PRINCIPLES OF STRATIGRAPHY

Roy R. Lemon

Florida Atlantic University

Merrill Publishing Company
Columbus   Toronto   London   Melbourne

Published by Merrill Publishing Company
Columbus, Ohio 43216

This book was set in Usherwood.

Administrative Editor: Stephanie K. Happer
Production Coordinator: Victoria M. Althoff
Art Coordinator: Jim Hubbard
Cover Designer: Russ Maselli
Photo Editor: Gail Meese

Library of Congress Catalog Card Number: 89–62283
International Standard Book Number: 0–675–20537–9
Printed in the United States of America
1 2 3 4 5 6 7 8 9—97 96 95 94 93 92 91 90

To Mary and Chris

PREFACE

Stratigraphy lies at the core of any geology curriculum because no investigation of earth history can proceed far without some grasp of stratigraphic principles. Stratigraphy is concerned with stratified rocks as the repository of a record of past events. It deals with sediments and sedimentary rocks, sedimentary environments, and fossils, and with the ways they can be described in terms of passing time. It provides the means of piecing together a calender so that world geologic records can be brought together into a coherent whole. The field of stratigraphy is broad. This causes considerable divergence of opinion as to what should or should not be included in a stratigraphic text. One of the earliest major texts was Grabau's great work published in 1913, and it is clear from a perusal of the contents that the author considered that virtually every topic from the atmosphere to vulcanicity should be embraced. Later texts emphasized the sedimentological aspects of stratigraphy, as did, for example, the classic work by Krumbein and Sloss published in 1950. Since that time, surprisingly few textbooks in stratigraphy have appeared, and virtually all of them have strongly stressed lithostratigraphy. On the other hand, the study of

sediments and sedimentary processes also has evolved rapidly and has seen a considerable growth in the number of texts available; nearly a dozen new ones have appeared in the past ten years. Because the stratigraphic record is contained within sedimentary successions, it is obvious that no treatment of stratigraphy would be complete without some explanation of this topic. In the present text, however, I have ventured to reduce those portions dealing with sediments and sedimentary rocks, as such, to a minimum. This approach is in no way intended to downgrade the importance of sedimentology, but merely to recognize that that field is itself increasingly served by its own separate literature. This treatment also emphasizes the fact that modern stratigraphy deals with magnetic stratigraphy, seismic stratigraphy, and an increasingly sophisticated numerical biostratigraphy. These topics are at least as important as those in the field of sedimentology and accordingly should be given considerable weight in a stratigraphy text.

The key role of biostratigraphy in all stratigraphic investigations is recognized by the special emphasis placed on this topic. Modern stratigraphy would be impossible without biostratigraphic control, and this justifies a more in-depth study of fossils than is found in more orthodox stratigraphic texts. Topics such as the geographic dispersal of fossils, the species concept, and mass extinctions are discussed. Also examined is the impact of the new approach to evolutionary theory in which punctualism is apparently replacing the gradualistic model and, in the process, explaining many puzzling features of the biostratigraphic record.

An adequate understanding of the subsurface data obtained from the many logging tools now available, and from seismic surveys, is also essential in stratigraphic studies. Because a large number, perhaps the majority, of geology graduates working in stratigraphy are in industry, much of their work will be based upon subsurface information.

Finally, at the ends of the geologic column are the Quaternary and Precambrian which, perhaps unexpectedly, have one thing in common: many of the so-called "normal" stratigraphic procedures cannot be applied to them. The special problems of Quaternary stratigraphy are accordingly dealt with in Chapter 15. Chapter 16, the closing chapter, ventures into what for many stratigraphers is a *terra incognita*. Attempts have been made to sort out Precambrian stratigraphy only relatively recently. Because we are here dealing with some seven-eighths of earth history, it is clearly a topic of immense importance with great opportunities for future work.

It is important to recognize the fact that, largely as a result of stratigraphic studies, the traditional uniformitarian view of earth history is being modified. It is clear, for example, that both in the accumulation of sediments and the evolution of life forms, the record speaks of constant interruptions, both of minor and major, literally earth-shaking, dimensions. Although some of the recent discoveries have raised important controversies yet to be completely resolved, some awareness of these ongoing discussions should be part of modern stratigraphy.

This text should stand in a corequisite relationship to those in the parallel fields of sediments, sedimentary rocks, and sedimentation. Within the geology program curriculum, it is expected that stratigraphy would, perhaps at the sophomore or junior level, occupy one semester and sedimentology another. Both areas could be covered simultaneously in corequisite courses, or alternatively, one would be the prerequisite of the other. Which should come first is open to debate and is, perhaps, not too important.

Acknowledgments

Throughout the writing of this book, I have benefited greatly from the helpful advice and encouragement of many colleagues. Of especial mention are Ian Watson, David Warburton, Edward Petuch, Rita Pellen, Sheldon Dobkin, and Ralph Adams. My sincere thanks and appreciation go also to the following who gave generously of their time and energy in reviewing the manuscript: Jon Avent, William Berry, Ronald Blakey, Mark Boardman, Karl Chauff, James E. Conkin, John Cooper, Brent Dugolinsky, Richard A. Flory, Lawrence A. Krissek, Fred Lohrengel II, Donald W. Neal, Paul Pausé, Charles Singler, Peter W. Whaley, and Grant Young. Their many helpful comments and suggestions have added enormously to the book.

My sincere thanks go also to Christina Soto, and most especially to Cynthia Mischler for typing the manuscript. The project could not have been possible without her hard work, infinite patience, and unfailing cheerfulness. Jean Brown's meticulous editing and many helpful suggestions have also contributed greatly to the book.

Finally, I offer my sincere thanks and appreciation to my wife, Mary, and son, Christopher. Without their help and encouragement over more years than I dare to count, the book would not have reached completion.

CONTENTS

PART ONE
TIME AND THE
TIMEKEEPERS

VERY LITTLE CAN BE SAID ABOUT STRATIGRAPHY WITHOUT REFERENCE TO THE TIME dimension. It is the record of sequential events that is of primary interest; so to gain some understanding of the relationship of sediments to time, and of the way geologic time is measured, are essential first steps. Perhaps the most prominent characteristic of the stratigraphic record is its lack of continuity, so it is with periodicity in earth history that this account begins.

1

TIME AND SEDIMENTS

There's something damn funny about the stratigraphical record.

Derek Ager

1.1 INTRODUCTION

So far as there is a written record, our modern understanding of rock strata can be said to begin with the writings of Nils Stenson, or as he is better known, Nicolaus Steno, a seventeenth-century Danish physician in the court of the Duke of Florence. Steno's studies of the rock strata of northern Italy were published in 1669 in a book that truly can be labeled as the first to deal with stratigraphic principles. Steno's observations of sedimentary rocks and their contained fossils led him to conclude that the land once had been covered by the sea and that layers of rock had formed by deposition of bed upon bed. The sequential ordering of beds from the oldest at the base to the youngest at the top of the succession followed what Steno called the *law of superposition.* A second law, which he labeled the *law of original horizontality,* was based on the observation that, although rock strata are to be seen often in an inclined attitude, they originally had been deposited in a horizontal position. A third law, that of *original continuity,* explained why strata exposed on one side of a valley could be seen also on the other. The implication here was that the edges of the strata seen in an outcrop must have been exposed by the breaking and removal, or the wearing away, of strata. In other words, the strata had once extended continuously across the area; the valley was, in fact, the result of their removal by surface agents of erosion. Steno's three laws are fundamentally simple and seem very obvious to us today. What is astonishing, perhaps, is that these first recorded observations occurred so late in history and that it had taken so long for a reasoned comment on the origin of sedimentary strata. It is very difficult to believe, for example, that the resemblance between sand deposits in a river or on a beach, and layers of sandstone in a cliff face, had not been noted by many people since the beginning of history. The lack of any written record of such observations during the hundreds of years before Steno's time can be explained only by assuming that no one was sufficiently interested! The origins of this lack of interest are not hard to find.

Much of what scholars were beginning to discover about the physical world during the seventeenth-century was, in fact, a rediscovery of many things known during the flowering of Greek science some fifteen hundred to one thousand years earlier. This promising beginning of science had come to an end, most scholars agree, with the destruction of the great library at Alexandria in the fifth century A.D. Probably the last in the great tradition of Greek scientists was Hypatia; her death in A.D. 415, at the hands of followers of Archbishop Cyril (who later was made a saint!), generally is accepted as marking the beginning of the so-called Dark Ages. During the thousand years that followed, and under the growing influence of Christianity, scholarship turned in a new direction. Intellectual debate dealt largely with the philosophical and mystical and was directed toward the spiritual rather than the real world.

Steno's law of superposition is the basis for what has come to be called "layer cake stratigraphy," in which each layer of sediment was laid down only to be immediately buried by the next layer above. This is what Ager (1981) called the "gentle rain from heaven" concept, in which at all locations and, more or less, at all times sediment is supposed to sift down from above. There is nothing intrinsically wrong with any of these ideas—the mistake lies in taking them too literally and applying them too widely. We can come to a proper understanding of the sedimentary record only if we realize that sometimes the processes do operate in this way and sometimes they most definitely do not. It is quite clear, for example, that sediment accumulation is not continuous over long periods of time. Even in deep-ocean sediments there are gaps in the record marking intervals of nondeposition or erosion, whereas in shelf and platform areas there is more gap than record. Most typically, the intervals of nonrecorded geologic time exceed by many times those time intervals for which there is a sedimentary record. This is what Charles Darwin referred to as the "imperfection of the geologic record." Also, implicit in an understanding of sediments and sedimentary rocks is the idea that the individual layers of sediment were deposited over a span of time and that the thickness of each layer is related in some way to the length of time it took to form. At a most elementary level of understanding, this is simply the "hourglass" idea. It is quite true that, other things being equal and under uniform conditions, 10 m of sediment will probably take twice as long to accumulate as 5 m. Time and sediment thickness are linked sometimes. In deep-sea cores, for example, quite reliable ages can be estimated by extrapolation of thicknesses downward on the basis of measured thicknesses higher in the core where there are some biostratigraphic or other age indicators. Actually, of course, there exists a tremendous variability in sedimentation rates. Sir Archibald Geikie in 1886 described large fossil trees (some of them more than 12 m tall) in their positions of growth (Fig. 1.1) and rightly concluded that the sandstone entombing them had accumulated before they had had time to rot away. At the other extreme, on certain parts of the ocean floor, sediments may accumulate at a rate of only 1 mm per 1000 years.

FIGURE 1.1
Sediment accumulation can be extremely rapid, as seen at this Pennsylvanian section in Nova Scotia, where a large tree trunk was buried in sediment before it had time to rot away. (Photograph courtesy of the Geological Survey of Canada, Ottawa.)

1.2 EPISODIC NATURE OF THE STRATIGRAPHIC RECORD

Balance between Deposition and Erosion

On many present-day marine shelves the bottom sediment consists, for the most part, of a transient veneer of clastic material, ranging from sand in the shallower areas to silt-size and finer material in deeper or more sheltered areas. One of the most important differences between shallow- and deeper-water regions is the frequency with which the bottom sediment is disturbed by wave or current action. In relatively shallow water—above what is called **wave base**—normal (fair-weather) wave action keeps the bottom sediment stirred up and in a state of constant shifting to and fro. Ripple marks, sand waves, submerged bars, and other features constantly form and reform. Below wave base, bottom sediment is moved less frequently, but is still shifted occasionally by storm waves and tidal currents. Obviously, the frequency of this disturbance bears a direct relationship to the depth of the water and/or the energy input to the water column. In such environments, where sediments are eventually disturbed

again, it is difficult at first to understand how any accumulation of shallow-water sediment can ever be formed and permanently preserved. The fact that sediments are preserved is due to several factors. Probably the most important is that the shallow shelves and platforms are slowly sinking. Causes of such sinking will be discussed in a later section. Thus, there will be a net accumulation of sediment, no matter how much redistribution takes place in the short term.

It is important to examine some of these short-term processes and local influences because they have a profound effect on the kind of sediment that is eventually preserved. Despite the essentially ephemeral nature of shallow-water sediments, there are numerous ways in which net accumulation does occur:

1. In the case of finer sediments, particularly clays, there is a natural cohesiveness—once deposited, fine sediments are relatively difficult to erode and reentrain.
2. Plants and benthic animals often play an important role as sediment-trapping devices by reducing current velocity at the sediment–water interface (Fig. 1.2). Other

FIGURE 1.2
A. Red mangroves (*Rhizophora mangle*) are effective in stabilizing and entrapping sediment in the supratidal zone. B. Black mangrove (*Avicennia germinans*) with nematophores growing up out of the water.

FIGURE 1.3
Cracked surface of drying mudflats covered by a mat of cyanobacteria.

organisms are sediment binders. Algal mats, for example, can spread over the sediment surface in a week or two of quiet-water conditions and thereafter tend to protect the underlying sediment from erosion (Fig. 1.3).

3. Many plants (*Halimeda, Penicillus,* etc.) and animals (crinoids, foraminiferans, etc.) are themselves active contributors to the local sediment.

4. Corals and encrusting organisms associated with them in **reefs** are efficient rock builders and flourish under the most turbulent conditions (Fig. 1.4), provided other ecologic parameters are met.

5. Shallow water is not necessarily turbulent; in protected bays and lagoons sediments accumulate rapidly, including evaporites and fine lime muds.

6. Much shallow-water deposition takes place not in a vertical direction but laterally by outbuilding in a horizontal direction. In a shallow, exposed, nearshore location, deposition is only slightly greater than erosion. Sediment is constantly being moved by bottom currents until it is finally swept out into deeper water, where accumulation significantly exceeds erosion. In such cases the sediments become progressively younger basinward rather than in a vertical direction (Fig. 1.5). This is typical of a deltaic environment.

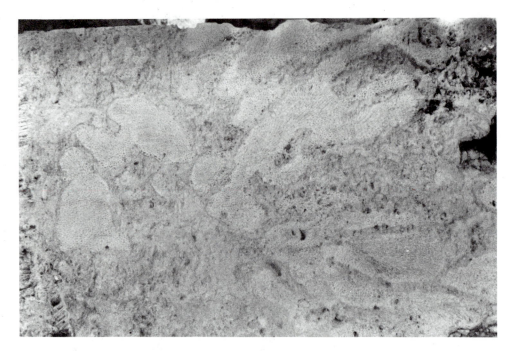

FIGURE 1.4
Massive limestone made up of scleractinian coral heads, Pleistocene Key Largo formation, Windley Key, Florida.

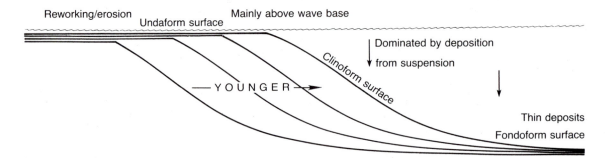

FIGURE 1.5
Much of the sediment deposited in many basins accumulates laterally, so the succession becomes progressively younger basinward rather than upward.

Beds and Bedding Surfaces

In a typical succession of sedimentary rocks, the most obvious feature is usually the bedding. The **bedding planes** separating the beds may be widely or closely spaced, regular or irregular. These characteristics are controlled by the physical and biologic features of the depositional environment. Bedding planes mark changes in the depositional environment: perhaps a pause in deposition, a brief interval of removal of sediment by bottom currents, or a change in the character of the sediment being laid down. Many bedding planes are seen to be marked by features such as ripple marks, miniature scour features, and desiccation cracks, together with many trace fossils, such as the tracks, trails, and burrows of bottom-living creatures. Such features can be seen on many modern seafloors or lake bottoms; it is clear that a bedding plane represents what was, for a short period of perhaps days, months, or even years, the sediment–water interface (Fig. 1.6). In effect, it is an ancient seafloor (or lake bed) preserved in much the same way as are certain fossils; indeed, fossils, particularly of benthic organisms, commonly are found as flattened impressions on bedding planes.

In describing the stratigraphy of a sequence of sediments, the geologist is always concerned with possible gaps in the succession; every effort is made, therefore, to find the most complete succession of strata. If there is no physical evidence of erosional breaks, and, more important, if there seems to be continuity in the contained faunas and/or floras, it has been generally assumed that there exists a more or less complete and satisfactory record of the time interval involved in the formation of the succession of strata. In recent years, this assumption is being increasingly questioned. It is now thought that, at least in terms of the time represented, the beds are far less important than are the bedding planes between them. There is nothing very continuous about the accumulation of sediment; it is, in fact, markedly episodic. At any one locality, there is little doubt that the succession of sediment preserved is a record of only a fraction of the total elapsed time represented from the bottom to the top of the succession. In other words, the sedimentary record is, at best, fragmentary. On the other hand, there is no doubt that even if sediment accumulation ceased, or erosion occurred, at one place, there was almost certainly sedimentation going on somewhere else. Piecing together and correlating many local sections within a sedimentary succession will go a long way toward improving the understanding of the overall record, even though, at the scale of individual beds and bedding planes, no long-distance correlation, as such, is possible. Even though bed-by-bed correlation is usually not possible beyond very short distances, the multiplicity of beds that constitute what we call a formation—a stratigraphic unit with a measure of lithologic homogeneity—can be correlated from place to place. As will be described in Chapter 3, a formation can be considered a fundamental mappable unit.

One of the main reasons the sedimentary record, at any one place, has so many gaps is that the delivery of sediment to any particular area of the seafloor is markedly episodic. The episodes may be quite random, with no obvious external cause, or they may be controlled by some rhythmic or cyclical mechanism, perhaps seasonal, monthly, or tidal. It is rare that this cyclicity can be discerned, for several reasons. Most commonly, there is the likelihood of contemporaneous erosion. Currents transporting sediment may also reentrain it, and the net accumulation may represent such a small part of the whole that no meaningful periodic signal is preserved. Equally important is **bioturbation,** the churning and mixing of sediment by burrowing organisms to varying depths below the sedi-

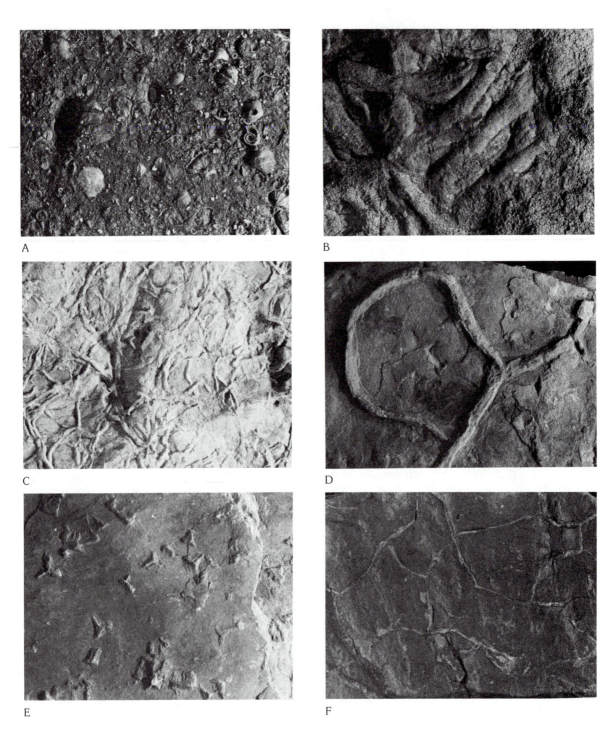

FIGURE 1.6
Bedding plane features. A. Broken mollusc and brachiopod shells and bryozoans make up a typical "shell hash." B, C, and D. Various trails and burrows are typical trace fossils (ichnofossils). E. Casts of salt crystals in tidal-flat mud deposits. F. Desiccation cracks in muds.

ment–water interface (Fig. 1.7). This activity often destroys finely laminated structures and obscures or blurs cyclical or rhythmic patterns. Only in sedimentary environments where bottom-dwelling organisms are rare or absent, such as under **anoxic** or hypersaline conditions, is there good correlation between a succession of beds and regular cyclical controls. Despite the effects of erosion and bioturbation, some sort of pattern may be preserved. The sedimentary boundaries may not be sharp, but depositional and dissolution cycles on the deep seafloor are still often discernible. Statistical analyses of bed thicknesses, bedding-plane intensity, and other characteristics have proved useful in restoring a meaningful periodic signal. In one study by Schwarzacher and Fisher (1982), statistical analyses were used to show that in both Cretaceous and Carboniferous limestones a cyclical pattern was present that was apparently climatic in origin and controlled by the Milankovitch mechanism, discussed in Chapter 4.

It is obvious from the above discussion that to assume that a succession of bedded sedimentary rocks will necessarily retain any readable account of what actually happened is sometimes optimistic. To make matters worse, it is apparent that parts of some sedimentary formations were laid down under unusual, rather than average (day-to-day) conditions. Certain successions have been interpreted as **"tempestites,"** in which sediments laid down earlier had been stirred up and then redeposited during an occasional storm surge. Such sediments are probably not representative of the total depositional environment at all. The message that emerges is clear. Even at the small scale of events like the deposition of layers of sediment, a uniformitarian approach, in the traditional sense, may not always be feasible. Reconstructing an ancient depositional environment by simply comparing the sedimentary rock with "similar" modern sediment may possibly result in a distorted picture of what actually was going on at the time the

FIGURE 1.7
Bioturbation. Conical mounds of sediment on an exposed tidal flat, pierced by exhalant holes of *Callianassa* burrows, Abu Dhabi, Persian Gulf. Black segments on rod are 10 cm long. (Photograph courtesy of Gerald M. Friedman.)

particles in the sedimentary rock were laid down.

Bedding and Time

In formal stratigraphic nomenclature, a bed is a lithostratigraphic unit, but over a given area it also can be considered a **chronostratigraphic unit**—a sedimentary unit bounded by time horizons. As Campbell (1967) pointed out, this fact has sometimes been overlooked. Because a bedding surface represents the actual seafloor or lake bed that existed for a time before it was covered by the next increment of sediment, it has a measure of synchroneity. Ideally, an **isochronous** (synchronous) surface represents an instant in time, but this is rarely, if ever, discernible. In practice, perhaps the shortest major event likely to leave a noticeable record within the sedimentary succession is a volcanic ash fall. Fallout from a single eruption event may spread a layer of ash over tens of thousands of square kilometers in the space of a few hours or days. There are many examples in the geologic succession of ash bands and **bentonite** beds that have proved to be invaluable time-horizon markers. Similar short-lived events are seen in the **graded beds** in a turbidite sequence, each one of these beds representing deposition during a single turbidity current flow event. A single bed may in places be traceable for many kilometers and, within its area of occurrence, can also be considered a chronostratigraphic unit because the bounding bedding surfaces are more or less synchronous over that area.

If we try to visualize the situation on the seafloor over a relatively large area it is obvious that while sedimentation takes place in one area no deposition is taking place in another. In other words, the bedding surface itself may have a wide area of distribution at a given point in time, but the time that has elapsed since the moment of arrival of the sediment forming that bedding surface (that

is, the age of the latest lamina) may differ from place to place. The intervals during which sediment is settling to the bottom, and the area over which it is deposited, might be analogous to localized and brief passing rain showers; long dry intervals between the showers might be compared with periods of hiatus during which no sediment is being deposited (Figs. 1.8, 1.9). This is a different concept from that of the gentle rain from heaven mentioned earlier. The time spans involved in these alternations of deposition and nondeposition are irregular, but there is little doubt that the times actually recorded by sedimentary increments are exceeded many times by the intervals for which no sedimentary record is preserved. This imbalance has long been suspected, but, as Dott (1983) pointed out, has only recently come to be appreciated.

The age of the uppermost layer, or lamina, in a bed differs from place to place, just as does the age of the initial layer of the overlying bed. The age variation is, however, likely to be encompassed within a range of time of considerably less magnitude than that of the interval of nondeposition separating the two beds. The scale of all these time parameters differs, depending upon the supply of sediment and the depositional environment. This model does seem to be supported by the evidence of seismic stratigraphy, in which reflections, identified as stratal surfaces approximating bedding surfaces, are traceable over considerable distances and seem to have all the attributes of at least quasi-isochronous surfaces.

1.3 FACIES CONCEPT

On a modern seafloor, local variations in energy input (turbulence), water depth, sediment supply, bottom faunas, and so on will result in different sediment types being laid down simultaneously in different areas. A glance at any distribution map of modern bottom sediments will confirm this (Fig.

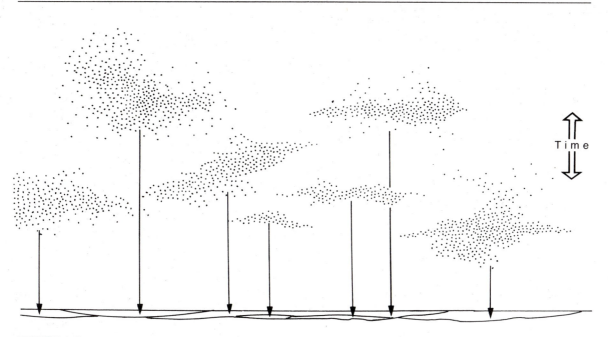

FIGURE 1.8
Sediment is delivered in pulses over time and settles to the seafloor or lake bed. Thus, a single bedding plane varies in age from place to place (ignoring penecontemporaneous erosion and reentrainment).

1.10). The correlation between sediment type and the environment in which it was deposited is recognized in the **facies** concept. At its most basic, the term facies simply means aspect and refers to the character of the rock. Most commonly, it is used in discussing the lateral variations seen in a particular stratigraphic unit such as a formation. So, for example, a sandstone formation that, traced laterally, becomes more shaly might be said to pass into a shaly facies. Additionally, certain implications regarding the depositional environment might be introduced. So, the sandstone might be described as being a shallow-water facies passing into a deeper-water (the shaly) facies, if indeed, this was how the lateral change in lithology had been interpreted. It was such lateral changes that were implicit when the term facies was first introduced in 1838 by Gressley. Since that time, however, it has come to have a variety of meanings. In some usages involving the time dimension,

changes in a vertical direction also are included. In discussing changes of a purely lithologic character, the term **lithofacies** is often used. When changes in the aspect of the contained faunas and/or floras are being discussed, the term **biofacies** is used. The facies concept is applied also to seismic data and even to igneous and metamorphic rocks. Facies are discussed again in Chapter 11.

Facies and Bedding

Facies changes often have considerable influence on the timing and vertical spacing of bedding surfaces (that is, on bedding thickness), but the bedding surfaces themselves will still be essentially isochronous. When we consider the relationship of bedding and facies through time as displayed in a vertical succession, it is inevitable that these isochronous bedding surfaces will cut across facies or formational boundaries (Fig. 1.11).

FIGURE 1.9

Types of bedding. A. Medium irregularly bedded limestones in the Mississippian of eastern Tennessee. B. Thin to medium regularly bedded limestones of Mississippian age, central Kentucky. C. Medium-bedded, cross-laminated quartzitic sandstones of Ordovician age, Cape Province, Republic of South Africa. D. Massive cross-bedded sandstone of the Upper Triassic Wingate Formation, southwest Colorado. E. Regularly bedded siltstones and shales in the Upper Cretaceous Menefee Formation, Durango, Colorado. F. Coals and associated shales and sandstones in the Menefee Formation, southwest Colorado.

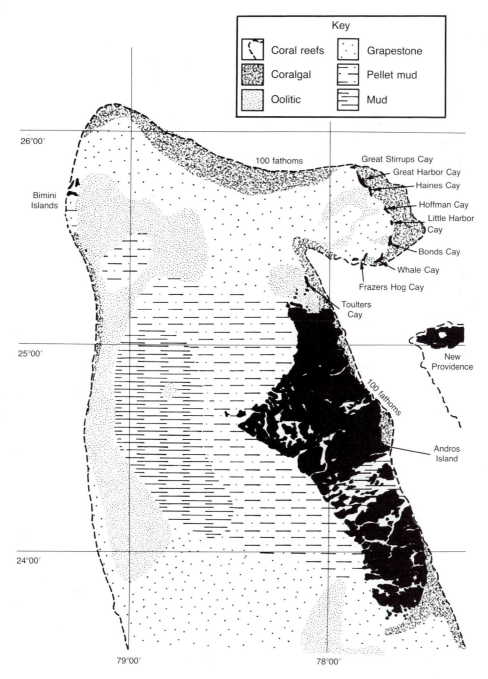

Key

Coral reefs   Grapestone

Coralgal   Pellet mud

Oolitic   Mud

26°00'

100 fathoms

Great Stirrups Cay
Great Harbor Cay
Haines Cay
Hoffman Cay
Little Harbor Cay
Bonds Cay
Whale Cay
Frazers Hog Cay

Bimini
Islands

Toulters
Cay

25°00'

New
Providence

100 fathoms

Andros
Island

24°00'

79°00'   78°00'

FIGURE 1.10

Variation in sea-bottom sediment type as shown in a facies map of the Andros Island area, Bahama Islands. (Redrawn after Purdy, E. G., 1963, Recent calcium carbonate facies of the Great Bahama Bank, 1, Petrography and reaction groups. 2, Sedimentary facies: Jour. Geology, v. 71, p. 334–335, 472–497, by permission of University of Chicago Press.)

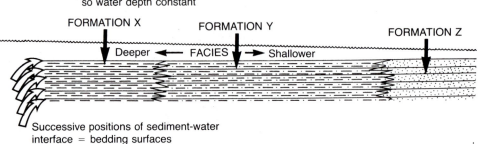

FIGURE 1.11

Relationship between bedding surfaces and formational/facies boundaries.

This is because such boundaries, except along depositional strike, are invariably diachronous as a consequence of the fact that sea level and sediment supply, the major controls of facies, are constantly changing through time. What this means is that a lithologic (formational) boundary marking a change in the depositional environment will, with time, inevitably move laterally. This is in response to a rising or falling sea level as it advances or recedes across a shelf or platform.

One of the important controls of bottom sediment type is water depth. The deeper the water, the smaller the input of turbulent energy on the seafloor. Unless there are other

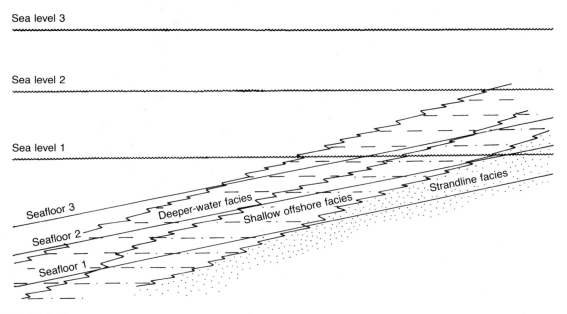

FIGURE 1.12

Lateral migration of lithofacies with rising sea level.

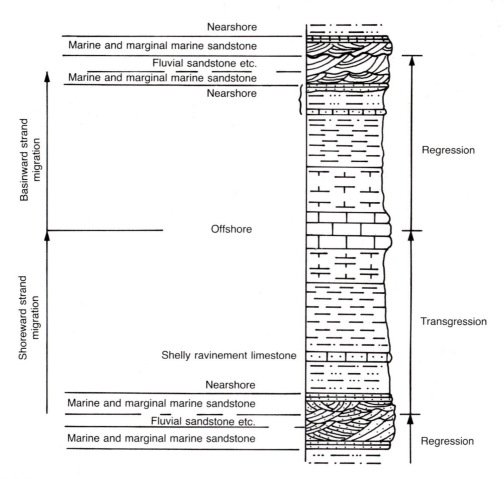

FIGURE 1.13

Interpretation of a typical cyclothem in the Upper Cretaceous of the Western Interior of
the United States in terms of transgression and regression. (Reproduced by permission
of the Geological Society from The great transgressions of the Late Cretaceous, Hancock,
J. M., and Kauffman, E. G., in Jour. Geol. Soc. London, v. 136, p. 175–186, 1979.)

factors involved, such as **relict sediments** remaining from an earlier regime, or deposition
by turbidity currents, deeper water is usually
characterized by finer sediments. The ideal
profile passes from sands in the shoreface region into silts and muds out on the shelf. The
various lithologies and faunal assemblages
associated with a particular depth of water
will migrate laterally up or down the regional
slope if sea level rises or falls (Fig. 1.12). In

the stratigraphic column at a given location,
a typical vertical succession, marking progressively deepening water during a marine
transgression, would pass upward from sandstone, through siltstone, to shales and perhaps limestones. Shallowing water would result in the reverse sequence (Fig. 1.13). This
is an illustration of an important principle in
stratigraphy known as *Walther's law of correlation of facies.*

Walther's Law

No discussion of facies change, particularly facies changes through time, can proceed very far without reference to the so-called *law of correlation of facies* proposed by Johannes Walther in 1894 (Walther, 1893–1894). The principle described by Walther is actually quite simple, but it was, as Middleton (1973) pointed out, misinterpreted, little understood, or ignored until relatively recently by many stratigraphers in the West. In its original form, the law can be stated as follows: "The various deposits of the same facies area, and similarly the sum of the rocks of different facies areas, are formed beside each other in space, though in cross section we see them lying on top of each other. As with biotopes, it is a basic statement of far-reaching significance that only those facies and facies areas can be superimposed primarily which can be observed beside each other at the present time" (trans. Middleton, 1973).

The concept of Walther's law can best be visualized by using the example of a marine transgression across a sloping shelf. On such a shelf, at any one time, the various sedimen-

tary facies, reflecting different water depths, would lie in contiguous zones roughly parallel to the bathymetric contours (Fig. 1.14). Under conditions of rising sea level, all the various depth-controlled zones would move up the slope and the sediments of one particular depth zone would become covered by sediments of deeper-water type. What this means is that, at any one place, the vertical succession of sediments that would eventually record the entire marine transgression would contain, from bottom to top, a succession of sedimentary facies whose vertical ordering would be the same as the lateral ordering of the adjacent depth zones on the sloping shelf.

Perhaps the most important single element in this concept is that of geologic time. The law is a statement of change and recognizes that all facies are diachronous and that facies boundaries cross timelines. What this means in practical terms is that, if at one location a time horizon within a facies of a particular lithology is traced laterally, it eventually passes into the facies that was lying above or below at the original location (Fig. 1.14).

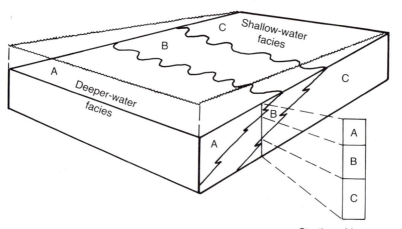

Stratigraphic succession

FIGURE 1.14
Walther's law illustrated by facies migration during a marine transgression.

1.4 GAPS IN THE RECORD

Effects of Erosion

Marine transgressions and regressions can be represented graphically in a zigzag diagram in which belts of depth-controlled facies move progressively backward and forward (Fig. 1.15). The situation becomes more complicated during a pronounced regressional phase, and progressive shallowing of the water finally leads to actual emergence. Subaerial erosion then begins to remove the sed-

iments of the earlier part of the cycle so that the basal beds initiating the next depositional cycle transgress an erosional surface (Fig. 1.16). Many fluctuations of sea level, such as those of glacio-eustatic origin, are short-lived, causing only brief periods of emergence and erosional beveling. Such breaks in the succession may not be very obvious because there is typically no angular discordance involved. As often as not, the breaks also occur within sands, and the basal transgressive sands of the new cycle may come to lie above

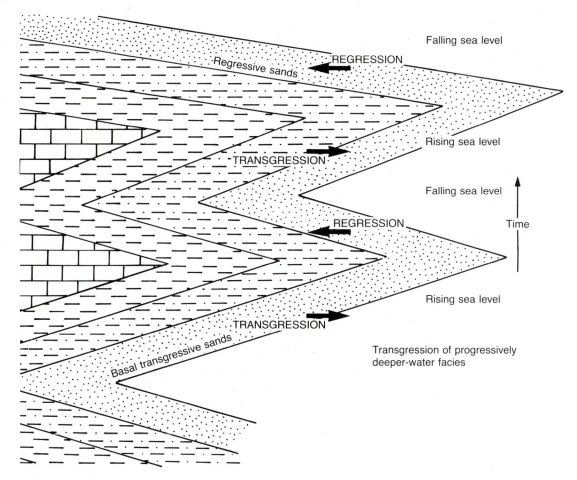

FIGURE 1.15
Shifting pattern of sedimentary facies during marine transgressions and regressions.

sands of the regressive phase preceding; indeed, they may be essentially the same sands reworked in the advancing shoreline. In such situations the hiatus is not easily detected, particularly in subsurface sections. The time span involved is short, so the break could probably be located only with sophisticated biostratigraphic control.

In most successions where continuing study of the stratigraphy has led to progres-

FIGURE 1.16
Balance between sediment accumulated and sediment lost to erosion during three transgressive-regressive cycles. In terms of the time recorded, the succession at stratigraphic section 1 contains a virtually complete record. In marked contrast, only a tiny portion of elapsed time is recorded at section 3—a good example of "more gap than record."

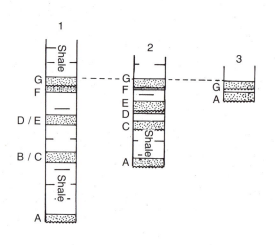

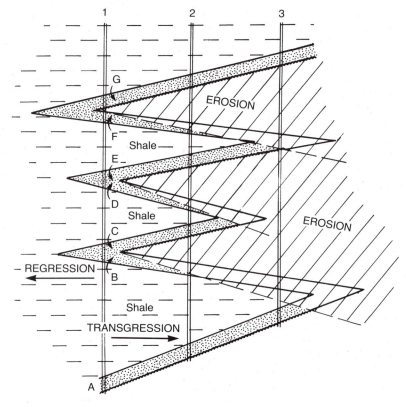

sively more sensitive control, the invariable result has been that more and more previously unsuspected breaks are revealed. When we speak of stratigraphic history as being more gap than record, it is not by any means due only to enormous missing slices involving whole geologic systems, although such do exist, but also to a large number of small gaps.

"More Gap Than Record"

Gaps in the stratigraphic record are due either to nondeposition or to later destruction of the sedimentary succession. If an area that was formerly submerged and the site of sediment accumulation becomes an area exposed to subaerial erosion, the earlier sedimentary record begins to be stripped away. This very elementary point is demonstrated throughout geologic history. For the most part, the history involves a balance between accumulation and erosion, although the word balance is perhaps often inappropriate because more sediment may be removed than is preserved. It is important to try to extract every bit of information possible from the preserved sedimentary record, but we must also attempt to assess the gaps in the record in terms of their causes and the magnitude of time they represent. Although the main concern here is with erosional gaps, it should be remembered that stratigraphic information is also lost as a consequence of metamorphism, melting, and lithospheric plate subduction.

Apart from nondeposition, major gaps in the stratigraphic record can result from many causes:

1. **Epeirogenic** upwarping of continental areas and loss of the sedimentary cover by erosion
2. Eustatic fall of sea level, causing emergence and erosion

3. Destruction at plate boundaries, due to
    (a) uplift and erosion during **orogenic** movement
    (b) metamorphism beyond recognition, or melting
    (c) loss by subduction; applies to some continental margin successions and ultimately to most ocean-basin sediments
4. Destruction of ocean-basin sediments, due to
    (a) erosion by deep-ocean currents
    (b) corrosion by bottom waters

The Nature of Hiatuses

A **hiatus** may be due simply to a cessation of deposition without any active erosion. Such gaps in the succession are likely to represent only brief time intervals, although they may be frequent. On a larger scale will be those hiatuses that represent not only cessation of deposition, but also erosion of previously accumulated strata. The smallest hiatus is that represented by a bedding plane, and the time involved may be only a few days. It may be due to nondeposition because of an absence of sediment delivered to the area or to bottom currents that prevent sediment from settling. In the second case, there is clearly a delicate balance between deposition and erosion. Currents moving rapidly enough to prevent sedimentation may, with little increase in velocity, begin to scour away existing sediment. Such penecontemporaneous and episodic events occur frequently and over very short time spans when considered within the geological context. Small-scale interruptions in the sedimentary succession of this nature are difficult to recognize; if they are to be considered as more than bedding planes, they usually are termed **diastems.** No biostratigraphic criteria are involved; the concept is purely lithostratigraphic in the physical sense and chronostratigraphically represents an insignificant hiatus. Where there is evidence of a concomitant faunal (or floral) discontinuity, Conkin and Conkin (1973, 1975) introduced the term *paracontinuity*. A **paracontinuity** is defined as a geographically widespread diastem manifested also by a faunal gap. The bed immediately above a paracontinuity is detrital and formed as a product of the process of erosion or abrasion that formed the paracontinuity surface.

When beds lying one above the other are all essentially parallel, and there is no evidence of erosion or structural disturbance, they are said to be **concordant** and, in a structural sense, **conformable.** The implication is that, in such a succession, sediment accumulation was more or less continuous and no hiatus exists. In fact, as we have seen, the very existence of bedding indicates that sedimentation was probably not continuous,

and so the terms are better used in a descriptive rather than a genetic sense. As will be discussed shortly, where there is evidence of an erosional break, those beds deposited after the erosional interval are described as being *unconformable* with the rocks below. In the case of breaks of major importance, there is usually a considerable loss of section, representing a relatively long interval of time. Typically, such **unconformities** show evidence of erosion of the older succession; if this was a consequence of emergence and subaerial exposure, such features as ravinements (Fig. 1.17), basal conglomerates, channeling and channel-fill deposits (Fig. 1.18), karstic features, and others may be present. In the case of submarine deposits, wave action, bottom currents, or slide and slump movements are responsible for erosion of the lower beds. In seismic reflection profiles un-

FIGURE 1.17
Small-scale ravinement in Pennsylvanian sandstones, eastern Tennessee.

FIGURE 1.18
Coarse gravels and sands filling a channel in Tertiary sediments, Mollendo, Peru.

conformities are usually discernible as such only if there is an angular discordance between older and younger sequences. **Angular unconformities** (Fig. 1.19) are the most obvious breaks in a succession, both in surface and seismic sections (Fig. 1.20). Such breaks indicate that a period of folding and/or faulting of the lower unit preceded its erosional beveling prior to the deposition of the upper unit. Angular unconformities are usually found in regions of tectonic activity; they were called *orogenic unconformities* by Chang (1975). If the upper unit lies over eroded and beveled igneous and/or metamorphic rocks the term **nonconformity** has come to have common usage. Major stratigraphic breaks without angular discordance of strata but with evidence of an erosional break of significant time magnitude are called **disconformities** (Fig. 1.20) or **nonangular unconformities** (Vail et al., 1977).

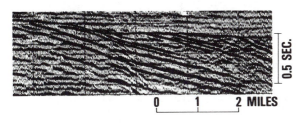

FIGURE 1.19
An angular unconformity as seen typically in a seismic record. (After Vail et al., 1977.)

Sometimes evidence of unconformities is subtle and may be present in the form of leached zones and other geochemical indicators of subaerial exposure. In limestones the effects of emergence into the vadose zone may be seen in dedolomitization, the forma-

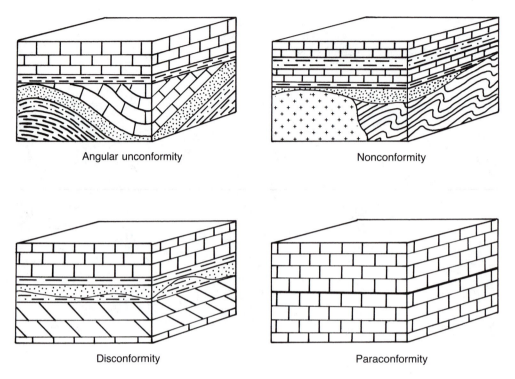

Angular unconformity

Nonconformity

Disconformity

Paraconformity

FIGURE 1.20
Block diagrams illustrating the major categories of unconformity.

tion of pisoliths, evidence of secondary porosity, and meniscus and pendant cements. On the other hand, there are some surprising examples of immense gaps in the section with no marked physical break and only the evidence of several missing biostratigraphic indicators to reveal the magnitude of the hiatus. Such breaks have been termed **paraconformities** (Dunbar and Rodgers, 1957) and are typically found in epeiric platform sequences of the mid-continent region. In northern Kentucky, for example, limestones of Middle Silurian age are overlain by late Early Devonian limestones; the break between them appears as little more than a bedding plane (Fig. 1.21), although it represents a hiatus of over 40 million years! Disconformities and paraconformities are typical

of cratonic successions and have been termed *epeirogenic unconformities* by Chang (1975). There is sometimes a tendency to suppose that major angular discordance is representative of a longer hiatus than breaks with no angular discordance, but there are cases in which only very small time intervals are involved. For example, at the opposite extreme from the paraconformity, there are, in areas of active tectonism such as California, beds of Middle Pleistocene age lying with pronounced angular discordance over beds of Late Pliocene age; the actual hiatus here is probably no more than one or two million years.

Considerable gaps exist also in deep-sea successions. Although the ocean basins would seem to be the logical place to look for

FIGURE 1.21
The paraconformity between the Louisville Limestone (uppermost lower Silurian) and Jeffersonville Limestone (lower middle Devonian). At Shanks quarry, eastern Louisville, Kentucky. (Photograph courtesy of James E. Conkin.)

the most complete stratigraphic record, the Deep Sea Drilling Project has, in fact, demonstrated the existence of hiatuses representing time intervals of up to tens of millions of years.

Although widely used in a general sense, the term hiatus was defined in a more restricted way by Wheeler (1964). The missing interval represented by an unconformity was defined as a **lacuna,** which consists of two separate parts; one part represents previously deposited strata later eroded (degraded), defined as the *degradation vacuity;* the other part represents the nondeposition-erosion cycle phase known as the *hiatus.* According to Wheeler (1964), and as Figure 1.22 shows, the essential boundary between areas of deposition and areas of erosion is the *base level,* the line showing the time at which deposition stops and erosion begins (the beginning of the hiatus) and the line showing the time at which erosion stops and deposition begins again (the end of the hiatus).

1.5 CONCLUSION

Gaps in the stratigraphic record have figured prominently in this chapter. It should not be thought, however, that such hiatuses represent a complete dearth of stratigraphic information. Later in this book some account will be given of the reasons for such gaps, and it should become apparent that stratigraphy is not to be compared with trying to solve a jigsaw puzzle with most of the pieces missing. The gaps, although real enough, often can be interpreted in terms of their nature, magnitude, and relative importance, and themselves are often useful sources of information. The theme that will be developed through this book is the essentially repetitive and cyclical nature of the stratigraphic record; it will be apparent that over spans of geologic time, virtually nothing is uniformly continuous.

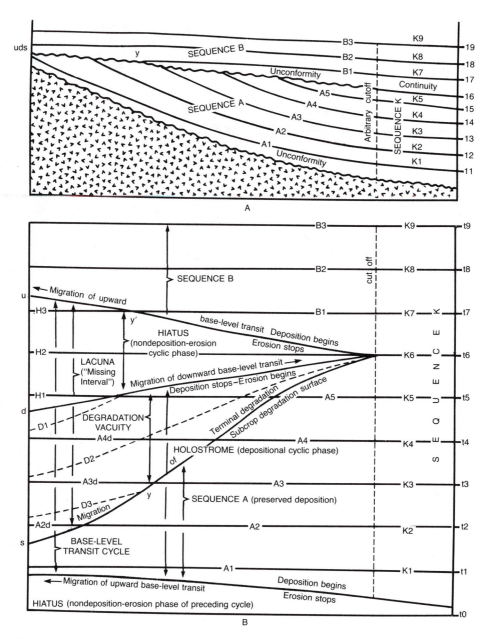

FIGURE 1.22
A. Section showing physical relationships of successive unconformity-bounded sequences. B. Area-time projections of A showing lithosphere surface-moment and base-level transit migration patterns in time stratigraphy. (After Wheeler, 1964.)

REFERENCES

Ager, D., 1981, The nature of the stratigraphical record, 2nd ed.: New York, Macmillan, 122 p.

Campbell, C. V., 1967, Lamina, laminaset, bed, and bedset: Sedimentology, v. 8, p. 7–26.

Chang, K. H., 1975, Unconformity-bounded stratigraphic units: Geol. Soc. America Bull., v. 86, p. 1544–1552.

Conkin, J. E., and Conkin, B. M., 1973, The paracontinuity and the determination of the Devonian-Mississippian boundary in the type Lower Mississippian area of North America: Univ. of Louisville Studies in Paleontology and Stratigraphy, no. 1, 36 p.

——1975, The Devonian-Mississippian and Kinderhookian-Osagean boundaries in the east-central United States are paracontinuities: Univ. of Louisville Studies in Paleontology and Stratigraphy, no. 4, 54 p.

Dott, R. H., Jr., 1983, Episodic sedimentation—How normal is average? How rare is rare? Does it matter?: Jour. Sed. Petrology, v. 53, p. 5–23.

Dunbar, C. O., and Rodgers, J., 1957, Principles of stratigraphy: New York, John Wiley & Sons, 356 p.

Geikie, A., 1886, Class book of geology: New York, Macmillan, 458 p.

Gressley, A., 1838, Observations Geologiques Sur le Jura Soleurois: Neuchâtel Neue Denkschr. Allg. Schweizer. Gesell. Gesteine Naturw., v. 2, p. 1–112.

Middleton, G. V., 1973, Johannes Walther's law of correlation of facies: Geol. Soc. America Bull., v. 84, p. 979–988.

Schwarzacher, W., and Fisher, A. G., 1982, Limestone-shale bedding and perturbations of the earth's orbit, in Einsele, G., and Seilacker, A., eds., Cyclic and event stratification: Berlin, Springer-Verlag, p. 72–95.

Vail, P. R., Mitchum, R. M., Jr., and Thompson, S., III, 1977, Seismic stratigraphy and global changes of sea level, part 4, Global cycles of relative changes of sea level: Am. Assoc. Petroleum Geologists Mem. 26, p. 83–97.

Walther, J., 1893–1894, Einleitung in die Geologie als historisch Wissenschaft: Jena, Verlag von Gustav Fischer, 3 v., 1055 p.

Wheeler, H. E., 1964, Base level, lithosphere surface, and time-stratigraphy: Geol. Soc. America Bull., v. 75, p. 599–609.

2
THE NEW UNIFORMITARIANISM

The jagged alligator, and the might
Of earth-convulsing behemoth, which once
Were monarch beasts, and on the slimy shores,
And weed-overgrown continents of earth,
Increased and multiplied like summer worms
On an abandoned corpse, till the blue globe
Wrapped deluge round it like a cloak; or some God
Whose throne was in a comet, passed, and cried
"Be not!" And like my words they were no more.

Percy B. Shelley

2.1 INTRODUCTION

As discussed in the previous chapter, traditional ideas that geologic processes are essentially continuous and ongoing have come to be replaced by the view that they may instead be episodic and markedly discontinuous. What is now becoming increasingly apparent is that the history of the earth as a whole also seems to be episodic and to have been interrupted by events of global magnitude. This awareness is leading to the emergence of a new kind of thinking about stratigraphy, a perspective in which periodicity rather than continuity is the underlying theme. Discussed in the following pages is the geological evidence for these major events, particularly those of apparently extraterrestrial origin, and their significance in stratigraphy. Inevitably, the emphasis here is on one of these events, that at the Cretaceous-Tertiary boundary, because it is the best documented. Mass extinctions are particularly important because it is largely in the biostratigraphic record that these major events are documented. Discussion is continued in Chapter 8.

2.2 UNIFORMITARIANISM AND CATASTROPHISM

As every student of the geological sciences knows, one of the most important steps forward in the history of geology was the realization of the great age of the earth and the extremely slow pace at which most geological processes seem to proceed. The publication of Hutton's *Theory of the Earth* in 1795 and later works by Charles Lyell in the nineteenth century (especially the 2nd volume of the 1st edition of the *Principles of Geology*) are generally considered to be among the most important milestones in the evolution of the science. Prior to this period, although there were notable exceptions, any genuine inquiry into the relationship between geological phenomena and time tended to be stultified by the general acceptance of the biblical account of creation. After all, if the earth and all its features had been created virtually instantaneously, then there was really no point in arguing about the relative age of this rock and that rock, or this valley and that mountain. Apart from the effects of this sort of mental straitjacket, it should be remembered also that, for the average person, the only real experience of geological and other natural phenomena comes in the form of catastrophes—earthquakes, volcanic eruptions, landslides, great floods, hurricanes, and the like. It was not surprising, therefore, that these awesome demonstrations of the power of natural forces were considered quite capable of molding the earth's surface features and causing the contortions and fractures of rocks so readily observed everywhere. Because the cause of these catastrophes was not understood, it was generally considered that they were somehow manifestations of divine displeasure and punishment for human wickedness. These ideas had a particularly powerful hold on the thinkers of the western world, raised as they were in the Judeo-Christian tra-

dition, which contained many historic precedents!

During the growth and evolution of geology as a science in the nineteenth century, one of the most important debates concerned the tempo at which geological processes operated. On the one side were proponents of what has come to be known as the "uniformitarian" school of thought and, on the other, those who were labeled "catastrophists." The uniformitarian view of earth history held that all geologic processes proceed continuously and at a very slow pace. The major features of the earth's surface could be accounted for as the result of the cumulative effects of very small changes operating over long periods of time. The catastrophists, on the other hand, believed that many large-scale changes took place rapidly and that many ongoing processes were cumulative until a sudden collapse or change brought the cycle to an end. Then, after a short period of violent change, the slow cycle would begin anew. Contrary to popular belief, this debate had little to do with the conflict between science and religion. The theologians, of course, attempted to explain everything in terms of divine acts. It is true that their dogma could be made to fit in somewhat better with a catastrophist view of the world, although, it must be admitted, even that must have taken some rather convoluted thinking! The so-called catastrophist interpretation of geological evidence was, in many ways, argued just as scientifically as was the uniformitarian view, as Rudwick (1972) has pointed out, but it was the uniformitarian view that eventually prevailed, largely because of the persuasive teachings of Charles Lyell. Although meticulous in his observations and reporting, there is little doubt that, overall, his opinions were biased against catastrophism, and he was not entirely objective in selecting his facts and examples. It should not be forgotten that he was by training a lawyer! The publication in

1859 of Charles Darwin's *The Origin of Species,* with its description of biological evolution, demonstrating slow change wrought by tiny advances over vast spans of time, provided a comforting confirmation of the rightness of the uniformitarian model.

Geologic catastrophes are, of course, happening all the time, but within the traditional uniformitarian context they were seen only as minor shakes and rattles in an otherwise smoothly running geological machine. The important point is that, although so-called catastrophic events are rare phenomena within human lifespans, many are relatively common within the larger span of geologic time. In the space of a few hours, a hurricane or flood may cause more changes than do the so-called "normal" geological processes of many years. As discussed in the previous chapter, the sedimentary record of unusual events may dominate the succession, whereas the record of "normal," day-to-day conditions may be quite negligible. Assuming that a hurricane or flood occurs at a particular locality, say, once every 50 years, even the geologically short time span of the past 1 million years provides us with 20,000 such catastrophes; it is in this context that they perhaps can be viewed in their true perspective.

Because many geologic processes do operate very slowly, at least in human terms, it is not surprising that most geologists until quite recently have been largely content with a uniformitarian view of their world. It is only in recent years that there has been a growing awareness that the catastrophic natural events that might be observed in even one human lifespan were not only more important than previously had been believed, but they were not necessarily representative of what could and apparently has happened in the geologic past. There has been no human experience of the kind of past catastrophe that has, as the geological record is beginning to reveal, occurred from time to time. Ager

(1981) suggested that what is emerging now is a new doctrine in geology, perhaps described as "catastrophic uniformitarianism" and summed up by the phrase "the history of any one part of the earth, like the life of a soldier, consists of long periods of boredom and short periods of terror."

To upgrade the importance of catastrophes as geological agents should not be construed as giving aid and comfort to fundamentalist doctrines which today still seek to promote a catastrophist view of earth history. Such absurd ideas are now appearing in the guise of "scientific" writings of creationist theorists and others. The works typically contain large numbers of references and quotations from the standard literature and, because such references are always meticulously acknowledged, there is, to the lay reader, the implication that the author of the particular work quoted supports the "facts" described. One of the more interesting catastrophes discussed in such literature is the explanation that the Grand Canyon was cut in a few days by the waters of Noah's flood!

2.3 MASS EXTINCTIONS

If the earth did experience catastrophic changes in the past, where is the evidence? It certainly is not to be seen in folding, faulting, or most erosional features. Instead, virtually all the evidence is biostratigraphic and is seen in the mass extinctions that have occurred periodically through the Phanerozoic. There have been, at times, marked reductions in the numbers and diversity of a wide range of species. The episodes are well-documented (e.g., Newell, 1967) and show that in the past 570 m.y. there have been at least six such major extinction events and numerous smaller ones (Fig. 2.1 and Table 2.1). Those at the ends of the Permian and Cretaceous were, of course, recognized by nineteenth-century geologists and selected as era

FIGURE 2.1
Percentage extinction of families of those invertebrate groups that are important as fossils during the Phanerozoic. (After Newell, N. D., 1967, Revolutions in the history of life: Geol. Soc. America Spec. Paper 89.)

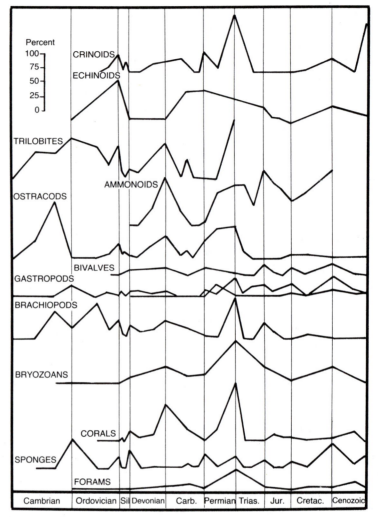

boundaries. According to Raup (1986), a 26-million-year periodicity is detectable in mass extinctions (see Fig. 4.12); this topic is discussed further in Chapter 4. The causes of these extinctions, and the question of just how sudden they were, have been the topics of lively debate for many years. Explanations have been many and varied and include climatic changes, salinity and temperature changes in the world's oceans, anoxic events, sea-level fluctuations, orogenies, and numerous other terrestrial phenomena. Other workers invoke extraterrestrial causes, such as radiation from a nearby supernova or meteoritic or cometary collisions with the earth. In considering the various possibilities, it should be remembered that several different changes may be linked in cause-and-effect relationships. Thus climatic change may cause (or be caused by) sea-level change, and that in turn would be responsible for an anoxic event, and so on. Even terrestrial and extraterrestrial mechanisms should not be divorced, because there is little question that a

TABLE 2.1
Major extinctions during the Phanerozoic (after Hallam, 1981)

Extinction Event	Animal Groups		Percentages of Families Extinct (after Newell, 1967)
	Extinct	Greatly Reduced	
End Cretaceous	Ammonites Rudistid molluscs Dinosaurs Large marine reptiles	Belemnites Corals Bryozoans Echinoids Sponges Planktonic foraminiferans	26
Late Triassic	Conodonts	Brachiopods Ammonites Fish Reptiles	35
Late Permian	Rugose corals Trilobites Blastoids Inadunate, flexibiliate and camerate crinoids Productid brachiopods Fusulinid foraminiferans	Bryozoans Reptiles	50
Famennian-Frasnian (Late Devonian)		Corals Stromatoporoids Trilobites Ammonoids Bryozoans Brachiopods Fish	30
Late Ordovician		Trilobites Brachiopods Crinoids Echinoids	24
Late Cambrian		Trilobites Sponges Gastropods	52

FIGURE 2.2
Examples of range charts.
A. Hippuroid bivalves.
(From Koogan, A. H., 1969,
Treatise on invertebrate paleontology, Pt. N, v. 3, Fig.
E234, p. N767, courtesy of
Geol. Soc. America and
Univ. of Kansas Press.) B.
Brachiopods. (From Williams, A., 1965, Treatise on
invertebrate paleontology,
Pt. H, Fig. 150, p. H240–
H241, courtesy of Geol.
Soc. America and Univ. of
Kansas Press.) Note the apparent coincidence of times
of extinction of different,
often only distantly related,
taxa.

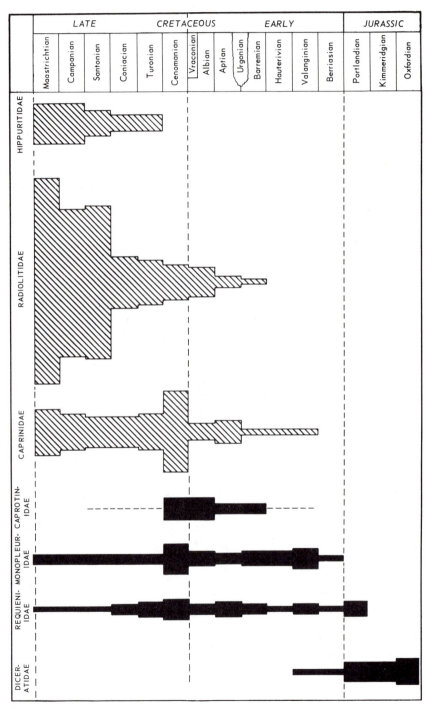

A

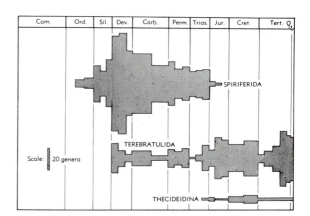

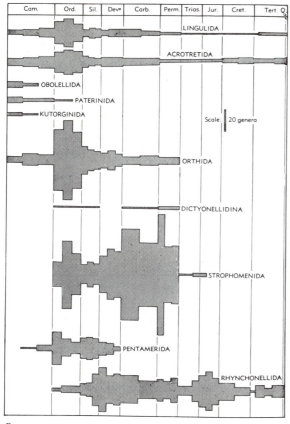

B

FIGURE 2.2
(*continued*)

major cosmic impact would have profound effects on many earth systems (for summaries and discussion, see Hallam, 1981, Ch. 10; Herman, 1981; Silver and Schultz, 1982; and Berggren and Van Couvering, 1984).

It is hardly surprising that after some 150 years of adherence to a uniformitarian doctrine, there are many geologists who would prefer to seek causes that were other than catastrophic. There are even some who would deny the existence of mass extinctions altogether, considering them artifacts of statistical analyses or an effect of that old favorite "the fragmentary nature of the fossil record." Circular thinking also is undeniably

sometimes a factor because it is obvious that if the end of a given time interval is defined by the disappearance of certain taxa, then clearly, those taxa will all, apparently, become extinct at the same time (Figs. 2.2 and 2.3). The true situation may be somewhat different. On the one hand, the time horizon established by an *apparent* coincidence of species extinction events, or other biostratigraphic markers, may not be confirmed by nonpaleontologic indicators such as magnetostratigraphic horizons. On the other hand, if extinction of a variety of species in different habitats occurs at approximately the same time horizon, it is not unreasonable to sus-

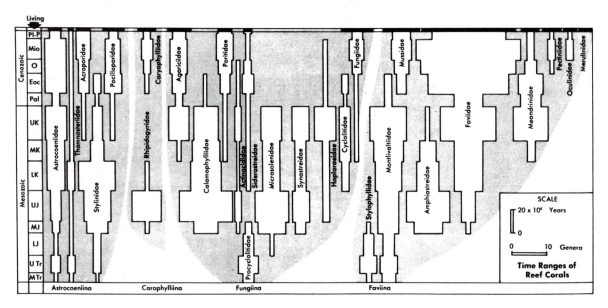

FIGURE 2.3
History of scleractinian reef corals, a group adapted to life in shallow, warm, and clear marine waters and particularly vulnerable to environmental change. (From Newell, N. D., 1971, Am. Mus. Nat. History Novitates, No. 2465.)

pect that the extinctions may have had a common cause and indeed occurred simultaneously. The search for such a cause is certainly a valid exercise; if nonpaleontologic evidence of a major event were discovered, it would be tempting to suggest that the extinctions and event were linked in some cause-and-effect relationship.

Even allowing for the many variables in biostratigraphic correlation and in nonpaleontologic dating methods, mass extinctions do seem to have been sudden. But how sudden is "sudden"? Clearly, sudden changes with global effect offer unique opportunities for stratigraphic correlation and time-scale calibration, so it is important to investigate the scale of the time intervals involved. As will be discussed in the next section, one of the explanations for the extinction of the dinosaurs is a meteorite impact. This obviously occurred in a matter of seconds and was "very sudden" by any standards, but it is certain that the dinosaurs did not disappear over-

night as a result. How long after the cosmic impact, if indeed this was the prime cause, the last dinosaur died, is unknown.

2.4 COSMIC IMPACTS

Of all the explanations for the causes of most mass extinctions, that of meteoritic or cometary impact seems to be finding increasing favor. There is impressive physical evidence that such collisions with the earth have occurred from time to time. The majority of meteorites probably originate in the asteroid belt that lies between the orbits of Jupiter and Mars. Certain of these bodies, known as Apollo objects, have orbits that cross that of the earth, and some inevitably collide. It has been calculated that objects as large as 10 km in diameter would be expected to impact about once every 60–100 m.y. (Shoemaker, 1982). Smaller meteorites would arrive at proportionately shorter intervals, those of 1-km diameter perhaps as frequently as every few hundred thousand years. The impact of a 10-km meteorite would, it has been calculated, have the explosive equivalent of 100 million one-megaton bombs. Whether it struck on land or in an ocean, it is likely that the crater would penetrate the crust and cause melting of crustal material. Enormous quantities of dust and water vapor would be injected into the atmosphere, causing a worldwide blocking out of sunlight for months or years and a consequent rapid cooling and climatic deterioration (Alvarez et al., 1980). Modeling of likely geochemical effects shows a compounding of the catastrophe: sudden heating of the atmosphere and the addition of meteoritic elements as aerosol particles at the time of impact could cause poisoning of the atmosphere and acid rains that could drastically alter the pH of the oceans (Hsü, 1980; McLaren, 1983; Mac-Dougall, 1988; Erickson and Dickson, 1987).

From even this brief summary, it would seem that the effects on the biosphere would be devastating and that extinctions would likely result across a broad spectrum of animal and plant species. Whether or not collision with large meteorites constitutes the prime cause of mass extinction, the physical evidence for periodic impacts is preserved on the earth's surface and within the sedimentary record in three ways: craters (**astroblemes**), tektites, and iridium anomalies.

Craters

More than one hundred craters or probable craters, ranging in age from Precambrian to Recent, have been documented (Fig. 2.4). One of the best known is Meteor (Barringer) Crater in Arizona; measuring 1.6 km across, it was made by an impact within the last 50,000 years (Fig. 2.5). The most recent ma-

FIGURE 2.4
Areal distribution of impact structures known up to 1982. Open symbols—craters with associated meteorite fragments; closed symbols—structures with shock metamorphic effects and in some cases siderophile anomalies. (After Grieve, A. F., 1982, The record of impact on Earth: Implications for a major Cretaceous/Tertiary impact event, Geol. Soc. America Spec. Paper 190, p. 25–39.)

FIGURE 2.5
View into what is perhaps the best-known meteorite impact crater, Barringer, or Meteor, Crater, near Winslow, Arizona.

jor cosmic event occurred in 1908 with the arrival of the so-called Tunguska meteorite in central Siberia. Damage extended to 40 km from the center of the impact area, although no crater was formed. This suggests that the meteorite disintegrated in the atmosphere or, alternatively, that a comet rather than a meteorite was responsible.

In the past, probably more than three-quarters of the earth's surface was covered by water, so it is obvious that the craters to be seen on the continents represent only a fraction of the possible impact sites. Inevitably, ocean impact sites are likely buried in sediment; in any case, the record of past im-

pacts is abbreviated by the lack of any oceanic crust older than Jurassic. On land, erosion quickly smooths out a crater—ancient impact sites are today typically quite subtle geologic features. For years many were described as old volcanic structures, ring dikes, or cryptoexplosion features. Some were not noticed until images from orbiting sensors became available (Figs. 2.6 and 2.7). Still others have no obvious surface features and were discovered by analysis of subsurface data. There is even oil production from supposed buried astrobleme structures in the Williston Basin (the Viewfield field in Saskatchewan and the Red Wing Creek field in

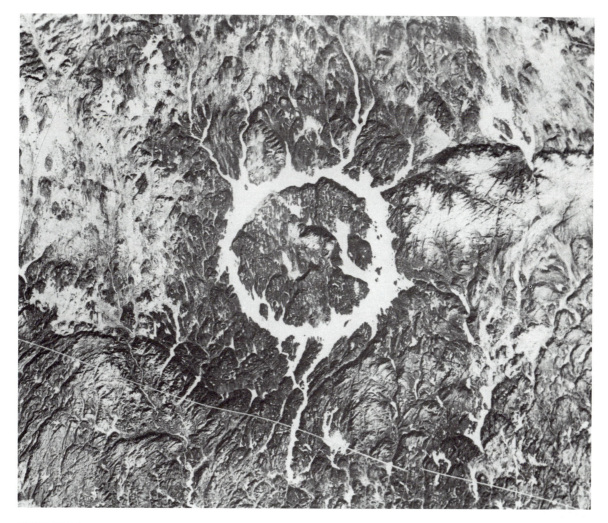

FIGURE 2.6
Manicouagan Lakes impact feature, Ontario, Canada. The visible structure is 70 km in diameter but fractures in the bedrock extend to 100 km. Radiometric dating indicates a late Triassic age. (Photograph courtesy Geological Survey of Canada, Ottawa.)

North Dakota) and possibly also the Lyles Ranch field in south Texas (Sawatzky, 1975; LeVie, 1986). Also in Texas, the Marquez Dome, in the East Texas salt diapir and long thought to be a salt dome, has recently been shown to be the central uplift of a 2–8-km impact structure (Shirley, 1989). Generally, proof of a meteoritic origin is based upon such criteria as the presence of nickel-iron fragments, shattercones, and high-pressure quartz (coesite).

Tektites

Tektites are small spheres of silicate glass, typically pea- to walnut-size and black, dark

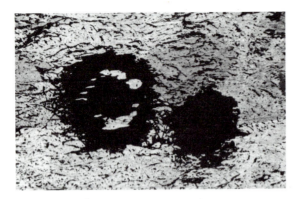

FIGURE 2.7
East and West Clearwater lakes, impact features in the glaciated surface of the Canadian Shield of northern Quebec, Canada. The craters are in crystalline rocks of Precambrian age, the western crater having a diameter of over 30 km, the eastern one measuring 26 km. The remnants of a central ring or peak are apparent in West Clearwater Lake. (Photograph courtesy Geological Survey of Canada, Ottawa.)

green, or sometimes yellowish in color. They have been found in surface sediments at various places around the world, and their source was, for a long time, a matter of conjecture. Many, with distinctive shapes and colors, were given local names; for example, those from Australia were termed australites, those from Czechoslovakia moldavites, and so on. One of the most characteristic features of tektites is a shape and/or surface texture that suggests that they were once molten. It was because of this that the name, derived from the Greek word *tektos,* meaning molten, was coined. More particularly, tektites show signs of ablation melting, indicating travel at hypersonic velocities through the atmosphere (Fig. 2.8). Once thought to have originated on the moon (Verbeek, 1897), it is now generally agreed that they represent crustal material melted and splashed out by large-body impacts on the earth's surface. The distribution,

composition, and ages of microtektites (typically less than 1 mm in diameter) found in deep-sea cores suggest association with known tektite strewnfields on land.

Four tektite strewnfields are known. The largest and youngest is the Australasian strewnfield that extends for some 8300 km in a great swath across southeast Asia and Australia (Fig. 2.9). It has been dated at 0.7 m.y. old and includes microtektites found in deep-sea cores. The impact crater of the meteorite responsible has not been found; this would be hardly surprising if, as Chapman (1971) suggested, australites are ejecta from the crater Tycho on the moon. The impact craters for two other strewnfields are known. The Ivory Coast tektites and associated deep-sea microtektites are thought to have come from the impact, about 1.5 m.y. ago, of a meteorite that formed what is now the 10-km diameter Lake Bosumtwi in Ghana. The Czechoslovak tektite field came from the Ries and Steinheim craters in Germany and has been dated at 14.8 m.y. old. The oldest strewnfield, known as the North American, has been dated at about 34 m.y. old. Microtektites from deep-sea cores in the Pacific and in the Caribbean are thought to belong to the same field. Samples from onshore locations in Barbados have been dated at 34.4 m.y. According to Petuch (1987), the impact site of the event is in the southern Everglades in Florida; a marked gravity anomaly there suggests that the meteorite, or part of it, may be buried beneath a thick cover of Cenozoic carbonates.

Iridium Anomalies

Iridium is a comparatively rare element in crustal rocks and is usually detectable only in background amounts of less than 100 ppm, in marked contrast to extraterrestrial material, which commonly contains iridium levels of thousands of parts per million. At some stratigraphic horizons unusually high con-

FIGURE 2.8
Typical tektites. A. Three views of flanged australite. B. Sculpturing of tektites; note deep meandrine grooves in this phillipinite. C. Pits and grooves are seen in this Ivory Coast tektite. D. A dumbbell shape is not uncommon.

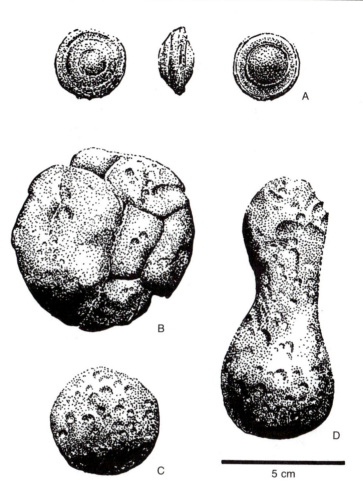

5 cm

centrations of iridium have been noted and several explanations have been proposed. It has been suggested that the known occurrence of iridium in volcanic ejecta is a likely source and that high concentrations in sediment indicate contamination by pyroclastic material. Another explanation has it that in areas of the seafloor with extremely slow sediment-accumulation rates, the iridium concentrations will be deceptively high when compared with parts of the section below and above, where sediment accumulation had proceeded at a more normal rate. Such mechanisms may be responsible for anomalies in special cases, but it is now generally believed that unusual iridium concentrations in sediment can best be explained as contamination by fallout from atmospheric dust

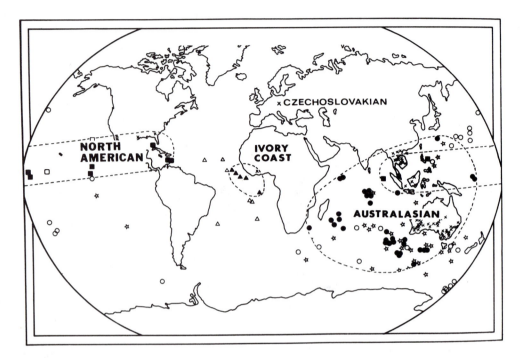

FIGURE 2.9

Tektite strewnfields. Tektite locations on land are indicated by Xs. The locations of cores containing Australasian, Ivory Coast, and North American tektites are indicated by solid circles, solid triangles, and solid squares, respectively. Cores with no tektites are indicated by open symbols. Boundaries of strewnfields (dashed lines) include all known tektite and microtektite occurrences. (From Glass, B. P., 1982, Possible correlations between tektite events and climatic changes?: Geol. Soc. America Spec. Paper 190, Fig. 1, p. 252.)

generated by a cosmic impact. Meteorites are known to have iridium concentrations in cosmic rather than terrestrial abundance. In the case of the Cretaceous-Tertiary boundary, discussed in the next section, the iridium anomaly has been found in Cretaceous-Tertiary boundary clays at more than 75 sites (Alvarez, 1986), both marine and nonmarine, around the world (Figs. 2.10, 2.11).

2.5 THE TERMINAL CRETACEOUS EVENT

Of the mass extinctions during the Phanerozoic, the one at the close of the Cretaceous has attracted the most attention and has come to be a sort of "type," or classical, example. It was marked by the disappearance of such diverse groups as the dinosaurs, most large marine reptiles, the flying reptiles, the ammonites, and the rudist molluscs. Many other organisms, such as the coccolithophorids and certain planktonic foraminiferans, suffered disastrous reduction in numbers and diversity at this time (Table 2.1). The suddenness of the faunal change seems to be well documented in numerous successions with good biostratigraphic control, as, for example, that reported by Smit and Hertogen (1980) for planktonic foraminiferans. In

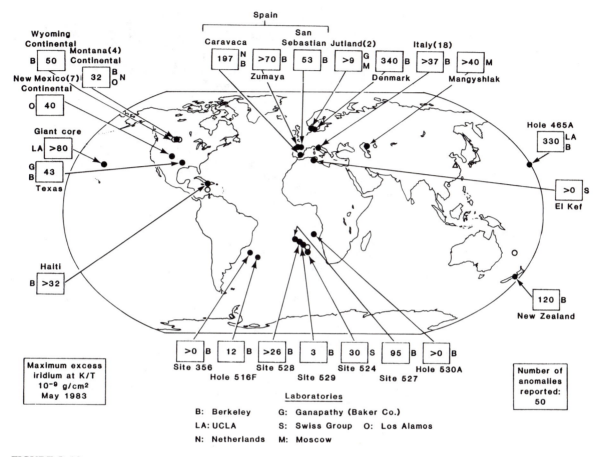

FIGURE 2.10

Worldwide occurrence of the iridium anomaly at the Cretaceous-Tertiary boundary. Values are given in terms of the excess iridium integrated beneath the curve of concentration (in parts per billion) as a function of depth. Sites marked >0 have anomalous iridium but not enough measured points to define the shape of the curve. Open circles indicate sites for which analyses were incomplete. (From Alvarez, W., Alvarez, L. W., Asaro, F., and Michel, H. V., The end of the Cretaceous: Sharp boundary or gradual transition?: Science, v. 223, Fig. 1, p. 1184. Copyright 1984 by the AAAS.)

all, according to Russell (1976), about 75 percent of all Late Mesozoic animal species disappeared by the beginning of the Cenozoic. There has been much speculation as to the cause of the mass extinction. One study (McLean, 1978) has suggested that rapid global warming was induced by a greenhouse effect triggered by a chain reaction involving increasing ocean temperatures and carbon dioxide expulsion. Other models proposed rely on sudden changes in the salinity of ocean surface waters (Gartner and Keaney, 1978, 1979; Clark and Kitchell, 1979; Gartner and McGuirk, 1979) or a massive increase in cosmic radiation consequent on a reduction in the earth's magnetic field or the explosion of a supernova (Waddington, 1967; Laster, 1968; Terry and Tucker, 1968). There is grow-

FIGURE 2.11
Clay layer at the Cretaceous-Tertiary boundary near Gubbio, northern Italy. The iridium
concentration is about 30 times higher in the clay than in the sediments below and
above. Fossil coccoliths are abundant immediately below the clay but only sparsely
present above it. (Photograph courtesy of Lawrence Berkeley Laboratories.)

ing evidence that the extinctions at the end of
the Cretaceous came as a result of the impact
of an Apollo object with a diameter of 10 km
or more, although some workers (Hsü, 1980)
prefer a comet. Numerous models describing
the effects of such an impact have been proposed. According to Emiliani (1980; see also
Fig. 2.12), the sudden extinction of so many
diverse animals and plants was due to atmospheric heating as a consequence of the impact. In Macdougall's (1988) view, such heating would have produced significant amounts
of nitrogen oxides, resulting in acid rains;

trace element toxification effects were described by Erickson and Dickson (1987). Alvarez et al. (1980), on the other hand, described a biological feedback mechanism
triggered by the dying off of land vegetation
and oceanic phytoplankton after the sunlight
was obscured by the injection of dust into the
atmosphere at the time of impact. The discovery at the Cretaceous-Tertiary boundary of
a clay layer with abnormally high levels of
iridium, an element rare on earth but common in meteorites, was regarded as strong
supporting evidence (Fig. 2.13). In another

FIGURE 2.12
Another view of the Cretaceous-Tertiary extinctions. (From Emiliani, C., EOS, v. 61, No. 26, Fig. 1, p. 505, 1980, copyright by the American Geophysical Union.)

study (Ganapathy, 1980), a whole range of noble metals considered sensitive indicators of meteorites (iridium, osmium, gold, platinum, etc.) was found in cosmic proportions in clays from the Cretaceous-Tertiary boundary, adding still further weight to the impact hypothesis. The geochemical evidence has been supported further by the discovery of shocked quartz (coesite and stishovite) in Cretaceous-Tertiary boundary clays at numerous sites (Bohor et al., 1987; McHone et al., 1989); these are considered unequivocal indicators of meteorite impact.

The site of the impact crater remains unknown. Clearly, it was nowhere on the continents, because a structure of that size could hardly have escaped attention, so an oceanic impact seems almost a certainty. No crater has been found and in any case, its absence can be conveniently attributed to subduction of oceanic crust or burial by sediments. On the other hand, some direct physical evidence of an oceanic impact is suggested by the presence of coarse breccias associated

with iridium anomalies at the Cretaceous-Tertiary boundary around the Gulf of Mexico. At several sites near the Brazos River, Texas, late Cretaceous-Paleocene quiet-water mudstones are interrupted by coarse clastics interpreted as deposited by a tsunami 50–100 m high. The most likely source, according to Bourgeois et al. (1988) was a **bolide**–water impact.

Inevitably, attention now has been focused on the stratigraphic horizons of other mass extinctions to see if these also can be explained by cosmic impacts. The search for evidence has met with mixed success (Table 2.2). At the Frasnian-Famennian boundary (Fig. 2.14), for example, marking a late Devonian mass extinction of major proportions, an iridium anomaly has been documented by Playford et al. (1984). On the other hand, at the Permian-Triassic boundary, marking the biggest mass extinction of them all, no anomaly occurs and iridium concentrations were reported by Zhou and Kyte (1988) as at or below normal. Geochemical studies of boundary clays from three Chinese locations suggested derivation from massive silicic volcanic eruptions. No iridium anomaly has been reported either from another mass extinction boundary, that at the end of the Triassic, although coesite has been found in sections at the top of the Rhaetic in Austria. In deep-sea cores, the Eocene-Oligocene boundary is said to be marked by an important mass extinction, although there is some controversy over this (Corliss et al., 1984), and it too apparently coincides with an iridium anomaly (Ganapathy, 1982). The Caribbean microtektites, mentioned previously, occur just above this iridium anomaly and so apparently do not record the same impact event (Sanfilipo et al., 1985). Yet another iridium anomaly, considered the signal of an impact, has been documented in late Pliocene cores from the Antarctic Ocean, some 1400 km west of Cape Horn and dated at 2.3 m.y.

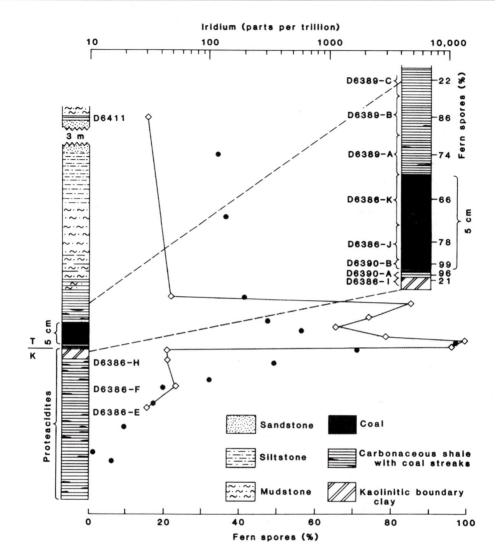

FIGURE 2.13
Diagram showing the lithology of the Cretaceous-Tertiary boundary at the Starkville North site, 5 km south of Trinidad, Colorado. The large black dots show the variation in iridium concentrations, the solid line and diamonds show the fern spore percentages, and the inset shows the detail of the boundary interval. The USGS location numbers are shown adjacent to the column. (From Tschudy, R. H., Pillmore, C. L., Orth, C. J., Gilmore, J. S., and Knight, J. D., Disruption of the terrestrial plant ecosystem at the Cretaceous-Tertiary boundary, western interior: Science, v. 225, Fig. 2, p. 1031. Copyright 1984 by the AAAS.)

(Kyte et al., 1981, 1988). This apparently was not attended by the alleged disastrous consequences of the earlier impacts because there is no known biostratigraphic disconti- nuity in the late Pliocene. On the other hand, there was a major climatic deterioration at this time and the impact may have been re- sponsible.

TABLE 2.2
Iridium anomalies at selected sites

Age	Site	Anomaly Value	X Local Background	Author
Eocene-Oligocene	Caribbean, DSDP Site 149	4.10 ppb	20	Ganapathy, 1982
Cretaceous-Tertiary	Gubbio, Italy Stebns Klint, Denmark Woodside, New Zealand Morgan Creek, Sask.	9.1 ppb 41.6 ppb 28.0 ppb 3.0 ppb	30 160 20 100	Alvarez et al., 1980 Brooks, et al., 1984 Nichols et al., 1986
Famennian-Frasnian (Devonian)	Western Australia	0.3 ppb	20	Playford et al., 1984

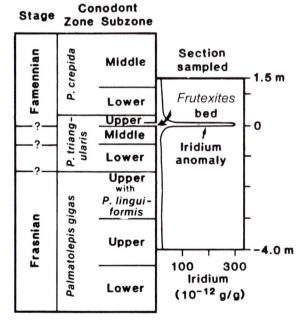

FIGURE 2.14
The iridium anomaly recorded near or at the Frasnian-Famennian boundary in the late Devonian. (From Playford, P.E., et al., Iridium anomaly in the Upper Devonian of the Canning Basin, Western Australia, v. 226, Fig. 2, p. 437. Copyright 1984 by the AAAS.)

Although the evidence for periodic cosmic impacts is strong, there are still many questions regarding their link with mass extinctions. In the continuing debate, opinions vary widely. Cosmic impacts are seen by some as the prime cause of mass extinctions and a major driving force in evolution and by others as only a minor influence, perhaps a "last straw" effect that sealed the fate of certain taxonomic groups already in decline. Although such a gradual decline has been suggested in the case of the dinosaurs, Russell (1975, 1984) has provided evidence that during latest Cretaceous time there was no decline in either dinosaur numbers or diversity. A study by Retallack et al. (1987) of paleosols across the Cretaceous-Tertiary boundary would support this view. It suggested that the decline in diversity and abundance of bone fossils toward the close of the Cretaceous was due to dissolution of bones and teeth and thus, a **taphonomic** effect.

But then again, in the case of the marine macrofossils, Kauffman's (1984) work suggests that rather than a wholesale catastrophe at the close of the Cretaceous, the evidence points to a complex pattern of extinctions that extended over millions of years. It is claimed also that the argument for a cosmic impact as the cause of the end of Cretaceous extinctions is weakened by the discovery of dinosaur remains (Fig. 2.15) in the San Juan basin of New Mexico at a horizon dated some 165,000 years younger than the iridium spike in the Cretaceous-Tertiary boundary clay (Morner, 1982). However, to suggest that the

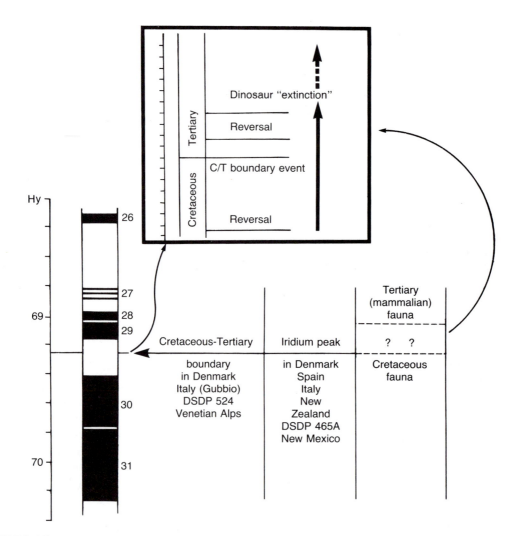

FIGURE 2.15

How long after the terminal Cretaceous impact did the dinosaurs survive? The data in this diagram indicate that dinosaur fossils have been found in the San Juan Basin of New Mexico up to the end of anomaly 29 of the magnetic polarity scale, whereas the Cretaceous-Tertiary boundary events took place at a time slightly earlier than the polarity switch at the start of anomaly 29. According to some authorities, the dinosaur fossils at the younger horizons were reworked from older deposits. (Based on data from Morner, N. A., 1982, Jour. Geology, v. 90, p. 564–573, by permission of the University of Chicago Press.)

disappearance of the dinosaurs and so many other forms at the close of the Cretaceous was virtually instantaneous and a *direct* result of a cosmic impact, is almost certainly an oversimplification. What is more likely is that an impact would set in motion a chain reaction of events involving collapsing food chains and deteriorating environments that continued over a protracted period of adjustment that might even have lasted for tens of thousands of years or longer. In human terms, these are hardly "sudden" events. However, in the context of a geological time frame, they are perceived as extremely rapid.

Before leaving the subject of mass extinctions, it is interesting to speculate on the effect that humans have had on the earth's surface. There is little doubt that some future stratigraphic succession will record the present time as also marked by a sudden extinction of many organisms. As with the Cretaceous terminal event, there likely will be a strong geochemical signal within the sedimentary record of humans' (probably brief) sojourn on this planet. The effects of a major nuclear exchange, described by Ehrlich et al. (1984) in the scenario for the "nuclear winter," bear a strong resemblance to those postulated by Alvarez et al. (1980) for the terminal Cretaceous cosmic impact. The discovery (Wolbach et al., 1985) of a carbon-rich layer at the top of the Cretaceous-Tertiary boundary clay, and interpreted as derived from the soot of wildfires on a global scale, adds a further point of similarity between the terminal Cretaceous event and what, in the context of terrestrial history, supposedly would be the "terminal Holocene event"!

2.6 EVENT STRATIGRAPHY

Global Events

If, as seems likely, the iridium spikes at the Cretaceous-Tertiary boundary and elsewhere were a direct result of major cosmic impacts, they record, as do such features as the widespread tektite fields, extremely short-lived events of global dimensions. Clearly, these features are excellent time horizons and more precisely defined than are biostratigraphic horizons. The recognition of the events themselves was considered by Ager (1981) in what he called "event stratigraphy." According to Ager, an event, such as a climate change or marine transgression, might be marked by a change in the lithology of a geological section at one location and discerned elsewhere by changes of a completely different nature. It is the event itself that provides the basis for correlation rather than correspondence of rock types or fossils. Apart from direct evidence for cosmic impacts, the geologic column contains many other indicators of rapid change, the nature of which usually can only be guessed. Ager was struck, as indeed many other geologists have been, by the remarkably sudden changes in lithology and the appearance of unusual rock types, often as quite thin formations, over very large areas of the earth at about the same time. For example, he cited a thin succession of black shales, red marls, and sandstones of latest Triassic age that are essentially the same in places as far apart as England, northern Italy, and Thailand. Successions such as these presumably record sudden changes in the sedimentary environment that may, on occasion, have reached global dimensions. What caused them and how rapid they were is a matter of debate. Some at least were climatic in origin, triggered in no small part by the movement of continental plates into new latitudes and the impacts of changing landmass configurations on ocean currents.

The Rare Event

Although natural catastrophes, by definition, are unusual happenings, the majority are repeated many times over. From large-scale happenings, like cosmic impacts, to local and regional phenomena such as earthquakes or volcanic eruptions, most recur at longer or shorter intervals, even if a complex interplay of different processes is involved. The catastrophic bursting of glacial dams, for example, such as that of Pleistocene Lake Missoula in Montana (Bretz, 1969; Baker, 1983), was repeated several times as the position of the Pleistocene ice fronts fluctuated. One of these floods is estimated to have involved a 600-m head of impounded water and to have run for two weeks. In places the torrent was thought to have been over 250 m deep and 14 km wide. The deep canyons and considerable hills of alluvial gravels of the channelled scablands remain today as testimony of what must have been an awesome spectacle.

The truly rare event, the one that perhaps occurs only once in history, probably should be considered separately. It is often said that "given time, anything can happen." In the sciences this must be stated more rigorously to read "given time, anything that can possibly happen will happen" because it is clearly necessary to exclude impossible phenomena that involve the breaking of natural laws. As Gretener (1967) pointed out, the simple principle involved in games of chance means that given a sufficient number of trials, the improbable becomes probable and eventually approaches certainty. There is no question that this principle has important implications in geology and is seen to operate in the case of random happenings such as the impact of large meteorites. Other examples occur when there is a chance combination of events, referred to by Simpson (1940) as the "sweepstakes effect," in the dispersal of faunas and floras when geographic barriers are eventually crossed as a consequence of a random set of circumstances. Leaving aside the obvious cases of the appearance and extinction of species, truly unique, large-scale events (events that happened only once in history), although difficult to find, did occur. One example is the breaching of the Straits of Gibraltar when the Atlantic Ocean poured in to refill the Mediterranean Basin after the so-

called Messinian salinity crisis of the Miocene; at that time the Mediterranean Basin became an evaporite basin below sea level, similar to the modern Caspian and Dead seas (Adams et al., 1977). Other rather obvious examples of unique events were the origin of planet Earth and the origin on earth of life itself.

2.7 CONCLUSION

Some of the discussion in this chapter has been concerned with hypotheses. It should be remembered that this is all they are—hypotheses, which only later work will prove or disprove. This is the way science works. The present status of these ideas can be summarized as follows: There seems little doubt that a major meteorite (or cometary) impact occurred at the end of the Cretaceous. There also seems to be an overwhelming body of evidence for numerous mass extinction events during the history of life on earth. One of these occurred, apparently, at about the same time as the end-of-Cretaceous meteorite impact. The timing of these two events is highly suggestive of a cause-and-effect relationship; the possibility that it was pure coincidence seems unlikely.

What then of other meteorite impacts and other mass extinctions? Did the other major meteorite impacts, for which there is strong evidence, also cause mass extinctions? Conversely, were other mass extinctions also caused by meteorite impacts? The search for the answers to these questions undoubtedly will be pursued vigorously in the years to come. The question of periodicity in both meteorite impacts and mass extinctions also has to be addressed. Can it be explained by the Nemesis star hypothesis, to be discussed in Chapter 4? Is there really an Oort cloud? There is clearly no shortage of questions; just as during the formative years of the plate tectonic paradigm, geology is again poised at the beginning of a period of exciting new discoveries.

REFERENCES

Adams, C. G., Benson, R. H., Ridd, R. B., Ryan, W. B. F., and Wright, R. C., 1977, The Messinian salinity crisis and evidence of Late Miocene eustatic changes in the world ocean: Nature, v. 269, p. 383–386.

Ager, D. V., 1981, The nature of the stratigraphical record, 2nd ed.: New York, John Wiley & Sons, 122 p.

Alvarez, L. W., Alvarez, W., Asaro, F., and Michel, H. V., 1980, Extraterrestrial cause for Cretaceous-Tertiary extinction: Science, v. 208, p. 1095–1108.

Alvarez, W., 1986, Toward a theory of impact crises: EOS, v. 67, p. 649, 653–655, 658.

Baker, V. R., 1983, Late Pleistocene fluvial systems, in Porter, S. C., ed., The Late Pleistocene, v. 1, in Wright, H. E., ed., Late Quaternary environments of the United States: Minneapolis, Univ. of Minnesota, 1984, p. 115–124.

Berggren, W. A., and Van Couvering, J. A., eds., 1984, Catastrophes and Earth history: Princeton, NJ, Princeton Univ. Press, 464 p.

Bohor, B. F., Modreski, P. J., and Foord, E. E., 1987, Shocked quartz in the Cretaceous-Tertiary boundary clays: Evidence for a global distribution: Science, v. 236, p. 705–709.

Bourgeois, J., Hansen, T. A., Wiberg, P. L., and Kauffman, E. G., 1988, A tsunami deposit at the Cretaceous-Tertiary boundary in Texas: Science, v. 241, p. 567–570.

Bretz, J H., 1969, The Lake Missoula floods and the channeled scabland: Jour. Geology, v. 77, p. 505–543.

Brooks, E. R., Reeves, R. D., Yang, X-H., Ryan, D. E. Holzbecher, J., Collen, J. D., Neall, V. E., and Lee, J., 1984, Elemental anomalies at the Cretaceous-Tertiary boundary: Science, v. 226, p. 183–195.

Chapman, D. R., 1971, Australasian tektite geographic pattern, crater and ray of origin, and theory of tektite events: Jour. Geophys. Research, v. 76, p. 6309–6338.

Clark, D. L., and Kitchell, J., 1979, Comment on the terminal Cretaceous event: A geologic problem with an oceanographic solution: Geology, v. 7, p. 228.

Corliss, B. H., Aubry, M., Berggren, W. A., Fenner, J. M., Keigwin, L. D., and Keller, G., 1984, The Eocene/Oligocene boundary event in the deep sea: Science, v. 226, p. 806–810.

Ehrlich, P. R., Sagan, C., Kennedy, D., and Roberts, W. O., 1984, The cold and the dark: New York, Norton, 229 p.

Emiliani, C., 1980, Death and renovation at the end of the Mesozoic: EOS, v. 61, p. 505–506.

Erickson, D. J., III, and Dickson, S. M., 1987, Global trace-element biogeochemistry at the K/T boundary: Oceanic and biotic response to a hypothetical meteorite impact: Geology, v. 15, p. 1014–1017.

Ganapathy, R., 1980, A major meteorite impact on the earth 65 million years ago: Evidence from the Cretaceous-Tertiary boundary clay: Science, v. 209, p. 921–923.

———1982, Evidence for a major meteorite impact on the earth 34 million years ago: Implications for Eocene extinctions: Science, v. 261, p. 885–888.

Gartner, S., and Keany, J., 1978, The terminal Cretaceous event: A geologic problem with an oceanographic solution: Geology, v. 6, p. 708–712.

———1979, Reply to comment on the terminal Cretaceous event: A geologic problem with an oceanographic solution: Geology, v. 7, p. 229.

Gartner, S., and McGuirk, J. P., 1979, Terminal Cretaceous extinction, scenario for a catastrophe: Science, v. 206, p. 1272–1276.

Gretener, P. E., 1967, The significance of the rare event in geology: Am. Assoc. Petroleum Geologists Bull., v. 51, p. 2197–2206.

Hallam, A., 1981, Facies interpretation and the stratigraphic record: San Francisco, Freeman, 291 p.

Herman, Y., 1981, Causes of massive biotic extinctions and explosive evolutionary diversification throughout Phanerozoic time: Geology, v. 9, p. 104–108.

Hsü, K. J., 1980, Terrestrial catastrophe caused by cometary impact at the end of Cretaceous: Nature, v. 285, p. 201–203.

Kauffman, E. G., 1984, The fabric of Cretaceous marine extinctions, in Berggren, W. A., and Van

Couvering, J. A., eds. Catastrophes and earth history: Princeton, NJ, Princeton Univ. Press, p. 151–246.

Kyte, F. T., Zhou, L., and Wasson, J. T., 1981, High noble metal concentrations in a late Pliocene sediment: Nature, v. 292, p. 417–420.

———1988, New evidence on the size and possible effects of a Late Pliocene oceanic asteroid impact: Science, v. 241, p. 63–65.

Laster, H., 1968, Cosmic rays from nearby supernovae: Biological effects: Science, v. 160, p. 1138.

LeVie, D. S., Jr., 1986, South Texas' Lyles Ranch field: Production from an astrobleme?: Oil and Gas Jour., v. 84, p. 135–138.

Macdougall, J. D., 1988, Seawater strontium isotopes, acid rain and the Cretaceous-Tertiary boundary: Science, v. 239, p. 485–487.

McHone, J. F., Nieman, R. A., Lewis, C. F., and Yates, A. M., 1989, Stishovite at the Cretaceous-Tertiary boundary, Raton, New Mexico: Science, v. 243, p. 1182–1184.

McLaren, D. J., 1983, Bolides and stratigraphy: Geol. Soc. America Bull., v. 94, p. 313–324.

McLean, D. M., 1978, A terminal Mesozoic "greenhouse": Lessons from the past: Science, v. 20, p. 401–406.

Morner, N. A., 1982, The Cretaceous-Tertiary boundary: Chronostratigraphic position and sequence of events: Jour. Geology, v. 90, p. 564–573.

Newell, N. D., 1967, Revolutions in the history of life: Geol. Soc. America Spec. Paper 89, p. 63–91.

Nichols, D. J., Jarzen, D. M., Orth, C. J., and Oliver, P. Q., 1986, Palynological and iridium anomalies at Cretaceous-Tertiary boundary, south-central Saskatchewan: Science, v. 231, p. 714–717.

Petuch, E., 1987, The Florida Everglades: A buried pseudoatoll?: Jour. Coastal Research, v. 3, p. 189–200.

Playford, P. E., McLaren, D. J., Orth, C. J., Gilmore, J. S., and Goodfellow, W. D., 1984, Iridium anomaly in the Upper Devonian of the Canning Basin, Western Australia: Science, v. 226, p. 437–439.

Raup, D. M., 1986, The Nemesis affair: New York, Norton, 220 p.

Retallack, G. J., Leahy, G. D., and Spoon, M. D., 1987, Evidence from paleosols for ecosystem changes across the Cretaceous/Tertiary boundary in eastern Montana: Geology, v. 15, p. 1090–1093.

Rudwick, M. J. S., 1972, The meaning of fossils: London and New York, MacDonald-Elsevier, 287 p.

Russell, D. A., 1975, Reptilian diversity and the Cretaceous-Tertiary transition in North America: Geol. Assoc. of Canada Spec. Paper 13, p. 119–136.

———1976, The biotic crisis at the end of the Cretaceous Period: Syllogeus, Natl. Mus. Canada, p. 11–23.

———1984, The gradual decline of the dinosaurs—Fact or fallacy?: Nature, v. 307, p. 360–361.

Sanfilippo, A., Riedel, W. R., Glass B. P., and Kyte, F. T., 1985, Late Eocene microtektites and radiolarian extinctions on Barbados: Nature, v. 314, p. 613–615.

Sawatzky, H. B., 1975, Astroblemes in Williston Basin: Am. Assoc. Petroleum Geologists Bull., v. 59, p. 694–710.

Shirley, K., 1989, Structures have impact on theories: A.A.P.G. Explorer, v. 10, p. 14–17.

Shoemaker, E. M., 1982, Bombardment of the Earth from late stages of accretion to modern times: Geol. Soc. America, Abs. with Programs, v. 14, p. 616.

Silver, L. T., and Schultz, P. H., eds., 1982, Geological implications of impacts of large asteroids and comets on the earth: Geol. Soc. America Spec. Paper 190.

Simpson, G. G., 1940, Mammals and land bridges: Jour. Washington Acad. Sci., v. 30, p. 137–163.

Smit, J., and Hertogen, J., 1980, An extraterrestrial event at the Cretaceous-Tertiary boundary: Nature, v. 285, p. 198–200.

Terry, K. D., and Tucker, W. H., 1968, Biologic effects of supernovae: Science, v. 159, p. 421–423.

Verbeek, R. D. M., 1897, Glaskogels van Billiton:

Jaarb. mijnwezen ned. Oost Indie, Jahrb. 36, p. 235–272.

Waddington, C. J., 1967, Paleomagnetic field reversals and cosmic radiation: Science, v. 158, p. 913–915.

Wolbach, W. S., Lewis, R. S., and Anders, E., 1985, Cretaceous extinctions: Evidence for wildfires and search for meteoritic material: Science, v. 230, p. 167–172.

Zhou, L., and Kyte, F. T., 1988, The Permian-Triassic boundary event: A geochemical study of three Chinese sections: Earth and Planetary Sci. Letters, v. 90, p. 411–421.

3
STRATIGRAPHIC CLASSIFICATION

As soon as it was recognised that the rocks which compose the earth's crust were not a mere congeries of rock-masses without order or sequence, but that when properly interpreted they arranged themselves in an orderly succession of formations, then it was seen to be necessary that names should be given to all parts of this wonderful succession of stratified deposits.

Jukes Brown

3.1 INTRODUCTION

As in all fields of science, stratigraphy has become increasingly sophisticated; it was obvious many years ago that rules and procedures were necessary in defining the many and varied kinds of units used in stratigraphy. Accordingly, in 1952 an International Subcommission on Stratigraphic Terminology (later called the International Subcommission on Stratigraphic Classification, or ISSC, and more recently the International Commission on Stratigraphy, or ICS) was created by the 19th International Geological Congress. The results of the labors of the many workers involved appeared in 1976 as the *International Stratigraphic Guide* (Hedberg, 1976). Although this publication represents a consensus of the members of the subcommission and contains valuable recommendations, it was never adopted as a statutory policy document by ICS. The most recent guidelines are to be found in the ICS statutes (Cowie, 1986), which, in many ways, can be considered as playing the same role as does the International Code of Zoological Nomenclature in zoology. Naturally, the rules set out in any stratigraphic code are subject to change as work continues. In 1979, for example, it was found necessary to publish a supplementary chapter to the *Stratigraphic Guide* to deal with magnetostratigraphic units (ISSC, 1979). In North America, the work of the American Commission on Stratigraphic Nomenclature had produced stratigraphic codes in 1961 and 1970 (ACSN, 1961, 1970), but more recently a new code has been written by the North American Commission on Stratigraphic Nomenclature; this was published in 1983 (NACSN, 1983).

Because all of stratigraphy deals with time and rocks, it follows that these are the two fundamental entities requiring division and classification. Thus, on the one hand are **lithostratigraphic** and similar units, based upon physical limits or internal attributes of bodies of rock, and on the other **geochronometric** units, measured in years. It has been suggested that these two kinds of divisions are the only really essential elements needed in stratigraphy (Ager, 1984), but this is a minority view. Be that as it may, from the earliest days of geology, there have been so-called **chronostratigraphic** and **geochronologic** units (Table 3.1), the first being rock units whose boundaries are defined by limiting time horizons, and the second being time units defined by the time span of the equivalent chronostratigraphic unit (for example, Silurian System and Silurian Period). There is something of a chicken-and-egg aspect about these two kinds of units, but they were the

TABLE 3.1

Summary of categories and unit terms in stratigraphic classification (after Hedberg, 1976)

Stratigraphic Categories	Principal Stratigraphic Unit Terms
Lithostratigraphic Defines a body of rock strata unified by overall homogeneity of lithology or combination of lithologies; may be sedimentary, metamorphic, or igneous	Group Formation* Member Bed(s)

Biostratigraphic
    Defines a body of rock unified by its fossil content
	Biozones:
	  Assemblage-zones
	  Range zones (various kinds)
	  Acme zones
	  Interval zones
	  Other kinds of biozones

Chronostratigraphic
    Defines a body of rock unified by being formed during a specific interval of geologic time; represents all rocks formed anywhere during a certain segment of earth history
 	
Geochronologic
    Defines a unit of geologic time determined by geologic methods; may correspond to the time span of a stratigraphic unit
	(Equivalent units)
	
	Eonothem Eon
	Erathem Era
	System Period
	Series Epoch
	Stage Age
	Chronozone Chron

*Fundamental unit (other categories do not have fundamental units)

natural outcome of attempts to` construct working divisions of the stratigraphic column in terms of elapsed time.

The classifications set out in both the *International Stratigraphic Guide* and the North American *Code of Stratigraphic Nomenclature* are summarized in Tables 3.1 and 3.2, and the more important elements are discussed in the following pages. The North American Stratigraphic Code appears in its entirety in Appendix 3. Magnetostratigraphic classifica-

tion is discussed in Chapter 6 and biostratigraphic classification in Chapter 9.

3.2 ROCK UNITS

Lithostratigraphic or rock units are the most fundamental of all the units used in stratigraphy because obviously this is where the tangible record begins. Rocks can be sampled and measured quite objectively, and in their initial description and definition no time

TABLE 3.2
Summary descriptions of the units recognized in the North American Stratigraphic Code (1983)

Rock Units	Description
Lithostratigraphic unit	Body of rock delineated by lithology and stratigraphic position; usually stratified and conforms to law of superposition
Lithodemic unit	Body of intrusive or metamorphosed rock not conforming to the law of superposition
Magnetostratigraphic unit	Body of rock unified by distinctive remanent magnetic properties
Biostratigraphic unit	Body of rock defined on the basis of its fossil content
Pedostratigraphic unit	Body of rock consisting of one or more buried pedologic horizons (paleosols or fossil soils)
Allostratigraphic unit	Body of rock contained between bounding discontinuities

Time and Time-Rock Units	Description
Geochronometric unit	Time unit measured in years
Geochronologic unit	Time unit based on a given chronostratigraphic unit and involving the time during which that unit formed
Polarity chronologic unit	Time unit that embraces that part of geologic time during which the earth's magnetic field had a characteristic polarity or sequence of polarities; corresponds to a particular polarity chronostratigraphic unit and is the basis for its definition
Diachronic unit	Time unit in which one or both boundaries are diachronous and thus mark the unequal spans of time represented by specific lithostratigraphic, biostratigraphic, pedostratigraphic, or allostratigraphic units or an assemblage of such units
Chronostratigraphic unit	Body of rock established as the basis for defining the specific time interval of a given geochronologic unit
Polarity chronostratigraphic unit	Body of rock that contains the magnetic signature imposed at the time the rock was formed; i.e., during the specific time interval of a polarity geochronologic unit

connotation is implied. As in all classifications a hierarchy of divisions of different rank is recognized, the most important being the formation.

Formations

The first procedure in stratigraphy is to measure the thickness of the various strata and to group within the succession those contiguous strata having homogeneous or similar lithology into a number of units, each of which is lithologically distinct from the units above and below (Fig. 3.1). Such lithologically homogeneous units described at a given location can be traced laterally and their distributions plotted on a map (Fig. 3.2).

Clearly, on a small-scale map for reconnaissance purposes, only large distinctive rock units would be readily mappable. On the other hand, in an area with large-scale base maps, good exposures and/or subsurface control, and enough stratigraphic data points to assure a sufficiently large sample size, even small units of only a few meters thick might be mapped. Such mappable units are the basic elements in stratigraphy and are termed **formations.**

The homogeneity of a particular formation may be expressed in one physical feature

FIGURE 3.1
Formational boundaries in the Triassic-Jurassic succession, Durango, Colorado. D = Dolores Formation, E = Entrada Sandstone, W = Wanakah Formation, JC = Junction Creek Sandstone, M = Morrison Formation, D = Dakota Sandstone.

FIGURE 3.2
Mapping formational boundaries from air photographs. The section seen in Figure 3.1 is
located in the upper center. The Entrada and Junction Creek sandstones are seen as
pale bands. (Photograph courtesy of U.S. Geological Survey.)

such as lithology, color, weathering characteristic, or even fossil content as in the case
of the shelly limestones, but is more usually
attributable to several features that are mutually related. The upper and lower limits of
a formation may be marked by a sudden
change in lithology or even an erosional
break. On the other hand, there may be a

gradual change through a thickness of transitional beds, sometimes called **passage
beds.** For example, a massive sandstone overlain by a shale may become progressively
more argillaceous toward its top and pass
gradationally up into the shale. Alternatively,
it may come to contain increasingly more
prominent intercalations of shale. The diffi-

culty lies in picking the boundary between the sandstone and shale. This usually can be done by selecting an arbitrary horizon—perhaps the highest single sandstone bed greater than a given thickness or, in working downward, the lowest shale band of a given thickness. This would mean that the upper part of the sandstone formation would be described as argillaceous, and the lower part of the shale formation would be arenaceous (Fig. 3.3). It might be convenient to recognize these parts of the section as distinct subdivi-

sions of the formation or members; for example, "upper shaly sandstone member" (see discussion of members in a later section). If a considerable thickness of these transitional beds is present, and the scale of the map warrants it, it might be better to separate them as a distinct formation of shaly sandstone or alternating sandstone and shale, occurring between a sandstone below and a shale above. There might still be a problem of transitional intervals at top and bottom but it would have been reduced in scale. It is

FIGURE 3.3
Stratigraphic section showing passage beds.

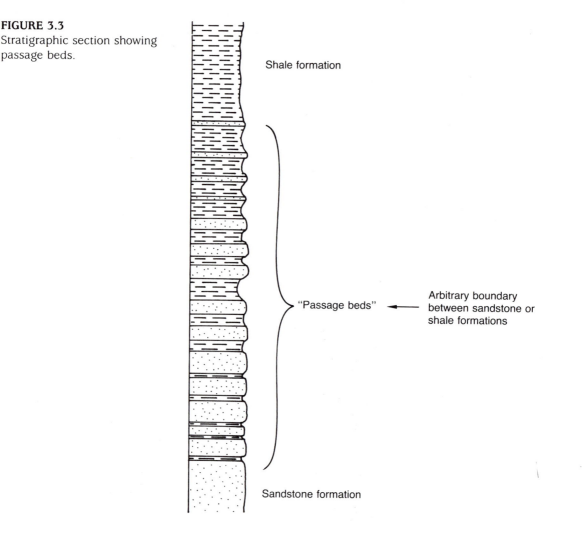

Shale formation

"Passage beds" ◄─── Arbitrary boundary between sandstone or shale formations

Sandstone formation

clear from this procedure that a formation can consist of alternations of two or more lithologies, the repetition of different lithologies being itself the distinctive feature of the formation. A formation also can consist of two or more distinct, nonalternating lithologies. For example, the Mississippian Sulphur Springs Formation of Missouri contains a basal shale, a middle oolitic limestone, and an upper sandstone. Each of these is too thin to map independently and too unlike the formations above or below to be included with them.

When a newly established stratigraphic unit, whether lithostratigraphic, chronostratigraphic, or biostratigraphic, is described, what is considered the most typical or representative section is selected as a **type section** (or type well); this is termed a **stratotype** (Table 3.3). The stratotype should include the full complement of the formation, although in areas of poor exposure it may be necessary to use outcrops at more than one location to provide a complete section of the formation; the stratotype is then described as a **composite section.** In naming a formation, a geographic name from the area of the stratotype is selected if possible (in the U.S., a geographical feature such as a river or mountain or the name of the nearest settlement with a post office is a reasonable guide). On maps and charts, this name precedes the word formation (or member, etc., as the case may be), although, quite frequently, a lithologic term is used instead, particularly in areas where traditional names have been long in use. Regional correlation charts contain many examples. For instance, the Upper Ordovician Plattin Limestone of western Kentucky is listed as the Plattin Formation in Indiana. Certain so-called "formations" are widely recognized and mapped, often without formal designations and stratotypes. In the mid-continent region, for example, Ordovician blanket sands overlying a pronounced unconformity are mapped over large areas, sometimes incorrectly, as "St. Peter Sandstone." The sands are sometimes referred to as the St. Peter Formation and, in some localities, are assigned group status.

As with taxonomic procedures used in the biologic sciences, care should be taken to avoid formation names already in use. In practice, this problem can be avoided by re-

TABLE 3.3
Stratotypes (after Hedberg, 1976)

1. *Primary Types*—originally designated and used in the original definition
   a. *Holostratotype*—original stratotype designated by the author at the time of establishment of a stratigraphic unit or boundary
   b. *Parastratotype*—supplementary stratotype used in the original definition by the original author to aid in elucidating the holostratotype
2. *Primary Types*—subsequently designated
   c. *Lectostratotype*—stratotype selected later in the absence of an adequately designated original stratotype
   d. *Neostratotype*—new stratotype selected to replace an older one that has been destroyed or nullified
3. *Secondary Types*
   e. *Hypostratotype* (reference section, auxiliary reference section)—stratotype designated to extend knowledge of the unit or boundary established by a stratotype to other geographical areas or to other facies; always subordinate to the holostratotype

ferring to the *Lexique Stratigraphique International,* published in Paris under the auspices of the International Union of Geologic Sciences (IUGS), which lists all published formation names. In the United States similar lexicons are published by the U.S. Geological Survey and periodically updated (for example, Keroher, 1968; Swanson et al., 1976). The 1966 revision was, in fact, published in Paris as volume 7 of the *Lexique Stratigraphique International.*

Within a formation or other lithostratigraphic unit, however it is defined, there are likely to be masses of lithologically simple and uniform rock that are recognizable as three-dimensional entities interfingering with and grading into the rocks that surround them. It is sometimes useful to separate such rock bodies from any role they may play in the hierarchy of lithostratigraphic units; the term **lithosome** was proposed for this purpose by Wheeler and Mallory (1956).

Members

Description of a stratigraphic section must begin with its division into suitable forma-

tions; further subdivision is not obligatory. Usually, however, it is convenient to recognize subdivisions of lower rank than formations, and groupings of formations into units of higher rank. Within a formation distinctive units may be established as **members.** These also are given names, as with the formations, and stratotypes are designated. Usually the formal names are geographic but may be informally descriptive; for example, "upper shaly sandstone member." A member may be less persistent than the formation containing it and is not necessarily recognizable over the entire geographic extent of the formation. Conversely, a distinctive member in one formation may continue laterally to become a member of an adjacent formation (Fig. 3.4). In some cases, a member may take on a formation status somewhere else. For example, the Dunderberg shale member of the Napah Formation in the southern Great Basin becomes the Dunderberg Formation in Utah and central Nevada. In any sequence of formations, only one or two formations may be subdivided into members. Because one formation has been divided, it is not necessary to divide the others. Similarly, within one for-

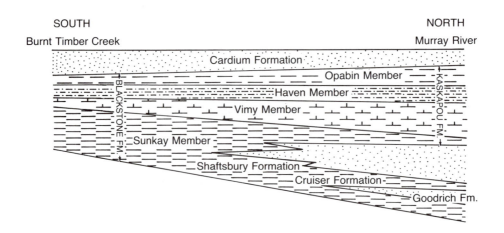

FIGURE 3.4
Formations and members in the Upper Cretaceous succession of the Rocky Mountain foothills, Alberta, Canada.

mation, the recognition of one or more distinctive units as members does not mean that the rest of the formation necessarily has to be divided into named members, although in some cases the presence of unnamed members between the named ones is implied. Not all members are necessarily stratified rock bodies; they may comprise irregular carbonate reefoid buildups (see Fig. 14.14), lens- or pod-shaped sand bodies, channel-fill sands (**shoestring sands**), tongues of gravel, and so on. A descriptive term— *lens, reef, tongue,* etc. (see Fig. 14.24b)—frequently replaces the term *member* in the name.

Groups

Moving up the hierarchy of lithostratigraphic units, two or more contiguous formations may be included in a unit known as a **group.** Again a suitable geographic name is used and a stratotype designated. The reason for bringing formations together in a group is usually to emphasize certain lithologic features these formations have in common and which set them apart from portions of the section above and below. Just as a formation does not have to contain everywhere all the members recognized at one location or another, so may a group differ from place to place in its constituent formations. Again, as with the member/formation relationship, a particular formation included in a group at one location may persist into an adjacent group. Where it is necessary to assemble groups into larger lithostratigraphic entities the term **supergroup** is used.

Changes in Rank

With improved knowledge and increasing refinement in the subdivision of the stratigraphic section, further subdivision may be called for. For example, the members of a formation may be themselves subdivided; a common procedure is to raise the members

to formation rank, with the new subdivisions as members. The original formation then may become a group. This may lead to the potentially confusing situation where the same name is used for a formation in one place and for a group in another. That the more sophisticated stratigraphic subdivision achieved in one place may not extend to an adjacent area may be due to several reasons. Sometimes it is because the character of the succession does not justify the same stratigraphic treatment. Commonly, it is due to facies changes or because the units become too thin. Sometimes, it is simply a case of poor exposure or because the area is less well known geologically.

Traditional Terminology

Although the various stratigraphic terms and naming procedures recommended by the ISSC have been widely adopted, in many places, notably in Europe, more traditional terminologies are still used. In Britain, for example, although stratigraphic procedures have been codified (Holland et al., 1978; Harland et al., 1982) many lithostratigraphic units have retained the names originally given them in the nineteenth century. As is still common in the U.S., a geographic name may precede a lithologic term with the implication that it is a unit of formational rank. Another common practice is to use the term "beds," as in "Ilfracombe Beds" of the Devonian of southwestern England. Sometimes subdivisions of such units also are labeled as "beds" and so, presumably, have member status. A growing practice is to add the term formation or other lithostratigraphic term to the earlier names. For example, correlation charts (House et al., 1977) for the Devonian of County Cork, Ireland, show the West Cork Sandstone Formation. In Killarney this unit becomes the Lower Purple Sandstone Formation, which, it must be admitted, is far more descriptive!

Correlation

In the mapping of the areal distribution of a formation, the most reliable method is by "walking out" the formation so that there is 100 percent certainty that its occurrence at one location can be established as having continuity with that at another. In areas with good exposure, such as in deserts or in the arctic, continuity can be established with some degree of confidence by air reconnaissance or by photogeologic surveys (Fig. 3.5), supplemented by spot checks on the ground, especially if the beds are flat-lying. The far more usual situation is one in which exposures are intermittent, either because of structural complexity, partial loss by erosion, or concealment by soils and surface deposits or vegetative cover. Establishing the continu-

ity of a single formation between two locations becomes increasingly unreliable with increasing distance but is strengthened greatly if a sequence of two or more adjacent formations can be correlated together. This applies particularly in the subsurface, where much of the correlation from well to well depends heavily upon data from electric, radioactivity, or sonic logs.

It should be remembered that beyond a very local area, or in special cases as with volcanic ash beds, lithologic correlation almost certainly does not mean time correlation. In comparing two widely separated stratigraphic sections, there may be a close *match* between the respective lithostratigraphic units, but to many workers this would not imply correlation. Instead, that term is reserved for strata that are demon-

FIGURE 3.5
Portion of a typical geophoto map. Prominent bands are limestones, shales, and sandstones of largely Cambrian, Mississippian, and Pennsylvanian ages, near Silverton, Colorado. (Photograph courtesy of D. S. Baars.)

strated as *coeval;* that is, of the same age. To the working stratigrapher, correlation in the time sense usually means biostratigraphic or magnetostratigraphic correlation. In practice, it is obviously wise to specify which kind of correlation is referred to.

Traced laterally from its stratotype a formation inevitably undergoes changes in its lithology and thickness. Eventually it becomes unrecognizable and, in fact, has become another formation, presumably with a stratotype somewhere else. In the area where one formation passes laterally into another, the stratigraphic section often shows an alternation of the lithologies of each of the formations concerned, so that they are seen to interfinger or intertongue. Because, after all, formations are part of (or entire) natural facies, some facies relationships literally defy subdivision into human-devised units of convenience. The problem of intertonguing formations and precisely what name to apply in the transitional zones can be handled by such devices as the arbitrary cutoff, as shown in Figure 3.6, or by using combination names. Sometimes map boundaries between intertonguing formations are controlled by localized nomenclature of regional geological surveys. In large countries like the U.S.A., Canada, Australia, etc., formation names applied by one state or provincial survey may not be always carried across state lines, and we have the phenomenon of "political unconformities" on geological maps.

For a particular area, usually a natural structural or other region—for example, eastern Tennessee, Gulf Coast Region, Ottawa Valley, Canning Basin—the commonly recognized formation names may be set out in a *standard section.* Correlation charts for areas of continental or subcontinental size bring these sections together to show their approximate geologic ages and age relationships to each other. Usually biostratigraphic data with local and standard worldwide zonal and stage boundaries are included also. Such charts are

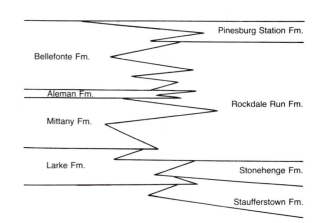

FIGURE 3.6
Interfingering of formations. An example from the Ordovician of southeast Pennsylvania.

not intended for detailed stratigraphic correlation at a local level and are only approximate indicators of age correlation (Fig. 3.7). In the U.S. the most recent stratigraphic correlation charts are those produced in 1985 by the American Association of Petroleum Geologists (AAPG, 1985) in their COSUNA project (for Correlation of Stratigraphic Units of North America). In Canada similar charts were published earlier by the Geological Survey of Canada (Douglas, 1967).

Facies Changes

Although it is important to define a stratigraphic unit at its stratotype as precisely as possible, as with all artificial classifications of natural objects there are always boundary problems. The facies concept discussed in Chapter 1 provides a useful degree of flexibility in dealing with the terminology of stratigraphic units and their lateral variation. Inevitably a stratigraphic unit will change in lithology as it is traced laterally but still may be recognized in some of its more important characteristics. If a massive limestone gradually passes laterally into a well-bedded lime-

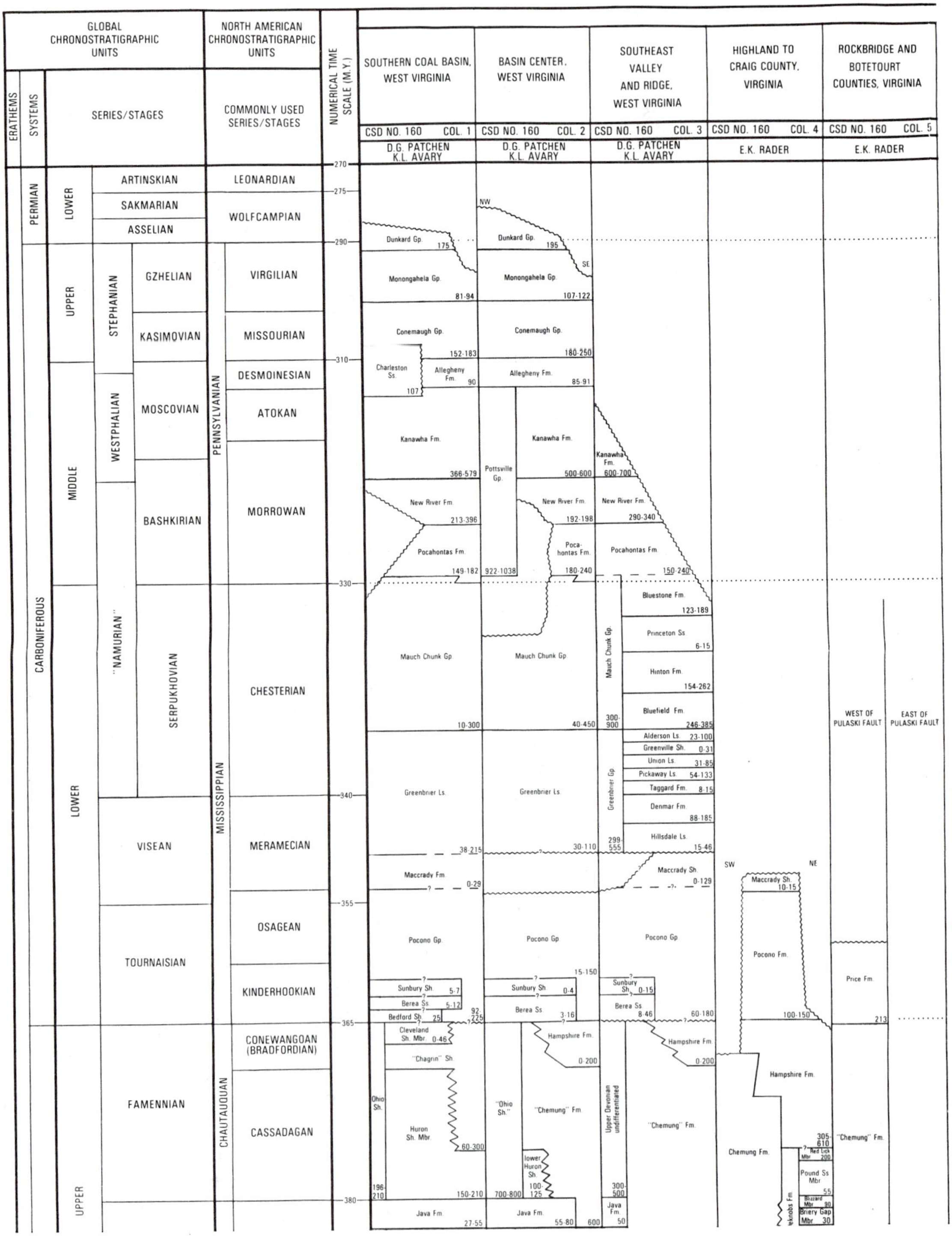

FIGURE 3.7
Portion of a COSUNA correlation chart. (Courtesy American Association of Petroleum Geologists, reprinted by permission.)

stone with shale partings, it is probably not sufficient reason to consider it a new formation. The existence of a shaly facies within that particular limestone formation undoubtedly would be included in the original description and definition of the formation, even though no shale is present at the stratotype. In other words, although shale occurs in the formation it is not a typical part of it.

The term *lithofacies* is applied when discussing the lithologic aspects of a stratigraphic unit. When dealing with a *biofacies* it is the environmental—that is, the paleoecological—aspect that usually is being considered. For example, in the classical descriptions of the Appalachian geosynclinal sequences the miogeosynclinal shelf successions often were referred to as the "shelly facies" (dominated by brachiopods) in contrast to the "graptolitic facies" of the deeper-water, eugeosynclinal sediments. For a paleontologically defined stratigraphic unit (for example, a biozone) there may be many changes in lithofacies within the total volume of the unit. Similarly, within a lithologically defined unit there may be numerous changes in biofacies.

Lithofacies Maps

In discussing lateral facies changes within a given stratigraphic interval a quantitative aspect is obviously involved. If a sandstone unit, followed along a line of section, becomes progressively more shaly until that interval is represented only by shale (that is, it "shales out"), there must be points on that section where the interval contains 75 percent sandstone and 25 percent shale, 50 percent sandstone and 50 percent shale, and so on. These points, plotted on a map and contoured by *isoliths* (lines joining points of similar lithologic ratio values), comprise a sand/shale ratio map (Fig. 3.8) for that particular stratigraphic interval. Similar maps can be constructed using three end members; sandstone, shale, and limestone are the common-

est. In such a map not only is the sandstone/shale ratio plotted but the ratio of the combined sandstone and shale to the limestone, known as the *clastic ratio* (Krumbein and Sloss, 1963), is plotted also. The proportions of the three end members contained in a given stratigraphic interval at any one location can be plotted as a point in a triangle in which the three end members, at 100 percent each, mark the apices. Similarly, the ratio values can be plotted on two separate facies maps, one showing the sandstone/shale ratio, the other, clastic ratio. These two maps then can be combined in various ways to demonstrate graphically the lateral variations in the proportions of the three end members. Virtually any combination of ratios of isolith values can be selected, so the maps are quite flexible (Fig. 3.9). Lithofacies maps of this type lend themselves well to subsurface mapping, in which all correlations from well to well are extrapolations. Lithofacies maps depict graphically the quantitative aspects of such correlations and often also can be useful in extrapolating regional trends beyond the area of control in paleogeographic reconstructions. For a discussion of quantitative mapping techniques, see Forgotson, (1960).

3.3 GEOCHRONOLOGIC AND CHRONOSTRATIGRAPHIC UNITS

Geochronologic, or geologic time, divisions and chronostratigraphic, or time-rock, divisions probably are best considered together because chronostratigraphic units are defined as bodies of rock bounded by isochronous surfaces and are supposedly the tangible record of the passage of specific time intervals. Chronostratigraphic units are included in all stratigraphic codes and have become the fundamental working units of the geologic time scale. Because of this, they are discussed later in their appropriate place. The fact remains, however, that they are often difficult to define, and it has been suggested that they be dispensed with as practical work-

FIGURE 3.8
Construction of a sand/shale ratio map.

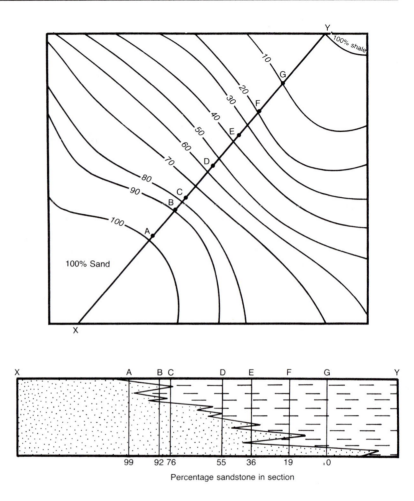

ing units. As Ager (1984) put it, "There are rocks which still remain, and there is time, which has passed and can never be recovered. All the rest is semantic confusion."

It was William Smith's careful studies of faunal successions in the Jurassic that opened the door to stratigraphic correlation, and the publication in 1799 of Smith's geologic table of strata marked the beginning of the geologic time scale as we know it today. By the middle of the nineteenth century, the main divisions had been established and stra-

tigraphic correlations were being rapidly expanded to all parts of the world. The divisions of the geologic time scale, although now becoming measurable in terms of real time, did not originate as quantitatively based temporal units. Seen as a calendar of earth history, the time scale began as a relative scale only. Strata and events were ordered in proper sequence, but the magnitude of the time intervals involved was unknown. Today, largely because of the data supplied by isotopic dating, the various divisions of the geo-

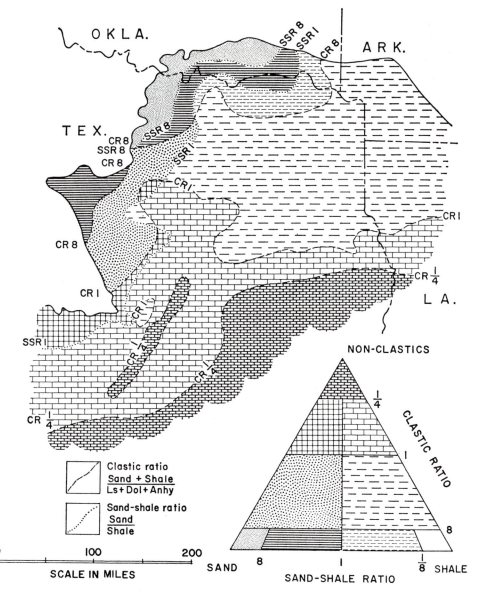

FIGURE 3.9
Triangle facies map of part of the Cretaceous Trinity Group in Texas. (From STRATIGRA-PHY AND SEDIMENTATION, 2/E, by W. C. Krumbein and L. L. Sloss, Fig. 12.11, p. 464. Copyright 1951, © 1963. Reprinted with the permission of W. H. Freeman and Company, San Francisco.)

logic time scale are being defined in terms of time measured in years. The evolution of the chronometric scale is discussed in a later section.

Geologic Time Scale

It was the objective measurement of sedimentary successions and the study of their contained faunas and floras that provided the foundation for the geologic time scale in use today (Table 3.4). The Silurian Period, for example, was not imagined or defined until the field studies of Roderick Murchison in the early 1830s. He demonstrated the presence of great thicknesses of sedimentary rocks containing a host of distinctive fossils that distinguished those strata from other parts of the geologic succession. The natural or supposed natural divisions distinguished in the total fossiliferous rock sequence came to be established as first-order units known as *systems.* A **system** is a basic time-rock or chronostratigraphic unit, recognized by its distinctive faunas and/or floras and including all rocks formed during the given time interval of a period. Although erosional unconformities and marked changes in lithology were often used as the boundaries of the original systems in their type areas, it was the distinctive features of the faunas and floras that set apart one system from another. Strictly speaking, therefore, the systems as originally defined were biostratigraphic units.

The geologic periods as many people originally visualized them marked major cycles of marine transgression and regression, with pronounced unconformities separating the systems. This is the reason for the twofold or threefold division into early and late, or early, middle, and late epochs (lower, middle, and upper series in chronostratigraphic terms).

Groupings of the geological periods from the Cambrian to the present into three **eras** was first suggested in 1841 by John Phillips.

There is no doubt that the divisions were fortunately chosen because both the Paleozoic-Mesozoic and Mesozoic-Cenozoic boundaries are marked by mass extinctions. It is believed that perhaps 95 percent of species became extinct at the close of the Permian and 75 percent at the close of the Cretaceous, as was discussed in Chapter 2. In the mid-nineteenth century there were still many unpersuaded to a uniformitarian view, and so the idea of catastrophic mass extinctions as marking divisions of geologic time seemed quite natural. Subdivisions of the systems into **series** (and *periods* into *epochs*) were usually based upon more-local stratigraphic discontinuities. The fact that originally in many of the systems a tripartite division into lower, middle, and upper (early, middle, and late in geochronologic terms) was recognized leads one to suspect a certain artificiality. In Murchison's view, the threefold division recognized natural phases of marine transgression, maximum inundation of the craton, and marine regression. In the Tertiary the original four divisions by Lyell were Eocene, Miocene, Older Pliocene, and Newer Pliocene, based upon percentages of extinct to extant molluscan species (see Table 9.5). However, this was later modified to include the Oligocene between the Eocene and Miocene, whereas the Newer Pliocene became the Pleistocene (Table 3.5). The origin of the geologic periods and systems is summarized in Table 3.6. The historical background of the evolution of the geologic time scale has been described by Berry (1968) and Harland et al. (1982).

Series are subdivided into *stages,* the geochronologic equivalents of which are *ages.* Theoretically, stages can be recognized worldwide but, in practice, many subdivisions of "stage" rank are of relatively local significance and can be correlated only approximately with the European stages, which, for the most part, set the world standard.

The progressive refinement of the divisions has proceeded rather unevenly. In the Trias-

TABLE 3.4

Geologic time scale (dates based on G.S.A. 1983 time scale; for more detailed subdivision see Tables 3.7, 3.8, and Appendix 1)

Eon	Era	Period		Epoch	Ma
Phanerozoic	Cenozoic*	Quaternary		Holocene Pleistocene	
		Tertiary	Neogene	Pliocene Miocene	
					23.7
			Paleogene	Oligocene Eocene Paleocene	
					66.4
	Mesozoic	Cretaceous		Late Early	
					144
		Jurassic		Late Middle Early	
					208
		Triassic		Late Middle Early	
					245
	Paleozoic Late	Permian		Late Early	
					286
		Carbon-iferous	Pennsyl-vanian	Late	
			Missis-sippian	Early	
					360
		Devonian		Late Middle Early	
					408
	Paleozoic Early	Silurian		Late Early	
					438
		Ordovician		Late Middle Early	
					505
		Cambrian		Late Middle Early	
					570
Proterozoic	Late	Precambrian: Further subdivision is based essentially on local successions, and no worldwide standard has, as yet, been agreed upon.			
	Middle				
	Early				2500
Archean	Late				
	Middle				
	Early				3500?

*Cainozoic or Kainozoic are variants.

71

TABLE 3.5
Evolution of division of the Tertiary and Quaternary (for details of Lyell's subdivisions on the basis of proportions of extinct to extant species see Table 9.5)

1833 Quaternary (Reboul)	1833 Newer Pliocene (Lyell)	1839 Pleistocene (Lyell)	1854 Pleistocene	1874 Pleistocene	Neogene (Hoernes, 1853)
T E R T I A R Y	Older Pliocene (Lyell)	Pliocene	Pliocene	Pliocene	
	Miocene (Lyell)	Miocene	Miocene	Miocene	
			Oligocene (Beyrich)	Oligocene	Paleogene (Naumann, 1866)
	Eocene (Lyell)	Eocene	Eocene	Eocene	
				Paleocene (Schimper)	

sic, Jurassic, and Cretaceous systems, for example, the recognition of the standard European stages, based upon the ammonite faunas, is possible virtually on a global scale (Van Hinte, 1976a, 1976b). In several of the other systems there is, as yet, no generally accepted world standard. Although a stage is now considered a chronostratigraphic unit, it was originally conceived as a biostratigraphic unit and is still considered as such by some workers. Stages are the subject of further discussion in Chapter 9.

Geologic Time Scale in Real Time

The geologic time scale began as a relative scale, and the time spans involved could only be estimated. The early attempts at determining the relative lengths of the various geologic periods sometimes have been shown to be surprisingly accurate. However, it was not until the development of radiometric dating that an actual time scale, often referred to, erroneously, as an "absolute time scale," became possible. Within the limits of accuracy of the various radiometric methods used, the divisions of the Phanerozoic are now coming to be fairly well established in terms of real time. In Table 3.7, several published geochronometric scales are compared and show, with a few exceptions, some measure of agreement. On such a scale, geologic events are located by measuring back in years from a datum selected as 1950 and with the notation B.P. (before present). Most commonly, the numbers are given in millions of years, abbreviated as Ma (mega ans). Although the abbreviations Ma and m.y. seem to be sometimes used interchangeably, m.y. should be used only for a specific time interval (with a beginning and an end).

Among the relatively recent time scales based upon radiometric dates have been those by Harland et al. (1964, 1982) and Odin (1982a, 1982b). The time scale of Palmer (1983) is shown in Table 3.8. For a discussion of time scales, see also Cohee et al. (1978). The scale by Harland et al. (1964), known as the Phanerozoic time scale (PTS), was long in common use but it was never intended as more than a state-of-the-art study which later work would progressively refine. However, as Harland and Francis (1971) pointed out, "The PTS had an effect opposite to that intended.

TABLE 3.6
Origin of the geologic periods and systems (data from Berry, 1968, and others)

Quaternary	Name first applied in 1829 by Paul A. Desnoyers, although incorrectly, to Tertiary sediments of the Seine Valley; redefined in 1833 by H.P.I. Reboul
Tertiary	First used as a period name by Charles Lyell in 1833; had been used earlier by Giovanni Arduino and others in the eighteenth-century simple division of rocks into Primary, Secondary, and Tertiary
Cretaceous	From the term *Terrain Cretace* used in 1882 by d'Omalius d'Halloy for chalk and greensand of northern France
Jurassic	Some Jurassic formations termed Jura Kalkstein by von Humboldt in 1799; modern use introduced as Jurassique in 1829 by French and Swiss workers for occurrences in the Jura Mountains
Triassic	Used in 1834 by Fredrich A. von Alberti for the three-division sequence of Bunter, Muschelkalk, and Keuper of Germany
Permian	System first defined by Roderick Murchison in 1841; based upon a type section near Perm, Russia
Carboniferous	Named in 1822 by W. D. Conybeare and William Phillips in their *Outlines of Geology of England and Wales*
Devonian	Named by William Lonsdale in 1837 after the county of Devon, England, for the marine facies of the Old Red Sandstone; first published by Sedgwick and Murchison in 1839
Silurian	System first defined in 1835 by Roderick Murchison and named for the Silures, an ancient tribe that had inhabited the type area in the Welsh borderland region
Ordovician	Name adapted by Charles Lapworth in 1879 from that of an ancient tribe, the Ordovices, to apply to a system of rocks in Wales and adjacent England that occurred between two unconformities within the Cambro-Silurian sequence of Sedgwick and Murchison
Cambrian	So-called after the ancient name for Wales and applied as a system name by Adam Sedgwick in 1835; no characteristic fossils were described so the system was not fully recognized until the faunas were described by Frederick McCoy and J. W. Salter in the 1850s
Pennsylvanian and Mississippian	In North America the marked twofold division of the Carboniferous; recognized as the Pennsylvanian, named in 1858 by H. D. Rogers, and the Mississippian, named by A. Winchell in 1870; both these divisions were given system/period status in 1905 in Chamberlin and Salisbury's *Textbook of Geology*

It sealed within the authority of hard covers a work intended to pinpoint and publicize the deficiencies of the current time-scale." In all geochronometric dates there is a range of accuracy that is variable from place to place within the scale. To avoid the difficulty of showing this graphically, in the scale adopted for the COSUNA stratigraphic correlation charts the values are rounded off (Salvador, 1985), as can be seen in Table 3.7 and Appendix 1. Refinements of the geochronometric scale continue to be made, with modifications due partly to new biostratigraphic data and partly to advances in sampling and preparation procedures used in isotopic dating methods.

TABLE 3.7
Comparison of geochronometric scales (After Salvador, 1985)

		Millions of Years Before Present (Ma)			
		COSUNA	Harland et al. (1982)	Odin (1982a, b)	GSA (Palmer, 1983)
Holocene					
		0.01	0.01		0.01
Pleistocene					
		1.7–2.8	2.0		1.6
Pliocene					
		5.3	5.1		5.3
Miocene					
		25	24.6	23	23.7
Oligocene					
		38	38.0	34	36.6
Eocene					
		55	54.9	53	57.8
Paleocene					
		67	65	65	66.4
Cretaceous	Late	100	97.5	95	97.5
	Early				
		140	144	130	144
Jurassic	Late	160	163	150	163
	Middle				
		180	188	181	187
	Early				
		200	213	204	208
Triassic					
		250	248	245	245
Permian					
		290	286	290	286
Carboniferous	Pennsylvanian	330	320		320
	Mississippian				
		365	360	360	360
Devonian					
		405	408	400	408
Silurian					
		425	438	418	438
Ordovician					
		500	505	495	505
Cambrian					
		570	590	530	570
Proterozoic					
		2500	2500		2500
Archean					

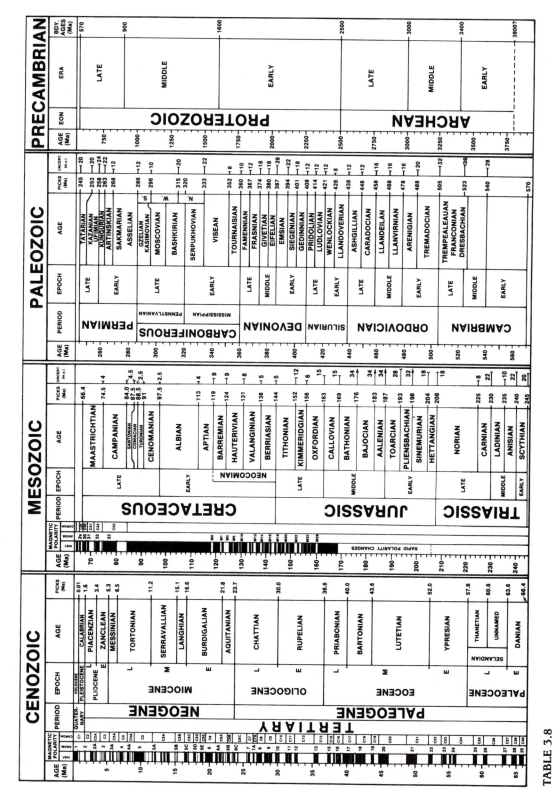

TABLE 3.8

Decade of North American geology, 1983 geologic time scale (from Palmer, A. R., comp., 1983, Geology, v. 11, p. 504)

Chronostratigraphic Units

A chronostratigraphic unit comprises a thickness of strata bounded below and above by isochronous surfaces. In other words, it includes all rocks formed everywhere during a fixed interval of time. It is easy to visualize such a unit, but it is usually difficult to determine one. Not only are true isochronous surfaces rarely discoverable in the stratigraphic record, but the rocks that are present probably record only a fraction of the time unit supposedly represented. Although strictly speaking an isochronous surface represents an instant in time, in practice, short-lived events lasting for days, months, or even thousands of years could leave a record in the rock strata that would be considered essentially isochronous within the broader context of geologic time. One of the best examples is seen in the fallout from a volcanic eruption, which can spread a layer of ash over hundreds of thousands of square kilometers in the space of a few days or weeks; wherever such a layer is preserved, it obviously provides an excellent time horizon. The ideal isochronous surface would, of course, be one recognizable on a global scale. Unless one is thinking in terms of some sudden worldwide climatic change or cosmic catastrophe that would leave a trace in the sedimentary record, it is difficult to visualize any way in which such a horizon could be formed. This topic was discussed in Chapter 2. Perhaps the closest approach to a global isochronous horizon is to be seen in the evidence of geomagnetic polarity reversals. The reversal process itself occurs over a time interval of a few thousand years and dating methods add further margins of error, but the times of reversals are still brief enough that the geologic record could be considered virtually isochronous. Even more rapid are changes in the isotopic composition of the oceans. The mixing time of the ocean waters is approximately 1000 years, so oxygen isotope records

also have great potential in global synchronous correlations. Magnetic reversals and oxygen isotope stratigraphy are discussed further in Chapters 6 and 15, respectively.

Even the largest chronostratigraphic units, representing hundreds of millions of years, began and ended—theoretically, at least—at an instant in time. This is not to say, as Ager (1981) put it, a great bell sounded in heaven to mark the end of one unit and the beginning of the next, but it should be remembered that erathemic or systemic boundaries are only as accurately defined as are the boundaries of the agreed-upon limiting **chronozones,** the smallest chronostratigraphic divisions (see Table 3.1). It is when we are dealing with these smaller units that the problems are brought into sharper focus. A chronozone, stage, or other chronostratigraphic unit should be defined in terms of limiting time horizons. These, in turn, are selected at a type section or in a type area by reference to particular events in geologic history. Such events may be indicated by features like a marked change in lithology (a formational boundary), erosional breaks, a bentonite layer, or the appearance or disappearance of a particular species.

A chronostratigraphic unit is normally defined by reference to an existing lithostratigraphic unit, such as a formation, or to a biostratigraphic unit, such as a biozone or grouping of biozones. If a given formation is the basis for defining the unit, as, for example with many stages, it should be made clear whether the time interval is that of the formation at its type section (stratotype) or that larger interval embracing the formation throughout its entire geographic extent—its full-complement interval. Similarly, in the case of biozones, and as discussed in Chapter 9, the time interval might refer to the local range of a fossil species, say at one location only (its teilzone), or to the entire life span of the species. In the latter case this might be difficult to ascertain because there is little likelihood of finding the very first and last specimens, a problem known as *collection failure*. It is clear that in defining chronostratigraphic units such potential ambiguities must be avoided. Traditionally, the recognition and definition of chronostratigraphic units, on anything other than a strictly local scale, has been possible only by reference to the presence or absence of certain species or assemblages selected as being of particular significance as zone markers. Biozones are the most important elements in the construction of a chronostratigraphic scale and are the basis for the recognition of stages (Table 3.9).

The evolution and increasing sophistication of the geologic time scale has proceeded largely on the basis of biostratigraphic data. The introduction of isotopically derived dates has undoubtedly assisted in giving a better grasp of rates of sediment accumulation, evolutionary change, and so on, but it is probably true to say that biostratigraphy op-

erates independently of the need for isotopic age data measured in real time. In other words, stratigraphic correlation in its purest form is concerned with the recognition of synchronous time horizons at different locations. This can be done on the basis of the recognition of biostratigraphic markers; the actual age of the event, in terms of real time, has no real bearing on the exercise. Indeed, when good biostratigraphic controls are available, the accuracy of even the best radiometrically derived age determination is such that it may provide only a general age for that part of the stratigraphic section in question. Such information may do no more than add an interesting, but possibly irrelevant, contribution to the accuracy of the stratigraphic correlation itself. As Conkin and Conkin (1984, p. 266) point out, radiometric dating can achieve an accuracy ranging from 0.1 to 2 percent; that is, about 3–5 m.y. in the Paleozoic and 1.6–3.2 m.y. in the Mesozoic.

TABLE 3.9
Devonian stages and biozones

System	Series	Stage		Biozone
DEVONIAN	MIDDLE DEVONIAN	Givetian		*Mucrospirifer mucronatus*
		Couvinian	Upper	*Calceola sandalina sandalina*
			Lower	*Schizospirifer(?) daleidensis, Fimbrispirifer divaricatus* and group, *Euryspirifer intermedius*
	LOWER DEVONIAN	Siegenian, Emsian		*Rhytistrophia beckii, Leptocoelia acutiplicata, Acrospirifer primaevus kasachstanika*
		Gedinnian		*Leptostrophia rotunda, Howellella mercuri*

Stratigraphic zonation based on fossils, on the other hand, can normally provide considerably more accurate age determination; for example, in the Mesozoic to around half a million years. In the Cenozoic, using biostratigraphic, magnetostratigraphic, and isotopic stratigraphic methods together, resolutions on the order of 100,000 years are regularly obtained. In practice, all available resources are brought to bear upon problems of stratigraphic correlation and any information on geologic age is potentially useful, even though data from some sources may only confirm existing data rather than add anything new. It should be obvious that the value of radiometric dating in stratigraphic studies is inversely proportional to the degree of available biostratigraphic control. In the Precambrian, only the vaguest outlines of geologic history are discernible without recourse to radiometric dating. As described in Chapter 16, many apparent stratigraphic relationships in Precambrian terranes proved, after radiometric dating, to have been interpreted entirely erroneously.

Stages

For most of the geologic column, the smallest chronostratigraphic unit that is now potentially recognizable on a global scale is the **stage.** The majority of stages were originally named after and defined in terms of formations by reference to a type section. Because, except along depositional strike, the formation boundaries are diachronous, long-distance correlations usually have to be made on the basis of fossils. So, for example, a succession might be described as containing a "typical Red River fauna" even though the Red River Formation itself is no longer present because of changes in facies. In effect, what is usually being described is an assemblage zone based on the recognition of species familiarly associated with the Red River Formation in its typical occurrences. The

stage in this case is essentially a biostratigraphic unit and consists of one or more biozones.

To cite two other examples among many, the Thanetian Stage of Late Paleocene age is named after the Thanet Sand, which occurs near London, England; in the Jurassic, the Oxfordian Stage is named after the Oxford Clay. Neither of these formations has more than a relatively local extent of a few kilometers, but on the basis of distinctive fossils the stages are recognizable in many other parts of the world. This topic is discussed in Chapter 9. With the gradual accumulation of data on the life spans of species, phylogenetic relationships, and the environmental controls of faunal provincial boundaries, it eventually becomes possible to define biozones very precisely; they have become the most useful criteria for the definition of chronozones. Implicit in the basic philosophy of biostratigraphic correlation is the acceptance of biostratigraphic units as quasi-chronostratigraphic units. Assuming that the boundaries of biostratigraphic units are isochronous is a compromise between the theoretical ideal and the practical. In general this compromise seems to work well.

It naturally takes many years and the efforts of a large number of workers to reach this level of sophistication, but it has been achieved for certain parts of the stratigraphic column particularly favored by abundant and well-preserved fossils. Because of the longer history of the geological sciences in Europe, faunas and chronostratigraphic relationships there are among the best understood in the world. It is not surprising, therefore, that world stratigraphy is based to quite a large extent on European standards. This is more a consequence of historical precedent than because European geologic sections are more complete or more fossiliferous. In the Tertiary, for example, the most complete stratigraphic sequence is found in New Zealand (Table 3.10). The main thrust in stratigraphy

TABLE 3.10

Stages in the Tertiary and Quaternary of New Zealand. Although stages are technically chronostratigraphic units, they are invariably defined on the basis of their contained faunas. Because virtually all faunal assemblages are more or less local or regional in their extent, it follows that stages, at least as long as they are based on fossils, can never be worldwide in their extent. This table provides a good example of stage classification on different sides of the world, and it should be emphasized that the correlation between the two classification schemes is only approximate. Radiometric dating and magnetic stratigraphy may be used in support, but, in general, are not accurate enough for detailed correlation.

	Series	New Zealand Stages		European Stages
UPPER PLEISTOCENE	Hawera			
LOWER PLEISTOCENE	Upper Wanganni	Castlecliffian	Putikian	Calabrian
			Okehuan	
		Nukumaruan	Marahauan	Astian
			Hautawan	
PLIOCENE	Lower Wanganni	Waitotaran (Mangapanian)		Piacenzian
		Waipipian		Zanclian
		Opoitian		
MIOCENE	Taranaki	Kapitean		Messinian
		Tongaporotuan		Tortonian
	Southland	Waiauan		Helvetian
		Lillburnian		
		Cliffdenian		Burdigalian
		Altonian		
	Pareora	Awamoan		Aquitanian
		Hutchinsonian		
		Otaian		Chattian
OLIGOCENE	Landon	Waitakian		
		Duntroonian		Stampian
		Whaingaroan		Sannoisian
EOCENE	Arnold	Runangan		Priobonian
		Kaiatan		
		Bortonian		Lutetian
	Dannevirke	Porangan		
		Heretaungan		Ypresian
		Mangaorapan		
PALEOCENE		Waipawan		Landenian
		Teurian		Montian
				Danian

now is to increase the level of understanding of regional stratigraphic relationships, to make as accurate a correlation as possible with the accepted world standard, and, if possible, establish workable, local, chronostratigraphic divisions in terms of that standard. As discussed in Chapter 9, there are several practical problems in defining and correlating stages because of the limitations of the faunas that characterize them. Stages, for all practical purposes, can be defined only biostratigraphically, so the distinction between such chronostratigraphic and biostratigraphic units tends to become rather obscure. The main difference, even though the data base may be the same, involves descriptive (biostratigraphic) versus interpretative (chronostratigraphic) units.

Chronostratigraphic Boundaries

One of the simplest ways of dividing up the stratigraphic column is, as discussed earlier, to use breaks in the succession. These might be marked by abrupt changes in lithology, erosional breaks, or the sudden disappearance or abrupt replacement of faunas or floras. In establishing lithostratigraphic units within a particular succession this approach is logical and, in fact, inevitable because the boundaries are usually very obvious. As McLaren (1970) pointed out, across such boundaries "something happened" and many workers have considered that the chronostratigraphic significance of what happened is sufficient to make the boundaries usable in establishing working chronostratigraphic units. It is conceded that erosional surfaces and other evidence of diastrophic changes are probably neither isochronous nor discernible everywhere, but, it is argued, such physical discontinuities reflect "happenings" that may have had a profound influence on faunas and floras also. It is, therefore, the paleontologic discontinuities that are virtually isochronous and of wide-ranging signifi-

cance. As is discussed in Chapter 13, erosional breaks do have some chronostratigraphic significance but should be treated somewhat differently.

There seem to be two approaches regarding the way in which chronostratigraphic boundaries should be selected. On the one side are those who favor using (biostratigraphic) boundaries across which "something happened," and on the other are proponents of the idea that boundaries across which "nothing happened" are the more desirable. The boundaries across which "something happened" certainly do exist but should really be considered usable in long-distance correlation only if they have widespread— ideally global—significance. Marked discontinuities in floras and faunas embracing a wide range of organisms can be due only to external, not phylogenetic, causes. If these are worldwide in their effect we may be looking at global climatic or extraterrestrial events as the cause. If cosmic, climatic, or other global events can be documented they are obviously invaluable isochronous horizons and should be selected as first-order chronostratigraphic boundaries. They are, however, far too infrequent to be used for detailed chronostratigraphic work. It is then that we must turn to the boundaries across which "nothing happened." Such a boundary selected within a continuous sedimentary succession clearly avoids the problem that might arise if, for example, an erosional surface is chosen as a boundary between two contiguous stages. If later work in other areas reveals strata equivalent in age to the hiatus represented by the erosion surface, revisions of the stage boundaries would be necessary. This topic will be discussed later.

Boundary Stratotypes

In defining and naming a stage, a stratotype is selected which shows the best possible and most complete section. In addition, it should contain the most diverse fauna and/or flora, with maximum representation of faunal and floral elements that might aid in its recognition and correlation elsewhere. Clearly this is not achieved if parts of the succession are known to be missing, so continuity of sedimentation is an important criterion in the selection of a stratotype. Successive stages are likely to possess stratotypes at widely differing localities; as more detailed studies are made it frequently happens that some overlap may occur. In other words, the top portion of stage A at one location may actually correlate with the lower part of stage B above it somewhere else (Fig. 3.10a). Alternatively, at another location there may be a situation in which strata above stage A may prove to correlate with strata below the base of stage B and thus be in a sort of stratigraphic limbo (Fig. 3.10b). This problem has proved to be very troublesome but the solution is surprisingly simple. Following a proposal by the British Mesozoic Committee (Ager, 1964), it was suggested that only the base of a stage will be defined so that the top will automatically occur at the base of the next stage above, wherever it is (Fig. 3.10c). This is what Ager (1981) called the topless fashion and it is embodied in the concept of the **boundary stratotype.** Although the whole stage is supposedly represented at the stratotype, it is only the base that is precisely designated. The task of determining the boundaries and selecting suitable boundary stratotypes is a difficult one because there clearly has to be considerable unanimity among geologists around the world. Several international committees have been and are working on such problems. The importance of such boundary stratotypes lies in the necessity for providing stable reference points that represent unique signals or instants of time within the geologic time scale. Such a designated boundary is referred to as a *global boundary stratotype section and point* (GBSSP); it must be defined precisely in terms of a particular stratigraphic

horizon, as seen at a single geographic locality (Cowie, 1986). For example, the GBSSP for the Silurian-Devonian boundary is located in a section at Klonk, near Suchomasty in Czechoslovakia. Here the boundary horizon occurs within a 7–10-cm-thick bed, designated as "bed No. 20," at the horizon where

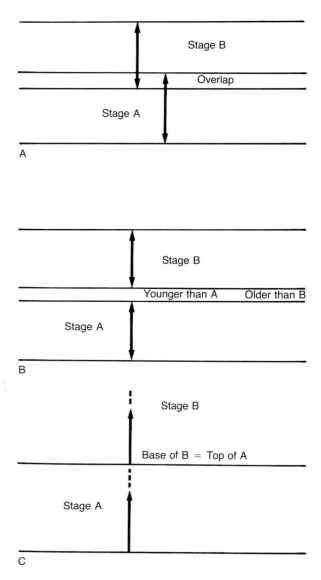

FIGURE 3.10
Boundary stratotypes

there is a sudden and abundant appearance of the graptolites *Monograptus uniformis* and *M. uniformis anguistidens.* Theoretically, at least, at the moment when those graptolite-bearing sediments were deposited, Ager's (1981) great bell in heaven sounded to mark the beginning of Devonian time! The Silurian-Devonian boundary stratotype, established in 1972 (McLaren, 1977), was the first of such "golden spikes." Since that time, other boundary stratotypes have been established, such as the Ordovician-Silurian boundary, the series and stage boundaries of the Silurian, and the Pliocene-Pleistocene boundary.

3.4 CONCLUSION

As a last word on the subject of classification, it should be remembered that classifications are human artifacts and, at best, are a compromise and a matter of convenience. Stratigraphy involves a unique array of entities, although there are really only two that might be considered fundamental. These are rock units and time units. Classifying them is usually no problem; it is when necessary to develop concepts involving both of them in combination that the exercise becomes more difficult. Further complexity stems from the many and varied tools now available to the stratigrapher, and much controversy attends the reconciling of scales and the divisions derived from the numerous techniques. As is to be expected, new ideas and new concepts appear on occasion; some are found useful and become incorporated in the stratigraphic edifice. For the future, the most fruitful field for new ideas in stratigraphy is in the Precambrian. It will be interesting to see if conventional schemes of stratigraphic classification can be adapted for use in the Precambrian, or whether a whole new approach in constructing a calendar will be necessary for some seven-eighths of earth history.

REFERENCES

ACSN (American Commission on Stratigraphic Nomenclature), 1961, Code of stratigraphic nomenclature: Am. Assoc. Petroleum Geologists Bull., v. 45, p. 645–665.

_____1970, Code of stratigraphic nomenclature, 2nd ed.: Am. Assoc. Petroleum Geologists, 45 p.

Ager, D. V., 1964, The British Mesozoic Committee: Nature, v. 203, p. 1059.

_____1981, The nature of the stratigraphical record, 2nd ed.: New York, Macmillan, 122 p.

_____1984, The stratigraphic code and what it implies, *in* Berggren, W. A., and Van Couvering, J. A. eds.: Catastrophes and Earth history, Princeton NJ, Princeton Univ. Press, p. 91–100.

AAPG (American Association of Petroleum Geologists), 1985, Correlation of Stratigraphic Units of North America (COSUNA) Project.

Berry, W. B. N., 1968, Growth of a prehistoric time scale: San Francisco, Freeman, 158 p.

Cohee, G. V., Glaessner, M. F., and Hedberg, H. D., eds., 1978, Contributions to the geologic time scale: Am. Assoc. Petroleum Geologists Studies in Geology, No. 6, 388 p.

Conkin, B. M., and Conkin, J. E., eds. 1984, Stratigraphy: Foundations and concepts: New York, Van Nostrand Reinhold Co., 363 p.

Cowie, J. W., 1986, Guidelines for boundary stratotypes: Episodes, v. 9, p. 78–82.

Douglas, R. J. W., 1967, Geology and economic minerals of Canada: Geol. Survey Canada Econ. Geology Rept. 5th ed., 838 p.

Forgotson, J. M., Jr., 1960, Review and classification of quantitative mapping techniques: Am. Assoc. Petroleum Geologists Bull. v. 44, p. 52–65.

Harland, W. B., Cox, A. V., Llewellen, P. G., Pickton, C. A. G., Smith, A. G., and Wilcock, B. eds., 1982, A geologic time scale: Cambridge, Cambridge Univ. Press, 131 p.

Harland, W. B., and Francis, E. H., eds. 1971, The Phanerozoic time scale—A supplement, Pt. 1, Supplementary papers and items: Geol. Soc. London, Spec. Pub. 5, p. 1–120.

Harland, W. B., Smith, A. G., and Wilcock, B. eds. 1964, The Phanerozoic time scale: Geol. Soc. London, Spec. Paper 1, 458 p.

Hedberg, H. H., ed., 1976, International stratigraphic guide: New York, John Wiley & Sons, 200 p.

Holland, C. H., Audley-Charles, M., Bassett, G., Cowie, J. W., Curry, D., Fitch, F. J., Hancock, J. M., House, M. R., Ingham, J. K., Kent, P. E., Morton, N., Ramsbottom, W. H. C., Rawson, P. F., Smith, D. B., Stubblefield, C. J., Torrens, H. S., Wallace, P., and Woodland, A. W., 1978, A guide to stratigraphical procedure: Geol. Soc. London, Spec. Rept. 11, 18 p.

House, M. R., Richardson, J. B., Chaloner, W. A., Allen, J. R. L., Holland, C. H., and Westoll, T. S., 1977, A correlation of Devonian rocks in the British Isles: Geol. Soc. London, Spec. Rept. 18, 110 p.

ISSC (International Subcommission on Stratigraphic Classification), 1979, Magnetostratigraphic polarity units—A supplementary chapter of the ISSC international stratigraphic guide: Geology, v. 7, p. 578–583.

Keroher, G. C., 1968, Lexicon of geological names of the United States: U.S. Geol. Survey Bull. 1350, 848 p.

Krumbein, W. C., and Sloss L. L., 1963, Stratigraphy and sedimentation, 2nd ed.: San Francisco, Freeman, 660 p.

McLaren, D. J., 1970, Presidential address: Time, life, and boundaries: Jour. Paleontology, v. 44, p. 801–815.

———1977, The Silurian-Devonian Boundary Commission: A final report, *in* Martinson, A.,

ed., The Silurian-Devonian boundary, IUGS Ser. A., Stuttgart, IUGS, p. 1–34.

NACSN (North American Commission on Stratigraphic Nomenclature), 1983, North American Stratigraphic Code, Am. Assoc. Petroleum Geologists Bull., v. 67, p. 841–875.

Odin, G. S. ed., 1982a, Numerical dating in stratigraphy: New York, John Wiley & Sons, 2 vols., 1040 p.

———1982b, The Phanerozoic time scale revisited: Episodes, v. 1982, p. 3–9.

Palmer, A. G., 1983, The decade of North American geology, 1983 geologic time scale: Geology, v. 11, p. 503–504.

Salvador, A., 1985, Chronostratigraphic and geochronometric scales in COSUNA stratigraphic correlation charts of the United States: Am. Assoc. Petroleum Geologists Bull., v. 69, p. 181–189.

Swanson, R. W., Hubert, M. L., Luttrell, G. W., and Jussen, V. M., 1976, Geologic names of the United States through 1975: U.S. Geol. Survey, Bull. 1535, 642 p. (computer list).

Van Hinte, J. E., 1976a, A Jurassic time scale: Am. Assoc. Petroleum Geologists Bull., v. 60, p. 489–497.

———1976b, A Cretaceous time scale: Am. Assoc. Petroleum Geologists Bull., v. 60, p. 498–516.

Wheeler, H. E., and Mallory, V. S., 1956, Factors in lithostratigraphy: Am. Assoc. Petroleum Geologists Bull., v. 60, p. 2711–2723.

4
GEOLOGIC TIMEKEEPERS

What seest thou else
In the dark backward and abysm of time?

William Shakespeare

4.1 INTRODUCTION

Time can be measured in two ways: either by reference to a regularly recurring cyclical phenomenon or to an ongoing nonreversible phenomenon. Obvious examples of cyclical phenomena would be the diurnal, lunar, and annual cycles, all related to the earth and moon in space. The most common artificial cycle is, of course, the passage of the hands around a clock. Among the ongoing, nonreversible phenomena, there is the ontogenetic development of an organism—the height and weight of a growing child can obviously be used as a rough estimate of the number of years since birth. Another example of a nonreversible phenomenon is the evolutionary change in faunas and floras over long spans of time, although the rate of progress of this phenomenon is certainly not constant. The decay of radioactive elements through a chain of daughter isotopes is ongoing, nonreversible, and constant in rate. Artificial examples include the burning down of a candle and the flow of sand in an hourglass, both used in the past as clocks. Although it is not difficult to think of many ways to measure the passage of time, only a very few time-de-

85

pendent phenomena leave any record preserved in the rocks. This chapter will deal with sedimentary and paleontological phenomena; radiometric dating, magnetic stratigraphy, and other topics in the general field of geochronology will be covered in later chapters.

4.2 CYCLICAL PHENOMENA

Rhythmites and Sedimentary Clocks

A feature sometimes encountered in sedimentary successions is the regular repetition of two or more sediment types. Typically, they are thin beds or laminae, often quite regular in thickness. Sometimes thousands of these rhythmically repeated layers, or **rhythmites,** as they are known, can be counted, and it is difficult to visualize anything other than a regularly repeating cyclical mechanism as the causative agent. Such rhythmic alternations of sediment type are seen in shale/siltstone sequences, clay deposits,

limestones, evaporites, and banded cherts, but there is rarely any direct evidence of the periodicity of the cycle. Annual climatic cycles, with their effect on water temperature, salinity, erosion, and sediment transport, are probably the most likely cause of many of these rhythmites.

Glacial Varves   The classic example of annual cyclical sedimentation is seen in the varved clays of glacial lakes. Typically a **varve** (Swedish for layer) measures about 50–150 mm in thickness and consists of two layers (a couplet). A thicker, light-colored layer is dominated by silt-sized material that washed into the lake during the spring and summer influx of glacial meltwater. The lighter-colored layer is overlain by a thinner, darker-colored layer of the clay-sized material that settled gradually during the winter, when discharge from glacial melting had stopped, the lake was frozen over, and the water was still (Fig. 4.1). The passage from the summer layer into the winter layer is not necessarily

FIGURE 4.1
Typical glacial varves; note banding within each varve. (Photographs courtesy of P. F. Karrow.)

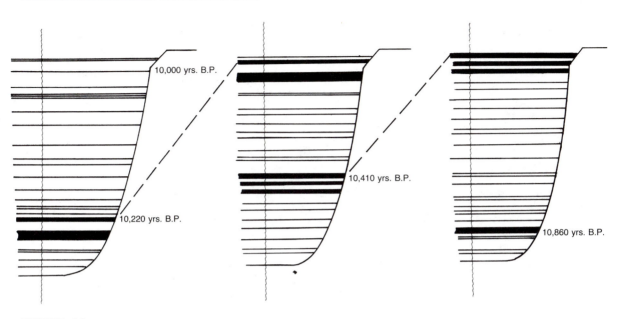

FIGURE 4.2
How glacial varve successions may be correlated between different glacial lakes, and
the timetable of the varve chronology extended back in time.

gradational. As Ashley (1972, 1975) pointed out, the two layers are genetically distinct and the summer layer may itself contain many thin laminations, often graded. Variations in the thickness of summer and winter layers reflect year-to-year differences in the length and kind of seasons experienced, so varve chronologies have provided useful information on past climatic fluctuations. It is the matching of distinctive groupings of thick and thin varves that allows the varve sequences from many different localities to be tied together into a composite succession and chronology (Fig. 4.2). The original Swedish varve chronology of DeGeer (1912) extended back to nearly 17,000 years B. P. (before present; i.e., the year 1950); since 12,000 years B. P., it is considered accurate to within 200 years. Varve chronologies in the Baltic region have been more or less tied in with the historical calendar and with pollen and dendrochronologic calendars. Similar attempts with varve chronologies from North America and

elsewhere have been less successful. Varved sediments of supposed fluvioglacial origin have been described from older glacial sequences of Permian, Ordovician, and Precambrian age (Fig. 4.3).

Nonglacial Varves In many lakes annual climatic cycles are reflected in both sedimentary and biological cycles. Although laminated sediments should be deposited as a consequence, they are rarely seen because of destruction by the churning action of mud-burrowing bottom organisms. Therefore, such sediments typically occur only in lakes where there is persistent stratification of the water, resulting in anoxic bottom conditions and a reduction or elimination of this **bioturbation** effect. Such conditions are thought to have existed in the lacustrine Green River Formation of Eocene age in Wyoming. Here calcareous silts alternate with thin, organic-rich shales considered by many to have been deposited during seasonal blooms of blue-

FIGURE 4.3
Probable glacial varves of Precambrian age. Dropstones in Gowganda "varvite," Well Township, Wakwekobi Lake map area, Ontario. (Photograph courtesy of Geological Survey of Canada, Ottawa.)

green algae (**cyanobacteria**) that formed mats on the lake floor (Bradley, 1929, 1964). Numerous other occurrences of varvelike rhythmites include those in many evaporite sequences. The repetitive laminae (Fig. 4.4) are presumably caused by cyclical changes in chemical equilibrium, with seasonal precipitation of carbonates, sulfates, and halite. In limestone and gypsum deposits of the Upper Jurassic Todilto formation in New Mexico,

Anderson and Kirkland (1960, 1966) described annual varves in which were recorded 11–13-year, as well as 60-, 85-, 170-, and 180-year cycles. Even when the rhythmic cycles can be shown as true annual varves, the causative mechanism is often in doubt. In certain evaporites, a halite/anhydrite repetition might be temperature-dependent; because anhydrite is more soluble at lower temperatures, it would represent the summer layer. On the other hand, it could be argued that dilution by rainfall and rivers during a cooler, wetter season and increased evaporation during the summer months would result in the precipitation of the more-soluble salts, such as halite, as the summer layer. In some cases evaporite rhythmites have been ascribed to seasonal mixing of stratified hypersaline water bodies, leading to complex salt-precipitation and solution cycles.

To discern the relationship between cyclical sedimentary features and external cyclical controls is, at best, a difficult task. As described above, all too often the same geologic evidence is open to interpretation in different ways. The problem is further complicated by the fact that more than one mechanism may be operating simultaneously or by the possibility that certain random effects (for example, tectonism) may exert some control also. In some cases no external controls are operating at all and the sedimentary mechanisms may be self-regulating. For example, one of the most prominent features of the stratigraphic succession in thick deltaic sequences is the cyclic repetition of sands and muds. Usually this can be attributed to the pulsatory nature of the growth of delta lobes. Seaward growth of the delta front and the flattening of the distributary stream gradient inevitably result in upstream crevassing and the abandonment of the delta lobe, the sands of which then come to be overlain by marine muds. Because the distributaries are constantly changing position, the overall succession at

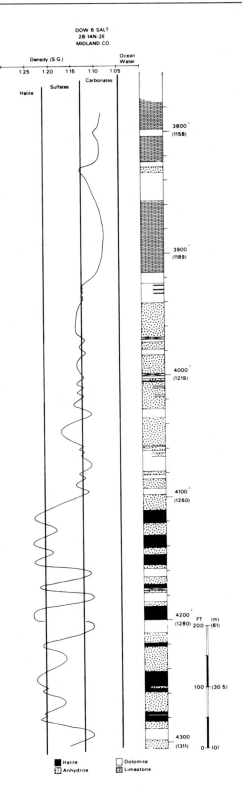

FIGURE 4.4
Description of core Dow 6 salt well showing cyclicity of evaporites. (From R. D. Matthews, 1977, Evaporite cycles and lithofacies in Lucas formation, Detroit River Group. Devonian, Midland, Michigan, *in* Reefs and evaporites—Concepts and depositional models, J. H. Fisher, ed., Am. Assoc. Petroleum Geologists Studies in Geology, 5, Fig. 2, p. 75. Reprinted by permission of AAPG, Tulsa.)

any one location—as, for example, in a well log—consists of alternating sands and muds (Fig. 4.5).

In certain rhythmites, as in some tree-ring chronologies, significant groupings of the annual laminae can be detected; it is clear that they reflect climatic cycles of some kind. Often the groupings indicate cycles of approximately 11 years duration, as in the Todilto formation mentioned earlier, strongly suggesting that they are linked in some way with the 11-year sunspot cycle. The possible terrestrial effects of sunspot cycles have long been the subject of debate; not only 11-year but 22-year and longer cycles have supposedly been revealed in a variety of phenomena: geological, geophysical, climatological, meteorological, and biological. Indeed, in some rhythmite sequences of doubtful origin, an apparent clustering of the beds in cycles of eleven has been used as the criterion by which they are considered to be annual layers. Apart from the fact that there has been no convincing explanation of the link between sunspot cycles and terrestrial phenomena, it has also been pointed out that sunspot cycles are not nearly as regular as is generally thought and vary from 8 to 16 years. Further, in many examples, it is only by statistical processing of data that the cyclicity is revealed at all. On the other hand, a study of glacial varves of late Precambrian age in South Australia (Williams, 1981) showed a well-marked cyclicity of 11, 22, and 90 years. These were equated with sunspot activity and indicated that solar cycles some 680 m.y. ago were similar to those observed at the present day.

In normal marine sequences, finely laminated rhythmites are unusual because small-scale sedimentary structures are typically destroyed by bioturbation. However, where bottom conditions are hostile to benthic organisms, usually because of oxygen deficiency, rhythmites have been recorded. Core samples from the Gulf of Aden, for example,

show regular rhythmites consisting of sandy layers alternating with organic-rich layers of mud. These are apparently caused by a seasonal fluctuation in windblown sand delivered to the area (Olausson and Olsson, 1969). Except in the case of the Late Pleistocene varve sequences, rhythmites are unreliable as anchors to the geologic time scale expressed in terms of real time. They remain as floating chronologies to faithfully record the passage of thousands or tens of thousands of years and reveal the presence of enigmatic longer cycles.

Other sedimentary cycles of much greater magnitude, and involving the repetition of up to ten or a dozen different lithologies, commonly occur in what are known as *cyclothems*. Sea-level fluctuations are responsible for many cyclothems and, as originally defined, a **cyclothem** is a succession of beds laid down during a single sedimentary cycle. The shortest of these cycles, those of glacio-eustatic origin, are measured in thousands of years (commonly 100,000 years) and, it is generally agreed, are linked to the so-called Milankovitch cycle of climatic change, described later in this chapter. Some cyclothems are clearly responses to other, larger cycles of both glacio-eustatic and other origins. These are described in Chapters 12 and 13.

Paleontological Clocks

In the majority of invertebrates that secrete hard skeletal parts, growth increments can be discerned in the form of growth lines. Typically, the lines are regularly spaced; in some cases they can be demonstrated as representing annual, monthly, or daily growth increments. In effect, these growth lines are evidence of what Thompson (1975) has called a bioclock, which controls the rhythm of growth. Determination of the actual ages of individuals and colonies is of interest to the biologist or paleontologist, but of much greater interest in the present context is the

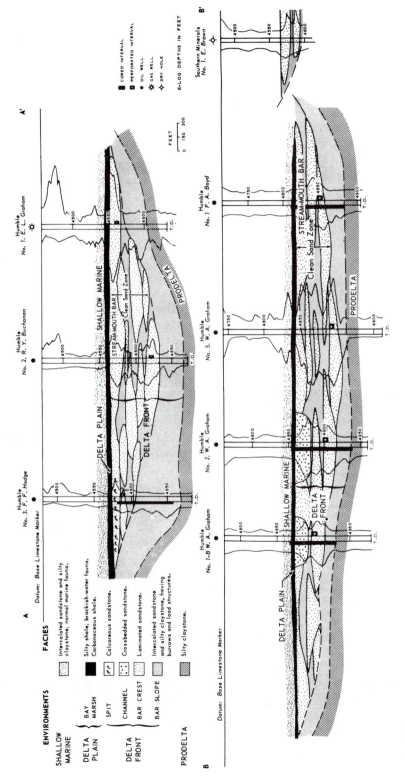

FIGURE 4.5

Alternating sands and muds in a typical deltaic sequence, seen in boreholes in the West Tuscola Field, Taylor County, Texas. (From Shannon, J. P., Jr., and Dahl, A. R., Deltaic stratigraphic traps in West Tuscola field, Taylor County, Texas: Am. Assoc. Petroleum Geologists Bull., v. 55, Fig. 7, p. 1200–1201. Reprinted by permission of AAPG, Tulsa.)

potential of such organisms as geochrono-metric tools. This is possible because in some organisms combinations of growth increments, daily and yearly or daily and monthly, can be discerned. Data from fossils of various ages have provided evidence of changes in the length of the day and the month through geologic time. Geophysical and astronomical theory has constructed a model in which lunar tidal torque has produced a gradual and uniform slowing of the earth's rotational spin by about 2 milliseconds per century. Paleontological evidence, although providing support for the concept by introducing some direct quantitative data on the length of the day and the month in the past, has suggested that the rates of change have not been constant, a phenomenon not predicted in the conventional model.

One of the first studies was by Wells (1963), who showed that in modern corals it was possible to count "around 360" small daily (circadian) growth lines in the space of one year's growth increment, whereas in corals from the Middle Devonian an average of about 400 daily growth lines per annum was counted. Two specimens from the Pennsylvanian gave counts of 385 and 390 daily increments. Wells compared these figures with astronomically derived data that gave a figure of 399 days for the Middle Devonian year. Following Wells, numerous workers have presented data derived from bivalves (Fig. 4.6), gastropods, cephalopods, brachiopods, and barnacles and also the organo-sedimentary structures known as stromatolites (Berry and Barker, 1968; MacClintock et al., 1968; Mazzullo, 1971). Curves have been con-

FIGURE 4.6
Daily growth increments in the shell of the mollusc *Mercenaria rileyi*, showing two cycles; the larger probably is annual, the smaller may be monthly.

structed to show that the rate of slowing of the earth's rotation through the Phanerozoic varied considerably. According to Creer (1975), there have been three periods of above-average deceleration of spin velocity, separated by two periods of moderate deceleration (Fig. 4.7). Evidence from the Precambrian is still fragmentary, but one record of growth rhythms in stromatolites from limestones of the Bulawayan Group in southern Africa (Pannella, 1972) indicates that tidal influences were affecting growth 2800 m.y. ago, so that the earth-moon system is at least that old. A tentative figure of 448 days in the year, derived from stromatolitic chert in the 1700–1900-m.y.-old Gunflint Formation of Lake Superior, Canada, suggests a very low rate of spin deceleration for that period of the Precambrian. The variations in the rate of change in the earth's spin were correlated by Pannella et al. (1968) with changes in the tidal torque consequent on the shifting of continental blocks and the expansion and contraction of epeiric seas. Creer (1975), on the other hand, described a correlation between the discontinuities in the spin deceleration curve and major events in the geomagnetic polarity chronology and suggested mass movement of material in the lower mantle as the common cause of both phenomena.

Apart from some lack of agreement between the paleontological data and astronomical calculations, there are also considerable problems concerning the accuracy of the bio-

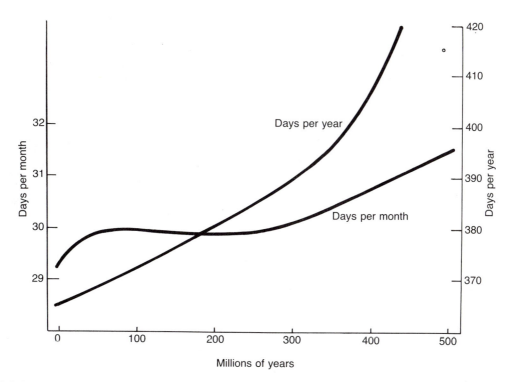

FIGURE 4.7
Variations in the number of days per month and days per year over geologic time, based on data from bioclocks. (After various sources.)

clocks themselves. Studies have shown that there can be considerable variations in the relationships between annual, daily, and other periodic growth increments. In certain bivalves, for example, the rate of growth slows with age so that, in young shells, growth lines may indeed be annual, whereas in mature shells they are not. Geographic variations introduce other factors. For instance, the season during which the growth increment is added varies systematically with latitude, as does the number of daily growth increments per growing season (Hall et al., 1974). Tidal effects, breeding cycle, and ecologic factors have also been seen by numerous workers as influencing growth patterns; human factors in the counting process introduce other variables (Crabtree et al., 1980).

It has been claimed that the history of earth-moon relationships can also be unraveled on the basis of paleontological clocks. Many astronomers believe that the moon was captured by the earth's gravitational field, but exactly when this event occurred is unknown.

As described earlier, there is paleontologic evidence of an earth-moon system in 2800-m.y.-old stromatolites. In even older stromatolites, those of 3100 m.y. age from the Pilbara region of Australia, it is possible that tidal rhythms may be detected. In one study by Kahn and Pompea (1978) it was suggested that changes in the length of the lunar cycle through geologic time could be determined from a study of cephalopod growth rhythms (Fig. 4.8). According to this study, daily growth lines can be detected in the shell of *Nautilus,* whereas the septa seem to be formed each lunar synodic month. The number of growth lines per chamber was counted as 30 ± 2 throughout the entire shell. This agrees closely with the 29.53 days of the present lunar synodic month, and the conclusion was that two developmental cycles were present, one daily and the other monthly. It was suggested that fossil nautiloids of several ages back to the Late Ordovician revealed a progressive reduction in the number of daily growth lines per chamber and that 420 m.y.

FIGURE 4.8
Growth lines on the outer shell of *Nautilus pompilius.* As can be seen, the counting of these lines is not easy, and opinions vary as to what constitutes a daily growth increment.

ago there were only about 21 days in the lunar month. The findings of this study were challenged in a work by Saunders and Ward (1979) in which a much more random relationship between daily growth lines and chamber-wall increments was reported. This applied to both living *Nautilus* and fossil cephalopods. Later work by Ward (1985) on living *Nautilus* showed that apertural growth rate was highly variable, although slowing down with approach to maturity. Chamber formation rates were found to vary from about three weeks in early chambers to nearly four months in later ones. Finally, it seems that astronomically determined data on the moon's recession and the speed of its rotation about the earth extrapolated back in time also give a different picture from that of Kahn and Pompea.

Paleontological clocks are potentially useful but many problems obviously remain; as Pannella (1975) argued, a more broadly based approach and, in particular, greater statistical control of data are needed. Numerous papers on growth rhythms and their application to geochronometric problems are referenced in Rosenberg and Runcorn (1975).

Dendrochronology

Dendrochronology is based on the fact that in many species of trees new rows of wood cells are formed annually, with larger cells in the spring and smaller cells in the summer and fall (Fig. 4.9). Particularly in marginal environments, fluctuations in seasonal averages of temperature and rainfall are likely to be reflected in variations in the thickness of these annual rings. Just as in the case of glacial varved clays, unique or distinctive groupings of growth rings can be used to make intertree correlation and to connect the record of living trees with stumps and logs in ancient peat bog deposits and with beams or posts in historic and prehistoric structures.

FIGURE 4.9
Marked contrast in the size of cells formed during the spring and those formed in the winter in the wood of *Pinus.*

Tree-ring analyses have provided the best means of studying climatic variations in the recent past (Fig. 4.10); chronologies from living bristlecone pines (*Pinus longaeva*) extend back some 6000 years, making cross correlation with even older material possible.

Deviations from the basic model are well known and have been described by numerous workers. For example, the effects of the seasonal climate on growth are not always evident in the same year as the climatic event; such effects may be delayed and spread over several subsequent years (Fritts

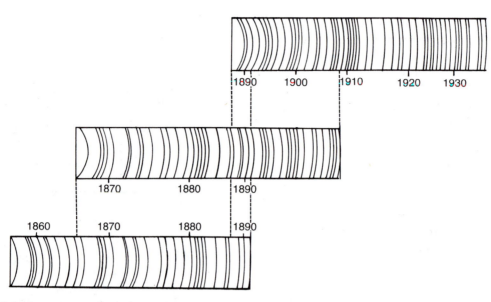

FIGURE 4.10
How tree ring patterns are correlated from recently cut to older trees and the chronology extended back in time.

et al., 1979). The response of trees in any one location to a change in temperature, sunshine, or precipitation may be different in one species than in another. Frost-damage zones can be detected in the annual rings of some species, and in still-growing trees the dating of the frosts can be determined to within a week or two. In some cases, an excessively cold or dry season or infestation by insects, rusts, blights, or other pathological organisms may result in the skipping of a growth ring altogether. Narrow or missing rings may be due also to the effects of ash fallout from volcanic eruptions. Ash bands dated in this way provide a key to the correlation of the **tephrochronologic** and dendrochronologic calendars. In forests adjacent to active volcanoes, it is possible to pinpoint the time of former eruptions with considerable accuracy, as has been shown by Yamaguchi (1985) at Mount St. Helens. This is an essential first step in understanding the eruption history and predicting future activity. Another link between dendrochronology and vulcanism is seen in the dating of unusually severe winters, some of which can be connected with volcanic activity. La Marche and Hirschboeck (1984), for example, demonstrated good agreement between frost-damaged rings and times of major volcanic eruptions in the historical past.

Although it is obvious that the very short time range of dendrochronology means that it is of only peripheral interest to geologists, the climatic cycles discerned in the postglacial period may provide us with the means to understand the mechanics that control the larger cycles over greater time spans. In addition, the unusual precision attainable with dendrochronologic dating has been the means of ascertaining the degree of drift noted in successively older C^{14} dates, consequent on past variations in the rate of atmospheric C^{14} production, and has enabled us to apply appropriate corrections. This is discussed further in Chapter 5.

Milankovitch Cycles

In addition to diurnal, lunar, annual, and sunspot cycles, the effects of certain longer astronomical cycles can be discerned in the geologic record by virtue of the influence of the cycles on world climate. Three cycles that affect the insolation in different parts of the earth can be detected in variations in the relationship of the earth and the sun (Fig. 4.11). These insolation effects were first described by Milankovitch (1941), although the cycles themselves had been calculated earlier by the mathematicians Leverrier in 1855 and Pilgrim in 1904 (Imbrie and Imbrie, 1979). For more recent reviews of the Milankovitch mechanism see Evans (1971) and Vernekar (1972).

The cycles are as follows:

1. Changes in the time of the year when the earth is closest (at perihelion) to and farthest (at aphelion) from the sun. This cycle, known as the precession of the equinoxes, takes place over a period of 19,000–23,000 years. At present the earth is nearest the sun during the southern hemisphere summer and farthest away during the northern hemisphere summer, with the result that the seasons now differ more markedly in the southern hemisphere than they do in the northern. The influence of this precession cycle is small at the poles but it becomes large toward the equator.
2. Variations in the obliquity of the earth's rotational axis. The variation ranges from 21.8° to 24.4°, going from minimum tilt to maximum and back to minimum over a period of 41,000 years. The greater the obliquity, the greater is the contrast between summer and winter, the effect being greatest at the poles and becoming small toward the equator.
3. Changes in the ellipticity of the earth's orbit around the sun. The variation in eccentricity ranges from a minimum of 0.017

(nearly circular) to a maximum of 0.053 (from less than 1 percent to about 6 percent); from minimum to maximum and back to minimum takes about 90,000–100,000 years. The cycle is rather irregular, as is the degree of ellipticity attained at maximum. Variations in ellipticity are thought to cause the difference between maximum and minimum insolation to vary by as much as 30 percent (Gribbin, 1978).

A relationship between the insolation cycle and the climatic cycle was suggested by Koppen (Koppen and Wegener, 1924), and the Koppen-Milankovitch hypothesis is now generally considered a valid explanation of the succession of glacials and interglacials that are well documented in the geologic record of the last 2.5 m.y. The climatic effects of each of the three cycles taken separately are small and could probably be masked by climatic fluctuations from other causes (volcanic ash in the atmosphere and solar cycles, to mention two examples). Taken together, the effects of the three cycles sometimes cancel each other out and sometimes augment each other so that the cumulative effect then becomes noticeable.

It is during glacial episodes that the effects of the earth's orbital variations, the so-called Milankovitch forcing, are particularly noticeable, although the response in terms of variations in ice volumes is not a simple one. Further discussion of this topic is postponed until Chapter 15. For a long time it was believed that orbital effects were restricted to glacial times, although, as long ago as 1929, Bradley had suggested that the cyclicity seen in the Eocene Green River shales was linked to the Milankovitch mechanism. Since that time, it has become increasingly clear that the effects of the Milankovitch cycles are to be seen in climate-sensitive facies throughout the stratigraphic column, during both glacial and nonglacial times. Herbert and

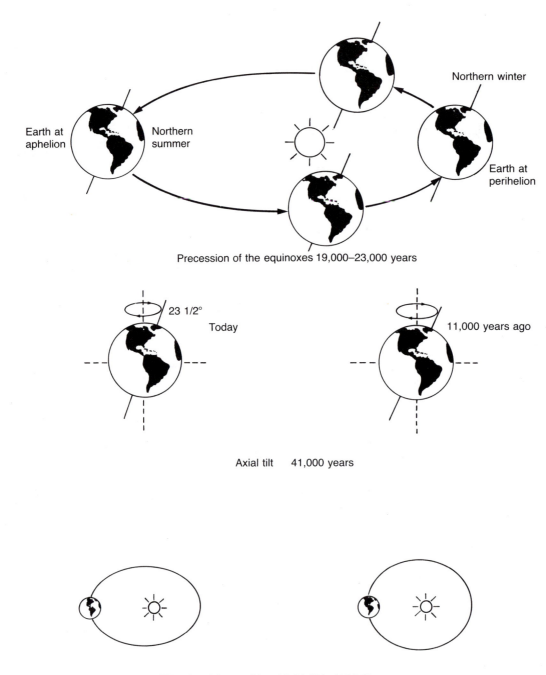

Precession of the equinoxes 19,000–23,000 years

Axial tilt   41,000 years

Ellipticity of the earth's orbit 90,000–100,000 years

FIGURE 4.11
The Milankovitch cycles.

Fischer (1986), working in Upper Albian (Cretaceous) sediments in Italy, described orbital signals in the Milankovitch frequency band (20, 41, 100, and 400 Ka) that are detectable as rhythmic variations in features such as sediment color and calcium carbonate content. As shown in Figure 4.12, limestone-shale couplets with a precession-related (21-Ka) cyclicity are bundled within a 100-Ka short eccentricity cycle, and these cycles in turn vary within a 400-Ka-long eccentricity cycle. Among other Milankovitch-type depositional cycles documented are those in the

Fort Hays limestone member of the Upper Cretaceous Niobrara Formation (Laferriere and Hattin, 1989). The cycles occur in rhythmically bedded pelagic and hemipelagic shales and limestones, again with limestone-shale couplets in high-frequency cycles grouped into bundles of lower frequency. They can be correlated over much of Colorado, Kansas, and northeastern New Mexico. Such recognizable time-equivalent cycles are, of course, isochronous entities and form the basis of what has come to be called cyclostratigraphy. As shown in Chapter 15, the

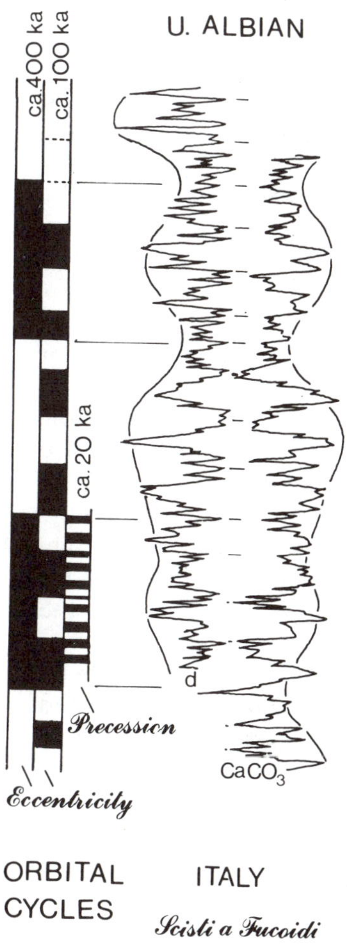

FIGURE 4.12

Signatures of Milankovitch cycles in nonglacial successions. The 400 Ka and 100 Ka eccentricity cycles contain "bundles" of high-frequency, precession-related (ca. 21 Ka) cycles seen in low-carbonate/high-carbonate bedding couplets. Core representing 1600 Ka from the Albian of Italy. Orbital signals are seen in both scan of sediment color (dark/light)—left, and $CaCO_3$ content—right curve. (From Herbert, T. D., and Fischer, A. G., 1986, Milankovitch climatic origin of mid-Cretaceous black shale rhythms in central Italy: Nature, v. 321, p. 739–743.)

ORBITAL   ITALY
CYCLES   Scisti a Fucoidi

Milankovitch curve has played an important role in the deciphering of Pleistocene stratigraphy. During older glaciations in the Permian, Ordovician, and Precambrian, there were also apparently fluctuations in the extent of the ice sheets, and there is no reason to doubt that they also were controlled by the Milankovitch mechanism.

Other Astronomical Cycles

The recurrence of ice ages and a supposed periodicity in mass extinctions are examples of two very long cycles that some workers have discerned in earth history. Both have been explained by terrestrial mechanisms of one kind or another; inevitably, however, extraterrestrial causes also have been invoked. The length of the cycles (measured in hundreds of millions of years for ice ages and around thirty million years for mass extinctions) suggests that if extraterrestrial influences indeed are responsible, they must be something beyond those ascribed to the celestial mechanics of the solar system as presently understood.

There have been at least six major glaciations since the early Precambrian at irregular intervals of from one hundred to several hundred million years. Although the episodic glacial advances and recessions during an ice age, at least in the most recent one, seem to be explained by the Koppen-Milankovitch hypothesis, there is considerably less unanimity on the subject of the basic cause of the ice ages themselves. As mentioned earlier, the thermal isolation of continents and oceans as a consequence of continental drift has been suggested, and it does seem to be supported by a certain amount of evidence; however, it is by no means the only mechanism proposed. Extraterrestrial causes have long been suggested, including the obvious one of a fluctuation in the output of energy from the sun, although the cause of this is not explained. Other ideas include that of Steiner

and Grillmair (1973), who suggested that a correlation can be demonstrated between the supposed periodicity of ice ages and variations in the central galactic force, a measure of the gravitational field of the galaxy. Through the galactic year, variously estimated at between 200 and 400 m.y., the solar system describes an elliptical orbit about the galactic center, and there is consequently a systematic variation in the value of the force. Exactly how variations in the central galactic force can trigger ice ages is not very clear. There are, of course, any number of cyclical phenomena in nature, and the search for matching cycles and cause-and-effect relationships has been an interesting pastime and the source of innumerable scientific papers from the early days of science. Given the large number of variables available in cosmic relationships, it usually is not difficult to find a match between an astronomical cycle and certain cyclical events in geology. The problem lies in trying to ascertain the mechanisms involved.

Turning to mass extinctions, the cyclicity supposedly discerned by different workers ranges from 26 to 32 m.y. Again, extraterrestrial mechanisms have been proposed and, as with the ice ages, one obvious approach is to find an astronomical cycle of similar magnitude. Perhaps the closest match is with a cycle of approximately 31–33 m.y. during which the solar system oscillates through the galactic plane. At such times, it has been suggested that increased dosages of cosmic radiation are the cause of mass extinction (Hatfield and Camp, 1970). In an alternative hypothesis, Raup and Sepkoski (1984, 1986) and Raup and Boyajian (1988) described a 26-m.y. cycle in mass extinctions (Fig. 4.13) and suggested that the earth is periodically bombarded by meteorites or, more likely, comets. To explain the periodicity, it has been postulated (Davis et al., 1984; Whitmire and Jackson, 1984) that the meteorites or comets originate in what is called the **Oort**

FIGURE 4.13

A. Record of percent extinction per million years from data set of 9773 marine animals. The best-fit 26 m.y. cycle is shown along the top. Ticks on abscissa denote stratigraphic stages. Dots indicate centers of sampling intervals. (From Raup, D. M. and Sepkoski, J. J. Jr., 1988, Testing for periodicity of extinction: Science, v. 241, p. 95, Fig. 2. B. Extinction profiles for ten samples of 1000 genera chosen at random from a data set of 19,897. (From Raup, D. M., and Boyajian, G. E., 1988, Generic extinction in the fossil record: Paleobiology, v. 14, Fig. 2, p. 116.)

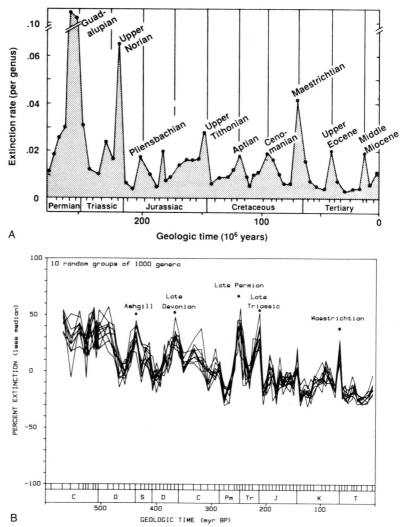

cloud, a swarm of millions of comets that is believed to lie in an orbit at the outer reaches of the solar system. At regular intervals another star, supposedly a small and dim companion of the sun, sweeps in on a highly eccentric orbit about the sun and passes through the Oort cloud (Fig. 4.14). Each time this happens, the gravitational influence of the star, which has been named Nemesis, sends many comets into new orbits and some eventually collide with the earth (Raup, 1986). This model finds some support in the considerable terrestrial evidence described in Chapter 2. Although the Nemesis theory is an interesting, and perhaps plausible, explanation, it should be emphasized that the existence of such a companion star or, for that matter, the Oort cloud, has yet to be demonstrated. That this hypothesis is taken seriously, however, is indicated by the fact that several astronomers now are actively engaged in a search for Nemesis.

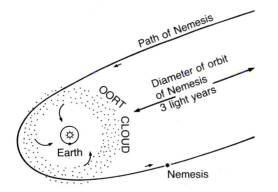

FIGURE 4.14
Oort cloud and the relationship between Sol and Nemesis.

4.3 ONGOING PHENOMENA

Apart from the cyclical phenomena just described, there are certain other geologic processes that are time-dependent. They are the ongoing processes, and their potential as timekeepers depends on their continuity and unvarying rate. Radiometric decay and evolutionary change are the two most important phenomena in this category; they are dealt with in later chapters.

Sediment Accumulation

Although, as described in Chapter 1, sediment accumulation is, for the most part, markedly episodic and usually a most unreliable basis for measuring time, there are some exceptions. In deep-sea environments, for example, estimates of age are sometimes made on the basis of sediment thickness because accumulation rates are thought to show little variation over long periods of time. They range from about 1–5 m/m.y. for siliceous oozes and clays to 5–40 m/m.y. or more for calcareous oozes. Provided some correction is made for compaction in deep-sea sediment cores, age can be plotted against the depth, and the age of each portion of the core can be read off directly from

a straight-line graphic plot. Once the upper part of the deep-sea stratigraphic column has been sampled and ages and thicknesses determined from cores, the ages of acoustic reflectors beyond the reach of the drill and down to acoustic basement can be estimated by extrapolation. The accumulation-rate method has been applied to ancient deposits with varying degrees of success. One study by Churkin et al. (1977) was based on the assumption that certain Ordovician and Silurian graptolitic shales were deposited at rates approximating those of modern pelagic siliceous oozes and clays.

Seafloor Spreading

The movement of the ocean floor away from a spreading center and the subsequent cooling and subsiding of the basement also provide approximate means of dating the ocean basement. This is done on the basis of linear extrapolations and assumes that these movements have been going on at a constant rate. Thus, the average depth to basement plotted against either age or distance from the spreading center shows a more or less constant relationship.

The pattern of magnetic anomaly stripes in the ocean basement, and the recognition of dated geomagnetic events within the pattern, provides another means of establishing a time scale. The most recent anomalies—that is, those occurring during the last 5 m.y.— have been quite accurately dated from both continental localities and deep-sea cores. When the distances of these seafloor anomaly stripes from the spreading center are plotted graphically against their ages, it can be shown that a linear relationship exists. The line joining the points is straight, indicating that seafloor spreading over this time period has apparently been proceeding at a constant rate (Fig. 4.15). If the distances from the spreading center to more-distant anomalies are plotted, their ages can be estimated by

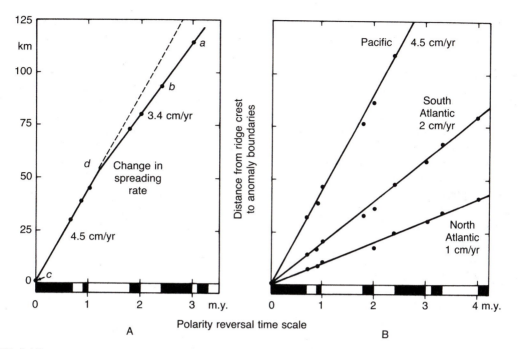

FIGURE 4.15

Rates of seafloor spreading estimated by plotting the distance of anomaly boundaries from the ridge crest against the calibrated age of each polarity reversal represented at each boundary. A. Average spreading rate given by slope of lines. B. Different spreading rates are recorded at different places along the oceanic ridges. (From Wyllie, P. J., 1976, The way the earth works, Fig. 10–8, p. 146. Reprinted by permission of John Wiley & Sons, New York.)

extrapolation. This is assuming, of course, that the ocean-floor spreading rate has remained constant during all this time. Comparisons of figures from different ocean basins provide some check on this. We return to this topic in Chapter 6.

4.4 CONCLUSION

In human experience the passage of time normally is measured by timekeepers that operate at a uniform rate. In the larger spans of geologic time, many of the timekeeping phenomena work in quite an irregular way; this does not detract from their value, provided they are nonerasable as applied to the geologic record itself. Clearly, the record of the passage of time must be preserved if it is to be of any use. An erasable phenomenon is one in which the evidence for it is likely to be destroyed not long after it formed. Many different time-dependent natural phenomena, operating at all scales, are of this nature, and it is only a tiny fraction that becomes part of the permanent record.

REFERENCES

Anderson, R. V., and Kirkland, D. W., 1960, Origin, varves, and cycles of Jurassic Todilto formation, New Mexico: Am. Assoc. Petroleum Geologists Bull., v. 44, p. 37–52.

_____1966, Intrabasin varve correlation: Geol. Soc. America Bull., v. 77, p. 241–255.

Ashley, G. M., 1972, Rhythmic sedimentation in glacial lake Hitchcock, Mass.-Conn.: Univ. Mass. Amherst, Geology Pub. 10, 148 p.

———1975, Rhythmic sedimentation in glacial lake Hitchcock, Massachusetts-Connecticut, in Jopling, A. V., and McDonald, B. C., eds., Glaciofluvial and glaciolacustrine sedimentation: Soc. Econ. Paleontologists and Mineralogists Spec. Pub. No. 23, p. 304–320.

Berry, W. B. N., and Barker, R., 1968, Fossil bivalve shells indicate longer month and year in Cretaceous than present: Nature, v. 217, p. 938–939.

Bradley, W. H., 1929, The varves and climate of the Green River epoch: U.S. Geol. Survey Prof. Paper 158-E, p. 87–110.

———1964, Geology of Green River Formation and associated Eocene rocks in southwestern Wyoming and adjacent parts of Colorado and Utah: U.S. Geol. Survey Prof. Paper 496A, 86 p.

Churkin, M., Jr., Carter, C., and Johnson, B. R., 1977, Subdivision of Ordovician and Silurian time scale using accumulation rates of graptolitic shale: Geology, v. 5, p. 452–456.

Crabtree, D. M., Clausen, C. D., and Roth, A. A., 1980, Consistency in growth line counts in bivalve specimens: Palaeogeog., Palaeoclimat., Palaeoecol., v. 29, p. 323–340.

Creer, K. M., 1975, On a tentative correlation between changes in the geomagnetic polarity bias and reversal frequency and the earth's rotation through Phanerozoic time, in Rosenberg, G. D., and Runcorn, S. K., eds., Growth rhythms and the history of the earth's rotation: New York, John Wiley & Sons, p. 293–318.

Davis, M., Hiut, P., and Muller, R. A., 1984, Extinction of species by periodic comet showers: Nature, v. 308, p. 715.

DeGeer, G., 1912, A geochronology of the last 12,000 years: Internat. Geol. Cong. XI, Stockholm, 1910, Compte Rendu, v. 1, p. 241–258.

Evans, P., 1971, Towards a Pleistocene time scale, in The Phanerozoic time-scale—A supplement: Geol. Soc. London Spec. Pub. 5, p. 123–356.

Fritts, H. C., Lofgren, G. R., and Gordon, G. A., 1979, Variations in climate since 1602 as reconstructed from tree rings: Quaternary Research, v. 12, p. 18–46.

Gribbin, J., 1978, The search for cycles, in Gribbin, J., ed., Climatic change, Sec. 3: Cambridge, Cambridge Univ. Press, p. 139–149.

Hall, C. A., Jr., Dollase, W. A., and Corbato, C. E., 1974, Shell growth in Tivela stultorum (Mawe, 1823) and Callista chione (Linnaeus, 1758) (Bivalvia): Annual periodicity, latitudinal differences, and diminution with age: Palaeogeog., Palaeoclimat., Palaeoecol., v. 15, p. 33–61.

Hatfield, C. B., and Camp, M. J., 1970, Mass extinctions correlated with periodic galactic events: Geol. Soc. America Bull., v. 81, p. 911–914.

Herbert, T. D., and Fischer, A. G., 1986, Milankovitch climatic origin of mid-Cretaceous black shale rhythms in central Italy: Nature, v. 321, p. 739–743.

Imbrie, J., and Imbrie, K., 1979, Ice ages: Hillside, NJ, Enslow Publishers, 224 p.

Kahn, P. G. K., and Pompea, S. M., 1978, Nautiloid growth rhythms and dynamical evolution of the earth-moon system: Nature, v. 275, p. 606–611.

Koppen, W., and Wegener, A., 1924, Die Klimate der geologishen vorzeit: Berlin, Gebrüder Borntraeger, 266 p.

Laferriere, A. P., and Hattin, D. E., 1989, Use of rhythmic bedding patterns for locating structural features, Niobrara Formation, United States Western interior: Am. Assoc. Petroleum Geologists Bull., v. 73, p. 630–640.

La Marche, V. C., Jr., and Hirschboeck, K. K., 1984, Frost rings in trees as records of major volcanic eruptions: Nature, v. 307, p. 121–126.

MacClintock, C., Pannella, G., and Thompson, M., 1968, Paleontological evidence of variations in length of synodic month since late Cambrian: Science, v. 162, p. 792–796.

Mazzullo, S. J., 1971, Length of the year during the Silurian and Devonian Periods: New values: Geol. Soc. America Bull., v. 82, p. 1085–1086.

Milankovitch, M., 1941, Kanon der Erdbestrahlung und seine Andwendung auf das Eiszeitenproblem: Royal Serb. Acad. Spec. Pub. 133, Belgrade, 633 p.

Olausson, E., and Olsson, I. U., 1969, Varve stratigraphy in a core from the Gulf of Aden: Palaeogeog., Palaeoclimat., Palaeoecol., v. 6, p. 87–103.

Pannella, G., 1972, Palaeontological evidence on the earth's rotational history since early Precambrian: Astrophys. Space Sci., v. 16, p. 212–237.

———1975, Palaeontological clocks and the history of the earth's rotation, *in* Rosenberg, G. D., and Runcorn, S. K., eds., Growth rhythms and the history of the earth's rotation: New York, John Wiley & Sons, p. 253–284.

Pannella, G., MacClintock, C., and Thompson, M. N., 1968, Paleontological evidence of variations in length of synodic month since Late Cambrian: Science, v. 162, p. 792–796.

Raup, D., 1986, The Nemesis affair: New York, Norton, 220 p.

Raup, D. M., and Boyajian, G. E., 1988, Patterns of generic extinction in the fossil record: Paleobiology, v. 14, p. 109–125.

Raup, D., and Sepkoski, J. J., Jr., 1984, Periodicity of extinction in the geologic past: Proc. Natl. Acad. Sci., v. 81, p. 801–805.

———1986, Periodic extinction of families and genera: Science, v. 231, p. 833–836.

Rosenberg, G. D., and Runcorn, S. K., 1975, Growth rhythms and the history of the Earth's rotation: New York, John Wiley & Sons, 538 p.

Saunders, W. B., and Ward, P. D., 1979, Nautiloid growth and lunar dynamics: Lethaia, v. 12, p. 172.

Steiner, J., and Grillmair, E., 1973, Possible galactic causes for periodic and episodic glaciations: Geol. Soc. America Bull., v. 84, p. 1003–1018.

Thompson, I., 1975, Biological clocks and shell growth in bivalves, *in* Rosenberg, G. D., and Runcorn, S. K., eds., Growth rhythms and the history of the earth's revolution: New York, John Wiley & Sons, p. 149–162.

Vernekar, A. D., 1972, Long-period global variations of incoming solar radiation: Meteorol. Mon., v. 12.

Ward, P. D., 1985, Periodicity of chamber formation in chambered cephalopods: Evidence from *Nautilus macromphalus* and *Nautilus pompilius:* Paleobiology, v. 11, p. 438–450.

Wells, J. W., 1963, Coral growth and geochronometry: Nature, v. 197, p. 948–950.

Whitmire, D. P., and Jackson, A. A., 1984, Are periodic mass extinctions driven by a distant solar companion?: Nature, v. 308, p. 713.

Williams, G. E., 1981, Sunspot periods in the late Precambrian, glacial climate, and solar-planetary relations: Nature, v. 291, p. 624–628.

Yamaguchi, D. K., 1985, Tree ring evidence for a two-year interval between recent prehistoric explosive eruptions of Mount St. Helens: Geology, v. 13, p. 554–557.

5

RADIOMETRIC DATING

If the rate of production of helium from known weights of the different radio-elements were experimentally known, it should thus be possible to determine the interval required for the production of the amount of helium observed in radioactive minerals, or, in other words, to determine the age of the mineral.

Ernest Rutherford

5.1 INTRODUCTION

The next two chapters are intended to bring together those phenomena that, although not directly related to sedimentary processes, nevertheless play an important role in the measurement of geologic time. In some cases, as with isotopic dating, they operate mainly, although not entirely, independently of sedimentary systems; in others—for example, paleomagnetism—they leave a record within the stratigraphic succession but do not play any direct part in the accumulation of the sediment itself.

It was the New Zealander Ernest Rutherford who, in 1905, was the first to suggest that the phenomenon of radioactivity could be used to measure time. If the rate of radioactive decay were known, and the proportions of the atoms, or **nuclides,** of both the parent and **daughter elements** could be determined, the time since the formation of the first **daughter nuclides** could be calculated. In minerals that incorporate radioactive elements, the crystallization process usually separates parent from daughter nuclides. This means that any daughter nuclides found in a mineral must have been formed since the mineral crystallized.

5.2   PRINCIPLES

The decay of radioactive elements began at the time of their creation within stellar masses, this nuclear synthesis occurring about 10 billion years ago (Stacey, 1969). The radioactive clocks did not actually start, however, until those elements became part of minerals in a rock. Once inside the mineral, daughter elements produced by decay usually cannot escape; it is the measurement of the proportions of parent and daughter elements that forms the basis for many isotopic dating methods. Because radioactive decay occurs only in the atomic nucleus (Table 5.1), changes involving the orbital electrons, such as chemical changes like oxidation, have no effect on the rate of decay. Radioactive decay occurs at a geometric rate; that is, over a given time the proportion of atoms that decay to the number remaining stays the same, no matter how long the decay process continues. The probability of decay is known as the **decay,** or *disintegration,* **constant,** usually expressed as λ (see Table 5.2). Because the reduction in the number of decaying atoms proceeds geometrically, it means that an infinite amount of time would be required for decay of all the parent atoms. However, the time taken for the decay of half the atoms of the parent element can be calculated and is known as the **half-life,** usually expressed as T. The value of the half-life and the decay constant are related and expressed as $T = 0.69315/\lambda$ (Table 5.2).

Isotopic ages are determined in real time, that is in thousands or millions of years before the present, and the time scale established by these methods is often referred to as an **absolute time scale.** Most isotopic dates are obtained from igneous rocks, the dates marking the time that the rock crystallized, or from metamorphic rocks, in which case it is the time of metamorphism that is dated—often this is a resetting of a radiometric clock started much earlier. Sedimentary rocks may be dated in three ways:

TABLE 5.1

Summary of decay processes

Decay Mode	Change	Result	Example
Alpha decay	Emits alpha particles	Reduces atomic number by 2; reduces mass number by 4	$_{92}U^{238} \rightarrow {}_{90}Th^{234}$
Beta decay	Emits negatively charged beta particle (electron)	Converts neutron into proton + electron; atomic number increased by 1; mass number unchanged	$_{37}Rb^{87} \rightarrow {}_{38}Sr^{87}$
Electron capture	Captures electron	Converts proton into neutron; atomic number decreased by 1; mass number unchanged	$_{19}K^{40} \rightarrow {}_{18}Ar^{40}$

TABLE 5.2
Decay constants and half-life values

Radioactive Isotope	Half-life (years)	Decay Constants
K^{40}	1.250×10^9	5.543×10^{-10}
Rb^{87}	48.8×10^9	1.42×10^{-11}
U^{238}	4.468×10^9	1.55125×10^{-10}
U^{235}	0.7038×10^9	9.8485×10^{-10}
U^{234}	2.47×10^5	2.806×10^{-6}
Th^{232}	14.010×10^9	4.9475×10^{-11}

1. They may be dated directly if they contain authigenic glauconite, which can be dated by the K-Ar method or, as in certain carbonate sediments, by a uranium-234–thorium-230 method. Direct dating of very young sediments—for example, peats—is possible by carbon-14 dating. These methods are described in more detail later.
2. The dating of lava flows and/or ash falls obviously provides a means of also dating the sedimentary succession with which they are interbedded, provided there is evidence of contemporaneity with the enclosing sediment. Ash bands incorporated within deep-sea sediments, and with indications of more or less uninterrupted sedimentation, would be an example.
3. The age of a sedimentary succession may be estimated by "bracketing." That is to say, a maximum age is indicated for sediments that overlie dated igneous rocks, and a minimum age when they are intruded by igneous rocks (see Fig. 5.1).

The main contribution of isotopic dating to stratigraphy has been to provide fixed dates within the succession of sedimentary rocks, thereby reducing the flexibility within what was initially a largely relative time scale. It should not be thought, however, that isotopic methods are intrinsically superior to or more accurate than the dating established by biostratigraphy (Conkin and Conkin, 1984, p.

266). Indeed, in successions with good controls, as for example the Jurassic ammonite zones, time intervals of a few hundred thousand years can be discerned. Older isotopic dates are often accurate only to several million years and, in the Precambrian, dates may be accurate to, say, only 50–70 m.y. This is still a very small percentage in the case of rocks 2 to 3 b.y. old; in the absence of biostratigraphic or other controls, radiometric dating is virtually the only way whereby Precambrian history can be deciphered.

5.3 SOURCES OF ERROR

Isotopic dates are normally expressed as a value followed by a plus and minus figure to signify a range in the precision of age determination. Thus, a figure of 245 ± 5 m.y. means that should the age determination process be repeated, it most likely would result in a date within a 10-m.y. time range. The magnitude of the ± range is simply a measure of the standard deviation of values obtained from numerous age determinations. There are two major kinds of uncertainty to be considered in isotopic age determination. One source of error lies in the possibility that the mineral used in the determination has not remained a closed system since crystallization, but rather has lost a proportion of the radiogenic daughter by leakage. This is one

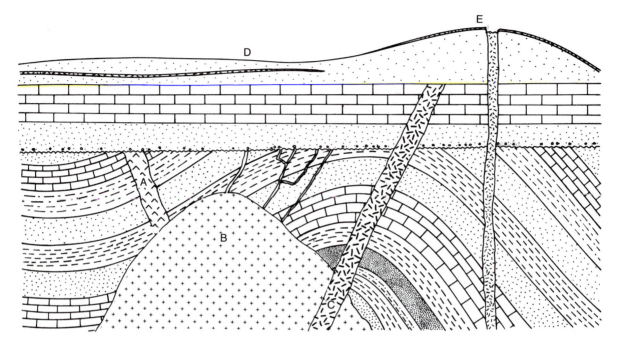

FIGURE 5.1
Relative time dating by bracketing, using the field relationships between country rock and igneous intrusions, mineralization, and lava flows. The folded sediments at the bottom of the section are pre-A in age; the limestone is post-B, pre-C in age; the lower sandstone is of post-C but pre-D age, the upper sandstone of post-D, pre-E age.

of the commonest problems encountered and eventually will result in a calculated date that is too young. Conversely, there is the possibility that, at the time of crystallization, not all preexisting radiogenic material was excluded; clearly, this distorts the determined age in the other direction. Secondly, there are certain errors that arise from the fact that the decay constants of the nuclides used in isotopic dating are known only to an accuracy of within 1 or 2 percent. This is a systematic error, but it must be added to the uncertainties that exist in calibrating the mass spectrometers used in this work and in the standards used. These are largely laboratory-generated variables, as also is what could be termed the skill factor. This is concerned with the accuracy of such data as the isotope ratio values.

Precision in this part of the process is increasing with the use of more sophisticated techniques and the growing use of computers in data acquisition. All of these sources of variation add up to interlaboratory differences of 3 percent or more.

5.4 POTASSIUM-ARGON METHOD

The potassium-argon method has come to be one of the most useful of the isotopic dating procedures. Potassium is the seventh most abundant element in the earth's crust and occurs in a wide range of rock-forming minerals. It can be used to date rocks from the oldest known to as young as late Pleistocene (Table 5.3). The decay of potassium-40 occurs in two ways: one decay scheme, involv-

TABLE 5.3

Common rock-forming minerals that are generally useful for potassium-argon dating; $\otimes$ = widely useful, x = sometimes useful (from POTASSIUM-ARGON DATING: PRINCIPLES, TECHNIQUES AND APPLICATIONS by G. Brent Dalrymple and Marvin A. Lanphere. Copyright © 1969. Reprinted by permission of W. H. Freeman and Company)

	Rock Type			
	Volcanic	Plutonic	Metamorphic	Sedimentary
Feldspars				
sanidine	$\otimes$			
anorthoclase	$\otimes$			
plagioclase	$\otimes$			
Feldspathoids				
leucite	x			
nepheline	x	x		
Mica				
biotite	$\otimes$	$\otimes$	$\otimes$	
phlogopite			$\otimes$	
muscovite		$\otimes$	$\otimes$	
lepidolite		x		
glauconite				x
Amphibole				
hornblende	$\otimes$	$\otimes$	$\otimes$	
Pyroxene	x	x		
Whole-rock	$\otimes$		x	

ing 89 percent of potassium-40, produces calcium-40 by beta decay; the remaining 11 percent decays to argon-40 by electron capture. Radiogenic calcium cannot be distinguished from common calcium, and because of the abundance of that element in crustal rocks of all kinds, the potassium-calcium decay scheme is unsuitable for dating. The potassium-argon method is the simplest of the isotopic clocks because it depends on the accumulation of the daughter isotope argon-40, none of which is normally present in the igneous rock at the time of formation. Only a minor correction is necessary for possible contamination by atmospheric argon (argon-36 and argon-35). Another advantage is that because argon is an inert gas it can be measured easily, even if present in only a minute

amount. It is for this reason that even rocks as young as 100,000 years can be dated. The chief source of inaccuracy in this method arises from the loss of radiogenic argon by diffusion. This is particularly likely to happen in the early stages of the cooling history of the rock until the temperature drops below what is called the "blocking temperature," or **diffusion threshold,** which has a value of around 200°C, but which varies with the mineral. The more slowly the rock cools, the longer the constituent minerals will remain above their diffusion thresholds, so that leakage is particularly likely in coarse-grained rocks cooled slowly at great depths. Potassium-argon dates from such sites may be erroneously too young by a considerable percentage. Lava flows, ashes, and rapidly cooled, fine-grained intrusives usually give a more accurate figure for the date of crystallization.

In marine sediments, potassium-argon dating can be used with the mineral glauconite. Glauconite, a complex potassium-iron-aluminum silicate, forms authigenically within the top few centimeters or meters of sea-bottom sediment as a replacement of skeletal debris, **ooliths,** fecal pellets, and mineral grains. As the glauconite grains grow within the sediment, they incorporate radioactive potassium-40 and, provided no leakage of the decay product argon-40 has occurred, the age can be calculated. Because glauconite grains grow within the sediment and are not deposited as clastic grains, the calculated age is, naturally, a little younger than that of the enclosing sediment. This discrepancy varies, depending upon the rate of sediment accumulation, the porosity and permeability of the sediment, and other factors, but, according to Odin and Dodson (1982), may range up to 25,000 years or more. Another potential source of dating error is a consequence of the geothermal gradient; it is suspected that a loss of argon may occur with heating of the glauconite as it becomes progressively more deeply buried. Cross checking with dates obtained by other means frequently shows glauconite dates to be as much as 20 percent too young. The percentage error decreases with age, however, and glauconite dates from pre-Tertiary sediments are considered more reliable than those from younger rocks.

5.5 URANIUM-LEAD METHODS

Uranium has three naturally occurring isotopes: uranium-238, uranium-235, and uranium-234, all of which are radioactive. Intimately associated with uranium is thorium which, although it occurs mainly as one isotope, thorium-232, is found also as five short-lived intermediate daughters of uranium-238,

uranium-235, and thorium-232. This chain of radioactive daughters ends with a stable isotope of lead. Uranium and thorium are found in numerous minerals, including zircon, sphene, monazite, apatite, and certain other less common species.

Uranium-238 decays to give rise to a number of daughter isotopes in what is termed the *uranium series.* Most of the daughters have fleeting half-lives (Fig. 5.2), and the series ends in an inert isotope of lead (lead-206). The decay involves eight alpha decay steps and six beta steps (Fig. 5.3); in practice, the intermediate members of the series are ignored and it is simply expressed as $U^{238} \rightarrow Pb^{206} + 8^4$ He. Uranium-235 decay gives rise to what is called the *actinium series,* again containing numerous daughter isotopes (Fig. 5.4) and ending in a stable lead isotope Pb^{207}. This series involves seven alpha steps and four beta steps and is usually summarized as $U^{235} \rightarrow Pb^{207} + 7^4$ He. Thorium-232 decays through relevant intermediate daughters (Fig. 5.5) to stable lead (Pb^{208}) with six alpha steps and four beta steps and is expressed as $Th^{232} \rightarrow Pb^{208} + 6^4$ He. Alpha decay produces helium, which is, therefore, a stable byproduct in addition to radiogenic lead in both uranium and thorium decay schemes. Many uranium-bearing minerals contain some original lead, and this must be detected; otherwise, an isotopic age will be calculated in excess of the true age. Fortunately, all common lead contains a certain proportion of the isotope lead-204, which is not radiogenic; if this is detected, it must have been present when the mineral formed. Its occurrence is an indication that a proportion of lead-206 and lead-207 was also present from the beginning, and an appropriate subtraction from the total lead can then be made.

The parallel decay schemes of uranium-235 and uranium-238 provide a means of cross-checking the accuracy of dates produced from them. Provided the mineral has remained a closed system and has neither

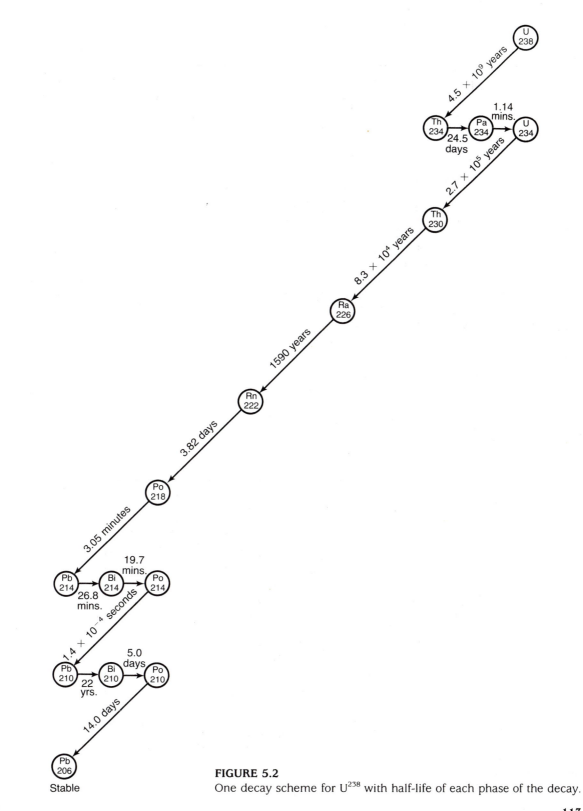

FIGURE 5.2
One decay scheme for U^{238} with half-life of each phase of the decay.

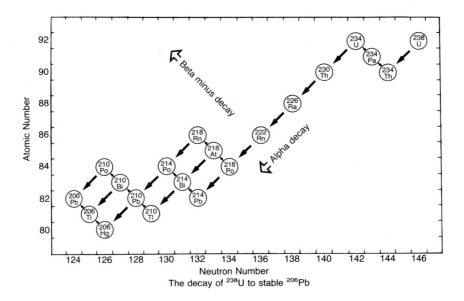

FIGURE 5.3
Decay scheme for U^{238}. (Redrawn after Faure, 1986.)

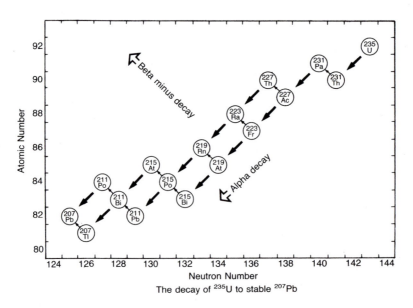

FIGURE 5.4
Decay scheme for U^{235}. (Redrawn after Faure, 1986.)

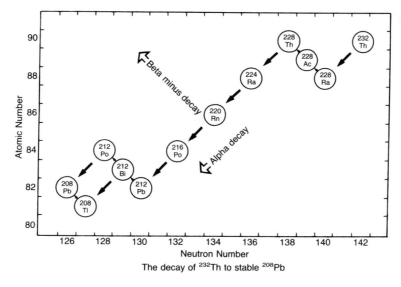

The decay of ^{232}Th to stable ^{208}Pb

FIGURE 5.5
Decay scheme for Th^{232}. (Redrawn after Faure, 1986.)

gained nor lost any uranium or daughter isotopes, and after some allowance is made for any original nonradiogenic lead (lead-204), the two decay schemes should provide dates that agree and which are then said to be concordant. The relationship between the uranium-235 and uranium-238 decay schemes can be shown graphically. The values of the lead-207/uranium-235 and lead-206/uranium-238 ratios as a function of age can be plotted as a curve known as the **concordia** (Fig. 5.6). In effect, it is a time curve defined by the loci of all concordant uranium-lead dates. Any dates that do not fall on the line are said to be discordant. For example, if the system has not been closed and some lead has been lost, the date will plot below the concordia. Many factors are involved in the loss of lead, but it is commonly caused by heat or stress during episodes of metamorphism and/or intrusion. If all the radiogenic lead is lost, the clock is, in effect, reset at zero. A further check on the accuracy of dates is provided by calculating the ratio of the two ra-

diogenic lead isotopes lead-207/lead-206 (the end members of the uranium-235 and uranium-238 decay series, respectively). This has been found to approximate quite closely the time of crystallization. Dates can be calculated also by comparing the rates of loss of lead, not only from the two uranium decay schemes but also from the thorium-lead scheme.

Uranium series dating can also be used with some calcareous skeletal material and limestone deposits. Corals, for example, apparently do not discriminate between calcium, strontium, radium, and uranium in the secretion of their skeletons and, consequently, incorporate some of these elements in their structure. As far as uranium is concerned, corals apparently have a uranium/calcium ratio similar to that of sea water, in which the dissolved uranium is enriched in uranium-234 by 15 percent over the parent uranium-238. Studies by Barnes et al. (1956) and Thurber et al. (1965) showed that living corals have a uranium-234/uranium-235 ratio

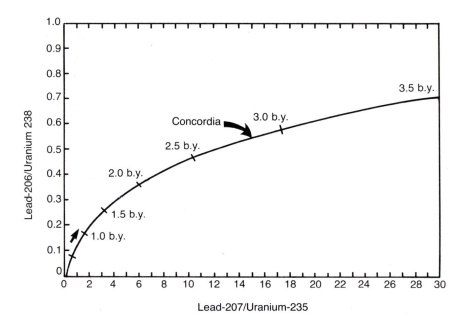

FIGURE 5.6

Concordia diagram for uranium-lead ratios. Provided the mineral has remained a closed system, the lead-207/uranium-235 and lead-206/uranium-238 dates should agree, and the ages are said to be concordant. The concordia curve is defined by the loci of all concordant ages.

similar to that of sea water, but with little or no thorium-230, a daughter isotope of uranium-238. Provided there has been no recrystallization and alteration to calcite, and the biogenic aragonite has remained a closed system, the age of the coral can be determined by measuring the growth of thorium-230 from alpha decay of uranium-234. Ages ranging up to about 350,000 years are potentially measurable. The same technique has been used in dating oolites (Broecker and Thurber, 1965) and also travertine deposits in caverns (Fornanca-Rinaldi, 1968). In **speleothems,** the uranium originates by leaching of uranium carbonates from the limestone above the cave and is carried down in solution by percolating rain water. Because thorium is relatively insoluble, it tends to remain behind and typically is found absorbed into clay minerals. Because the clays may also be

carried downward, there is the possibility of contamination of the speleothems, and extreme care is needed in the selection of only the purest calcite. In practice, stalagmites have been found the most suitable sites for calcite sampling.

5.6 THORIUM-230 METHOD

In dating relatively recent marine sediments, it often is found that potassium-argon dating is unreliable, and an alternative dating method that measures the uranium-238 present in sea water has come to be used. Uranium-238 decays into another unstable isotope, thorium-230, which typically is taken up readily by absorption into clays or by incorporation in authigenic minerals, presumably at a fairly constant rate. Thorium-230, in turn, decays to radium-226, with a half-life of

83,000 years. Progressive sampling for thorium-230 down a core shows a measurable decrease with depth and, provided that sedimentation rates have been uniform and accumulation continuous, the value of the ratio of thorium-230 at the sediment–water interface to that at depth provides a means of dating the core.

5.7 RUBIDIUM-STRONTIUM METHOD

Rubidium is found in two natural isotopes, rubidium-85 and rubidium-87. Rubidium 87 constitutes 25 percent of all rubidium, is radioactive, and decays by emitting a single beta particle to form strontium-87. Rubidium is not an abundant element but it is found in a variety of minerals. With an ionic radius close to that of potassium, rubidium is often found as a substitute in potassium-bearing minerals such as potassic feldspars, the micas, pyroxenes, amphiboles, olivine, and certain clay minerals. The Rb-Sr method is thus widely used in dating granites, pegmatites, and basic rocks as well as many metamorphic rocks, particularly in the Precambrian (Neumann and Huster, 1974).

Many rubidium-bearing minerals also contain common strontium, a mixture of strontium-86 and strontium-87, so that the isotopic proportions must be determined before the amount of radiogenic strontium-87 can be calculated. The proportion of strontium-86 to strontium-87 at the time of rock formation can be ascertained so that the amount of strontium-86 in a sample provides an indicator of how much strontium-87 was originally present. Just as in the case of the original lead in the uranium-lead method, this original strontium-87 must be subtracted in making the age calculation. In young rocks the amount of original strontium-87 relative to new radiogenic strontium-87 is clearly greater than in older rocks in which there has been time for the accumulation of a larger proportion of radiogenic strontium-87. Because there is always a small percentage error in calculating the strontium-86/strontium-87 ratio, it follows that Rb-Sr dating is more reliable when applied to very old rocks, such as those from Archean shield areas or the moon. So-called whole-rock dating often is used in the Rb-Sr method because it sometimes is found that the rock as a whole has remained a closed system, whereas the constituent minerals have not.

5.8 RADIOCARBON DATING

Radiocarbon dating differs from the methods described above in that it is concerned with carbon compounds formed by once-living organisms or by processes operating in equilibrium within the atmosphere and hydrosphere. Material dated by this method includes wood and charcoal, seeds, cloth, paper, peat, bones, shells (both organic carbon in conchiolin and inorganic carbon in aragonite or calcite), freshwater limestones, and cave deposits. Dates from wood, charcoal, and charred bones are usually reliable, but dates from other material may be less so because of problems of contamination, leaching by ground-water circulation, and so on.

Carbon has three isotopes: carbon-12 and carbon-13, which are stable, and carbon-14, which is radioactive. The radioactive isotope is produced in the upper atmosphere about 16 km above the surface by a reaction between cosmic-ray neutrons and nuclei of stable nitrogen-14 atoms. In the reaction, a nitrogen-14 atom absorbs a neutron and emits a proton, thereby changing to carbon-14. Carbon-14 is incorporated in the carbon dioxide molecules in the atmosphere and also quickly comes to be present in all carbon compounds in the oceans and in living plants and animals. Although carbon-14 eventually decays back to nitrogen-14 (Fig. 5.7), its loss is compensated by the production of new carbon-14 by neutron bombardment of nitrogen-14, and so the carbon-14 reservoir is

FIGURE 5.7

The decay of radioactive carbon (C^{14}) by the spontaneous emission of a β particle results in an N^{14} atom.

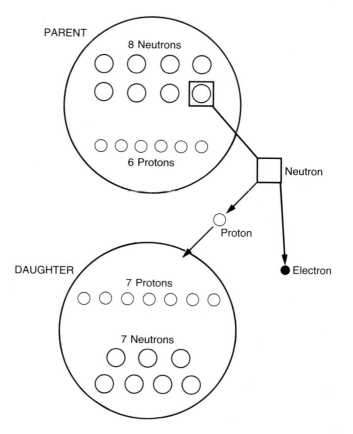

essentially in steady state throughout the atmosphere, hydrosphere, and biosphere. As long as a plant or animal is living and recycling carbon during its metabolism it is in equilibrium with the carbon-14 reservoir, but on death the recycling ceases and the carbon-14 existing at the time of death begins to decay. Radiocarbon dating depends upon the measurement of the amount of remaining carbon-14 in dead organic material and is thus termed a *decay clock,* as opposed to an *accumulation clock,* which depends upon the measurement of an accumulating daughter isotope.

The half-life of carbon-14 is now taken as 5730 ± 40 years, a figure adopted internationally at the 5th Radiometric Dating Conference of 1962. An earlier figure of 5568 ± 30 years was used in older studies and is still used for dates published in the journal *Radiocarbon.* In practice this is no problem—the dates can be easily corrected by a simple calculation. With such a brief half-life the biggest problem in radiocarbon dating lies in the detection of carbon-14 in older material. With standard techniques the age limit lies at about 10 half-lives, by which time the carbon-14 remaining is only about 1/1000 of that present originally. In practice, beyond about 40,000 years the amount of carbon-14 remaining is so small as to become virtually impossible to detect. A more recent technique (Grootes, 1978; Stuiver et al., 1978, 1979), in which carbon-14 is concentrated through a thermal diffusion process, has extended the range of radiocarbon dating back to about 75,000 years, but special precautions must be taken to avoid contamination, and extra shielding is needed to reduce background count. In addition, relatively large samples of 60–120 g of carbon are needed for the process, and the procedure is fairly time consuming, taking about five weeks for a sixfold enrichment. In a departure from conventional dating methods based on the counting of particles emitted during decay, a new technique, using a particle accelerator and an ultrasensitive mass spectrometer, was developed at the University of Toronto during the late 1970s (Litherland and Rucklidge, 1981). This technique has made possible the direct detection of carbon-14 atoms in the sample,

and, with sensitivities better than parts per quadrillion (10^{15}), dates as old as 70,000 years B.P. can be measured. Dating is rapid, taking only hours instead of days, as before. In addition, sample size is greatly reduced and can be less than a milligram.

At first it was thought that the production of carbon-14 in the atmosphere had occurred at an unchanging rate, at least throughout the recent geologic past. Later, as serious discrepancies were revealed between carbon-14 dates and dates obtained from tree rings, historically documented material, and so on, it was discovered that the production of carbon-14 has apparently fluctuated over time (Fig. 5.8). The variations are due to changes in the

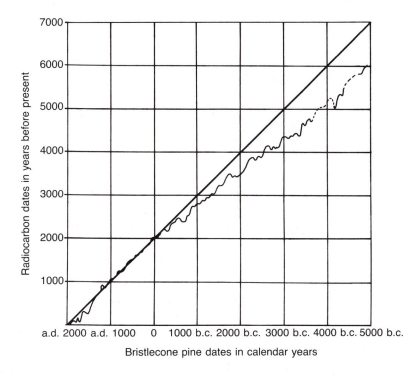

FIGURE 5.8

Calibration curve of radiocarbon dates based on dendrochronological records for the period from around 6000 B.P. to the present day. The C^{14} dates were obtained from groups of ten annual rings, and the curve is based on nearly 300 samples. Note that from beyond 2000 B.P. there is an increasing divergence between the C^{14} date and the calendar date based on tree rings. (Redrawn after Suess, H. E., 1979, Fig. 1, p. 778.)

cosmic-ray flux in the upper atmosphere and these, in turn, are linked to cyclical changes occurring in the sun and/or to changes in the intensity of the earth's magnetic field. There are apparently both short- and long-term variations in carbon-14 production. The short-term oscillations are of about 100 to 200 years and are related to the solar-cycle effects on the cosmic-ray flux. The longer-term variations follow a cycle of about 7000 years and are apparently related to changes in the earth's magnetic field intensity. The problem of these deviations has been the subject of considerable study (Suess, 1965, 1979; Stuiver and Suess, 1966; Michael and Ralph, 1970; and Stuiver, 1970) and conversion tables and calibration curves have been prepared. A further problem arises as a consequence of the burning of fossil fuels over the last 200 years or so since the beginning of the industrial revolution. The introduction of "dead" CO_2 into the atmosphere in that time has diluted the carbon-14 content. Conversely, the detonation of nuclear weapons in the period since 1945 has increased the amount of carbon-14 in the atmosphere. The bias introduced by these essentially random effects can be eliminated by calibrating results against a standard sample of wood dating from before the industrial revolution.

5.9 OTHER DATING METHODS

Isotope dates also can be obtained from certain other decay schemes such as samarium-neodymium, rhenium-osmium, potassium-calcium, and lutetium-hafnium. These methods are mainly new and less commonly used; they are reviewed in Faure (1986), so will not be discussed further. Worthy of mention, however, is another class of dating tools that uses radiation-damage phenomena. During radioactive decay, the fission fragments that are emitted leave a trail of damage as they pass through the surround-

ing mineral or volcanic glass. Suitably etched, these *fission tracks* can be detected under high magnification. Pioneers in this field were Price and Walker (1962), who later showed that the fission tracks of U^{238} in mica and other minerals could be used to date the host mineral by counting the density of the tracks. Clearly, the longer fission had been going on, the greater the density of the tracks, ranging up to several thousands of tracks per square centimeter in older samples. Fission tracks become annealed and fade if the sample is heated; this means that, in effect, the fission-track age gives the date since cooling rather than the true age of the mineral or glass. Episodes of metamorphism and heating are, therefore, the chief source of error in this method. To avoid certain other systematic errors, the fission-track dates are compared with calibration standards established from minerals whose ages have been determined by other methods. Zircons from rapidly cooled and subsequently unheated volcanic rocks are selected for this. For a review of fission-track dating, see Poupeau (1981).

Another kind of radiation-damage phenomenon is seen in *thermoluminescence*. In this case, the damage arising from bombardment by alpha, beta, and gamma rays or by other ionizing radiation is seen in the displacement of electrons from their parent atoms. Some of these electrons are trapped in certain disordered states in the crystal lattice and, therefore, retain the excess energy they received on collision. When the mineral containing the trapped electrons is heated, the electrons return to their stable configuration, releasing in the process the excess energy as light. This phenomenon has been used for dating, the technique involving the heating of the mineral at increasing temperature to produce a luminescence. Measurement of the luminescence produces values that are plotted on a so-called "glow-curve," the area enclosed by the curve being proportional to the amount of radiation damage. The method is used mainly on archeological material such as artificial glasses, ceramics, bones, and flints. Thermoluminescence is discussed in detail by McDougal (1968).

A second technique using displaced metastable electrons in radiation-damaged minerals and glasses is the *electron spin resonance* (ESR) method. In this technique a spectrometer is used to measure the absorption of microwave radiation by the trapped unpaired electrons when they are exposed to a powerful magnetic field (see Ikeya, 1978).

5.10 CONCLUSION

Geochronometry is one of the fastest-growing areas of research among the earth sciences, and, although a specialist field outside the areas normally associated with stratigraphy, its contribution to that field and to geology as a whole has been enormous. During the first 150 years of stratigraphy's evolution, there was little knowledge of the real ages of rocks and fossils. The main object of stratigraphic correlation was to establish the synchroneity of events; their ages in real time were often felt to be of secondary importance. The situation is very different today. It is now routine to incorporate a real-time scale in all stratigraphic tables, and isotopic dates now are used frequently as primary data in stratigraphic correlation. Recent years have seen the introduction of new techniques and increasingly sophisticated instrumentation; accuracy is improving and the number of dates derived from many different techniques is growing rapidly. Of particular value is the contribution radiometric dating is making to Precambrian studies. Without isotopic dates, our understanding of seven-eighths of the history of our planet would be vague, indeed.

REFERENCES

Barnes, J. S., Lang, E. J., and Potratz, H. A., 1956, Ratio of ionium to uranium in coral limestone: Science, v. 124, p. 175–176.

Broecker, W. S., and Thurber, D. L., 1965, Uranium-series dating of corals and oolites from Bahamian and Florida Key limestones: Science, v. 149, p. 58–60.

Conkin, B. M., and Conkin, J. E., 1984, Stratigraphy foundations and concepts: Princeton, NJ, Van Nostrand Reinhold, 365 p.

Faure, G., 1986, Principles of isotope geology, 2nd ed.: New York, John Wiley & Sons, 589 p.

Fornanca-Rinaldi, G., 1968, Dating of cave concretions: Earth and Planetary Sci. Letters, v. 5, p. 120–122.

Grootes, P. M., 1978, Carbon 14 time scale extended: Comparison of chronologies: Science, v. 200, p. 11–15.

Ikeya, M., 1978, Electron spin resonance as a method of dating: Archaeometry, v. 20, p. 147–158.

Litherland, A. E., and Rucklidge, J. C., 1981, Radioisotope detection and dating with accelerators: EOS, v. 62, p. 105–108.

McDougal, D. J., ed., 1968, Thermoluminescence of geological materials: Orlando, FL, Academic Press, 678 p.

Michael, H. N., and Ralph, E. K., 1970, Correction factors applied to Egyptian radiocarbon dates from the era before Christ, *in* Olsson, I. V., ed., Radiocarbon variations and absolute chronology: New York, John Wiley & Sons, p. 109–119.

Neumann, W., and Huster, H., 1974, The half-life of ^{87}Rb measured as a difference between the isotopes ^{87}Rb and ^{85}Rb: Physik, v. 270, p. 121–127.

Odin, G. S., and Dodson, M. H., 1982, Zero isotopic ages of glauconites, *in* Odin, G. S., ed., Numerical dating in stratigraphy: New York, John Wiley & Sons, p. 277–307.

Poupeau, A., 1981, Precision, accuracy and meaning of fission-track ages: Proc. Indian Acad. Sci., Earth Planet: Science, v. 90, p. 403–436.

Price, P. B., and Walker, R. M., 1962, Fossil tracks of charged particles in mica and the age of minerals: Jour. Geophys. Research, v. 68, p. 4847–4862.

Stacey, F. D., 1969, Physics of the Earth: New York, John Wiley & Sons, 324 p.

Stuiver, M., 1970, Long-term variations in radiocarbon variations and absolute chronology, *in* Olsson, I. V., ed., Radiocarbon variations and absolute chronology: New York, John Wiley & Sons, p. 197–213.

Stuiver, M., Heusser, C. J., and Yang, I. C., 1978, North American glacial history extended to 75,000 years ago: Science, v. 200, p. 16–21.

Stuiver, M., Robinson, S. W., and Yang, I. C., 1979, ^{14}C dating to 60,000 years B.P. with proportional counters, *in* Berger, R., and Suess, H. E., eds., Radiocarbon dating: Berkeley, Univ. California Press, p. 202–215.

Stuiver, M., and Suess, H. E., 1966, On the relationship between radiocarbon dates and true sample ages: Radiocarbon, v. 8, p. 534–540.

Suess, H. E., 1965, Secular variations of the cosmic-ray-produced carbon-14 in the atmosphere and their interpretation: Jour. Geophys. Research, v. 70, p. 5937–5952.

————1979, A calibration table for conventional radiocarbon dates, *in* Berger, R., and Suess, H. E., eds., Radiocarbon dating: Berkeley, Univ. California Press, p. 177–184.

Thurber, D. L., Broecker, W. S., and Potratz, H. A., 1965, Uranium series ages of Pacific atoll coral: Science, v. 149, p. 55–58.

6
MAGNETIC STRATIGRAPHY

The whole globe of the Earth is one great magnet.

Edmund Halley

6.1 INTRODUCTION

Of all the modern tools available to the stratigrapher, probably the most surprising to the nineteenth-century worker would be the various ways in which certain peculiarities in the earth's magnetic field can be used in stratigraphic correlation. Magnetic stratigraphy is based on the fact that the earth's magnetic field reverses itself at intervals and that the record of these changes, "frozen in," as it were, within the stratigraphic succession, is accessible to suitable laboratory techniques. A second paleomagnetic phenomenon that is time-related is seen in the so-called apparent polar wandering curve (APW curve). Although the APW curve is merely a reflection of continental drift, paleomagnetism is obviously a necessary ingredient in this phenomenon.

6.2 EARTH'S MAGNETIC FIELD

It generally is agreed that the earth's magnetic field is generated by movements in the liquid outer core. These movements are probably a combination of convective flow and flow generated by the earth's rotation.

Movement of this metallic core material through weak magnetic fields produced by a battery effect due to compositional differences within the earth, perhaps augmented by the solar magnetic field, generates an electric current. This current, it is believed, in turn generates the strong geomagnetic field we call the main field. In a model described by Elsasser (1965) and Bullard (1971), the earth behaves like a self-exciting disc dynamo (Fig. 6.1). The axial dipole of the field is linked to the rotation of the earth, although the north and south magnetic poles do not coincide with the rotational (geographic) poles because of a small nondipole component in the main field, caused possibly by some irregularity in the convective motion of the fluid outer core. At present, the north and south magnetic poles lie close to the geographic poles at a distance that changes by a small amount each year. This secular variation (magnetic declination—at present about 2° in a westerly direction per 100 years) is a

manifestation of long-term changes that take place over a cycle lasting about 7000 years and that are primarily produced by variability in the nondipole component of the field. Direct observations of magnetic declination date only to the sixteenth century, but further data for earlier years in the historical period can be gained from **archeomagnetic** evidence. Archeologic materials made of clay, such as pottery and bricks, invariably contain magnetic particles and, therefore, acquired a remanent magnetism as they cooled after firing. Provided the orientation of the object at the time of firing can be established (usually not difficult for both pots and bricks when a pottery kiln is excavated), the magnetic field direction can be determined. Although the original objective in archeomagnetic studies was to establish ancient pole positions, dated master curves of former pole positions have come to provide a means of dating artifacts of unknown age (Fig. 6.2).

Geological evidence for the prehistoric period confirms that the predominant values for angular deviation seem to lie between 6° and 27°, with a maximum of about 36° (Harrison, 1974). If the positions of the magnetic poles are plotted for a period of several thousand years they cluster significantly around the geographic poles (Fig. 6.3), and there is no reason to suppose that this has not always been the case in the geologic past.

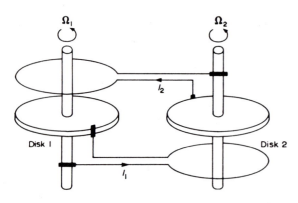

FIGURE 6.1

Self-exciting dynamo model. If rotated at sufficient speed, two interconnected disk dynamos become self-excited. Experiments have shown that, under certain conditions, the current generation oscillates and even undergoes field reversal. (From Stacey, F. D., 1969, Physics of the Earth, Fig. 5–10, p. 156. Reprinted by permission of John Wiley & Sons.)

6.3 REMANENT MAGNETISM

Magnetization Processes

Magnetite and, to a lesser extent, certain other iron-rich minerals are natural magnets. In such minerals, the tiny magnetic fields that are produced around atoms by the orbital spin of electrons reinforce one another and become aligned parallel to the ambient geomagnetic field. Many rocks contain a small percentage, perhaps up to 10 percent, of these magnetic minerals. In igneous rocks

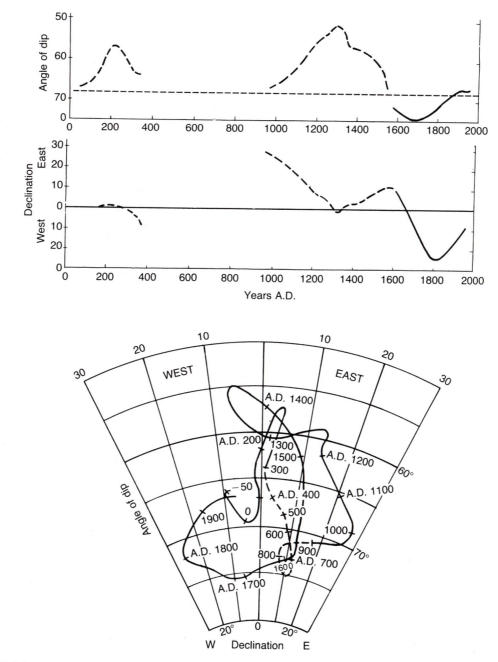

FIGURE 6.2
Secular variations of magnetic declination and inclination over the past 2000 years, as measured from Britain.

FIGURE 6.3
Paleomagnetic poles for the last 7000 years, plotted on polar projection. (From Tarling, D. H., 1983, Palaeomagnetism: Principles and applications in geology, geophysics and archaeology, Fig. 8–14, p. 193. Reprinted by permission of Chapman and Hall.)

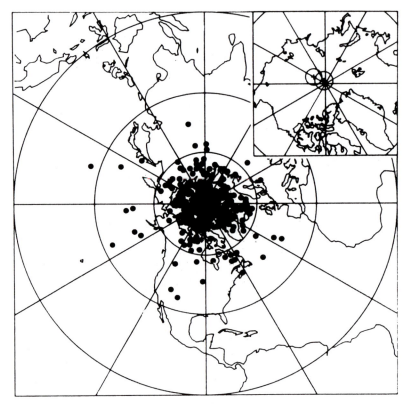

the magnetism appears during the cooling of the melt when the temperature drops below what is known as the **Curie point.** This varies with the mineral, but is on the order of 400°C–600°C. As the mineral cools a few tens of degrees below the Curie point, it passes through what is called the *blocking temperature.* Once this stage is reached, the magnetism is said to be "hard," and removal of, or change in, the external geomagnetic field does not alter the mineral's magnetism. In effect, the geomagnetic field with a characteristic direction and polarity at that location and at that time is "frozen in" and is preserved in the mineral. This type of acquired magnetism is termed **thermoremanent magnetism (TRM).** The magnetic properties of the mineral are preserved over geologically very long periods of time unless the rock is

subjected to later heating above the Curie point, at which time thermal agitation of the atoms destroys their magnetic alignment and the remanent magnetism is lost. When the rock eventually cools again, it will acquire a new magnetic signature appropriate to the time and place of the later cooling.

In sedimentary rocks, magnetic mineral grains that had acquired an earlier TRM will become oriented by the geomagnetic field as they settle to the seafloor or lake bottom and are incorporated within the sedimentary fabric. Such acquired magnetism is called detrital or *depositional remanence (DRM)* and is more stable in finer-grained sediments such as clays and shales than in coarse-grained rocks. This is because the larger clastic grains are influenced more by hydraulic forces and gravity, whereas finer particles are more apt to be aligned with the ambient geomagnetic field because they tend to settle under quiet, low-energy conditions. Although the optimum conditions for magnetic orientation within the ambient field occur at the moment of deposition, some DRM is postdepositional. This is seen particularly in finer sediments where water films around the grains allow them to rotate into alignment with the geomagnetic field. In coarser sediments, water films are of minor significance and the grains are mostly locked in contact with one another. A third type of remanent magnetism is acquired at the time of crystallization of certain minerals that form as a result of chemical reactions at relatively low temperatures, far below their Curie points. This *chemical remanence (CRM)* can also be acquired during recrystallization because of oxidation, dehydration, or other alteration processes. This type of remanent magnetism dates from the time of alteration and will, obviously, be younger than the date of original formation of the rock.

In comparing paleomagnetic data from lava flows (TRM) with those from sediments (DRM or CRM) it must be remembered that because lava flows cool and acquire their remanent magnetism quickly they provide, in effect, a spot reading of the earth's magnetic field virtually at an instant in time. In contrast, in deep-sea sediments the mixing of mineral grains by bioturbation processes in the upper 10 cm or so of the sediment homogenizes the record, and the evidence of secular variations is smoothed out. The resultant remanent magnetism would be expected to indicate an average figure for the time of deposition of the sediment layer as a whole. The samples used in these studies are about 2.0–2.5 cm thick. With a typical accumulation rate of 1 cm per 1000 years, the magnetic direction indicated by any given sample should be the average of secular variations over a 2500-year time span (Harrison, 1966, 1974).

Measuring Remanent Magnetism

Remanent magnetism is measured in the laboratory, so the first requirement is to obtain properly oriented samples. Field collecting must be done with great care, and in structurally disturbed rocks the orientation at the time of magnetization must be established. In the laboratory the remanent magnetism is measured with a magnetometer, of which there are several different types. One of the earliest developed was the astatic magnetometer (Fig. 6.4A). Developed soon after was the spinner magnetometer, so-called because the rock sample is spun near a pickup coil so as to induce a current in it. In the most recent version of this instrument, using what is known as a ring fluxgate pickup system linked to a microcomputer (Fig. 6.4B), measurements of both magnetic direction and intensity can be obtained in 20 minutes or less. Even faster are cryogenic magnetometers, in which measurements can be made in a few seconds (Fig. 6.4C).

One of the chief problems in paleomagnetic studies arises from the fact that magnetized minerals are influenced by the ambient geomagnetic field during the time that has elapsed since the mineral acquired its primary magnetism. The term "permanent" magnetism is something of a misnomer because over geologic time even so-called hard magnetism is slowly lost by a process known as *relaxation*. Relaxation time varies and is controlled mainly by mineral composition, temperature, and the strength of the main field at the time of primary magnetization. If the mineral has a long relaxation time, the primary remanence will likely still be retained but overprinted by a secondary or *viscous* remanent magnetization (VRM). Such secondary magnetism is derived from the ambient geomagnetic field and must be removed by a "cleaning" process before the primary remanence can be measured. For a fuller account of laboratory procedures see Tarling (1983).

FIGURE 6.4
Magnetometers: A. astatic type, B. spinner type with ring fluxgate system, C. cryogenic type. (From Tarling, D. H., 1983, Palaeomagnetism: Principles and applications in geology, geophysics and archaeology, Fig. 5–4, p. 85. Reprinted by permission of Chapman and Hall.)

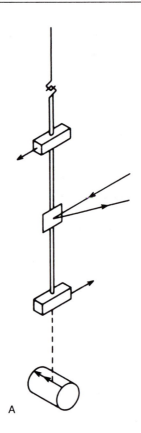

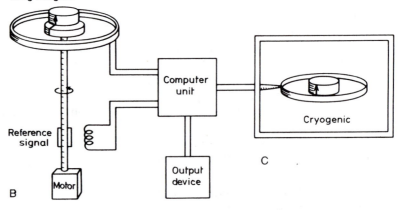

6.4 MAGNETIC REVERSALS

Not only do the positions of the magnetic poles change through time, but the strength, or intensity, of the geomagnetic field also fluctuates, presumably in response to irregularities in the movement of material in the earth's outer core. From time to time, the field strength drops briefly to zero and then recovers, but with an opposite, or reversed, polarity. The strength of the earth's magnetic field fluctuates markedly with time. It is thought that what happens during a reversal is that the field strength drops to a very low value, presumably briefly to zero, before increasing again with the opposite polarity. The total reversal process is thought to take from 1000 to 10,000 years, a period so short that only rarely has it been "caught in the act" within the stratigraphic record. The study of paleomagnetism revealed early on the existence of **magnetic reversals.** For example, roughly 50 percent of all rock samples measured showed a reversed magnetic polarity. Although self-reversal in magnetized minerals is known, it is not a common phenomenon, and it would indeed be a coincidence if it had occurred in 50 percent of the rocks. Several other lines of evidence (Cox, 1969) remove all doubt that the magnetically reversed rocks were formed during periods when the geomagnetic field was actually reversed (Fig. 6.5). Reversal apparently has occurred many times in the geologic past. Periods of one polarity lasting on the order of 1 m.y. were termed **epochs.** Over the past 5 m.y., for example, there have been four such epochs, two with normal polarity (including the present) and two with reversed polarity. Within each normal epoch were brief periods when the geomagnetic field switched again so that there have been short-lived reversed **events** lasting from 10,000 to 100,000 years. In the same way, the reversed epochs have been similarly interrupted by normal events (Fig. 6.6). Some workers distinguish between events of 50,000–100,000 years duration and "short events" of around 10,000 years. The term **excursion** has also been used for these shorter events. Counting from the present and moving backward through time, the early practice was to name the polarity epochs after well-known geomagnetists, with polarity events named after the locations where they were first discov-

ered. Beyond about 5.1 m.y. ago, however, a numbering system has been adopted, so that counting the Brunhes Normal, Matuyama Reversed, Gauss Normal, and Gilbert Reversed Epochs as 1 through 4, the normal epoch predating the Gilbert becomes number 5 and so on. Reversal events within epochs are lettered A, B, C, etc. with increasing age (Fig. 6.7), and smaller excursions within events are identified by a further letter or numerical subscript. In 1979 the International Subcommission on Stratigraphic Classification (ISSC) of the International Union of Geological Sciences (IUGS) ruled that the use of terms like epoch, event, interval, and so on is confusing and undesirable, and alternative terms have been suggested. These are discussed in a later section.

The dating of the boundaries between the various magnetic intervals has been based upon both radiometric and biostratigraphic data. Back to around 5 m.y. ago the geomagnetic time scale is considered to be fairly complete and accurate, although because of the brevity of some of the reversal events and the often discontinuous sedimentary record, they are not always easy to detect. Many dating methods are accurate only to a few tens of thousands of years and some polarity switches do not last that long. That events recorded in one place cannot always be detected in another may obviously be due to the problems just mentioned, but it has caused some workers to be somewhat skeptical. It has been suggested, for example, that some events may be due to relatively local nondipole magnetic activity occurring during a period of low main-dipole intensity (Watkins, 1976). Such occurrences would obviously not be of worldwide significance and should not be given "event" status. Watkins (1976) suggested the terms **deviation** or *excursion* for such apparent polarity reversals, when indeed they can be demonstrated. Some inaccuracies of the dating methods have resulted in miscorrelation. For example,

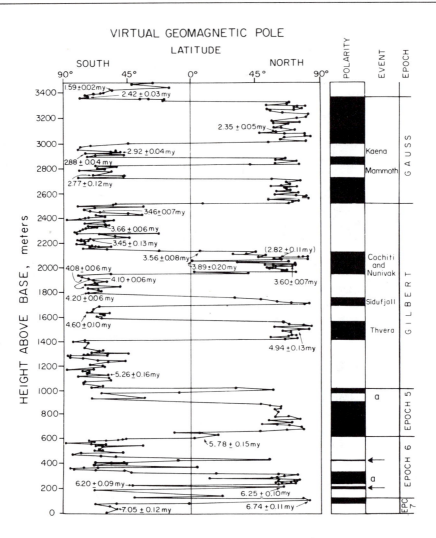

FIGURE 6.5

Typical magnetostratigraphic section. Determinations to the right show normal polarity and are depicted as black bars on the polarity scale; determinations to the left show reversed polarity. (From McDougall, I., Saemundsson, K., Johannesson, H., Watkins, N. D., and Kristjansson, L., 1977, Extension of the geomagnetic polarity time scale to 6.5 my: K-Ar dating, geological and paleomagnetic study of a 3,500 m lava succession in western Iceland: Geol. Soc. America Bull., v. 88, Fig. 4, p. 11.)

the Olduvai event in the Matuyama Epoch (Chron) was found to contain within its time range the interval of the Gilsa event, which had formerly been recognized as a separate entity. This controversy is still unresolved.

Because magnetic reversals are presumably worldwide, if they can be detected within sedimentary successions they provide a physical indication of isochronous boundaries between reversal and normal polarity

FIGURE 6.6

Time scale for the most recent polarity chrons.

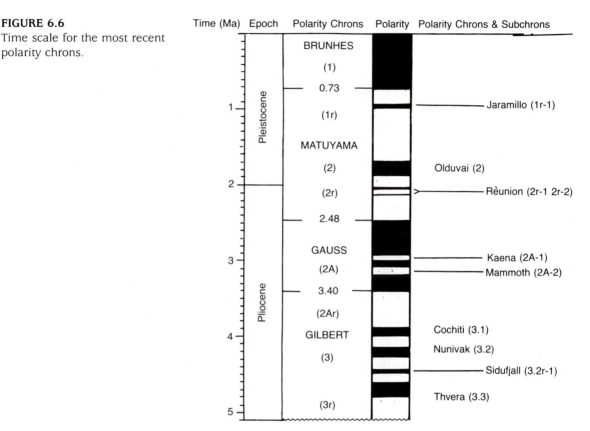

zones. Also, because they can be detected in a variety of rocks, both on the continents and in deep oceans, they provide a useful correlation tool with great potential in stratigraphy. It should be remembered, however, that the alternation of normal and reversed polarities in the rocks is only a flip-flop phenomenon and not a unique feature of the succession as would be, for example, the vertical range of a given fossil. Usually, direct correlation from one section to another is possible only by counting the alternations back through time. If complete sections are present and, as in many deep-sea sections, constant rates of deposition are assumed, it may be possible to match the spacings of various epochs and events. This procedure is often unreliable because the shorter-term events may not be recorded in some sections where the rate of ac-

cumulation of sediment was very slow or where the succession is blurred by bioturbation. Some occasional, short switches of polarity (events) or distinctive groupings of events within very long quiet intervals may be correlatable, but the crowding of reversals during mixed intervals makes direct correlation by comparison difficult. In practice, magnetostratigraphic correlations are made with the help of biostratigraphic indicators, volcanic ash bands, and/or radiometrically dated markers.

Magnetic Stratigraphy of the Ocean Floors

In the ocean basins, magnetic stratigraphy can be worked out in two ways. In the first, the detrital magnetism of sediments recov-

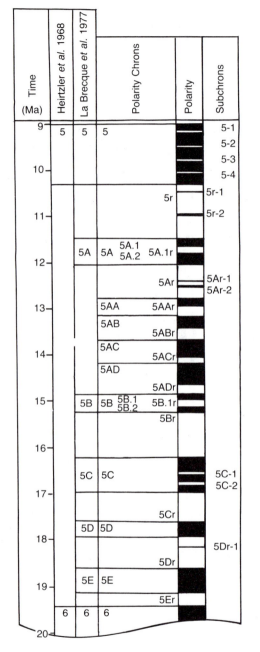

FIGURE 6.7
Magnetostratigraphic scale for the last 20 m.y.

ered in cores can be determined and, together with the biostratigraphic and/or oxygen isotope (discussed in Chapter 15) succession, a sequence of magnetic reversals can be established in the same way as in continental sequences involving lava flows; that is, in a vertical direction through the succession. Secondly, by virtue of the seafloor-spreading mechanism, whereby basaltic lavas originate along the mid-ocean ridges and are subsequently carried away on the surface of a moving lithospheric plate, there is obviously a progressive increase in age of the basalts with increasing distances from the spreading center (Fig. 6.8). Although the seafloor flows lie over each other at their margins, the large-scale sequential succession is not in a vertical direction as in other stratigraphic situations, but in a horizontal one. When the surveys from shipborne magnetometers are plotted on the charts, **magnetic anomalies** show up as stripes that mark alternations of normal and reversed polarities in seafloor basalts. In a traverse at 90° away from a spreading center, the pattern of relative width and spacing of the stripes is roughly the same as the pattern seen in a traverse from the spreading center in the opposite direction. The discovery in 1958 (Mason, 1958; Mason and Raff, 1961) of these anomalies and the later discovery that these stripes represented magnetic reversals (Vine and Matthews, 1963; Morley and Larochelle, 1964) provided convincing proof of the hypothesis of seafloor spreading.

As magnetic surveys of the world's oceans proceeded through the late 1960s and 1970s a numbering system of anomalies came to be adopted, counting from the spreading center outward and applying numbers to anomalies formed during periods of dominantly normal polarity (Fig. 6.9). Later studies turned up new anomalies between previously numbered ones; these were accommodated by adding a letter or a second number; for example, the anomaly discovered between

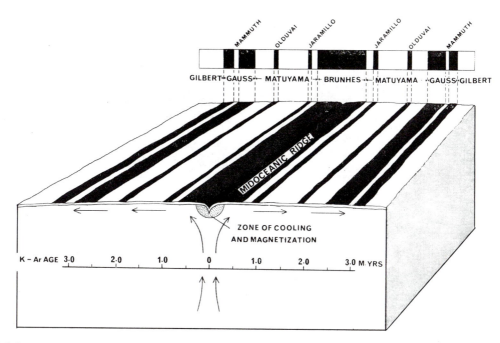

FIGURE 6.8
Generation of magnetic anomalies at a spreading center. This is an oversimplified picture and at depth the magnetic polarity may vary. (From Allan, T. D., 1969, A review of earth magnetism: Earth-Science Reviews, v. 5, Fig. 7, p. 230. Reprinted by permission of Elsevier Science Publishers B.V.)

numbers 3 and 4 was labeled 3^A or 3^1. Because of these procedures, there is no direct correlation between the epoch numbers, discussed previously, and anomaly numbers. Many published marine magnetostratigraphic successions show both systems, although there is by no means a consensus yet on correlation between them. Certain other number systems have also been used for anomalies on the seafloor, such as the M numbers (M standing for Mesozoic) of Larsen and Chase (1972), Larsen and Pitman (1972), and Larsen and Hilde (1975), in which the reversed rather than the normal anomalies are numbered (Figs. 6.10, 6.11). In an attempt to head off the proliferation of numbers and avoid the possibility of confusion in correlation between epoch numbers and anomalies, Hsü et

al. (1984) proposed that the anomaly numbering system be used as the basis for the magnetostratigraphic scale. The divisions are chrons, the term recommended by the ISSC (1979) to replace epoch, and each one is numbered the same as the corresponding anomaly, boundaries being set by the youngest reversal boundary (Fig. 6.12). The prefix C was suggested to remove a possible source of confusion with other magnetostratigraphic numbering systems. The succession of polarity switching as revealed by anomalies on the seafloor has been demonstrated as far back as the early Jurassic, which is the age of the oldest oceanic crustal rocks. The time scale of geomagnetic events is based on counting backward through time and is, therefore, to some extent still a relative scale. The task of

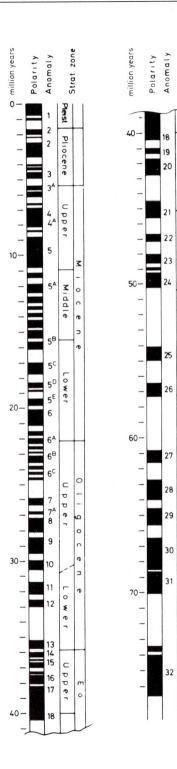

assigning ages in terms of real time is proceeding rapidly and becoming progressively more refined.

In cores of sediment the geomagnetic succession is determined from detrital magnetism in clastic grains. Even in oozes with 95 percent calcareous material the magnetism, though weak, is usually still detectable. Because the original geographic orientation of the sediment in the core is lost in the recovery process, the magnetic direction in terms of north and south poles (that is, the polarity as such) is not measured but rather the inclination, or dip angle. This is down for normal polarity and up for reversed in the northern hemisphere and the opposite in the southern hemisphere. In many parts of the ocean floor the rate of accumulation of sediment is more or less constant over several million years so that there is a direct correlation between the depth within the core and age. Thus the sequence of polarity boundaries determined within the core can be plotted as a series of points on the straight line representing age vs. sediment thickness (Fig. 6.13). The steeper the inclination of the line, the more rapid the sedimentation rate. Such graphs provide a useful check on the ages of the briefer magnetic intervals, which are frequently in doubt because of the tolerances in the radiometric dating methods used.

In the case of older polarity events the sedimentary and detrital magnetic record is often beyond reach of the drill, but ages can be calculated by reference to anomalies on the seafloor and by extrapolation from younger events. This is done not on the basis of rates

FIGURE 6.9

Cenozoic polarity time-scale. (From Tarling, D. H., 1983, Palaeomagnetism: Principles and applications in geology, geophysics and archaeology, Fig. 9.2, p. 204. Reprinted by permission of Chapman and Hall.)

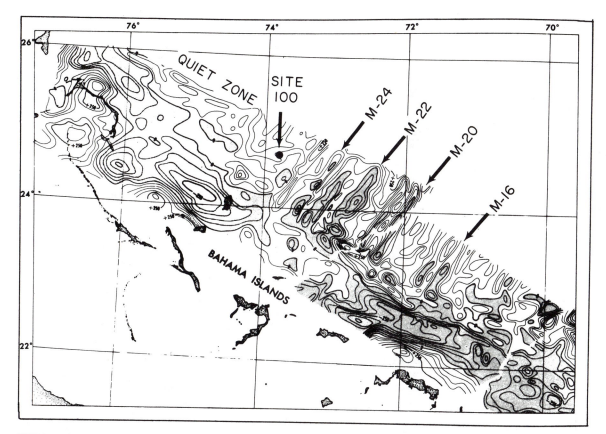

FIGURE 6.10
Magnetic anomaly contour chart in the vicinity of the Bahama Islands and DSDP site 100. (From Larsen, R. L., and Hilde, T.W.C., Jour. Geophys. Research, v. 80, Fig. 7, p. 2592, 1975, copyright by the American Geophysical Union.)

of constant accumulation of sediment but on the basis of constancy of spreading rates of ocean floor over periods of several million years. There is, as we saw earlier, a direct correlation between the age of a magnetic anomaly and its distance from a spreading center. If the younger anomalies, whose ages are known, are used to plot a straight-line distance/age graph, this line can clearly be extended and the age of older, more-distant anomalies determined by extrapolation (Fig. 6.14). This is, of course, assuming a constant rate of spreading. Actually, spreading rates

do vary not only from ocean to ocean and from place to place along a particular mid-ocean ridge, but also with time. This, however, does not pose too much of a problem within a single ocean basin and over moderate time spans. If the spreading rate of the oceanic floor over several million years is assumed to be constant, the correlation between age of oceanic crust and distance from spreading center clearly provides a means of mapping the age of the ocean floor. Maps have been prepared for all the world's oceans, with the exception of the Arctic

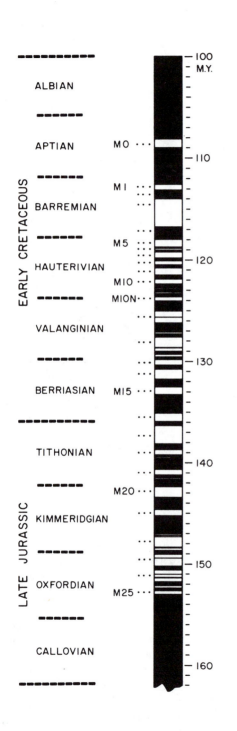

Ocean, showing the age of the ocean floor by means of isopleths that are, in fact, **iso-chrons**—lines of equal age (Figs. 6.15 and 6.16).

In constructing diagrams of geomagnetic polarity and isochron maps of seafloor anomalies, the age of the anomaly is frequently determined on the basis of micropaleontology of the sediments above the basaltic basement. To translate this into actual time requires selection of a time scale, of which there are several. The Phanerozoic time scale (Harland et al., 1964) has been accepted for many years but, increasingly, different workers tend to use their own favorite time scale, so there may be variations from one study to another. Thus, for example, an increase in the spreading rates of the ocean floor during the Cretaceous has been claimed by Larsen and Pitman (1972); this can be demonstrated adequately on the Phanerozoic time scale. If, on the other hand, other equally accepted time scales are used (and, as described in Chapter 3, there are several), the apparent range in rate of spreading can be reduced in magnitude or can be made to disappear entirely (Baldwin et al., 1974). Ultimately, of course, the age in real time of both biostratigraphic zones and magnetic anomalies depends on isotopic dating and, as can be seen in the numerous figures and tables in this and other chapters, a scale (in m.y.) is invariably incorporated.

FIGURE 6.11
Magnetic reversal time scale for the Early Cretaceous and Late Jurassic. Stage names and radiometric dating from Geological Society of London (1964) time scale (except for Berriasian and Tithonian stage names). In this scheme, the reversed anomalies are numbered. (From Larsen, R. L., and Hilde, T.W.C., Jour. Geophys. Research, v. 80, Fig. 8, p. 2593, 1975, copyright by the American Geophysical Union.)

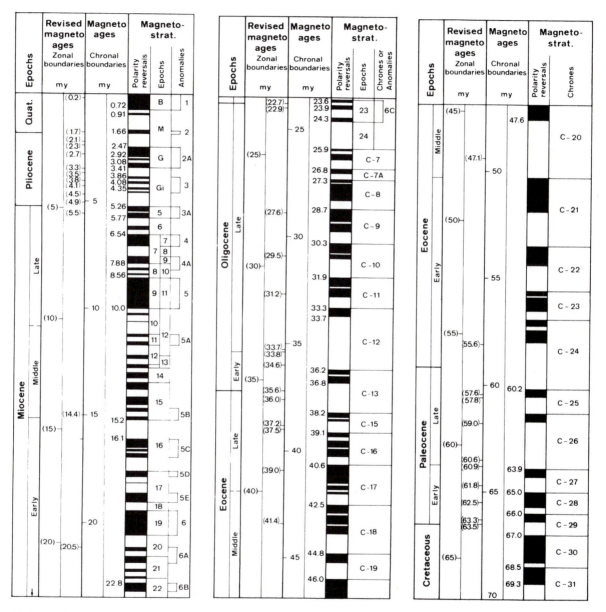

FIGURE 6.12

Magnetostratigraphic scale for the Eocene to Recent. This shows the correlation between epoch and anomaly numbers. Note the alternative interpretations for epochs 7–13. After epoch 24, chron numbering is used. (From Hsü, K. J., et al., 1984, Numerical ages of Cenozoic biostratigraphy datum levels: Results of South Atlantic Leg 73 drilling: Geol. Soc. America Bull., v. 95, Fig. 9, p. 871–873.)

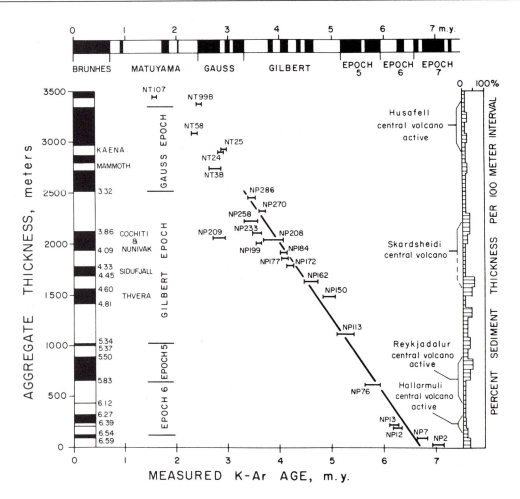

FIGURE 6.13

Aggregate stratigraphic thickness plotted against age. Horizontal bars indicate precision
limits of K-Ar ages. The polarity time scale at top of diagram is after Talwani et al.
(1971). The regression line is least-squares fit with the Gilbert Gauss epoch boundary
fixed at 3.32 m.y. at 2520 m above base of sequence. (From McDougall, I., Saemundsson, K., Johannesson, H., Watkins, N. D., and Kristjansson, L., 1977, Extension of the
geomagnetic polarity time scale to 6.5 my: K-Ar dating, geological and paleomagnetic
study of a 3,500 m lava succession in western Iceland: Geol. Soc. America Bull., v. 88,
Fig. 5, p. 57.)

Magnetic History of the Phanerozoic

As new paleomagnetic data have been accumulated, an increasing number of short-term
events has been discovered, so the original
sharp distinctions between the magnetic epochs have become somewhat blurred. Not
only has the validity of some of the shorter
events been open to question but, as mentioned, the dating of others has not been accurate enough to preclude the possibility of

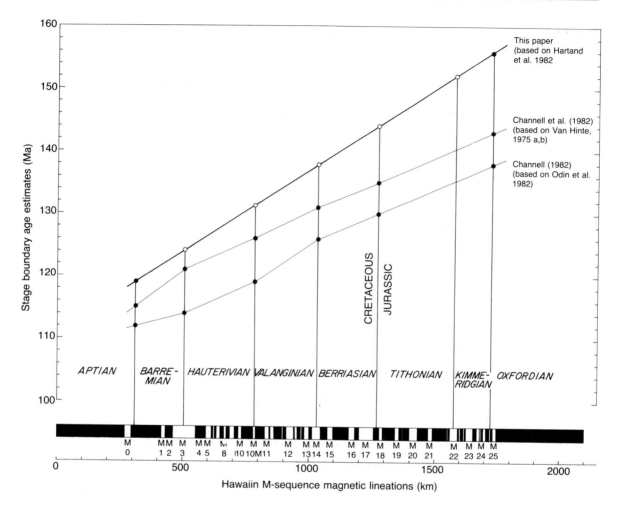

FIGURE 6.14

Calibration plot of age versus distance of seafloor magnetic anomalies. An example from the Hawaiian lineations in the central Pacific. A constant spreading rate of 3.836 cm/yr is assumed on the basis of age estimates of 119 m.y. for the Barremian/Aptian stage boundary and 156 m.y. for the Oxfordian/Kimmeridgian boundary. (From Kent, D. V., and Gradstein, F. M., 1985, A Jurassic and Cretaceous geochronology: Geol. Soc. America Bull., v. 96, Fig. 2, p. 1422.)

miscorrelation. The problem becomes progressively greater with increasing age of the rocks. The apparent conflict of data, which sometimes seem to indicate both normal and reversed polarity at the same time, is simply a consequence of the uncertainty inherent in

methods of isotopic dating. A factor of ±5 percent in a 2-m.y. date gives a variation of ±100,000 years; some short magnetic events last only that long. With progressively older rocks it becomes difficult to separate even epochs from each other. Despite these short-

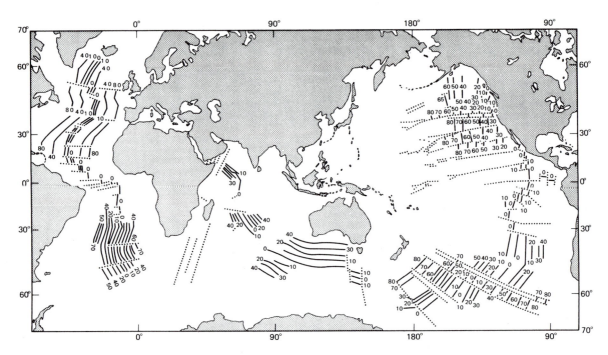

FIGURE 6.15
Linear magnetic anomalies on the seafloor. The lines are isochrons showing the age of
the seafloor. (From Heirtzler, J. R., Dickson, G. O., Herron, E. N., Pitman, W. C., and
Le Pichon, X., Marine magnetic anomalies, geomagnetic field reversals and motions of
the ocean floor and continents: Jour. Geophys. Research, v. 73, Fig. 4, p. 2124, 1968,
copyright by the American Geophysical Union.)

comings, however, it is possible to detect
marked changes in the magnetic behavior of
the earth at intervals of roughly 30-50 m.y. or
so. It seems that the Cenozoic as a period of
mixed polarities is not typical of the Phanerozoic record as a whole. Figure 6.17 shows
that from around 70 m.y. ago to about 120–
125 m.y. ago polarity was relatively stable,
with a long normal interval interrupted by
only two short reversals. This is the Cretaceous quiet interval (referred to as KN by Irving and Pullaiah, 1976), which is seen as a
magnetically quiet zone in the anomaly pattern of the central Atlantic and Pacific. In the
early Cretaceous and latest Jurassic, there
was a disturbed or mixed interval and this, in

terms of seafloor anomalies, is known as the
Keithley sequence in the Atlantic and the
Phoenix, Japanese, and Hawaiian sequences
in the Pacific. In middle and late Jurassic
time, there was another largely normal interval (the JN of Irving and Pullaiah, 1976),
which in the North Atlantic is seen as a marginal quiet zone. The major differences between various interpretations are due to the
different time scales used (Fig. 6.18). Prior to
the Jurassic, the record is rather more obscure for the simple reason that we can no
longer rely on anomalies on the ocean floor
because there is no ocean floor old enough.
The most striking feature of the earlier record, however, is the long period of reversed

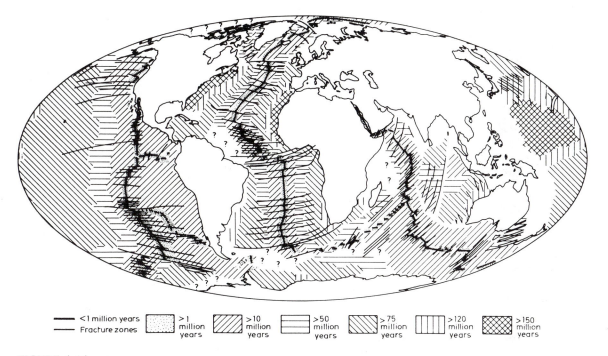

Key:
— <1 million years / Fracture zones
>1 million years
>10 million years
>50 million years
>75 million years
>120 million years
>150 million years

FIGURE 6.16
Magnetic age of the ocean floors. Many of the ages are based on extrapolation from known areas. No ocean floor is known that is older than about 200 m.y., so there is a considerable restraint on the possible variations in plate reconstruction. (From Tarling, D. H., 1983, Palaeomagnetism: Principles and applications in geology, geophysics and archaeology, Fig. 9.5, p. 210. Reprinted by permission of Chapman and Hall.)

polarity in the Permian and Pennsylvanian (the PCR of Irving and Pullaiah, 1976). The Paleozoic record as a whole shows roughly equal numbers of mixed normal and reversed intervals, with the mixed intervals perhaps dominant in terms of total time (Fig. 6.19). Figure 6.20 represents the compilation of data by Soviet geologists.

Concerning polarity reversals in the Paleozoic, one rather important point should not be overlooked; namely, that as a consequence of apparent polar wandering, the north magnetic pole (the pole indicated by north-seeking magnetism during normal polarity) may at times appear to have lain south of the equator. A possible source of confusion here is avoided if the pole indicated is

located with respect to the APW curve for that particular crustal plate.

6.5 CLASSIFICATION

Terminology

Magnetostratigraphy is a relatively new kind of stratigraphy, and during the period from the later 1950s to the 1970s it acquired its own hierarchy of terms. There was little attempt to formalize these terms and equate them with the usage in conventional stratigraphy, so more recently certain changes were proposed by the ISSC (ISSC, 1979; NASCN, 1983). The chief point of criticism by the ISSC is that terms like epoch, event, and interval

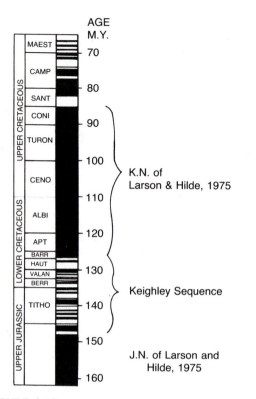

FIGURE 6.17
Polarity sequence through the Mesozoic. Note the long normal sequences in the Upper Jurassic and Mid-Cretaceous.

already have been long in use in geology with other meanings. Accordingly, virtually all the familiar terminology that has grown with magnetic stratigraphy has been set aside and replaced by new terms (Table 6.1). Although it may be some time before there is general acceptance and usage of the proposed terminology, there is no question that the arguments in support of the recommendations are both logical and correct. The basic **magnetostratigraphic unit,** according to the ISSC proposals, is a *magnetostratigraphic polarity zone* (or simply **polarity zone**). Such a polarity zone consists of a thickness of strata distinguished by a particular magnetic character. Most often this would be a single or

predominant polarity direction throughout the unit. On the other hand, in some situations it might be that the unit is characterized by alternating (mixed) polarities, setting it off from predominantly normal or reversed units above and below within the magnetostratigraphic sequence. This is reminiscent of the relationship between certain lithostratigraphic units in which a formation, for example, might be set apart from formations above and below by a distinctive homogeneous lithologic character or, alternatively, by virtue of its rapid alternation of lithologies—heterogeneous in contrast to homogeneous. A magnetostratigraphic polarity unit is based on the magnetic polarity identified in the rock and is therefore an objective descriptive unit, just as is a lithostratigraphic or biostratigraphic unit. It should be remembered, however, that its lower and upper boundaries are, in theory at least, marked by isochronous surfaces potentially recognizable on a global scale. Consequently, if it were possible to establish the precise horizons marking the polarity switches, magnetostratigraphic units would, in effect, be chronostratigraphic units. True chronostratigraphic units—units whose boundaries are defined by time horizons—are rarely if ever discoverable. This topic is discussed elsewhere in the book. In practice, the worldwide nature of polarity changes makes them virtually unique among the various tools available to stratigraphers. However, the reversals themselves are not unique events, and correlation from one location to another frequently relies upon other criteria such as biostratigraphic or radiometric data.

The original terms epoch, event, interval, etc. were used as time units within a hierarchy in which a rather loosely defined scale in years was used as the criterion of rank (Table 6.2). For the past 5 m.y. or so, the dating of reversals is reasonably accurate and probably most of even the short-term events have been detected. With older rocks, however, the inherent inaccuracies of dating methods means

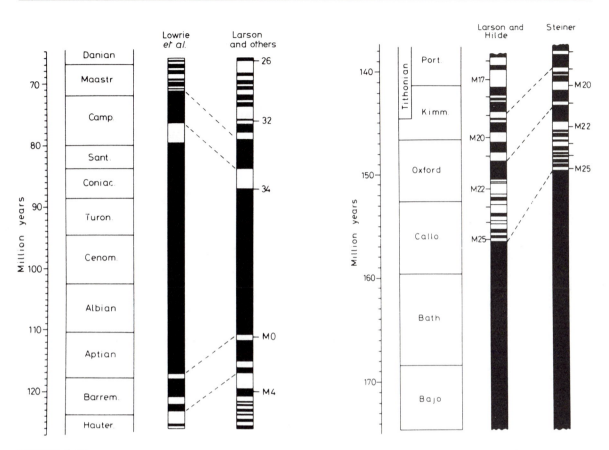

FIGURE 6.18

Mesozoic polarity scales. Much of the history of magnetic reversals is based on studies of oceanic magnetic anomalies. There is general agreement on the relative length and ordering of the reversal intervals, but uncertainties about the dating and the use of different time scales leads to the differences shown here. Left: Mainly Cretaceous polarities after Lowrie et al. (1980a, b), Larsen and Pitman (1972), and Larsen and Hilde (1975). Right: Mainly Jurassic anomalies based on Larsen and Hilde (1975) and Steiner (1980). (From Tarling, D. H., 1983, Palaeomagnetism: Principles and applications in geology, geophysics and archaeology, Fig. 9.4, p. 208. Reprinted by permission of Chapman and Hall.)

that these units of lower rank are often difficult to detect. Geologic time and geochronologic units are obviously not influenced by the method whereby chronostratigraphic units are established, so no new terms are necessary when discussing time as measured by magnetostratigraphy. Accordingly, both the existing terms (chron, etc.) and chrono-

stratigraphic terms (chronozones, etc.) can be used; for example, the Gilbert Reversed Epoch becomes the Gilbert Chron and the body of rock deposited during that time is known as the Gilbert Chronozone. It should be noted that the virtual isochroneity of the horizons marking switches of polarity means that polarity zones (magnetostratigraphic units)

FIGURE 6.19
Sequence of polarity reversals
through the Phanerozoic. (From
Wyllie, P. J., 1976, The way the
earth works, Fig. 9–4, p. 125.
Reprinted by permission of John
Wiley & Sons.)

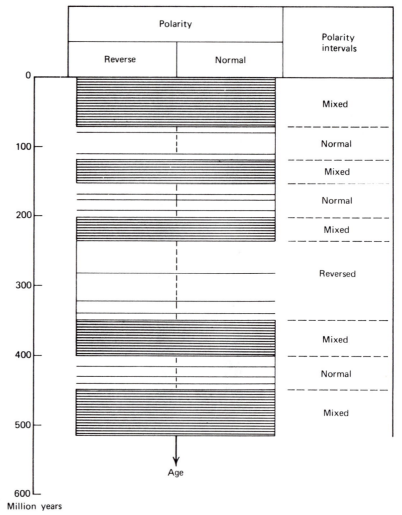

come very close to coinciding exactly with chronozones (chronostratigraphic units). However, to reiterate, one is an objective unit, the other theoretical. For detailed discussion of magnetostratigraphic terminology and stratigraphic terminology as a whole, the reader is referred to the *International Stratigraphic Guide* (Hedberg, 1976) and the supplementary chapter (ISSC, 1979), also Oriel et al. (1976), Watkins (1976), Irving (1971, 1972), and NACSN (1983).

Type Sections

As with other formal stratigraphic schemes, it has been recommended that type sections should be established when the correlation between the magnetic stratigraphic time scale and biostratigraphic and other scales can be established with confidence. This has been done, for example, at the important Cretaceous-Tertiary boundary, and a magnetostratigraphic type section has been proposed (Alvarez et al., 1977) at Gubbio in the Um-

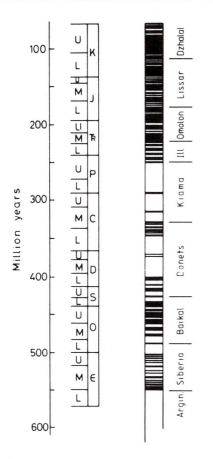

FIGURE 6.20
Russian polarity scale for the Phanerozoic. (After Molotovsky et al., 1976, reproduced in Tarling, D. H., 1983, Palaeomagnetism: Principles and applications in geology, geophysics and archaeology, Fig. 9–6, p. 211. Reprinted by permission of Chapman and Hall.)

brian Appenines, Italy. Here, in the Scaglio Rossa Limestone, a coccolith-foraminiferal limestone, there is thought to be a continuous succession across the Cretaceous-Tertiary boundary. The boundary can be very precisely located (Premoli Silva, 1977) by the disappearance of large foraminiferans of the genus *Globotruncana* (at the top of the *Abathomphalous mayaroensis* foraminiferal zone and the *Pachydiscus neubergicus* ammonite zone); this means that it can be biostratigraphically correlated with the deep-sea succession. In terms of the magnetostratigraphic succession, the faunal boundary occurs near the top of the reversed interval preceding anomaly 29 of the deep-sea magnetic polarity sequence; this has been labeled (Lowrie and Alvarez, 1977) in the Gubbio section as reversed polarity zone Gubbio G_1 (Fig. 6.21).

6.6 POLAR WANDERING CURVES

In addition to magnetic reversals there is another paleomagnetic phenomenon that is time-dependent and possibly usable as a means of dividing geologic time, albeit on a much vaster scale than is done with polarity reversals. Time division in this case is based on variations in the shape of **apparent polar wandering curves (APW curves).** These are derived from plotting the position of a pole as it appears to vary with time relative to an assumed fixed point. From the earliest days of

TABLE 6.1
Terminology used in magnetostratigraphy (ISSC, 1979)

Traditional Term	Geochronologic	Chronostratigraphic	Magnetostratigraphic
Interval	Chron (or superchron)	Chronozone (or superchronozone	Polarity superzone
Epoch	Chron	Chronozone	Polarity zone
Event	Chron (or subchron)	Chronozone (or subchronozone)	Polarity subzone

TABLE 6.2
Hierarchy of magnetostratigraphic divisions in terms of length in years

Traditional Unit	Magnitude	New Unit
Interval	30–50 m.y. +	Chron (or superchron)
Epoch	1 m.y. +	Chron
Event	10,000–100,000 yrs. (of some authors) or 50,000–100,000 yrs.	Chron (or subchron)
Excursion (short event)	10,000 yrs. or less	Chron

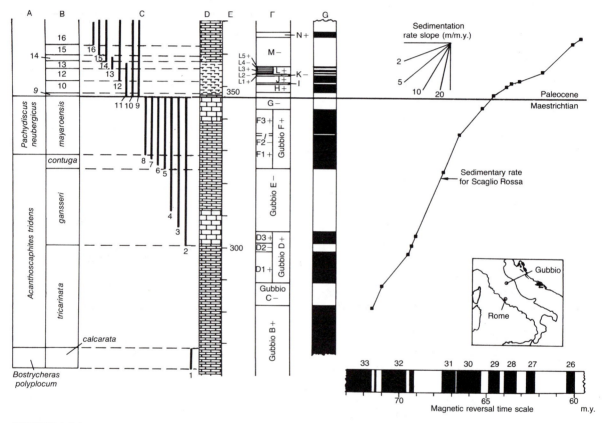

FIGURE 6.21
Magnetostratotype at the Cretaceous-Tertiary boundary in the Gubbio section, Italy. Column A, ammonite zones; B, foraminiferal zones (after Premoli Silva, 1977); C, foraminiferal ranges; D, lithology; E, stratigraphic level (in meters); F, polarity zone; G, magnetic polarity. Key to foraminiferal ranges: 1, *Globotruncana calcareta*; 2, *G. gansseri*; 3, *Pseudoguembelina excolata*; 4, *Abathomphalus intermedius*; 5, *Globotruncana contusa*; 6, *Rugoglobigerina rotundata*; 7, *Racemiguembelina fructicosa*; 8, *Abathomphalus mayaroensis*; 9, *Globigerina eugubina*; 10, *Globorotalia pseudobulloides*; 11, *Globigerina daubjergensis*; 12, *Globorotalia trinidadensis*; 13, *G. uncinata*; 14, *G. angulata*; 15, *G. pusilla pusilla*; 16, *A. pseudomenardii*.

FIGURE 6.22
Apparent polar-wandering curves for the Precambrian as plotted from Europe (top) and North America (bottom). (From Tarling, D. H., 1983, Palaeo-magnetism: Principles and applications in geology, geophysics and archaeology, Fig. 9.10, p. 241. Reprinted by permission of Chapman and Hall.)

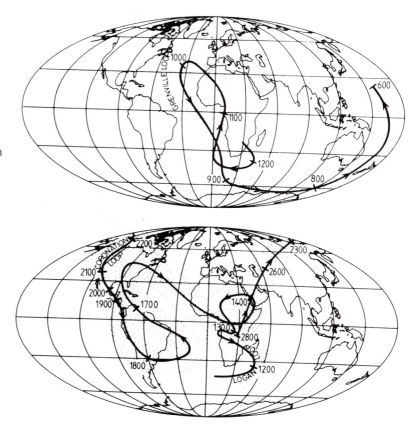

paleomagnetic studies it was known that the apparent position of the magnetic poles has changed through geologic time. It was found also that in rocks of a given age paleomagnetic indicators on two continents were likely to point to widely disparate locations for one magnetic pole. According to the axial dipole theory the magnetic poles coincide more or less with the rotational poles. The theory also eliminates the possibility of multiple poles at any one time. Finally, there are marked constraints on any model that suggests that the earth's axis of rotation has changed appreciably in the past. With these points established, the wide scatter of positions for the pole in the geologic past, as measured by paleomagnetic data, must clearly be due to relative movements of the continental blocks containing the magnetized rocks. These ob-

servations provided perhaps the most persuasive arguments in support of continental drift at a time when plate tectonic mechanisms were not understood as well as they now are.

With growing numbers of paleomagnetic observations it has become possible to plot pole positions from all the continents at many points in time from the Precambrian to the present day. These points joined comprise apparent polar wandering curves. It has usually been found more convenient to plot the APW curve for a single continental block and to compare it with curves for other continents rather than to plot what actually happened—namely, the movements of the continental blocks relative to a fixed pole. The APW curves provide an important insight into the direction and speed of movement of the various continental blocks (Fig. 6.22). The com-

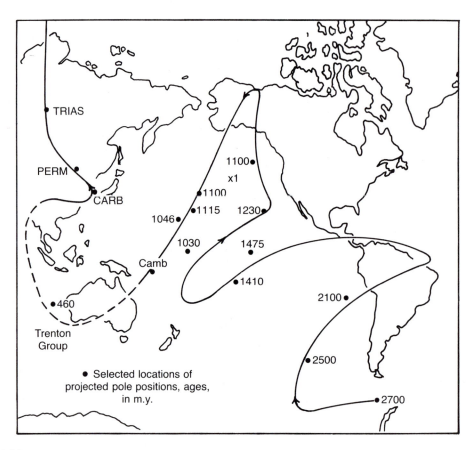

FIGURE 6.23
Coming together of the apparent polar-wandering curves for the Slave and Superior tectonic provinces of the Canadian Shield, indicating their earlier independent movement as protocratons. (After Cavanaugh, M. D., and Seyfert, C. K., 1977, Apparent polar wander paths and the joining of the Superior and Slave provinces during early Proterozoic time: Geology, v. 5, Fig. 1, p. 207.)

ing together and blending of two APW curves derived from two continental blocks, for example, records the collision of these blocks and their later movement as one (Fig. 6.23). Conversely, the dividing of an APW curve into two divergent traces marks the splitting of a continental block. Although clearly the initial construction of an APW curve depends upon the correct dating of the magnetized rocks, a well-documented curve may serve to indicate the age of a magnetic sample of unknown age if that sample's pole position plots on the

curve between points that are dated. This procedure is analogous to the archeomagnetic dating from a master curve, mentioned in an earlier section.

When the APW curve for a given continental block is traced from the Precambrian to the present, it is seen to follow a highly irregular course with relatively smooth segments known as *tracks,* alternating with loops or kinks called **hairpins.** The hairpins mark sharp changes in the direction of drift of a continental block (Fig. 6.24). The alternation

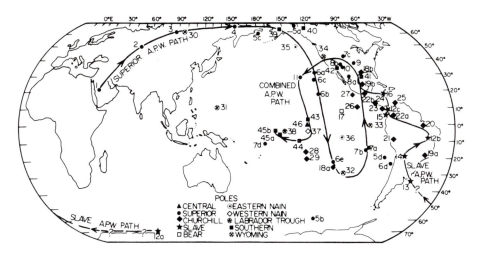

FIGURE 6.24
Apparent polar-wandering curve for the Canadian Shield, Archean to Triassic. (Redrawn from Irving and Park, 1972.)

of the long periods represented by the tracks and the much briefer periods marked by the hairpins suggested to Irving and Park (1972) that they might be used as a basis for dividing geologic time into what they termed *superintervals*. The magnitude of these time divisions, measured in hundreds of millions of years, means that they are applicable only to the Precambrian. Further discussion of this topic is postponed until Chapter 16.

6.7 CONCLUSION

It should be clear by now that what stratigraphers really are looking for is an isochronous marker of global extent. With the discovery of magnetic reversals, it almost might seem that this "holy grail" had been discovered! In fact, magnetic stratigraphy has made an invaluable contribution in solving problems of stratigraphic correlation. Although subject to many problems of measurement, dating, and correlation, magnetic stratigraphy both enhances and is enhanced by the various other stratigraphic tools available, such as biostratigraphy, oxygen isotope stratigraphy, and te-

phrochronology. It is, therefore, in a complementary role rather than by itself that its true value lies. As explained, the precision and applicability of magnetic stratigraphy drops off markedly in rocks of pre-Jurassic age, indicating very clearly how important seafloor anomaly successions are in magnetostratigraphic classification.

REFERENCES

Alvarez, W., Arthur, M. A., Fischer, A. G., Lowrie, W., Napoleone, A., Premoli Silva, I., and Roggenthen, W. M., 1977, Upper Cretaceous-Paleocene magnetic stratigraphy at Gubbio, Italy, V. Type section for the Late Cretaceous-Paleocene geomagnetic reversal time scale: Geol. Soc. America Bull., v. 88, p. 383–389.

Baldwin, B., Coney, P. J., and Dickinson, W. R., 1974, Dilemma of a Cretaceous time scale and rates of sea-floor spreading: Geology, v. 2, p. 267–270.

Bullard, E. C., 1971, The earth's magnetic field and its origin, *in* Gass, I. G., Smith, P. J., and Wilson, R. C. L., eds., Understanding the earth: Cambridge, MA, M.I.T. Press, p. 71–78.

Cox, A., 1969, Geomagnetic reversals: Science, v. 163, p. 237–245.

Elsasser, W. M., 1965, Hydromagnetic dynamo theory: Rev. Mod. Physics, v. 28, p. 135–163.

Harland, W. B., Smith, A. G., and Wilcock, B., eds., 1964, The Phanerozoic time scale: Geol. Soc. London Quart. Jour., v. 120S, 458 p.

Harrison, C. G. A., 1966, The paleomagnetism of deep-sea sediments: Jour. Geophys. Research, v. 71, p. 3033–3043.

———— 1974, The paleomagnetic record from deep-sea sediment cores: Earth-Science Rev., v. 10, p. 1–36.

Hedberg, H. D., ed., 1976, International stratigraphic guide: New York, John Wiley & Sons, 200 p.

Hsü, K. J., La Brecque, J. L., Percival, S. F., Wright, R. C., Gambos, A. J., Pisciotto, K., Tucker, P., Peterson, N., McKenzie, J. A., Weissert, H., Karpoff, A. M., Carman, M. F., Jr., and Schreiber, E., 1984, Numerical ages of Cenozoic biostratigraphy datum levels: Results of South Atlantic Leg 73 drilling: Geol. Soc. America Bull., v. 95, p. 863–876.

Irving, E., 1971, Nomenclature in magnetic stratigraphy: Royal Astron. Soc. Geophys. Jour., v. 24, p. 529–531.

———— 1972, Paleomagnetic stratigraphy—Names or numbers? Comments on earth sciences: Geophysics, v. 2, p. 125–130.

Irving, E., and Park, J. K., 1972, Hairpins and super-intervals: Can. Jour. Earth Sci., v. 9, p. 1318–1324.

Irving, E., and Pullaiah, G., 1976, Reversals of the geomagnetic field, magnetostratigraphy, and relative magnitude of paleosecular variation in the Phanerozoic: Earth-Sci. Rev., v. 12, p. 35–64.

ISSC (International Subcommision on Stratigraphic Classification), 1979, Magnetostratigraphic polarity units—A supplementary chapter of the ISSC international stratigraphic guide: Geology, v. 7, p. 578–583.

Larsen, R. L., and Chase, C. G., 1972, Late Mesozoic evolution of the western Pacific Ocean: Geol. Soc. America Bull., v. 83, p. 3627–3644.

Larsen, R. L., and Hilde, T. W. C., 1975, A revised time scale of magnetic reversals for the early Cretaceous and late Jurassic: Jour. Geophys. Research, v. 80, p. 2586–2594.

Larsen, R. L., and Pitman, W. C., III, 1972, Worldwide correlation of Mesozoic magnetic anomalies and its implications: Geol. Soc. America Bull., v. 83, p. 3645–3662.

Lowrie, W., and Alvarez, W., 1977, Upper Cretaceous-Paleocene magnetic stratigraphy at Gubbio, Italy, III. Upper Cretaceous magnetic stratigraphy: Geol. Soc. America Bull., v. 88, p. 374–377.

Lowrie, W., Alvarez, W., Premoli Silva, I., and Monechi, S., 1980a, Lower Cretaceous magnetic stratigraphy in Umbrian pelagic carbonate rocks: Royal Astron. Soc. Geophys. Jour., v. 60, p. 263–281.

Lowrie, W., Channel, J. E. T., and Alvarez, W., 1980b, A review of magnetic stratigraphy investigations in Cretaceous pelagic rocks: Jour. Geophys. Research, v. 85, p. 3597–3605.

Mason, R. G., 1958, A magnetic survey off the west coast of the United States between latitudes 32° and 36° N., longitudes 121° and 128° W.: Royal Astron. Soc. Geophys. Jour., v. 20, p. 320–329.

Mason, R. G., and Raff, A. D., 1961, A magnetic survey off the west coast of North America 32° N to 42° N.: Geol. Soc. America Bull., v. 72, p. 1259–1265.

Molotovsky, E. A., Pevzner, M. A., Petchersky, D. M., Rodionov, V. P., and Khramov, A. N., 1976, Phanerozoic magnetostratigraphic scale and geomagnetic field inversions regime: Geomagnetic Researches, v. 17, Moscow, Nanka, p. 45–52.

Morley, L. W., and Larochelle, A., 1964, Paleomagnetism as a means of dating geological events: Roy. Soc. Canada Spec. Pub. 8, p. 512–521.

NACSN (North American Commission on Stratigraphic Nomenclature), 1983, North American stratigraphic code: Am. Assoc. Petroleum Geologists Bull., v. 67, p. 841–875.

Oriel, S. S., MacQueen, R. W., Wilson, J. A., and Dalrymple, G. B., 1976, Stratigraphic Commission Note 44, Application for addition to code concerning magnetostratigraphic units: Am. Assoc. Petroleum Geologists Bull., v. 60, p. 273–277.

Premoli Silva, I., 1977, Upper Cretaceous-Paleocene magnetic stratigraphy at Gubbio, Italy, II. Biostratigraphy: Geol. Soc. America Bull., v. 88, p. 371–374.

Steiner, M. B., 1980, Investigations of the geomagnetic field polarity during the Jurassic: Jour. Geophys. Research, v. 85, p. 3572–3586.

Talwani, M., Windisch, C. C., and Langseth, M. G., Jr., 1971, Reykjanes ridge crest: A detailed geophysical study: Jour. Geophys. Research, v. 76, p. 473–512.

Tarling, D. H., 1983, Palaeomagnetism: Principles and applications in geology, geophysics and archaeology: London, Chapman and Hall, 379 p.

Vine, F. J., and Matthews, D. H., 1963, Magnetic anomalies over ocean ridges: Nature, v. 199, p. 947–949.

Watkins, N. D., 1976, Polarity Subcommission sets up some guidelines: Geotimes, v. 21, no. 4, p. 18–20.

PART TWO
THE FOSSIL
RECORD

FOSSILS PLAY A UNIQUE ROLE IN THE STUDY OF STRATIGRAPHY BECAUSE THEY INTRO-
duce into sedimentary successions the record of an ongoing time-dependent phenomenon—evolutionary change. The sediments that contain this
record can be thought of as the pages of a book; the record of passing time
is written largely by the fossils. Sedimentary environments can be interpreted from a study of lithofacies and often in remarkable detail, but the
record is typically repetitive. A Cambrian shale can look very much like
one of Cretaceous age and may have been deposited under essentially the
same conditions. It is the overprint of the fossil record that enables geologic history to be read as a sequential account.

7
FOSSILS AND BIOGEOGRAPHY

To eyes that have learned to see, fossils are very much alive.

George Gaylord Simpson

7.1 INTRODUCTION

The essential units in biostratigraphy are **biozones**—successions of strata defined on the basis of their contained faunas and/or floras—within which certain taxa are selected as index fossils. The ideal taxon for stratigraphic correlation, an index fossil, should (1) belong to a species that has a short temporal span or is part of a rapidly evolving lineage; (2) be nektonic, planktonic, or nekroplanktonic, so that it is widely dispersed and will not be facies-restricted; (3) be robust enough to be preserved in a variety of depositional environments and still be identifiable. In a special category are microfossils that are small and abundant enough to be recognizable in well cuttings. The value of biostratigraphic units in long-distance correlation of strata is clearly related to the geographic range of the taxon or fauna upon which the biozone is based. The mechanisms whereby different species become distributed and extend their ranges, and the barriers that circumscribe their expansion, form the topic of discussion in the first section.

7.2 GEOGRAPHIC DISPERSAL

Oceanic Dispersal Mechanisms

Ocean currents have the most important influence in dispersing marine organisms of all types. Organisms can be placed in several different categories as follows.

Plankton Plankton comprise animal and plant species, many of microscopic size, that passively float in the water. Temperature and salinity are the chief controls in determining the geographic range of living species. The fossil record is dominated by the Foraminifera, Radiolaria, and Diatomacea; in Paleozoic successions graptolites were important.

Pseudoplankton and Rafted Organisms

Pseudoplanktonic or rafted organisms are those that attach themselves, either in larval or adult stages, to floating objects such as seaweeds. In modern seas, patches of floating *Sargassum* weed (gulfweed) play host to a variety of organisms, such as bryozoans, gastropods, polychaetes, and hydroids (Fig. 7.1). It is clear that some of these species find widespread distribution. Certain seeds remain viable even after long immersion in sea water, and many, notably tropical plant species such as *Ipomoea pescapra* (railroad vine), come to have a global distribution.

Necroplankton Necroplanktonic dispersal affects organisms that possess skeletal parts that float for a time after death. Modern *Nautilus* shells, for example, are known to float, sometimes for years, after the death of the animal. Many ammonites and graptoloid graptolites were widely distributed in this manner before finally sinking as a consequence of becoming waterlogged, punctured, or weighted by commensal organisms.

Planktonic Larvae of Benthic Organisms

Many benthic species produce larvae with limited swimming ability (for example, the veliger larvae of many molluscs), but dispersal potential is still largely dependent upon ocean currents. In the case of shelf species the migration and settlement of larval stages is controlled on a local scale by temperature, salinity, and bottom-sediment type. On a larger scale the barriers to dispersal are likely to be deep oceans and climatic zonal boundaries. Whether or not such major barriers can be crossed depends largely upon the longevity of the larvae (Table 7.1). Long-lived larvae may well survive considerable ocean voyages, as the occurrence of the same shelf species on both sides of an ocean will testify (Fig. 7.2). Some species, such as the tonnacean gastropods *Tonna galea* and *Cymatium pileare,* are even found to have global distribution. Transoceanic migration of larvae may be very rapid because oceanographic surveys have indicated, for example, that the 5000 km between the Bahamas and the Azores are covered in from 128 to 300 days by the North Atlantic Drift. The South Atlantic can be crossed via the South Equatorial current in from 60 to 154 days (data from U.S. Hydrographic Office, quoted in Scheltema, 1977). On both the local and transoceanic scales, transportation of larval stages by currents enables environmentally unfavorable areas to be bypassed. This process is typically very rapid in the geologic context, even if major oceanic barriers are involved. Migration along continental margins may also be very rapid, as in the case of the living gastropod *Littorina littorea,* documented by Kraueter (1974). Originally introduced from Europe, presumably by humans, this species has extended its distribution southward along the coast from Canada to New Jersey, a distance of 1200 km in the last 130 years.

In older successions, the wide distribution, sometimes of almost global extent, of certain shallow benthic species has frequently been noted. Rather than resulting from the migration of long-lived larvae, this **pandemism** is often more apparent than real and may be caused by continental drift. Kauffman (1975) pointed out that the trans-Atlantic distances that constitute a barrier to the migration of many modern bivalve larvae did not exist in the early Cretaceous. At that time, the proto-Atlantic was relatively narrow (Fig. 7.3), ranging from some 2000 km to less than 1000 km at its narrowest and still, therefore, crossable even at a slow 1 km/hr current velocity (the modern Equatorial Current has a velocity of 3 km/hr). The presence of stepping stones in the form of islands and shoals was likely also a factor in the rapid dispersal of many forms. By the late Cretaceous, with continual and accelerated plate movement, the separation of North American and European shelf areas had grown too great for much transoceanic larval exchange; from this time on, there is a natural increase in endemism in European and North American faunas. Because of these plate tectonic factors, the pandemism of many benthic bivalves makes them good index taxa for long-distance correlation. As Kauffman (1975) pointed out, their value in solving problems of Cretaceous stratigraphy is in many ways superior to that of ammonites. It is well known that the mobility and dispersal potential of ammonites makes them the oft-cited ideal index fossils, but as Kauffman (1975) showed, when comparing trans-Atlantic Cretaceous faunas, between 70 and 80 percent of benthic bivalve species are common to North America and Europe as far east as Russia. This compares with less than 10 percent for ammonite species. The ammonites also appear to have been considerably less tolerant of temperature changes and are markedly more restricted in their distribution by climatic zonal boundaries.

Terrestrial Dispersal Mechanisms

Terrestrial dispersal mechanisms can be classified broadly under two headings: active, in which animals move of their own volition, and passive, in which an external agency, such as the wind, is involved. The active migration and dispersal of terrestrial organisms apparently is controlled largely by climatic changes and their influence on the vegetation. Clearly, a whole range of organisms is linked within complex food chain relationships. Passive dispersal is chiefly seen in the plants, where seeds are transported by gravity, wind, running water, and animals, both internally (edible fruit and seeds) and externally (sand spurs and other burrlike fruit). The migration and spread of flying insects, although obviously not entirely passive, is influenced markedly by winds. The wide migration of the monarch butterfly, for example, was studied by Urquhart (1960), who noted a close correlation between routes traveled and weather patterns.

Continental Drift

Although dispersal mechanisms can be logically discussed in terms of living organisms, there remains one mechanism that can apply only in the case of fossils: postfossilization transport by virtue of continental drift. Long-distance correlation from continent to continent is frequently simplified by the fact that the strata involved were once laid down in a single localized basin, perhaps with a highly endemic fauna, and later split between two or more continental fragments. In the days when continental drift generally was not accepted, such distant connections were very puzzling. Many ingenious land bridges and ancient seaways were invented to explain the observed distributions (Fig. 7.4). Even in Arkell's immensely important and influential work, *The Jurassic Geology of the World,* published in 1956, there is only the briefest reference to the possibility of continental drift. The conclusion reached, on the evidence of Jurassic faunal distributions, was that it could be discounted. On the other hand, it should be remembered that many years before the great unifying concept of plate tectonics

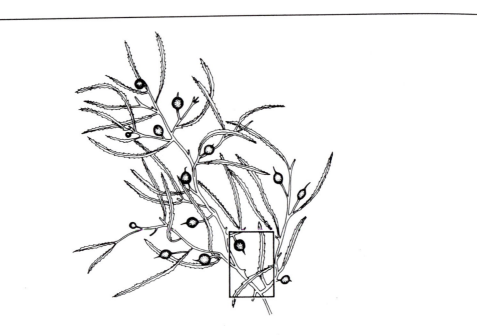

A. *Litiopa* (gastropod mollusk)
B. *Amphithoë* (amphipod crustacean)
C. *Luconacia* (caprellid amphipod)
D. *Anemonia sargassensis* (sea anemone)
E. *Platynereis* (nereid polychaete)
F. *Sertularia* (hydroid cnidarian)
G. *Clytia* (hydroid cnidarian)
H. *Scyllaea* (nudibranch gastropod)
I. *Spirorbis* (serpulid polychaete)
J. Copepod crustacean
K. *Zanclea* (hydroid cnidarian)
L. *Ceramium* (an epiphytic red alga)
M. *Membranipora* (bryozoan)
N. *Doto* (nudibranch gastropod)
O. *Gnescioceros* (polyclad flatworm)
P. *Obelia* (hydroid cnidarian)
Q. *Anoplodactylus* (pycnogonid) feeding on a nudibranch
R. *Fiona* (nudibranch gastropod)

FIGURE 7.1
Invertebrates that live on floating *Sargassum* weed. See detail opposite. (From Barnes, R. D., 1980, Invertebrate zoology, 4th ed., Fig. 18–9, p. 894, 895. Reprinted by permission of Saunders/Holt, Rinehart and Winston, Philadelphia.)

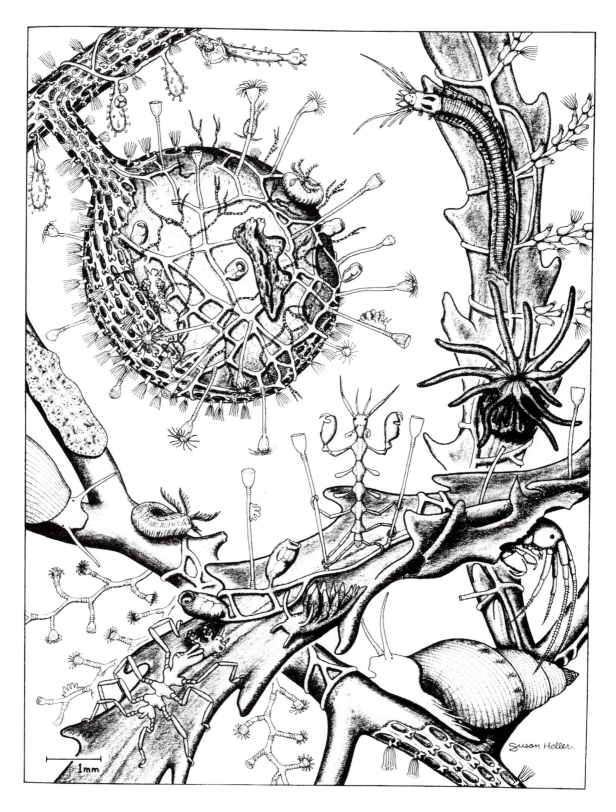

1mm

Susan Heller.

TABLE 7.1

Normal dispersal time and distance of some common living Bivalvia with long-lived planktonic larvae in 1 and 3 km/hr currents typical of modern oceanic systems (From Kaufmann, E. G., 1975, Table 1, p. 170; reprinted by permission of Geological Association of Canada, Ottawa)

Taxa		Temperature or Climatic Zone	Planktonic Larval Duration in Days	Distance Traveled in km/mi	
Family	Genus/Species			Current 1/km/hr	Current 3/km/hr
Long-lived larvae					
Veneridae	*Paphia*	N. TEMPERATE	42	1008/625	3024/1875
	Venus sp.	18°C	30	720/446	2160/1338
		21°C	28	672/417	2016/1251
		24°C	22	528/327	1584/981
		27°C	20	480/298	1440/894
		30°C	16	384/238	1152/714
Ostreidae	*Crassostrea*				
	virginica	17°–18°C	21	504/313	1512/939
		20°C	17	408/253	1224/759
		22°–24°C	13	312/193	936/579
		24°–27°C	7	168/104	504/312
	Ostrea				
	edulis	16°C	17–18	406–432/253–269	1224–1296/759–807
		17°C	13–14	312–336/193–208	936–1008/579–624
		18°–21°C	9–10	216–240/134–149	648–720/402–447
		22°–23°C	6	144/89	432/267
Mytilidae	*Mytilus*				
	crassitesta	Temperate	?–86	2064/1280	6192/3840
	edulis	<18°C-LAB.	TO 120	2880/1786	8640/5358
		18°C	28	672/417	2016/1251
		20°C	21	504/313	1512/939
Myidae	*Mya* SP.	18°C	14	336/208	1008/624
		20°C	21	504/313	1512/939
Pectinidae	*Pecten*				
	maximus	16°C	33–38	792–912/491–565	2376–2736/1473–1695
General for family		Temperate	14–21	336–504/208–313	1008–1512/624–939
		Tropical	14–15	336–360/208–223	1008–1080/624–669
Arcidae	*Anadara* sp.	Temperate	27–37	648–888/402–551	1944–2664/1206–1653
Teredinidae	*Teredo* sp.	Temperate	1.5–15	36–360/22–223	108–1080/66–669
Pholadidae	*Barnea*	Temperate	35	840/521	2520/1563
General for bivalvia					
Average		Temperate	14–21	336–504/208–313	1008–1512/624–939
Range for 90%		Temperate	14–42	336–1008/208–625	1008–3024/624–1875
Total recorded range		All zones	0–120	0–2880/0–1786	0–8640/0–5358

FIGURE 7.2

A. Distribution of bivalve veliger larvae in the north and tropical Atlantic Ocean. Large circles indicate collections containing larvae: large black circle, family Pinnidae; large circle containing cross, family Teredinidae; large circle containing small black circle, both families Pinnidae and Teredinidae; large open circle, other bivalve families. Small open circles are locations where bivalve larvae were absent. B. Distribution of gastropod veliger larvae in the north and tropical Atlantic Ocean. Large black circle *Cymatium parthenopeum;* large half-blackened circle, *Charonia variegata;* large circle containing small black circle, both *Cymatium parthenopeum* and *Charonia variegata;* large circle containing cross, other species of family Cymatiidae; large open circle, other gastropod families. Small open circles are locations where gastropod larvae were absent. Arrows show direction of surface currents. (From Scheltema, R. S., 1977, Dispersal of marine invertebrate organisms: Paleobiologic and biostratigraphic implications, *in* Kauffman, E. G., and Hazel, J. E., eds., Concepts and methods of biostratigraphy, Figs. 5 and 6, p. 83, 84. Reprinted by permission of Dowden, Hutchinson and Ross, Stroudsburg, PA.)

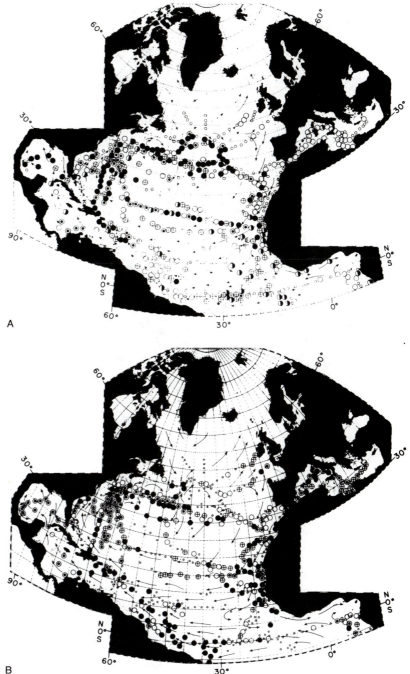

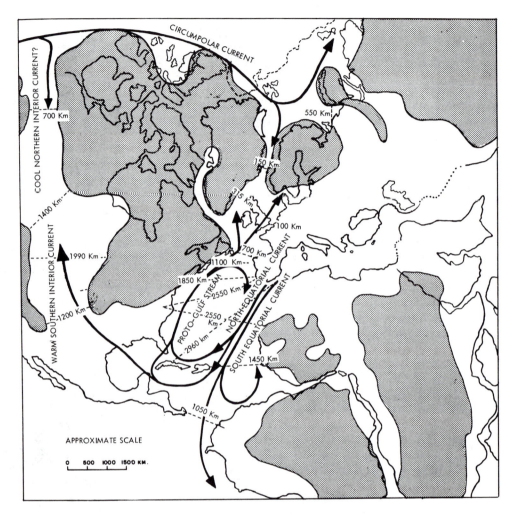

FIGURE 7.3

Cretaceous plate reconstruction showing distribution of proposed Cretaceous current systems and approximate maximum distances in kilometers across selected points of the proto-Atlantic and its marginal seaways and across the Western Interior seaway during maximum late Cenomanian–earliest Turonian transgression, from shallow shelf area to shallow shelf area. Compare these figures with those for larval dispersal (Table 7.1) of common bivalves on 1 and 3 km/hr current systems. At 3 km/hr, the larvae of Cretaceous bivalves should have been able to disperse throughout the Euramerican region and across the Atlantic extension of Tethys in short, "geologically instantaneous" periods of time. (From Kauffman, E. G., 1975, Dispersal and biostratigraphic potential of Cretaceous benthonic Bivalvia in the Western Interior, *in* Caldwell, W. G. E., The Cretaceous System in the Western Interior of North America, Spec. Paper 13, Fig. 2, p. 183. Reprinted by permission of Geological Association of Canada, Ottawa.)

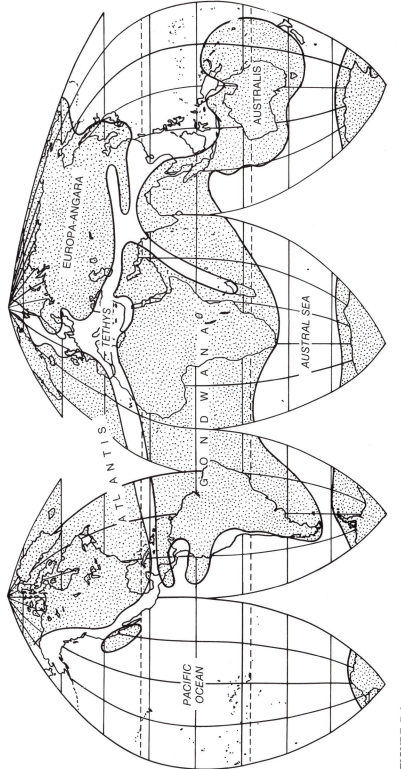

FIGURE 7.4

Typical paleogeographic reconstruction, using land bridges and seaways to explain the distribution of fossil faunas and floras, without recourse to continental drift. This particular one dates from 1949 and depicts the supposed distribution of land and sea in the early Jurassic.

163

threw light on the mechanisms responsible for continental drift, many paleontologists were the most vociferous supporters of the idea. The apparent absence of any conceivable mechanism did not in the least dampen their enthusiasm!

7.3 GEOGRAPHY AND THE GENE POOL

Geographic Variability

In describing an extant animal species of the present day, one of the important factors to consider is the geographic range of the species and the amount of morphologic variability that occurs within the total population. If the geographic range is extensive, there are likely to be several local varieties, races, or subspecies recognized. Occasionally, the extreme variants within the total range may be sufficiently different from one another to have been originally mistaken for separate species. In fact, there are presumably numerous local variants between the two extremes which serve to connect them by gradational variation within a common gene pool. This would be an example of what the biologist calls a **cline** (Fig. 7.5). As Mayr (1971) put it, a cline represents a balance between selection on the one hand and gene flow on the other. Selection by itself would result in every local population being unique within its own ecologic niche, whereas gene flow tends to smooth out local differences and to make all populations similar within a given species throughout its entire range. Clearly, the degree of morphologic variation within a cline will depend upon the relative intensity or effectiveness of these two opposing influences. A cline may be seen in one or more physiological or morphologic attributes of the species: overall size, coloration, limb proportions, and so on. The cline in each feature may or may not show a gradient similar to that of the cline in any one of the other fea-

tures. The continuity within a cline might be seen in a graduated series of local subspecies, but according to Wright (1967), it is more likely that, ultimately, there would emerge only a few well-differentiated species, these being derived from the more successful of the subspecies. An alternative explanation for this apparent stability of form over an extended geographic range is offered by Thom's fold catastrophe model, described in Chapter 8.

Among the most important mechanisms in the evolutionary process, leading to the emergence of new species, is geographic isolation of portions of the gene pool, a process known as **cladogenesis.** Within a small isolated population, genetic drift and an interplay of local environmental stress factors and natural selection are presumably sufficient to cause a progressive statistical shift in various physiological and morphological characters. Eventually the cumulative effect of this shift would result in significant differences from the parent population—differences sufficiently great that, even if the geographic barriers were removed and the separate gene pools came into contact, no hybridization would occur. A species that emerges as a consequence of geographic isolation is termed an **allopatric species,** or *vicariant.* The *vicariance paradigm,* as it is known, is illustrated best when the distribution of species is viewed on a global scale. Some strange relationships are revealed. There are, for example, and not surprisingly, species of freshwater fishes in the rivers of Africa different from those in South America. However, when the taxonomy of these fishes is examined in detail, there are some marked resemblances at higher levels. These clearly point to the origin of both African and South American freshwater fishes from a common ancestral stock that must have existed in Gondwana. This is a clear case where an original parent population had been split by continental drift. This splitting was a prerequisite for

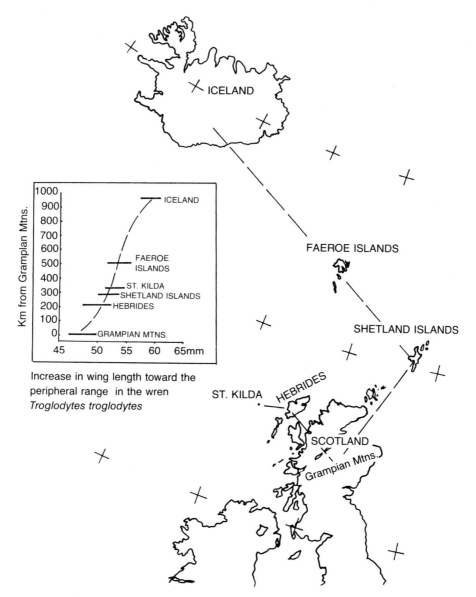

FIGURE 7.5
Example of a cline in a living bird species. (Data from Salomonsen, F., 1931, Diluviale isolation und arten bildyng: Proc. 7th Internat. Ornith. Kongr Amsterdam, p. 413–438.)

evolutionary divergence and must have preceded the speciation process that produced the vicariant biota.

In the case of the shallow-water neritic organisms that dominate the fossil record,

sea-level and climatic fluctuations have introduced a constantly shifting pattern of environmental changes. Climatic changes are superimposed upon the effects of sea-level changes and may in part be due to the same underlying causes. A marine regression affects the population of shallow neritic environments in two ways: first, in isolating portions of the gene pool behind geographic or salinity and temperature barriers and second, by increasing interspecific and intraspecific competition within contracting habitats. It should be noted that marine transgressions have similar effects on the terrestrial populations of areas cut off to form islands.

The total region occupied by a given species may change with time, either because of shifts in climatic or other geographic boundaries or because of a gradual expansion of the species into new areas which, while only marginally favorable to the original stock, might come to be occupied by better-adapted variant genotypes. It has been suggested that it is only in such marginal areas that the genetic potential of species (that is, the potential of genotypes) can be exploited and that this is also an important factor in the evolutionary mechanisms responsible for the appearance of new species. Local populations at the extreme limits of the geographic range of the species, what Eldredge and Gould (1977) have called **peripheral isolates,** are likely to exhibit some of the more extreme morphological variations within the species. Also, such peripheral populations stand a greater probability of being cut off and isolated from the main gene pool.

Essentially the same mechanism operates in the case of colonization of isolated habitats, notably islands. What is known as the "founder principle" describes the effects of the colonization of an island by a single fertilized female. Clearly this "founder" will have carried with her only a tiny fraction of the genetic potential of the main gene pool. Descendant generations, isolated from the stabilizing gene flow of the parent population, will likely contain unusual gene combinations, and natural selection will result in rapid genetic change and the appearance of new species. Although not proven, this founder effect undoubtedly has some influence on isolated populations and, as such, plays a role in the mechanism of speciation.

Evolution and the Environment

Environmental changes play a vital and possibly even an essential role in the process of speciation. It is generally agreed that the rate of evolutionary change tends to be more rapid in the case of organisms in high-stress environments such as rocky shores or tidal flats than in those in more stable habitats. In a completely static and geographically extensive environment a large population with unrestricted gene flow would, as Mayr (1969) put it, remain "totally conservative." There are some striking examples of evolutionary conservatism in the fossil record. These are generally thought to occur when the organisms were so well adapted to an environment that remained virtually unchanged that no environmental stress existed and selection pressures were absent or much reduced. The brachiopod *Lingula,* for example, is generally cited as demonstrating evolutionary conservatism, and species from the Ordovician are not significantly different from modern forms. The monoplacophoran *Neopilina* is another example. Thought to have become extinct in the Devonian, living specimens were discovered in 1953 in deep-water dredges from the eastern Pacific and Caribbean (Fig. 7.6). The modern specimens are so similar to the Silurian and Devonian *Pilina* that some taxonomists would have preferred not to have introduced another generic name. The persistent habitats in which such "living fossils" are found are commonly quite localized in extent and the populations are small.

7.4 BARRIERS

Barriers to the migration of animals and plants are of two kinds. Some are uncrossable in the sense that they mark a threshold in a continuous gradient such as temperature change. The expansion of a tropical species

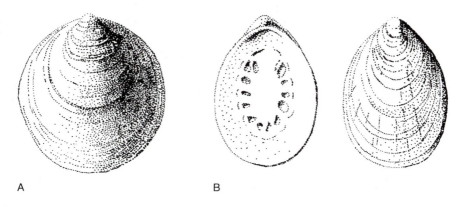

FIGURE 7.6
Example of a "living fossil." *Neopilina galatheae,* (A) a Recent species of monoplacophoran, compared with *Pilina unquis* (B) from the Middle Silurian.

into higher latitudes, for example, clearly is
blocked unless there is movement of the
threshold itself. Most of the major latitudinal
controls are of this nature. The second type
of barrier is potentially crossable, depending
upon its effectiveness. This is the idea of the
"sweepstakes route" in species migration described by Simpson (1952). Presumably,
given enough time, the barrier will eventually
be crossed. Within the human or historical
context it might take a long time before a
combination of favorable circumstances resulted in the crossing of major barriers and
the occupation of all suitable habitats. However, even a period involving thousands of
years can, in the geological context, be considered as virtually instantaneous. It should
be remembered that an apparent diachronism in the spread of a species may be simply
linked to a progressive expansion of suitable
habitats. Clearly, if a sedimentary formation
or facies is diachronous, so also will be the
fossil species contained in it, particularly in
the case of facies-linked taxa. In the case of
the short-lived species selected as zonal indicators, this effect is usually not important,
but it is frequently noticeable in the case of
long-lived species.

In modern faunas the barriers between adjacent populations are sometimes very subtle

and are seen to be behavioral rather than environmental. In dealing with fossil faunas,
such barriers are naturally an unknown
quantity; even the nature of some physical
barriers sometimes only can be guessed at.
Thus, water temperature and salinity barriers
may have been the causes of a pronounced
change in a fossil fauna, but no direct evidence is to be seen in the enclosing sediment. The marked contrast between the Tethyan and Boreal ammonite faunal provinces in
the Jurassic Volgian stage apparently marks a
temperature barrier.

Shifting temperature barriers can be considered under two headings: short-term and
long-term. Short-term changes, occurring
over periods of the thousands of years of the
Milankovitch cycle, are apparent mostly during ice ages, whereas long-term cycles, measured in millions of years, are most likely a
consequence of continental drift. Temperature changes have been especially pronounced since the beginning of the late Cenozoic ice age, and the effects on both marine
and terrestrial biota are very apparent. For
example, numerous marine molluscan species, such as *Merceneria merceneria, Polinices
duplicata,* and *Oliva sayana,* which range
along the U.S. east coast, are also found in
the Gulf of Mexico. They do not occur today

in southern Florida, however, because it is apparently too warm. The range was presumably once continuous during glacial periods, a supposition confirmed by the presence of these species as fossils in Pleistocene faunas of southern Florida (Fig. 7.7). Terrestrial faunas and floras provide some striking instances of populations isolated by temperature barriers. In both North America and Europe are numerous examples of so-called **glacial refugia** in mountainous areas to which relict arctic faunal and floral elements retreated as postglacial climates ameliorated (Fig. 7.8). Refugia of various kinds have undoubtedly played a role during times of environmental stress and increased likelihood of extinction. Jablonski and Flessa (1986) de-

scribed the effects of eustatic lowering of sea level on species-area relationships of shallow-marine benthic faunas. They suggested that, although decimated on the continental shelves, representatives of many families might find refuge on oceanic islands which, because of their conical shape, actually gained slightly in perimeter during a sea-level drop.

Although physical barriers of one kind or another are the usual boundaries to natural population ranges, there may be other, more subtle, influences involved. Interspecific competition between two ecologically similar species where their ranges overlap may cause individuals to be at a disadvantage. As pointed out by Valentine (1977), the effect of

FIGURE 7.7
Faunal provinces in the marine molluscs of the southeastern Atlantic seaboard and the Gulf of Mexico. C = Carolinian, CA = Caribbean, CG = Carolinian, gulf component, P = Panamic, B = Brazilian. (After Petuch, E. J., 1988, Neogene history of tropical American mollusks: Charlottesville, VA, Coastal Education and Research Foundation, 217 p.)

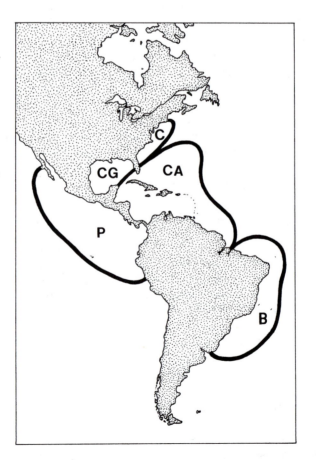

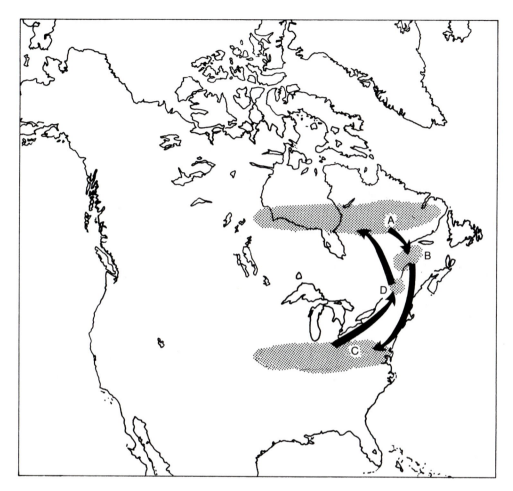

FIGURE 7.8
Map showing how populations became stranded during the advance and retreat of the Pleistocene ice cap in eastern North America. Area A marks the original distribution of a subarctic species. Area C marks the distribution of the same species in a more southerly area after the advance of the ice, with a population left isolated at B. After retreat of the ice, populations returned to the former range A, leaving a population relict on high ground at D. (Based on a diagram by Darlington, P. J., Jr., 1957, Zoogeography: The geographical distribution of animals: John Wiley & Sons, New York.)

this is to emphasize the partition between the ranges of the two forms. In considering the fossil record over spans of geologic time, geographic barriers appear and disappear in response to the changing configuration of the continents and to sea-level changes. One of the best documented of these barriers is the Panamanian isthmus. Its emergence and the formation of a land bridge between Central and South America occurred between 3.5 and 3.1 m.y. ago (Fig. 7.9). As Vermeij and Petuch (1986) and Petuch (1982) point out, this event had important repercussions that extended far beyond the immediate region.

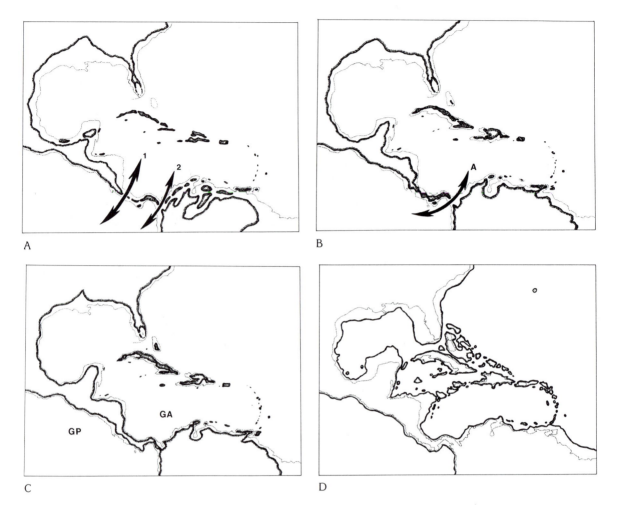

FIGURE 7.9
History of the Panamanian land bridge. A. Configuration of the American continents and Caribbean islands in the middle Pliocene (Zanclian Stage). The main seaways were (1) the Balboa seaway and (2) the Atrato seaway. B. Configuration of coastlines in the Late Pliocene (Piacenzian Stage) showing Atrato seaway (A). C. Configuration of coastlines in the Plio-Pleistocene. The Gatunian faunal province is bisected by the Panama land bridge, resulting in an Atlantic faunal component (GA) and a Pacific component (GP). D. Configuration of coastlines during a Pleistocene glacial sea-level lowstand. Sea levels dropped approximately 200 m, exposing the wide platforms of Florida, the Bahamas, Yucatan, and Honduras. (From Petuch, E. J., 1988, Neogene history of tropical American mollusks, Fig. 11, p. 109; Fig. 12, p. 111; Fig. 13, p. 113; Fig. 14, p. 125. Reprinted by permission of Coastal Education and Research Foundation, Charlottesville, VA.)

Apart from the obvious effect of the large-scale two-way migration of land mammals, the severance of the Caribbean-Pacific connection had profound effects on the marine environment. The north-south circulation became intensified and the chemistry of the Atlantic was altered (Keigwin, 1982), these changes, perhaps, playing a role in the onset of the Late Cenozoic glaciations. The effects on marine biota were particularly striking on the Atlantic side, where, according to Vermeij and Petuch (1986), extinction of molluscs was far more severe than on the Pacific side.

The effect of the removal of a major geographic barrier also makes an interesting study. One that actually has been observed is the cutting of the Suez Canal. As described by Vermeij (1978), a major episode of biotic interchange began after 1869, although it appears to have been very much a one-way flow, with some thirty species of fish, twenty decapod crustacean species, and forty molluscan species invading the Mediterranean. The paucity of Mediterranean forms making the return journey is explained by the hypersalinity of the Red Sea and of the Bitter Lakes along the line of the canal. The elevated salinities were clearly less of a hindrance to Red Sea euryhaline species than they were to Mediterranean species.

7.5 FAUNAL PROVINCES

Biogeography is concerned with the areal distribution of living animals and plants and with the controls that influence that distribution. An animal or plant community is an association of species living within what biologists call a **biotope**—a **habitat** having a certain range of environmental parameters. The boundaries between adjacent biotopes are often set by subtle controls such as slight differences in water temperature or salinity, food supply, or interspecific competition or even by differences between a north- or south-facing hillslope.

In considering the geographic range of a species, Valentine (1977) has distinguished between the *prospective range,* which includes the geographic areas where all the environmental requirements are met and where the species *could* flourish, and the *realized range,* which is the region that is, in fact, occupied by the species. For example, a freshwater mollusc inhabiting a series of lakes in Europe undoubtedly would be able to colonize similar environments in North America given the opportunity. There are, of course, numerous examples of human interference with natural distributions, usually with unhappy consequences. What Valentine called the *local range* comprises that region which is occupied by a species and which is circumscribed by ecologic barriers marking the limits of the biotope. Thus, the realized range of a species comprises a number of local ranges, the barriers between which are occasionally crossed by individuals. If interchange between different local-range populations is frequent enough, gene flow is maintained. However, there clearly exists the possibility that over longer periods of time the barriers might become increasingly effective until portions of the gene pool become isolated within one or more local-range populations.

Communities are grouped within larger entities known as **provinces,** the boundaries of which are typically set by major ecologic barriers. In the marine environment these are climatic (water-temperature) zonal boundaries, ocean currents, deep and extensive oceans and, of course, land barriers. In the case of terrestrial organisms they include oceans, climatic zonal boundaries, and physiographic barriers such as mountain ranges and deserts. The role and significance of these various kinds of barriers has varied through geologic time as the configuration of continents and oceans, together with the climatic zonal gradients, have changed. For example, during the existence of Pangaea (roughly through the late Paleozoic), major barriers were few and provincialism in the faunas is seen to be very low. In contrast, at the present day, provincialism is high because many major physiographic barriers trend longitudinally with the continental masses; at the same time, there are unusually high climatic zonal gradients that are latitudinal in their effect. Climatic zonation is undoubtedly the chief control of provincialism. This is very obvious in the case of terrestrial faunas and floras, but in the marine environment ocean currents, water depths, and salinity variations are also important influences.

Just as in the biosphere at the present day, fossil faunas and floras may show marked provincialism in the distribution of species. Thus, a given faunal province will contain a particular assemblage of species that distinguishes it from an adjacent province. Those species limited to a particular province are said to be **endemic.** Sometimes marked geographic or other barriers separate the provinces. Often, however, the boundaries are more or less gradational and certain faunal or floral elements in one province may persist for a short distance into the territory of the adjacent province. There then will be a transition zone in which the more tolerant (**eurytopic**) species of the respective provinces are intermingled. Still other species appear to ignore the barriers altogether and are equally abundant throughout several provinces. Such pandemic species within the fossil record are clearly useful in long-distance correlation. Sometimes the widespread distribution is due to necroplanktonic effects and has little to do with ecologic factors. Modern marine faunal provinces are usually distinguishable on the basis of characteristic species assemblages (Fig. 7.10). For fossils, faunal provinces are usually based on the distribution at the family or occasionally the generic level and so are more broadly defined. A study by Campbell and Valentine (1977), however, has shown that the majority of modern marine

faunal provinces can still be discerned at the generic and usually also at the family level. This means that modern and ancient provinces are clearly compatible in significance.

The boundaries of fossil marine provinces are seen to shift with time, usually because of changes in the configuration of oceans and continents and paleolatitudinal shifts resulting from lithospheric plate movements. The degree of provincialism also varies with time (Fig. 7.11). Periods during which climatic gradients were shallow, and/or there was continuity of shelf seas, were characterized by a large proportion of pandemic species and provincialism was less marked. For such time intervals long-distance biostratigraphic correlation would tend to be easier, although, as is discussed later, the more uniform environments often resulted in lower speciation rates and consequently a tendency toward a reduced sensitivity in zonation. Although the concept of faunal provinces is well rooted in the literature, there is considerable controversy as to the definition and validity of such provinces. Ager (1971) argued that the distribution of faunas should be considered only in terms of ecologic factors operating at various levels and expressed reservation that the faunal provinces, recognized by many workers, exist at all. He also pointed out that, in terms of worldwide distribution of certain taxonomic groups, the apparent provincialism seen in comparing faunas at the generic and specific levels may often be more an artifact of nationalistic and parochial taxonomy in which members of what is, in fact, a single species are given a multiplicity of local names.

7.6 BIOSTRATIGRAPHY AND TAXONOMY

Biological and Paleontological Species

It cannot be emphasized too strongly that biostratigraphy, in common with all biologic

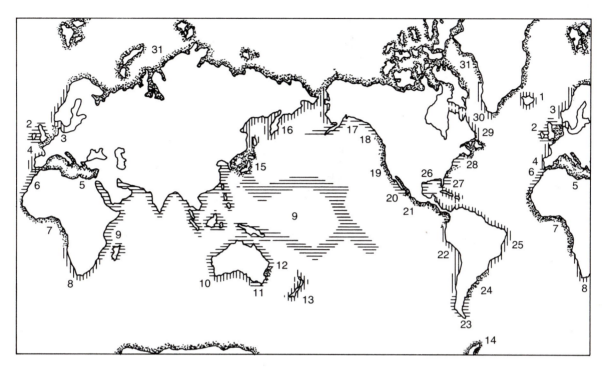

FIGURE 7.10
Marine molluscan provinces of the continental shelves of the world. 1—Norwegian; 2—Caledonian; 3—Celtic; 4—Lusitanian; 5—Mediterranean; 6—Mauritanian; 7—Guinean; 8—South African; 9—Indo-Pacific; 10—South Australian; 11—Maugean; 12—Peronian; 13—Zealandian; 14—Antarctic; 15—Japonic; 16—Bering; 17—Aleutian; 18—Oregonian; 19—Californian; 20—Surian; 21—Panamic; 22—Peruvian; 23—Magellanic; 24—Patagonian; 25—Caribbean; 26—Gulf; 27—Carolinian; 28—Virginian; 29—Nova Scotian; 30—Labradorian; 31—Arctic. (From Valentine, J. W., EVOLUTIONARY PALEOECOLOGY OF THE MARINE BIOSPHERE, © 1973, p. 356. Reprinted by permission of Prentice-Hall, Inc., Englewood Cliffs, N.J.)

and paleontologic studies, begins with taxonomy. An accurate determination of species is fundamental. When speaking of species, it is very obvious that the paleontologist's definition cannot be the same as that of the biologist. The criteria used with living organisms are many and varied, but among the numerous ways species are defined, in theory at least, emphasis is placed upon the necessity for demonstrating reproductive isolation of interbreeding natural populations. With fossil material, there is clearly no way of establishing this; the paleontologist has to rely solely on the morphology of the skeletal parts that have been preserved. It is assumed that the differences between any two such **morphospecies** should be of the kind and degree to be expected if they did belong to two reproductively isolated populations; it must be admitted, however, that this is only an educated guess.

Systematics long has tended to be a battleground between "lumpers" and "splitters." Both approaches when carried to extremes naturally tend to obscure the real relationships between different taxa. In the study of

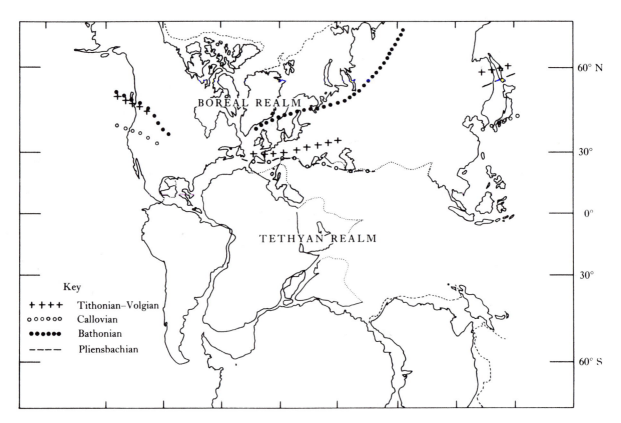

FIGURE 7.11
Approximate boundaries of the Tethyan and Boreal realms at different times in the Jurassic. (From Hallam, A., 1975, Jurassic environments, Fig. 10.1, p. 200. Reprinted by permission of Cambridge University Press, Cambridge, England.)

many groups of animals, particularly fossil groups, there was an early phase of extreme splitting. This led to the recognition of vast numbers of species and subspecies based upon very small morphological differences. Often such determinations were based upon analyses of very few individuals, sometimes a single specimen. A better understanding of geographic variations, ecologic controls, and phylogenetic relationships has usually resulted in a great reduction in the number of genera and species. Many of the early-described "species" of ammonites, for example, have proven to be merely examples of infraspecific variation and sexual dimorphism.

Westermann (1966), for instance, was able to show that no fewer than 78 previously described "species" and "subspecies" of ammonite were actually variants of a sexually dimorphic single species.

Homeomorphy

A not uncommon phenomenon in the evolutionary history of many animal groups is parallel evolution, in which two groups that originally arose from a common ancestor, and initially diverged, later evolved along approximately the same lines. Sometimes this process went a step further: the two separate lines tended to converge and, with time, began to look more and more alike. This was presumably a response to similar environmental influences acting upon what was often a relatively limited potential for morphologic variability. Examples occur in many invertebrate groups, including the corals, brachiopods, cephalopods, and crinoids. Species that look alike but which are not related below the family level have been termed **homeomorphs.** A potential source of confusion to taxonomists is the case of isochronous homeomorphs, which appear at roughly the same time. In biostratigraphy, particularly troublesome are cases of heterochronous homeomorphy. Here the dating and correlation of strata have been based upon supposed zonal indicator species that were actually unrecognized homeomorphs of species having ranges through earlier or later time. For example, the nautiloid genus *Permoceras,* from the late Permian, is an almost perfect homeomorph of *Pseudonautilus* of late Jurassic age (Sweet, 1964), and *Liroceras,* of Mississippian to Permian age, is virtually identical to *Eutrephoceras,* of late Jurassic to Miocene age. Among the brachiopods, Cloud (1948) described *Tetractinella* from the middle Triassic as a homeomorph of the late Jurassic *Cheirothyris.*

7.7 CONCLUSION

Biostratigraphic studies are inevitably much concerned with the range through time of given species. Range charts figure prominently in all studies. It should be remembered, however, that such charts are only a means to an end—the real object is to use them for correlation purposes. The geographic range of zonal species and the factors affecting their dispersal become of major importance in both local and regional stratigraphy. The establishment of time-stratigraphic divisions that are of global significance may prove difficult or even impossible in some

parts of the geologic column; nevertheless, this is what stratigraphers seek as the ultimate goal.

REFERENCES

Ager, D. V., 1971, Space and time in brachiopod history, *in* Middlemass, F. A., Rawson, P. F., and Newall, G., eds., Faunal provinces in space and time: Liverpool, Seel House Press, p. 4–110.

Arkell, W. J., 1956, Jurassic geology of the world: London, Oliver and Boyd, 608 p.

Campbell, C. A., and Valentine, J. W., 1977, Comparability of modern and ancient marine faunal provinces: Paleobiology, v. 3, p. 49–57.

Cloud, P. E., Jr., 1948, Some problems and patterns of evolution exemplified by fossil invertebrates: Evolution, v. 2, p. 322–350.

Eldredge, N., and Gould, S. J., 1977, Evolutionary models and biostratigraphic strategies, *in* Kauffman, E. G., and Hazel, J. E., eds., Concepts and methods of biostratigraphy: Stroudsburg, PA, Dowden, Hutchinson and Ross, p. 25–40.

Jablonski, D., and Flessa, K. W., 1986, The taxonomic structure of shallow-water marine faunas: Implication for Phanerozoic extinctions: Malacologia, v. 27, p. 43–66.

Kauffman, E. G., 1975, Dispersal and biostratigraphic potential of Cretaceous benthonic Bivalvia in the Western Interior, *in* Caldwell, W. G. E., The Cretaceous System in the Western Interior of North America: Geol. Assoc. Canada Spec. Paper 13, p. 163–194.

Keigwin, L. D., Jr., 1982, Isotopic paleooceanography of the Caribbean and East Pacific: Role of Panama uplift in Late Neogene time: Science, v. 217, p. 350–353.

Kraueter, J. N., 1974, Offshore currents, larval transport, and establishment of southern populations of *Littorina littorea* along the U.S. Atlantic Coast: Thalassica Jugoslavica 10, p. 159–170.

Mayr, E., 1969, Principles of systematic zoology: New York, McGraw-Hill, 428 p.

———— 1971, Populations, species and evolution: Cambridge, MA, Harvard Univ. Press, 453 p.

Petuch, E. J., 1982, Geographical heterochrony: Contemporaneous coexistence of Neogene and Recent molluscan faunas in the Americas: Palaeogeog., Palaeoclim., Palaeoecol., v. 37, p. 277–312.

—— 1988, Neogene history of tropical American mollusks: Charlottesville, VA, Coastal Education and Research Foundation, 217 p.

Scheltema, R. S., 1977, Dispersal of marine invertebrate organisms: Paleobiogeographic and biostratigraphic implications, *in* Kauffman, E. G., and Hazel, J. E., eds., Concepts and methods of biostratigraphy: Stroudsburg, PA, Dowden, Hutchinson and Ross, p. 73–108.

Simpson, G. G., 1952, Probability of dispersal in geologic time: Am. Mus. Nat. History Bull., v. 99, p. 163–176.

Sweet, W. C., 1964, Nautiloidea-Orthocerida, *in* Moore, R. C., ed., Treatise on invertebrate paleontology, part K, Mollusca 3: Geol. Soc. America and Univ. Kansas Press, p. K216–K457.

Urquhart, F., 1960, The monarch butterfly: Toronto, Univ. of Toronto Press, 361 p.

Valentine, J. W., 1977, Biogeography and biostratigraphy: Stroudsburg, PA, Dowden, Hutchinson and Ross, p. 143–162.

Vermeij, G. J., 1978, Biogeography and adaptation: Patterns of marine life: Cambridge, Harvard Univ. Press, 332 p.

Vermeij, G. J., and Petuch, E. J., 1986, Differential extinction in tropical American molluscs: Endemism, architecture, and the Panama land bridge: Malacologia, v. 27, p. 29–41.

Westerman, G. E. G., 1966, Covariance and taxonomy of the Jurassic ammonite *Sonninia adicra* (Waagen): Neues Jahrb. Geologie u. Paläontologie Abh., 124, p. 289–312.

Wright, S., 1967, Comments on the preliminary working papers of Eden and Waddington: Nistar Inst. Symp. Mon. 5, p. 117–120.

8

SPECIES THROUGH TIME

Race after race resigned their fleeting breath—
The rocks alone their curious annals save.

Timothy Abbott Conrad

8.1 INTRODUCTION

The fundamental taxonomic unit used in any refined biostratigraphy is the species. This chapter will be concerned with the temporal parameters of the species as discerned by the paleontologist. Of particular concern is the relationship between the species and time because an understanding of this is fundamental to biostratigraphy. It is inevitable that evolutionary mechanisms must also be discussed, but it should be clear that any more than a cursory review of that subject is beyond the scope of this text.

Recent years have seen an apparent resolution of an old conflict between what might be called the classical neo-Darwinian model of gradual phyletic change and the fossil record, which has conspicuously failed to turn up the evidence for it. The doctrine of gradualism has always found strong support from biologists who saw the emergence of new species and higher taxa as inevitably the results of extrapolation of the microevolutionary changes that could be observed in extant populations or even manipulated in the genetics laboratory. The lack of direct fossil evidence for populations gradually changing over long spans of time was always ascribed to the fragmentary nature of the stratigraphic record. That there are gaps in sedimentary successions has been freely admitted. However, largely because of the gradualistic view of the speciation process, a rather extreme notion was held by some workers that the paleontological record contained in the stratigraphic column must be so fragmentary as to be virtually useless. This is not true. In fact, it is becoming increasingly apparent that speciation is a punctuated process.

8.2 DECLINE OF GRADUALISM

In discussing the principle of the cline and continuity through a common gene pool, we have been considering a distribution through space. It is now necessary to apply the same principle within the time dimension. Just as the cline is a geographic entity, so the **lineage** can be considered a temporal one. A lineage also connects extreme variants within a common gene pool, but the connection is through time (Fig. 8.1). That is to say, variant B at the end of the lineage is a descendant of ancestral variant A at the beginning. Taken separately, these two variants might, although often do not, have morphological differences from each other that would normally be accepted by a paleontologist as sufficient to denote separate species. In fact, they are genetically connected through time from an ancestral to a descendant gene pool, and if variant B has completely replaced variant A, **anagenesis,** or *phyletic evolution,* has occurred.

At any given time, the variability in any morphologic character in a population can be expressed as a normal statistical distribution about a mean. A lineage may show a progressive change through time of this mean. A lineage in which one species or subspecies gradually evolves from another by a progressive shift in one or more morphological parameters is a process of speciation by **phylo-**

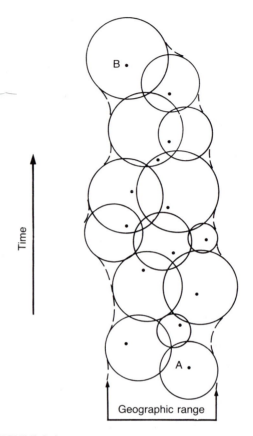

FIGURE 8.1
Gene pool in space and time. Circles represent local populations, black dots are individuals within successive generations. Descendant B is descended from ancestor A; no morphologic difference between A and B is implied or necessary.

genesis, what Eldredge and Gould (1972) have called **phyletic gradualism,** or anagenesis, in current terminology. If such a lineage were preserved in its entirety in a sedimentary succession, the statistical data on the morphological trends could be plotted graphically and numerous arbitrarily defined species might be recognized (Fig. 8.2). Such a succession would clearly represent a long period of more or less continuous sedimentation. Any progressive evolutionary changes seen in a lineage could be graphed as a

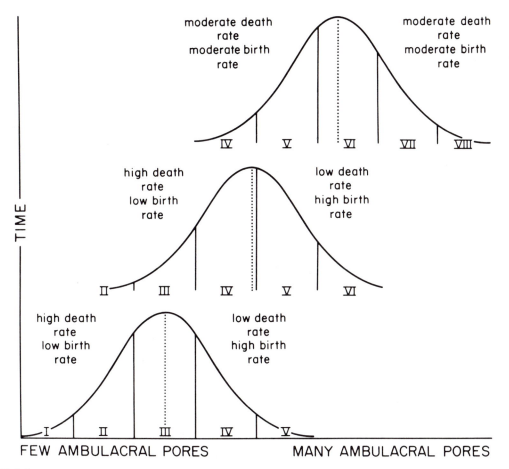

FIGURE 8.2

Arbitrary definition of a species. An example using the heart urchin *Micraster*. Each successive population has a different mix of individuals and a different average when counting the ambulacral pores. Although each population might be considered morphologically distinct enough to be considered a separate species, there is, in fact, some overlap between the three populations. Note that individuals described as type IV or V are found in all three populations. (From Beerbower, J. R., SEARCH FOR THE PAST: An introduction to paleontology, 2nd ed., © 1968, Fig. 6.4, p. 141. Reprinted by permission of Prentice-Hall, Inc., Englewood Cliffs, NJ.)

regression line, with changes in a particular feature plotted against time. Such a regression line, depicting, for example, a progressive increase in size, would be a continuous line only if a more or less complete sample of individuals within the lineage were present. Any significant hiatus would obviously be indicated by a break in the regression lines plotted for the various morphological shifts manifested in the lineage. If there is no significant time lag in the spread of a species through its geographic range, then a similar lineage, or even parts of it, found at a distant location could be correlated precisely, provided that local environmental factors did not introduce new effects.

Lineages clearly are potentially very precise biostratigraphic tools; unfortunately they are rare. The very refined zonal subdivision of the Cretaceous, using different morphologic stages in evolving lineages of irregular echinoids, particularly in the phyletic series of *Micraster, Echinocorys,* and *Offaster-Galeola* (Ernst and Seibertz, 1977), is an example

(Fig. 8.3). Numerous other examples of lineages that supposedly demonstrate phyletic evolutionary changes have been described at one time or another, but in virtually all cases later work has thrown doubt on their validity. Among the relatively recent studies was one by Hallam (1978), working with Jurassic pelecypods. This work shows that, although a mechanism of phyletic gradualism was apparently operating, it seemed to be control-

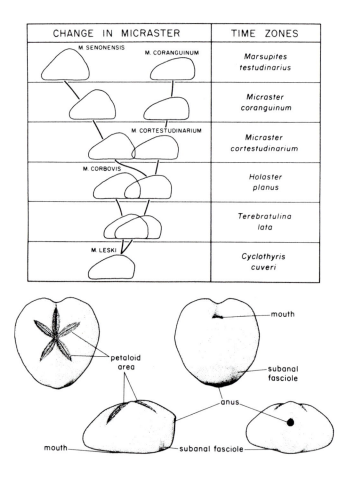

FIGURE 8.3

Evolutionary trends in the echinoid genus *Micraster* from the Cretaceous chalk; top, changing shape of the test with time; bottom, general anatomical features of *Micraster*. (From Beerbower, J. R., SEARCH FOR THE PAST: An introduction to paleontology, 2nd ed., © 1968, Fig. 6.2, p. 136. Reprinted by permission of Prentice-Hall, Inc., Englewood Cliffs, NJ.)

ling only the overall size, and a phyletic size increase was not accompanied by anything other than minor shape changes (Fig. 8.4). A similar conclusion was reached in a study of *Hyopsodus,* a small Eocene condylarthran mammal, by West (1979), who saw virtually no morphologic evolution except in size increase for over 2 m.y. (Fig. 8.5). It should be added that this is not the conclusion of Gingerich (1977); his findings apparently demonstrated gradualistic phyletic change within several lineages. In other cases, the morphological changes appear, on closer inspection, to be not so much unidirectional as oscillatory. In the short term, portions of the lineage might demonstrate some progressive shift in a morphological character, but extrapolated over a longer span, the descendant variant in the end looks very similar to the ancestral form.

The gradual change from one species to another, and the sorting out by the taxono-

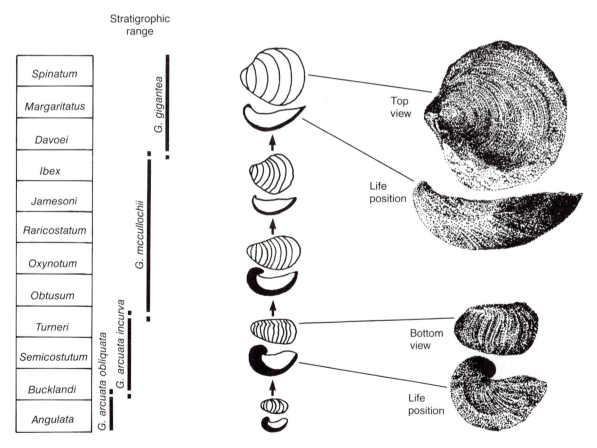

FIGURE 8.4

Lineage in the Early Jurassic coiled oyster *Gryphaea.* There is an apparently gradual trend toward increased size; at the same time, the shell becomes thinner and flatter, probably as an adaptation to a softer substrate and increased water movement. (Redrawn after Hallam, A., 1968, Morphology, paleoecology and evolution of the genus *Gryphaea* in the British Lias: Philos. Trans. Royal Soc., London, B. v. 254, p. 91–128.)

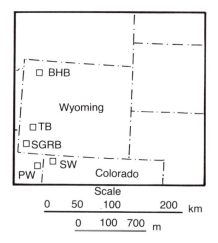

Location map for *Hyopsodus* assemblages in Wyoming, Utah and Colorado. (BHB - Bighorn Basin, SGRB - Southern Green River Basin, TB - Tabernacle Butte. PW - Powder Wash, SW - Sand Wash).

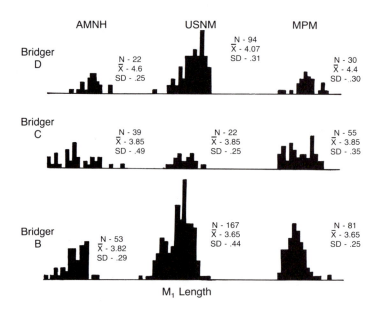

M₁ Length

FIGURE 8.5

Apparent stasis in early Eocene condylarthran *Hyopsodus*. In vertebrates, the teeth are usually the most diagnostic and (fortunately) the best preserved part of the skeleton. In this diagram, the length of the lower first molar is measured in three populations collected at three different horizons within the Bridger Formation. The time span represented between Bridger B, the lowest of the lithostratigraphic units, and Bridger D, the highest, is estimated at about 1 m.y. Note the similarity in size distribution between all three horizons, indicating no significant morphologic change over the time span involved. (After West, 1979, Fig. 1, p. 253, Fig. 5, p. 256, and Gingerich, 1977, Patterns of evolution in the mammalian fossil record, *in* Hallam, A., ed., Patterns of evolution as illustrated by the fossil record, Fig. 12, p. 492. Reprinted by permission of Elsevier Science Publishers B. V.)

mist of subspecies and species by arbitrary statistical boundaries is certainly not the usual way the biostratigrapher encounters the fossil record! Instead, when a new species appears within a given stratigraphic succession it usually appears suddenly, persists for a given interval of strata (Fig. 8.6), and then as suddenly disappears, never to be seen again. In many cases no perceptible evolutionary changes can be detected in a species during its entire temporal existence. Eldredge (1971, 1972), for example, found that certain trilobite species and even subspecies remained essentially unchanged throughout their stratigraphic ranges. Sometimes this involved an extended period, as with *Phacaps rana rana,* which apparently persisted for 2 m.y. Numerous other examples of species that lasted for considerably longer periods are cited later in this chapter.

Appearance and Disappearance of Species

However a species is defined and whatever morphological parameters are assigned to it, there must have been a moment in time when it first appeared and another when it disappeared. In stratigraphy these moments are rarely discoverable. The first occurrence of a new species within a sedimentary succession, assuming we are not dealing with an arbitrarily defined morphospecies within an evolutionary lineage, is not only abrupt but is unlikely to represent the time of its first appearance on earth. Very commonly the appearance of a species coincides with a change in lithology; this may be because the species is facies-linked and migrated from elsewhere as the favorable habitat expanded. In other cases the coincidence of appearance (or disappearance) with a change in lithology may simply reflect a change in conditions of burial, fossilization, and/or preservation (Fig. 8.6). Sometimes there is no apparent change in lithology. In such cases, it may be that

more subtle alterations in the environment, such as changes in water temperature or salinity, might be sufficient to cause the appearance or disappearance of a particular species, but not to affect the character of the sediment being deposited.

Punctuated Nature of the Fossil Record

The rather compelling evidence from the rocks, together with the fact that certain evolutionary lineages cited in early biostratigraphic studies have been shown to be invalid, have led some workers to doubt that such lineages exist at all. Consequently, the mechanism of speciation by phylogenesis demonstrated in a lineage has also been called to question. Supporters of phylogenesis (Schopf and Hoffman, 1983; Gingerich, 1985), on the other hand, argue that the abrupt appearances and disappearances of species are explained as due to the fragmented nature of the fossil record. They suggest that the appearance of stasis rather than gradual change is an artifact of the way stratigraphic data are sequentially grouped together. It is true that the preservation of a complete evolutionary lineage would require a long period of uniform environment and continuity of deposition. Such conditions rarely occur; more commonly, there are gaps in the fossil record due to nondeposition, erosion, nonrepresentation in unsuitable facies, or nonpreservation. The gaps would effectively break up a lineage into artificially separate temporal populations, each of which usually would be considered a separate species by the paleontologist (Fig. 8.7). Thus, it is suggested, the lineage is obscured but could be demonstrated if the gaps between the fragments were bridged by extrapolation.

In considering these "gaps," the problem is largely one of scale. As described in Chapter 1, sediment accumulation, in virtually all depositional environments, is markedly episodic, but it is difficult to ascertain the mag-

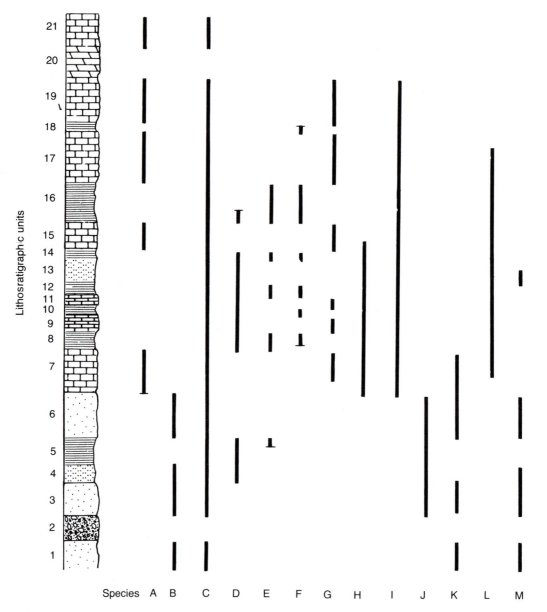

FIGURE 8.6

Appearance and disappearance of species within a stratigraphic succession. Species A,B,D,E,F,J, and M are all markedly facies-controlled. A appeared at the base of unit 7, but this is only a first appearance for this particular locality. Species E and F first appeared in units 5 and 8, respectively, and may be true first appearances because they are not present at lower horizons with similar facies. Species D's disappearance in unit 16 may be a true extinction event. Species C is not facies-controlled, being found in all facies except that of unit 20, where its absence is probably due to destruction by dolomitization. Species I's nonappearance in unit 21 might be due to its extinction. Species L and M are found in various facies, have apparently short time ranges, and might be good zonal species. Species K's occurrence seems to indicate that it also could be a zonal species. Its absence in units 4 and 5 is fortuitous. The absence of fossils in unit 2 is likely due to unsuitable facies and/or postdepositional destruction. The absence of fossils in unit 20 is likely due to destruction during dolomitization.

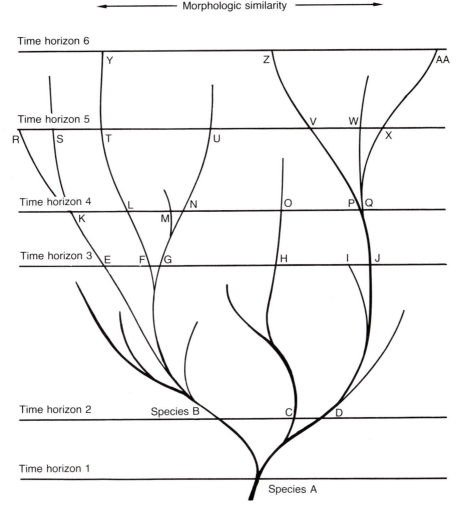

FIGURE 8.7

How gaps in the stratigraphic record can be used to define species within a hypothetical gradualistic lineage. This is the traditional "family tree" diagram widely used in demonstrating phylogenetic relationships in fossil lineages. The six time horizons depicted here are represented in one particular succession or area, and only the base and the upper branches of the tree can be determined from the assemblages collected there. The evolutionary radiation shown between time horizons 2 and 3 is based on evidence from faunas collected elsewhere. The tree's various branches are drawn by connecting those species that have the most morphological features in common and separating those that are least similar. For example, species C and D are more similar to each other than either of them is to species B, but they are all similar enough to species A, which is found in older strata, to be considered as originating from that species as the common ancestor. Species F and G, M and N, and P and Q are also closely related. Further taxonomic grouping into genera, families, etc. is done in the same way. Thus, all the species on the branches above species B might be grouped into a genus, separated from those on the C and D branches. The C and D branches might, in turn, be considered subgenera. These procedures are highly subjective; using the same collections, different workers might construct different trees. Even when the sedimentary record is relatively complete or contains, perhaps, a lot of small gaps, examples of an actual fossil record of the gradational changes supposed, by the gradualist, to occur along one of the "branches" of the tree are so rare as to be virtually nonexistent.

nitude of the intervals between the episodes. What is not clear is the scale involved when comparing the time recorded with the time not recorded. By analogy with a book, is it a case of, say, every fifth page preserved, every fifth word, or every fifth paragraph? Because of this uncertainty, the sedimentologists have usually been unable to provide a satisfactory answer *per se* when faced with the obvious gaps in the fossil record. Supporters of phyletic gradualism have, in effect, supplied their own answer to the question of scale: the gaps in the stratigraphic record obviously must be big—clearly, whole chapters are missing! Translated into real time, such gaps are so large as to be unacceptable in many well-documented sedimentary successions. It is difficult to accept them, for example, in thick epicontinental limestone successions, in which an impressive degree of stasis in fossil lineages in turn demonstrates long-continued uniform bottom conditions, with only short-term pulsatory variations in sediment accumulation. The discontinuities in the fossil record may be real enough, but what is difficult to find is matching evidence in the sedimentary succession for gaps of the magnitude required to support a gradualistic view of the biostratigraphic record. From a purely geologic standpoint, the physical evidence indicates that sedimentary successions, in general, seem to contain rather more small gaps and fewer very large ones. The fact remains that, when all the fossil evidence is assembled and all the allowances made, the stratigraphic record as a whole does not do a very good job of supporting phyletic gradualism.

An alternative mechanism in macroevolution has been proposed (Eldredge and Gould, 1972; Gould and Eldredge, 1977; Gould, 1980), suggesting that the sudden appearances and disappearances of species within the stratigraphic succession can be only partially explained by the fragmented fossil record. Rather than external causes like facies changes, nondeposition, or erosion, the abrupt appearances and disappearances can be ascribed to a mechanism of speciation that Eldredge and Gould called *punctuated equilibrium.*

8.3 THE PUNCTUATED EQUILIBRIUM MODEL

In this model, the parent stock of a species exists within a particular territory practically unchanged through what may be protracted intervals of time. A descendant species of the original population arises suddenly from a peripheral variant genotype that has become isolated from the parent gene pool. The new species eventually migrates back into the ancestral range after the parent species has become extinct and vacated its ecologic niche (Fig. 8.8). Alternatively, under changing environmental conditions, the new species may successfully compete with the ancestral species. It is in the small, geographically isolated populations, probably under environmental stress, that rapid evolutionary changes likely occur. Isolation from the stabilizing influence of the parent gene pool, described earlier in the founder principle, is probably an important factor. It has been suggested (Williamson, 1980) that the geographic isolation also serves to protect the emerging transitional species at a time when it would be vulnerable to competition from the unaltered descendants of the parent stock. The fossil record of emerging populations produced by this mechanism would show species existing for long periods with no change, then being replaced abruptly by new species (Fig. 8.9). This is exactly what the biostratigrapher normally finds, so, on that basis alone, the punctuated equilibrium model has much to recommend it.

Even under relatively stable and slowly changing conditions, a punctuated rather than a gradualistic mechanism may be likely. Dodson and Hallam (1977), in a model based

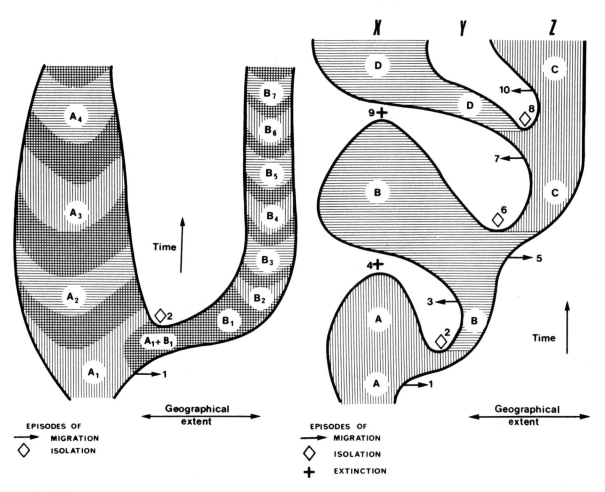

FIGURE 8.8

Punctuated equilibrium and gradualistic evolutionary models compared. Model A (left): speciation by phyletic gradualism. Migration begins at time level 1 and geographic isolation is established by time level 2. The two branches ultimately become genetically separated. Each time-separated grade (A_1, A_2, etc. and B_1, B_2, etc.) is a chronological subspecies. Model B (right): speciation by punctuated equilibrium. Species A in region X gives rise at time level 1 to a branch that migrates to region Y; when it becomes isolated at level 2, it changes morphologically into species B. This species then migrates back to region X at time level 3, replacing species A, which becomes extinct at level 4. Species B also migrates to region 2 at level 5 and changes to species C after isolation at level 6. Similarly, a branch of species C migrates back to region Y at level 7, changes to species D after isolation at level 8, and then invades region X, displacing species B. Region X is thus characterized by three successive species A, B, and D, which abruptly replace each other. (From Sylvester-Bradley, P. C., 1977, Biostratigraphical tests of evolutionary theory, *in* Concepts and methods of biostratigraphy, Kauffman, E. G., and Hazel, J. E., eds., Concepts and methods of biostratigraphy, Fig. 1, p. 43, and Fig. 2, p. 45. Reprinted by permission of Dowden, Hutchinson and Ross, Stroudsburg, PA.)

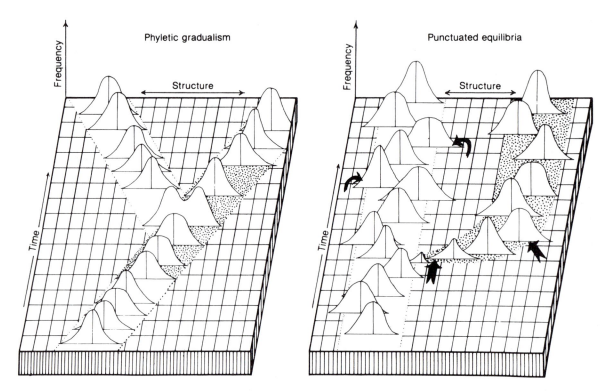

FIGURE 8.9
Punctuated equilibrium model. (From Vbra, E., 1980, Evolution, species and fossils:
How does life evolve?: South African Jour. Science, v. 76, Fig. 1, p. 62. Reprinted by
permission.)

upon Thom's "fold catastrophe" theorem, showed that along a slowly changing environmental gradient, the evolutionary response is likely discontinuous. New species arise as a result of quantum evolution in peripheral populations living beyond a given environmental threshold. The new species then invades the range of the ancestral population and replaces it within a radius limited by its fitness.

As explained by Pielou (1979, p. 85), the response of a given species to its environment can be shown statistically as a *fitness curve,* which will be a bell-shaped curve showing the distribution of a given morphologic feature through the population, as shown in Figure 8.10. If the species expands

its range, it will eventually occupy areas with a less favorable environment, but a stabilizing selection factor will prevent progressive displacement of the peak in the fitness curve, even though it becomes progressively lower and eventually bimodal. Along an environmental gradient, a stacked series of fitness curves becomes a *fitness surface,* also shown in Figure 8.10. A sudden change occurs at the point along the environmental gradient when the secondary peak of the bimodal curve becomes the primary peak. This is the "catastrophe" of the fold catastrophe theory. The stabilizing factor is what causes the sudden switch and explains why a series of isolated subpopulations produce just two species rather than a completely gradational

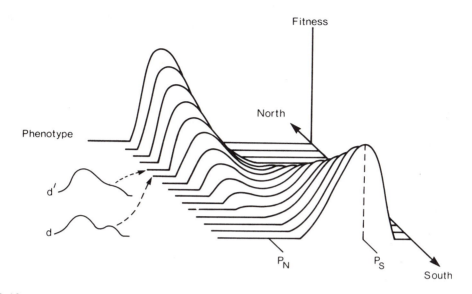

FIGURE 8.10

Fold catastrophe model of speciation. Each curve shows fitness as a function of the phenotype at a particular latitude. A species spreading north from a range initially confined to the far south will continue to have its southern phenotype P_S until there is no longer even a local maximum at P_S on the fitness curve. This point, somewhere between d and d', marks the boundary at which its phenotype switches abruptly to the northern form, P_N. (From Pielou, E. C., Biogeography, Fig. 3.7, p. 86. © 1979 by John Wiley & Sons, New York.)

series of subspecies. This is essentially a process of allopatric speciation, although no geographic barriers, isolation, or marked environmental discontinuities are invoked.

Strong support for the punctuated equilibrium model is seen in one of the best-documented examples of a virtually complete fossil record of speciation, studied by Williamson (1981). He described a sequence of late Cenozoic lacustrine deposits in northern Kenya, where some 400 m of sands and silts contain a rich fauna of freshwater molluscs, including numerous lineages, both sexual and asexual. The pattern of morphologic variation through time is the same in all cases, with long-term stasis punctuated by very short episodes of marked change and the appearance of new species (Fig. 8.11). The time spans of intermediate forms between ancestral and descendant species were found to range from 5,000 to 50,000 years. Because living relatives of the snails can be studied, this is known to represent, in very general terms, some 20,000 to 30,000 generations. Within the stratigraphic succession are recorded several episodes when lake levels fell markedly. At such times, the lake environments were under stress and local populations would have been isolated. The fact that such episodes coincide in time with speciation events within the molluscan lineages is hardly surprising, because it is generally accepted that environmental stress and geographic isolation seem to function as "triggers" in the speciation process. Although there is an unusually complete sedimentary record and conditions would seem to be ideal, evidence for gradualistic phyletic

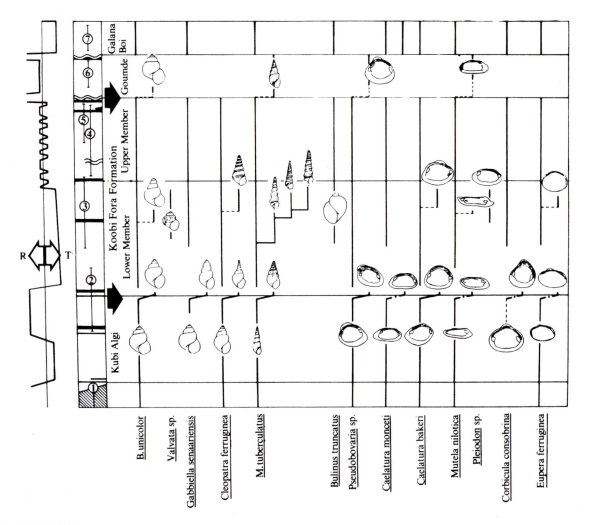

FIGURE 8.11
Patterns of evolutionary change in the Turkana Basin (Kenya) molluscan sequence.
The heavy line on the extreme left is a generalized indication of lake-level change
(T = transgression, R = regression). Note particularly (1) the simultaneous speciation
events in all lineages at the horizons indicated by the heavy arrows and (2) the adaptive
radiation of several stocks in the middle part of the sequence. (From Williamson, P. G.,
Paleontological documentation of speciation in Cenozoic molluscs from Turkana Basin.
Reprinted by permission from *Nature,* v. 293, Fig. 4, p. 442. Copyright © 1981, Macmillan Journals Limited.)

change is conspicuously absent. Instead, the
mechanisms documented fit the punctuated
equilibrium model; there is no doubt that the
Williamson study makes a very strong case
for its validity.

In terms of the fossil record as a whole, the
fragmentary nature of the stratigraphic column obviously accentuates the existing
punctualistic discontinuities in the fossil
succession. Thus, it is hardly surprising that

the record contains little evidence to support the traditional gradualistic evolutionary model in macroevolution. In particular, three features should be noted:

1. The extreme scarcity of gradualistic lineages in the geologic record, even those documented, seem to demonstrate only microevolutionary shifts and/or phyletic size increase.
2. Many fossil species typically show no morphologic changes from the beginning to the end of their stratigraphic ranges, even if long time intervals are involved. In other words, *stasis* rather than change seems to be the most characteristic feature of the fossil record.
3. "Missing links" connecting a species to a descendant species are virtually absent, and fossil evidence of intermediate forms connecting ancestral taxa to new higher taxa is lacking.

The fossil record is admittedly often incomplete, but there are enough examples of relatively complete fossiliferous sections, not to mention the very numerous cases of long-lived species, to demonstrate that stasis and not continuous change is an important feature in evolutionary history.

8.4 STASIS

There may be several explanations for stasis. It is possible, for example, that the apparent lack of microevolutionary changes during a species life span may reflect a relatively static environment over long periods of time. Our view of climatic and other environmental changes is, after all, colored by our Pleistocene experience. The rapid changes in climate and sea level, and steep climatic zonal gradients, are all very atypical of much of geologic time. It could be argued also that the stasis seen through the life of many fossil species is not a true reflection of genetic change. The paleontologists have no way of

determining this and would always prefer to assume a correlation between genetic and morphologic characteristics, if only to lend some measure of reality to the morphospecies. Ornithologists and icthyologists are two groups who should appreciate this particular point because their species are often defined on the basis of color, behavior, and other features that are simply not reflected in the skeletal morphology at all.

It is possible also that, as pointed out by Lewin (1980), the minor morphologic and genetic modifications seen by the biologist as microevolution may be simply oscillations about a mean. In the long term, and with the added factor of the fragmentary fossil record, this would be seen by the paleontologist as stasis. Perhaps the problem is again one of scale. The sudden appearance of a new species (the "instant" speciation so unpalatable to the gradualist) appears to be sudden only within the context of the fossil record. If, for example, a new species emerged over a period of 50,000 years this would probably be seen by the biologist as long enough for speciation by anagenesis. On the other hand, this time span, in many cases, represents less than one percent of the life of the species.

The punctualistic model would seem also to provide an adequate explanation for the lack of "missing links" in the fossil record. As Simpson (1953) suggested, the appearance of a new type often reflects an adaptive breakthrough with rapid evolution and radiation expansion into new territory. Because most major changes occur initially in small populations in relatively restricted geographic areas, this would clearly reduce the likelihood of discovery in the fossil record. As already noted, times of higher rates of evolutionary change, in shallow-water marine organisms, for example, frequently coincide with marine regressional phases that are typically marked by erosional breaks in the stratigraphic record. This obviously further increases the odds against preservation of transitional forms.

8.5 SPECIES LONGEVITY

There have been numerous studies of species longevity. Kennedy and Cobban (1976) and Kennedy (1977), for example, have reported that certain endemic species of the ammonite genus *Baculites* had life spans of between 500,000 and 700,000 years; species of *Scaphites* of between 600,000 and 700,000 years. More widely occurring (pandemic) species, on the other hand, apparently had considerably longer life spans; for example, *Baculites undatus,* 3.3 m.y., and *Hoploscaphites constrictus,* nearly 4.0 m.y. Even longer time ranges were cited for certain other ammonite species: 8.0 m.y. for *Austeniceras austeni,* over 20 m.y. for *Phylloceras serum,* and 25 m.y. for *Phylloceras thetys.* These figures, however, are certainly not typical of the ammonites as a whole—it is very obvious that the sophisticated zoning of the Mesozoic based on ammonites would not have been possible if they were. On the basis of average zonal durations, an estimate for ammonite species longevity in the Triassic and Jurassic is 1.2 m.y. and in the Cretaceous nearly 2.0 m.y. (Kennedy, 1977). Close agreement is seen in an estimate of 1.0 m.y. in Jurassic ammonite zones given by Hallam (1976). It is interesting to note that in the graptolites, another animal group that has provided sensitive zonal indicators, the species longevity is comparable. For the British Silurian graptolites, a mean duration of 1.9 m.y. has been calculated by Rickards (1977). Data for mean species longevity in several orders of Jurassic pelecypods show a range from under 1 m.y. to 63 m.y., with an overall mean of 15 m.y. (Hallam, 1976). In another study, the average species life of Cenozoic pelecypods was cited as 7 m.y. by Stanley (1977). In the case of mammals, Kurten (1959) calculated that certain Cenozoic species had life spans ranging from 0.3 to 7.5 m.y., whereas Stanley (1976) gave a figure of 1.2 m.y. for mean species duration for Plio-Pleistocene mammals of Europe.

Within one faunal assemblage, presumably sharing common environmental stresses, the speciation rate has been found to differ from species to species; this is apparently linked to variations in the response of different species to stress and to generation time. Work by Jackson (1977) has shown that in some groups the more geographically widespread and environmentally tolerant species tend to persist for relatively long time spans, whereas short-lived species seem to be the ones that are most susceptible to environmental stress. This is what Vrba (1980) called the *effect hypothesis,* in which the emphasis is on internal parameters rather than the environment, in influencing speciation rates and species longevity. In general, it is the highly specialized species that have shorter life spans because they occupy relatively restricted habitats and are thus likely to suffer rapid extinction in the event of even small changes in the environment. Speciation is frequent, with many different but closely related species coexisting in habitats that differ only in subtle details. In contrast, a generalist species—one with a tolerance for a wide range of environmental variables—will obviously be much less affected by changes in the environment. Speciation is infrequent and at any one time there is likely only one species occupying that particular habitat.

Environmental stress in terms of fluctuations of, for example, salinity, oxygen, and temperature is not always itself the sole factor; what is often important is the organism's ability to cope with the stress. Those species well-adapted to high-stress situations, as for example many brackish-water species, apparently are numbered among those with longer-than-average species life, as also are those species not subject to undue interspecific competition. Kauffman's (1967, 1977) studies of the Cretaceous of the western United States and Canada have indicated a close correlation between the timing of major marine transgressions and regressions and the speciation rates of several molluscs. During a marine transgression, when shallow-marine environments are expanding, and presumably environmental stress for many species is decreasing, it might be supposed that speciation rates would decline. This would be the intuitive model suggested by many studies. In fact, the actual situation has been shown to be much more complex. For example, in some cases the speciation rate actually increased during the early stages of the transgression, precisely at the time when enlarged habitat areas for marine organisms would presumably be reflected in a lessening

of environmental stress. This increase was due to the fact that certain species, what Kauffman (1977) has called *opportunistic taxa,* were able more rapidly to exploit those ecological niches made larger by the transgression.

The findings of these workers are of considerable value to the biostratigrapher because they introduce a new degree of sophistication into biostratigraphy, emphasizing paleoecological controls, as interpreted from lithofacies, and linking them directly to biostratigraphy. As can be seen in Figures 8.12 and 8.13, provided the various sedimentary

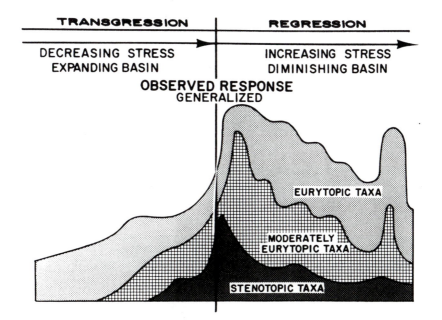

FIGURE 8.12

Evolutionary response of diverse Cretaceous molluscan lineages with different stress-tolerance limits to environmental changes consequent on sea-level changes in the Western Interior Cretaceous basin. The height of the curve represents speciation rates (species per m.y.) and also reflects species duration. During a marine transgression, there is a slow rise in speciation rate as a result of radiation among opportunistic species in expanding environmental niches and also because of competition. As regression proceeds and habitats contract, environmental stress increases. The peak of evolution in moderately eurytopic species near the maximum regression is in nearshore clastic facies faunas. (From Kauffman, E. G., 1977, Evolutionary rates and biostratigraphy, *in* Concepts and methods of biostratigraphy, Fig. 7, p. 137. Reprinted by permission of Dowden, Hutchinson and Ross, Stroudsburg, PA.)

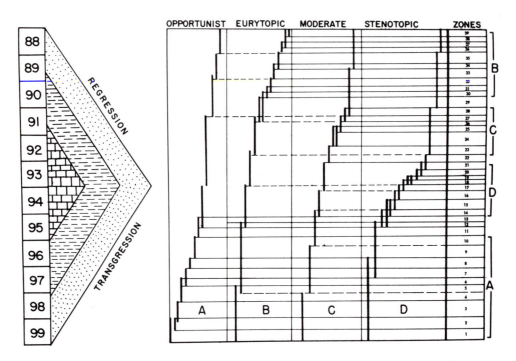

FIGURE 8.13

How the best biostratigraphic indicators may be selected within a transgressive/regressive succession. For example, during the late phases of the transgression, stenotopic species provide the means for a more sensitive zonation than do eurytopic species. (From Kauffman, E. G., 1977, Evolutionary rates and biostratigraphy, *in* Concepts and methods of biostratigraphy, Fig. 8, p. 139. Reprinted by permission of Dowden, Hutchinson and Ross, Stroudsburg, PA.)

facies of the transgressive/regressive cycles can be recognized (in practice usually not difficult), the biostratigrapher can, in making zonal divisions, predict those species that are likely to have the shortest range through time and, therefore, be the most sensitive zone fossils. Thus, during the initial stages of the transgression, the best zonal indicators would be found among species of the opportunistic group, whereas in the late transgressive phase, **stenotopic** species become more useful.

8.6 MASS EXTINCTIONS

Very large punctuations of the biostratigraphic record are seen in the periodic mass

extinctions of organisms. Not only are such catastrophic changes important in directing the pathway of evolution—which is, after all, the essential component in biostratigraphy—but the events themselves are potentially first-order time-horizon markers. Mass extinctions were once thought to be only quantitatively different from the "normal" or background extinctions that presumably have been a feature of evolutionary progress since life began. It is now realized, however, that they are qualitatively different and that they have played a special role in shaping large-scale evolutionary patterns. In effect, a modified evolutionary model is emerging in which the ongoing progress of change, involving a background extinction regime, is periodically interrupted by mass extinctions that are different in both kind and degree. Background extinctions are an essential part of evolutionary progress in that they are responsible for the elimination of those species that are less well adapted. This is the traditional "species selection" of Darwin. It is a type of constructive or positive selection process because, in theory at least, survivors are by definition "better" than those species that succumbed. The essential point is that the potential for survival of species is constantly being tested; this is what might be called the driving force in evolution. What makes the system work is that potentially advantageous traits can emerge by this selection process because they are heritable.

Mass extinctions are qualitatively different in that the features or traits that might be potentially advantageous to a species during such an extinction event are probably *not* heritable. It follows, therefore, that mass extinctions represent a selection process that is not constructive. It seems, at least to some workers, an entirely random process, and survivability depends on chance. As Raup (1986) pointed out, only if major biotic crises occurred frequently would the natural selection process be effective. It seems that the selectivity that operates during a mass extinction applies at a level above the species level and, as Jablonski (1986) suggested, ignores the individual adaptations that had enhanced the species during periods of background extinction. At the same time, other traits that might have had no survival value during "normal" times suddenly became important in mass extinction crises. If such advantageous features or traits do exist, they are certainly not very obvious. Mass extinctions, at times, seem to have eliminated almost the entire spectrum of organisms, in virtually every ecologic niche. As it is now becoming clear to paleontologists, if we are to understand mass extinctions, it is not the extinction event or those species that became extinct that we should be studying, but rather the survivors.

As well as considering the attributes of individual species, recent studies have turned to an examination of **clades,** or groups of species related phylogenetically. At this level, there is some suggestion that a pattern may be discernible. Preliminary studies by Raup (1986) and Jablonski (1986), for example, suggest that vulnerability to mass extinction events may have something to do with provinciality. That is to say, those clades that have a wide geographic extent are likely to do better during a mass extinction than those with a more restricted or provincial distribution, regardless of the number of species in the clade. A species-rich clade does not seem to have any particular advantage over a species-poor clade, either during background or mass extinction. On the other hand, it has been noticed that tropical species tend to suffer greater decimation during mass extinction than do those from higher latitudes. The bias may be partially a statistical artifact because tropical faunas tend to contain more species-rich clades, but it may be due also to the generally greater provinciality of tropical clades. Because mass extinctions alternate with periods of background extinction, selection through several mass extinctions would clearly favor those species with a combination of traits that would increase their survivability through both background-type and mass extinctions. As Lewin (1986) put it, "such a combination would be a sure route to success through the history of life." Whether future work will reveal such combinations of heritable traits is open to debate. At the moment, the geographic distribution factor that seems to be the only obvious trait with survival value during mass extinctions does *not* seem to be heritable (Jablonski, 1986).

In considering the survivors of a mass extinction, many have proved immensely successful in the post-event period. However, it should be remembered that this success

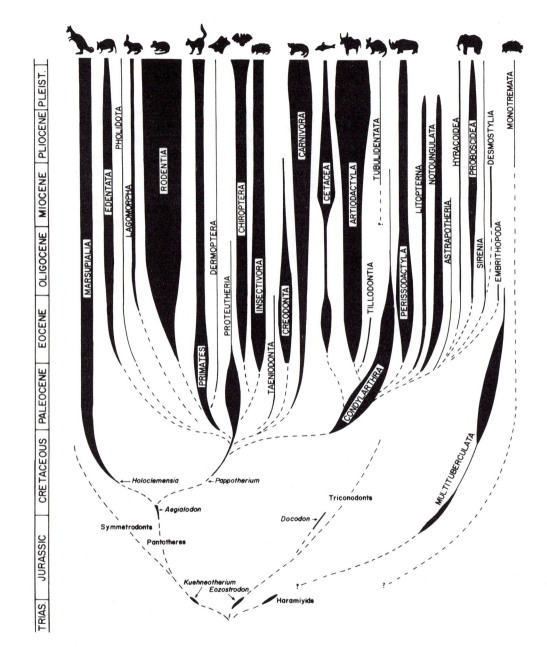

FIGURE 8.14
Evolutionary radiation of mammals in the Cenozoic era. (From Gingerich, P. D., 1977, Patterns of evolution in the mammalian fossil record, *in* Hallam, A., ed., Patterns of evolution as illustrated by the fossil record, Fig. 1, p. 471. Reprinted by permission of Elsevier Science Publishers B. V.)

might not necessarily reflect any particular superiority, but be due to the elimination of competition. After all, the mammals and the dinosaurs had existed side by side for some 100 million years; it was only after the dinosaurs had been removed from the scene that the great evolutionary radiation of the mammals began (Fig. 8.14). It also should not be thought that those species that survived a mass extinction event simply continued through the stratigraphic succession after the disappearance of their less fortunate fellows. Not all those species that disappeared at an extinction boundary became extinct. Many survivors may have been absent because of causes other than extinction, such as unsuitable facies or preservational factors. The eventual reappearance of these species illustrates what has been called the Lazarus effect. One example is seen at the Cretaceous-Tertiary palynological boundary in western North America, marked by the disappearance of most angiosperms. At the same time, the proportion of fern spores showed an abrupt increase (Tschudy et al., 1984; Nichols et al., 1986) before decreasing again as the angiosperm flora began to recover and recolonize the devastated areas (Fig. 8.15).

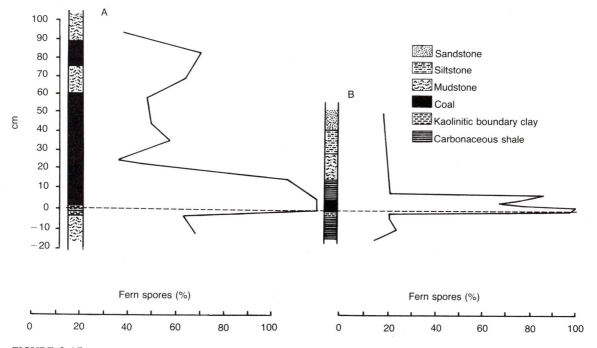

FIGURE 8.15

Percentage of fern spores at the Cretaceous-Tertiary boundary at two sites. A. Morgan Creek, Saskatchewan. B. Starkville North, Colorado. (A after Nichols, D. J. et al., 1986, Palynological and iridium anomalies at Cretaceous-Tertiary boundary, south-central Saskatchewan: Science, v. 231, 14 February 1986, p. 714-716, Fig. 1, p. 715. Copyright 1986 by the AAAS. Reprinted by permission; B after Tschudy R. H. et al., 1984, Disruption of the terrestrial plant ecosystem at the Cretaceous-Tertiary boundary, Western Interior: Science, v. 225, 7 September 1984, p. 1030–1032, Fig. 2, p. 1031. Copyright 1984 by the AAAS. Reprinted by permission.)

8.7 TIERS OF TIME

In viewing the history of life as a whole, Gould (1985) has emphasized the importance of mass extinctions. As he put it, "mass extinctions are more *frequent,* more *rapid,* more *extensive in impact,* and more *qualitatively different in effect* than our uniformitarian hopes had previously permitted most of us to contemplate." Gould's suggestion is that the progress of life is occurring at three different levels, or "tiers," with very different time scales (Fig. 8.16). Each one has a special mechanism operating virtually independently of the other two. At the first tier are those microevolutionary changes observed by biologists and demonstrated by studies in population genetics. They occur over time periods typically measured in thousands or tens of thousands of years. It is now widely believed that such microevolutionary changes are not necessarily cumulative and that there is no simple extrapolation from microevolutionary to macroevolutionary trends. Beyond the microevolutionary changes and probably overriding them, are those events taking place on the second tier and produced by the mechanism of punctuated equilibrium. These changes occur during time frames measured in millions of years. They seem to be decoupled from microevolutionary mechanisms and arise from a higher-order selection process affecting groups of species. Once the individual species appear on the scene, they may undergo little further change. Third-tier phenomena, occurring in a time frame of tens of millions of years, are the mass extinctions. Whatever the mechanism responsible, such catastrophes force the whole biosphere through narrow "bottlenecks." As new faunas and floras come to occupy the vacated ecologic niches, they are often seen to have compositions profoundly different than those populations that had preceded them.

The fact that a new form is "different" does not necessarily mean that it is "better." As we saw earlier, the survival of its ancestral stock past the bottleneck of the extinction event may have been entirely fortuitous. In the model that inevitably emerges, there is no obvious place for the concept of "evolutionary progress" as traditionally it has been defined. Lest this be misconstrued by those who harbor a certain lack of affection for Charles Darwin, it should be emphasized that this in no way denies the underlying theme of evolution and the efficacy of natural selection. In simplest terms, the punctuational model demonstrates that evolutionary progress does not proceed smoothly. As Gould (1985) pointed out, natural selection requires transitional forms, but they do not have to be part of an insensibly graded series. Mass extinctions, in particular, play a vital role. As Raup (1986) suggested, they may have a greater impact on evolution than do the processes operating between extinctions.

8.8 CONCLUSION

The punctuated equilibrium model and mass extinctions clearly introduce important modifications of the traditional Darwinian view of the history of life, but there is, in addition, a third modification that now may be emerging. This is concerned with interpretation rather than with new discoveries and questions one of the main tenets of Darwin's thesis. It commonly is believed, especially by paleontologists, that Darwin's discoveries were based to a large extent on fossil finds. This is not true. Darwin, in fact, spoke rather disparagingly of "the imperfections of the geological record"; what made a far greater impact on him was, according to Oldroyd (1980), Malthus's views on population. Given the socioeconomic climate of the Victorian age, this hardly is surprising, and, as Hsü (1986) suggests, it led Darwin to place too much emphasis on "the struggle for survival" as the main driving force in evolution. From this arose a common misconception that the "fit-

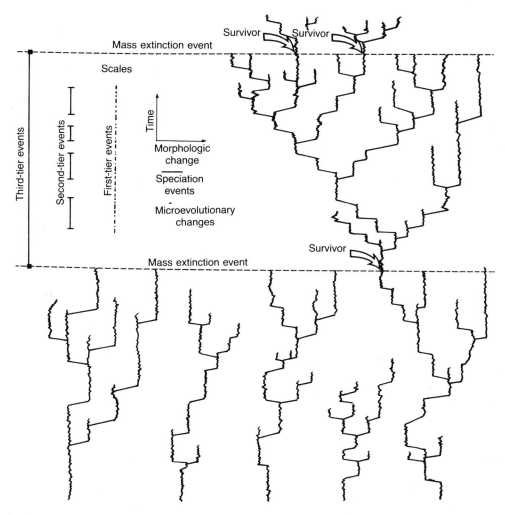

FIGURE 8.16

Tiers of time. A diagrammatic representation of the three sets of evolutionary events or changes that occur on markedly different time scales, as described by Gould (1985). The short-term (microevolutionary) changes of the first tier are seen as the wiggly lines. The occasional wiggly line leaning to left or right represents a case of so-called "genetic drift." The sudden shifts to left or right, of much greater magnitude, represent specia-tion events as described in the punctuated equilibrium model and are events in the second tier. Third-tier events, occurring at very infrequent intervals (tens of millions of years) are the mass extinctions. These represent the "bottlenecks" in the larger flow of evolutionary change.

test" who survived in this struggle were somehow winners in a kind of direct confron-tation between species, the survival of one species entailing the extinction of others. This view is held still by those biologists who are orthodox Darwinians. As we saw earlier, the fossil record contains little evidence for the conventional Darwinian model; instead, it suggests that, rather than interspecific competition, extinctions are caused by the destruction of habitats. In this model, the fittest species are those better able to adapt to or survive environmental change, the rate of change probably being a key factor. What the paleontologic record also shows is that coevolution of species rather than competition between species may be the more likely pattern in life history. In coevolution, the relationships between species are seen in varying degrees of symbiosis. As Hsü (1986) put it, with the exception of humans, one species does not seek to eliminate another—"even predators do not seek the extermination of their prey."

The concept of coevolution has been a topic of considerable discussion in recent years. A selection of papers on the subject is to be found in Futuyma and Slatkin (1983).

It could be argued that evolutionary mechanisms are of little concern in stratigraphy. On the other hand, environmental selection pressure is an important aspect that may be understood better through an analysis of ancient environments. This is where lithostratigraphy and the efforts of the sedimentologists can make a big contribution. Although stratigraphers can only use fossils as they find them, it should be remembered that biostratigraphic units, as entities defined by appearances and disappearances of species, perhaps, do mirror the real world of a punctuational rather than a gradational fossil record.

REFERENCES

Dodson, M. M., and Hallam, A., 1977, Allopatric speciation and the fold catastrophe: The American Naturalist, v. 111, p. 415–433.

Eldredge, N., 1971, The allopatric model of phylogeny in Paleozoic invertebrates: Evolution, v. 25, p. 156–167.

———— 1972, Systematics and evolution of *Phacops rana* (Green, 1832) and *Phacops iowensis* Delo, 1935 (Trilobita), *in* The Middle Devonian of North America: Bull. Am. Mus. Nat. History, v. 47, p. 45–114.

Eldredge, N., and Gould, S. J., 1972, Punctuated equilibria: An alternative to phyletic gradualism, *in* Schopf, T. M. J., ed., Models in paleobiology: San Francisco, Freeman, Cooper, p. 82–115.

Ernst, G., and Seibertz, E., 1977, Concepts and methods of echinoid biostratigraphy, *in* Kauffman, E. G., and Hazel, J. E., eds., Concepts and methods of biostratigraphy: Stroudsburg, PA, Dowden, Hutchinson and Ross, p. 541–563.

Futuyma, D. J., and Slatkin, M., 1983, eds., Coevolution: Sunderland, MA, Sinauer, 555 p.

Gingerich, P. D., 1977, Patterns of evolution in the mammalian fossil record, Chapter 15, *in* Hallam, A., ed., Patterns of evolution as illustrated by the fossil record: Amsterdam, Elsevier, 591 p.

———— 1985, Species in the fossil record: Concepts, trends and transitions: Paleobiology, v. 2, p. 27–41.

Gould, S. J., 1980, Is a new and general theory of evolution emerging?: Paleobiology, v. 6, p. 119–130.

———— 1985, The paradox of the first tier: An agenda for biology: Paleobiology, v. 11, p. 2–12.

Gould, S. J., and Eldredge, N., 1977, Punctuated equilibria: The tempo and mode of evolution reconsidered: Paleobiology, v. 3, p. 115–151.

Hallam, A., 1976, Stratigraphic distribution and ecology of European Jurassic bivalves: Lethaia, v. 9, p. 245–259.

———— 1978, How rare is phyletic gradualism and what is its evolutionary significance? Evidence from Jurassic bivalves: Paleobiology, v. 4, p. 16–25.

Hsü, K. J., 1986, Sedimentary petrology and biologic evolution: Jour. Sed. Petrology, v. 56, p. 729–732.

Jablonski, D., 1986, Background and mass extinctions: The alternation of macroevolutionary regimes: Science, v. 231, p. 129–133.

Jackson, J. B. C., 1977, Some relationships between habitat and biostratigraphic potential of marine benthos, *in* Kauffman, E. G., and Hazel, J. E., eds., Concepts and methods of biostratigraphy: Stroudsburg, PA, Dowden, Hutchinson and Ross, p. 65–72.

Kauffman, E. G., 1967, Cretaceous *Thyasira* from the Western Interior of North America: Smithsonian Misc. Coll. 152, p. 1–159.

—— 1977, Evolutionary rates and biostratigraphy, *in* Kauffman, E. G., and Hazel, J. E., eds., Concepts and methods of biostratigraphy: Stroudsburg, PA, Dowden, Hutchinson and Ross, p. 109–141.

Kennedy, W. J., 1977, Ammonite evolution, *in* Hallam, A., ed., Patterns of evolution: Amsterdam, Elsevier, p. 251–304.

Kennedy, W. J., and Cobban, W. A., 1976, Aspects of ammonite biology, biogeography, and biostratigraphy: Palaeontol.-Assoc. Spec. Paper in Palaeontology, No. 77, 94 p.

Kurten, B., 1959, Rates of evolution in fossil mammals, Cold Spring Harbor: Symp. Quant. Biology, 24, p. 205–215.

Lewin, R., 1980, Evolutionary theory under fire: Science, v. 210, p. 883–887.

—— 1986, Mass extinctions select different victims: Science, v. 231, p. 219–220.

Nichols, D. J., Jarzen, D. M., Orth, C. J., and Oliver, P. Q., 1986, Palynological and iridium anomalies at Cretaceous-Tertiary boundary, south-central Saskatchewan: Science, v. 231, p. 714–717.

Oldroyd, D. R., 1980, Darwinian impacts: Milton Keynes, England, Open Univ. Press, 398 p.

Pielou, E. C., 1979, Biogeography: New York, John Wiley & Sons, 351 p.

Raup, D. M., 1986, Biological extinctions in earth history: Science, v. 231, p. 1528–1533.

Rickards, R. B., 1977, Patterns of evolution in the graptolites, *in* Hallam A., ed., Patterns of evolution: Amsterdam, Elsevier, p. 333–358.

Schopf, T. J. H., and Hoffman, A., 1983, Punctuated equilibrium and the fossil record: Science, v. 219, p. 438–439.

Simpson, G. G., 1953, The major features of evolution: New York, Columbia Univ. Press, 434 p.

Stanley, S. M., 1976, Stability of species in geologic time: Science, v. 192, p. 267–269.

—— 1977, Trends, rates and patterns of evolution in the Bivalvia, *in* Hallam, A., ed., Patterns of evolution: Amsterdam, Elsevier, p. 209–250.

Tschudy, R. H., Pillmore, C. L., Orth, C. J., Gilmore, J. S., and Knight, J. D., 1984, Disruption of the plant ecosystem at the Cretaceous-Tertiary boundary Western Interior: Science, v. 225, p. 1030–1032.

Vrba, E., 1980, Evolution, species and fossils: How does life evolve?: South African Jour. Science, v. 76, p. 61–84.

West, R. M., 1979, Apparent prolonged evolutionary stasis in the primitive Eocene hoofed mammal *Hyopsodus:* Paleobiology, v. 5, p. 252–260.

Williamson, P. G., 1980, Evolutionary implication of late Cenozoic freshwater molluscs from the Turkana Basin, North Kenya. Ph.D. thesis, University of Bristol.

—— 1981, Paleontological documentation of speciation in Cenozoic molluscs from Turkana Basin: Nature, v. 293, p. 437–443.

9
BIOSTRATIGRAPHY

Scarcely any palaeontological discovery is more striking than the fact, that the forms of life change almost simultaneously throughout the world.

Charles Darwin

9.1 INTRODUCTION

Now that some of the spatial and temporal relationships of fossils have been reviewed, this chapter will be concerned with the use of fossils as stratigraphic tools. It was the study of the biostratigraphic features in rocks that laid the foundations for modern stratigraphy, and it is biostratigraphy that still provides the most reliable means for correlation between sedimentary successions around the world. Possibly, this is the most important chapter in the book.

The basic principle of biostratigraphy is that evolutionary changes in faunas and floras are nonreversible. Any given species has only a finite existence in time, perhaps reminiscent of the idea conveyed by Shakespeare's Macbeth when he speaks of "a poor player that struts and frets his hour upon the stage, and then is heard no more." A recognizable species or other taxon that is found only within a finite succession of strata is the basis of the zone concept. **Biozones** are the fundamental units used in biostratigraphy; they can be described as bodies of rock defined by the presence, absence, or relative abundance of certain species or other taxa or of assemblages of species or other taxa. Although taxa of any rank, or groupings of taxa, can be used in defining the various kinds of biozones, in practice it is the species that is the fundamental taxon used in biostratigraphy because it is the fundamental unit in taxonomy. Biozones based upon genera or taxa of higher rank often lack the necessary refinement for meaningful stratigraphic studies.

9.2 ORIGINS OF BIOSTRATIGRAPHY

The earliest works on stratigraphy, such as those of Steno, mentioned in Chapter 1, Lehmann, Arduino, and others, showed an awareness of stratigraphic sequences, but only at a most rudimentary level. Little account was taken of either fossils or of lithotratigraphic classification. This was due largely to the dominant place held by the teachings of Gottlob Werner (1759–1812). Werner's assignment of all rocks into four different ages: Primary, Secondary, Tertiary, and what he called Flotsgebirge was not only crude, but grossly inaccurate and full of ambiguities. In addition, Werner's pedantic insistence that all rocks were formed in an aqueous environment was an even greater cause for contention. It was William Smith (1769–1839; Fig. 9–1) who was the first person to observe and make use of the fact that fossil species occurred in an orderly succession within strata and that a sequence of sedimentary rocks in one place could be correlated with that in another on the basis of their contained fossils. For that period of history this was an astonishing discovery, and it laid

FIGURE 9.1
William Smith, 1769–1839. (Courtesy Geological Society of London.)

the groundwork for the rapid developments that were to follow during the nineteenth century. In 1799 Smith drew the first geologic map. It showed the distribution of rock formations around Bath, in southwestern England, and was accompanied by a table of strata. Later versions drawn by Smith, in collaboration with Reverends Benjamin Richardson and Joseph Townsend, were published in 1813 and 1815, the latest table appearing with a geologic map of England and Wales.

As described in Chapter 2, not everyone in the early years of geology subscribed to the uniformitarian doctrine; indeed, it was the evidence of biostratigraphy as much as anything that persuaded some that a catastrophist view of earth history was the correct one. Among these pioneers was Alcide d'Orbigny (1802–1857), and it is to him that we owe the conception of the original biostratigraphic units, the *stage* and the *zone*. To d'Orbigny, a stage represented a world that had existed in the past. His belief was that the present-day earth and its inhabitants were simply the most recent of a whole series of creations. He interpreted the geologic column and its contained fossil faunas and floras in terms of 27 stages, each of which was the record of a world during a temporary resting stage, or, as Monty (1968) put it, a "paleo-today," complete with oceans and mountains, animals and plants. The interval between one stage and the next was marked by a period of geologic upheaval. Despite his, by modern standards, unorthodox views, d'Orbigny's stage concept found an echo in the uniformitarian approach to the division of the geologic column. It is interesting to note that not only was his term adopted, but that many of the Jurassic and Cretaceous stage names used today can be seen in d'Orbigny's original list of 27 *etages* (Table 9.1; see also Table 3.8).

Although the zone concept is also considered to have originated with d'Orbigny, nowhere does he define the term as such. However, from his writing, it is clear that he considered zones as subdivisions within stages and that they referred only to fossils. Each stage contained hundreds or even thousands of species, but from them d'Orbigny made a selection of only those he considered especially useful and characteristic. These were grouped into zones. Thus, the zone emerged, at least in one sense, as a division of a stage and characterized by certain index species, or *zone fossils*. Some of d'Orbigny's views were certainly strange in modern terms, but be that as it may, and as with the stage concept, the idea of the zone and "zonal species" was readily adopted into the developing biostratigraphic edifice as it grew during the nineteenth century. The first person to

visualize the zone in the modern sense was Albert Oppel (1831–1865). Again, it is only from his writing that we are able to determine what he meant because, like d'Orbigny, he nowhere defined a zone. It could be argued that in referring to zones, his statement "which through the constant and exclusive occurrence of certain species, mark themselves off from their neighbors as distinct horizons" would seem clear enough, but it was left to later workers to provide a definition. Marr (1898), for example, defined zones as "belts of strata, each of which is characterized by an assemblage of organic remains of which one abundant and characteristic form is chosen as index." Other definitions were less explicit and tended to confuse the issue. Woodward (1892) described zones as "assemblages of organic remains" but, as someone pointed out, a cemetery would as well fit that particular definition. As new kinds of zones were defined so were time units to correspond with them. For example, the time encompassed by a biozone was termed a **biochron**. Other terms are listed in Table 9.2, but they are not widely used today. The species or other taxa that are selected as characteristic of a biozone are known as *index* taxa. Ideally, they should have the attributes described in Chapter 7, such as a short time range and widespread geographic dispersal.

9.3 KINDS OF BIOZONES

The evolution of the zone concept has resulted in numerous kinds of zones and quite a varied terminology, as can be seen from Table 9.2. The present discussion is based mainly on the recommendations set out in the North American Stratigraphic Code (NACSN, 1983).

There are three principal kinds of biozone recognized: *interval zones, assemblage zones,* and *abundance zones* (Fig. 9.2).

TABLE 9.1
Original 27 stages of D'Orbigny (After Monty, 1968, Table 1, p. 694.)

Terrains	Ages Geologiques
	Etages
Tertiaires	27. Subapennin 26. Falunien Supérieur ou Falunien Inférieur ou Tongrien 25. Parisien 24. Suessonien
Crétacés	23. Danien 22. Sénonien 21. Turonien 20. Sénomanien 19. Albien 18. Aptien 17. Néocomien
Jurassiques	16. Portlandien 15. Kimmeridgien 14. Corallien 13. Oxfordien 12. Callovien 11. Bathonien 10. Bajocien 9. Toarcien 8. Liasien 7. Sinémurien
Triasiques	6. Saliferien 5. Conchylien
Paleozoiques	4. Permien 3. Carboniférien 2. Dévonien 1. Silurien Supérieur et Inférieur

Interval Zones

Interval zones are of several types: taxon range zones, concurrent and partial range zones, lineage zones, and interzones.

Taxon Range Zones A **taxon range zone** can be defined as the body of rock representing the maximum stratigraphic and geographic range of specimens of a species or other taxon. As Shaw (1964) pointed out, such a biozone is an abstract concept because its true extent through space and time is usually unknowable. It is obvious that the time of first appearance of a species within a

TABLE 9.2
Terminology of biozones

(North American Stratigraphic Code)	(International Stratigraphic Guide; Hedberg, 1976)	Synonyms	Time Term
Interval Zone	Interval Zone		
Taxon range zone	Taxon range zone	Acrozone (of some authors)	Biochron
Concurrent range zone	Concurrent range zone, or Oppel zone	Overlap zone, range overlap zone	
Partial range zone		Local range zone, teilzone, topo-zone	Teilchron
Lineage zone	Lineage zone (phylozone)	Evolutionary zone, morphogenetic zone, phyloge-netic zone	
Abundance zone	Abundance zone	Epibole, peak zone, flood zone, acro-zone (of some authors), acme zone	Hemera
Assemblage zone	Assemblage zone	Faunizone, ceno-zone	Secule, moment
Interzone, barren interzone, in-trazone	Interzone, barren inter-zone		

sedimentary succession occurs at some un-known interval after its actual appearance on earth. The same principle holds true for the final disappearance of a species because, with the possible exception of cases of mass extinction, it undoubtedly lingered on some-where for some time after the time of its last recorded occurrence. Within a particular sec-tion or locality, the range of a species has only an essentially local significance and the term **teilzone** (part of zone) is used to define it. It follows that the true range (or a reason-able percentage of it) can be determined only by a compilation of data from innumerable teilzones. Why a species is not present in a particular sedimentary succession can be due to numerous reasons; these are dis-cussed later.

Concurrent Range and Partial Range Zones A **concurrent range zone** is defined as the interval of overlap between the first ap-pearance of one species or other taxon and the last appearance of a different taxon (Fig. 9.3). A modification of this is seen where the ranges of the two taxa do not overlap so that there is, in effect, a gap between the last ap-pearance of one taxon and the first appear-

FIGURE 9.2
Types of biozones.

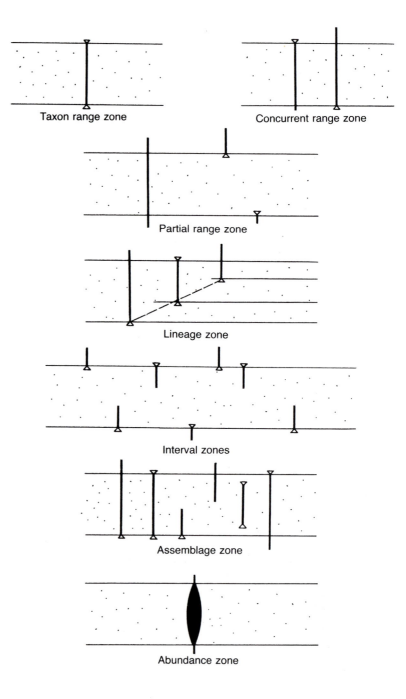

Taxon range zone

Concurrent range zone

Partial range zone

Lineage zone

Interval zones

Assemblage zone

Abundance zone

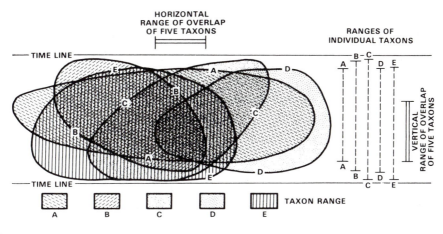

FIGURE 9.3
Variations in extent of concurrent-range taxon zone depending on number of taxa.
(From Hedberg, H. D., International stratigraphic guide, Fig. 7, p. 57, © 1976, John
Wiley & Sons, Inc., New York.)

ance of the next. This interval is used to define a portion of the range of a third taxon and is known as a **partial range zone.**

Lineage Zones If an evolutionary lineage can be documented, the interval between the first appearance of one taxon and the first appearance of a descendant taxon is known as a **lineage zone.** When the taxa involved are unrelated or when the bounding horizons are of last occurrences of either related or unrelated taxa, the term *interval zone* is used.

Interzones Unfossiliferous intervals (presumably also strata containing unrecognizable fossils or those of no zonal value) that are delineated below and above by defined biozones, are termed **barren interzones,** or **intrazones.**

Assemblage Zones

An **assemblage zone** is defined as a biozone characterized by the association of three or more taxa. The grouping of species or other taxa within it is unique and distinguishable

from the assemblages in strata below and above, although certain elements of the assemblage may range through lower and higher strata. The taxa constitute a natural assemblage that lived (and died) together in a particular habitat—a **biocoenosis**—in which case they are usually good environmental indicators as well. If the fossils were brought together after death by sedimentary processes (in a **thanatocoenosis**), they still represent a natural contemporaneous succession, unless it can be demonstrated that fossils derived by reworking of older strata are included.

Abundance Zones

An **abundance zone** is defined as a body of strata that contains the maximum abundance in numbers of individuals of one or more species or other taxa. Theoretically, it represents the time during which the taxa involved were at their acme, and the term **acme zone** is a widely used synonym. When one species is being considered, the term **epibole** is the traditional term. In practice, abundance zones are difficult to use in anything other than local situations. The marked abundance of a particular species, when compared with a sparser occurrence at lower and higher horizons, may be entirely an artifact of the sedimentary environment, preservation, or local paleoecologic factors. As such, it may have little, if anything, to do with a temporally significant event affecting the population size. Long-range correlation based upon abundance zones is particularly risky because the succession of such zones at one place may appear in different order from that in another. In an example cited by Arkell (1933, p. 31), the sequences of ammonite species epiboles recorded in two different localities are markedly dissimilar, and it is clear that the acmes of particular species simply marked brief intervals of favorable conditions that were quite local in their effect.

9.4 BIOSTRATIGRAPHIC AND CHRONOSTRATIGRAPHIC UNITS

Biozones and Stages

The relationship between biostratigraphic units and chronostratigraphic units is a curious one. Chronostratigraphic units are abstract concepts, yet they are used all the time and at all scales. The smallest chronostratigraphic unit to have global significance is, according to Hedberg (1976), the stage. However, unless one is thinking of stages in the abstract sense, this is untrue. As we have seen (Chapter 3), stages were originally conceived as biostratigraphic units, and some workers (e.g., Hancock, 1977; Ludvigsen and Westrop, 1985) still consider them as such because for all practical purposes they can be recognized only on the basis of their contained fossils.

Because a stage is defined in terms of numerous biozones of greater or lesser local significance, inevitably its recognition is limited by whether the biozones are present. As

Ludvigsen and Westrop (1985) point out, zonal and stadial successions in one region will likely be different from those in another. Following this reasoning, stages could be global divisions only if all the component biozones also could be recognized on a global scale, which, of course, they cannot. The recognition of a stage at a distant locality is based on an aggregate of biozones that are judged to be approximately coeval with those of the stratotype. This approximation is based, in turn, on correlations through numerous local zonal schemes that presumably link up eventually with the zonal divisions at the stratotype. This is discussed further later in this chapter. Stages are universally used in all detailed stratigraphic charts and are sometimes further subdivided into substages. In most cases, however, smaller-scale subdivisions of the stratigraphic column are shown as zones, which, within the chronostratigraphic hierarchy, are supposedly chronozones. Because in reality they are biozones, there is a switch from one kind of stratigraphic division to another (see, for example, Table 3.9).

The relationship between biostratigraphic units, which are practical, objectively defined units, and chronostratigraphic units depends upon how we view the relationship between the speed of dispersal of a zonal species through its geographic range and the rate of accumulation of sediments. As discussed in Chapter 7, the rate of dispersal of many species, even over long distances, is often measured in only thousands or even hundreds of years. In any practical geological sense, dispersal at this speed could be considered virtually instantaneous. In effect, therefore, a biostratigraphic horizon marked by the first appearance of a zonal species could be considered an isochronous horizon. Assuming that both the lower and upper boundaries of the biostratigraphic unit were properly defined, it is clear that the only difference between the unit as a chronostratigraphic or as a biostratigraphic entity is the difference between theory and practice. In other words, it is the difference between interpretation and description.

It is common to discern an *apparent* diachronism in biozonal boundaries, but this is invariably due to external factors. Except along depositional strike, formational and facies boundaries are typically diachronous and so, often, will be the contained faunas. Although good index fossils are supposed not to be affected by such ecologic influences, the distribution of even the most widely disseminated pandemic species may still be affected by factors inherent in the depositional or postdepositional environment. Clearly, the time of arrival of a marine species must await a marine transgression. In other cases, the absence of a species may be simply a consequence of it not being preserved. These external influences as they affect the presence or absence of biozonal indicators are discussed in a later section. It is argued that if diachronism were a significant factor in species dispersal, the vertical and temporal sequence of different species at one locality would be markedly dissimilar from that at another. Species originating in other regions, both distant and close by, would be expected to migrate and arrive at different times and in random order. In fact, this seems not to be the case and, as Hallam (1975, p. 3) has confirmed, even with the Jurassic ammonite zones, probably as sensitively defined as any, the same sequence of genera and species is found in different parts of the world.

All of this sounds very convincing, but it would be more acceptable if there were some independent means of cross-checking these suppositions. Magnetostratigraphy might provide the answer in special cases, but it should be remembered that detailed magnetostratigraphic correlations themselves invariably depend upon biostratigraphic data. Further, not only does the dating of magnetostratigraphic horizons become progressively

less accurate with increasing age, but between a third and a half of the Phanerozoic is characterized by quiet intervals of one polarity or the other and so is essentially devoid of magnetostratigraphic markers.

Biomeres

The possibility that certain faunal boundaries might be diachronous was raised by Palmer (1963) in studying trilobite zones in the Cambrian. Certain well-marked faunal changes are abrupt and nonevolutionary and are thought to be due to the sudden replacement of faunal elements invading from outside the area. In the proposed model, there were successive invasions of species derived from a slowly evolving basic stock, and the invaders apparently appeared earlier in the west than the east (Fig. 9.4). This suggested that the source area was along the craton margin in the ancestral Pacific basin. Because the zonal boundaries were shown to be diachronous, so also would be any higher-order stratigraphic units defined by them. This would mean that they could not, for example, be used as the basis for stadial divisions because, by definition, stages are bounded by isochronous horizons. Accordingly, Palmer (1963) proposed a new class of biostratigraphic unit termed a *biomere* (literally segment of life), defined as "a regional biostratigraphic unit bounded by abrupt non-evolutionary changes in the dominant elements of a single phylum" and with boundaries that "may be" diachronous. It would seem that, rather than the nature of the faunal turnover, the important element in this definition is the diachronous nature of the boundaries. It is on this feature, as Ludvigsen and Westrop (1985) rightly point out, that the uniqueness of the biomere, as a distinct class of biostratigraphic unit, rests. More recent work has thrown some doubts on the biomere concept because Palmer (1982) suggested that the apparent diachronous biomere boundaries

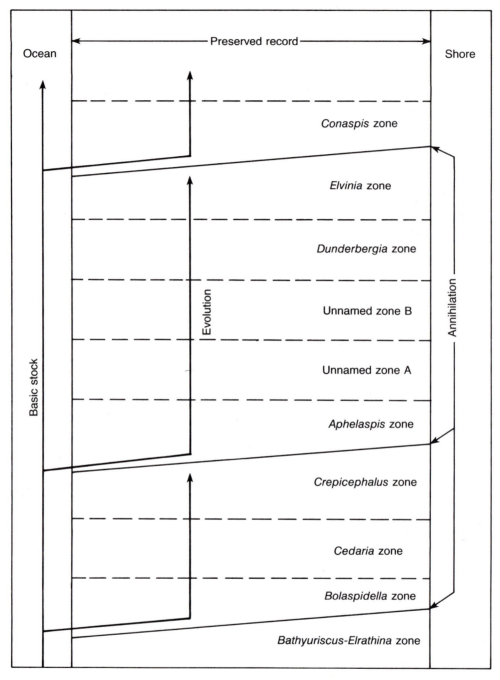

FIGURE 9.4
Apparent diachronism in trilobite zonal boundaries from west to east in western North America. In this interpretation, faunas apparently evolved from a basic stock in deeper water to the west and progressively migrated eastward into shallower water. (Redrawn from Palmer, A. R., 1963, Jour. Paleontology, v. 39, p. 151.)

could be explained by biofacies changes. If this is indeed the case, then the abrupt faunal changes, probably a manifestation of a punctuated equilibrium mechanism, are essentially isochronous and biomeres become, in effect, stages.

Defining and Naming Biozones

Biozones cannot normally be defined in terms of stratotypes, but Hedberg (1976) has recommended that reference localities be named where the particular zone can be studied under optimum conditions of exposure, continuity, and preservation. In establishing a range biozone, the ideal index species or other taxon has a short temporal existence, is facies-independent, and has as wide a geographic distribution as possible. The zone takes its name from the taxon; in the case of zones based upon species, many authors drop the generic name and refer to the zone by the trivial name, but capitalized; for example, the *Cardioceras cordatum* zone of the Jurassic Oxfordian stage is simply known as the Cordatum zone. In the case of assemblage and concurrent range zones, two or more characteristic (index) species are selected and the names hyphenated. In well-studied and fossiliferous sections, zones may be further subdivided into **subzones** (Table 9.3), and these are occasionally broken down still further into **zonules** (or horizons in British terminology), which are defined as the smallest biostratigraphic subdivision, commonly a single stratum. Subdivision of one zone does not necessarily mean that adjacent zones can be similarly broken down. Although there are exceptions, the smaller the subdivision, the more local is its significance. It is common practice to assign the names of index species to the zone or subzone, but in some cases, particularly in zonation schemes based on microfossils, a shorthand system using letters, numbers, or combinations is often used (Table 9.4); for example, P for Pa-

leogene foraminiferans, N for Neogene, and so on. One obvious disadvantage to numbering systems is that when new zones are established, they cannot be easily inserted.

Alternative Zonal Schemes

In any given part of the stratigraphic succession there are usually several schemes of zonal subdivision in use, based upon different kinds of organisms. Sometimes these are mutually exclusive, as in the case of facies-linked faunas. The Silurian, for example, has been subdivided on the basis of graptolites (Fig. 9.5), but these fossils are found almost exclusively in shale facies; thus in sandstone and limestone facies a zonation depending upon brachiopods has been established. In some cases, different zonal schemes are more or less in competition with each other and one or another may be used, depending upon expediency. Subsurface studies in which almost all data are obtained from well cuttings have inevitably resulted in a dependence on microfossils such as foraminiferans, and these have come to provide some of the most sophisticated biostratigraphic tools of all. Within one region, the boundaries of zones based on different fossil groups may not match up at all. On the other hand, they sometimes fall at closely similar horizons, particularly in the case of benthic-facies–linked organisms, and it is likely that the faunal changes are linked to sea-level fluctuations.

9.5 CORRELATION OF ZONES

A biozone, as it is originally defined, is usually of more or less local significance. When traced laterally, the faunal components gradually change so that the complete suite of species in an assemblage zone and the index taxon and associated species in a range zone, at the original reference section, are no longer present and certain species may be re-

TABLE 9.3
Division of ammonite zones into subzones and subzones into zonules, examples from the Lower Jurassic of southern England

Lower Pliensbachian	*Davoei*	*figulinum*
		capricornus
		maculatum
	Ibex	*luridum*
		valdani
		masseanum
	Jamesoni	*jamesoni*
		brevispina
		polymorphus
		taylori
Upper Sinemurian	*Raricostatum*	*aplanatum*
		macdonnellii
		raricostatoides
		densinodulum
	Oxynotum	*oxynotum*
		simpsoni
	Obtusum	*denotatus*
		stellare
		obtusum
Lower Sinemurian	*Turneri*	*birchi*
		brooki
	Semicostatum	*sauzeanum*
		scipionianum
		reynesi
	Bucklandi	*bucklandi*
		rotiforme
		conybeari
Hettangian	*Angulata*	*complanata*
		extranodosa
	Liasicus	*laqueus*
		portlocki
	Planorbis	*johnstoni*
		planorbis

A

Zones	Dorset (W. D. Lang)	Radstock, Avon (J. W. Tutcher and A. E. Trueman)
Davoei	*Oistoceras* spp. *striatum* *lataecosta*	*?brevilobatum* *lataecosta* *Beaniceras* sp. *cheltiense* *sparsicosta*
Ibex	*centaurus* *actaeon* *maugenesti* and *valdani*	*actaeon* *maugenesti* *ibex*
Jamesoni	*masseanum* *pettos* *jamesoni* *obsoleta* *brevispina* *polymorphus* *Tetraspidoceras* spp. *peregrinum* *taylori*	*masseanum* *jamesoni* *obsoleta* *brevispina* *polymorphus* *Phricodoceras*
Raricostatum	*leckenbyi* *exhaeredatum* *raricostatoides* *aeneum* *obesum* and *armatum* *bispinigerum* *densinodulum*	*leckenbyi* *lorioli* *aplanatum* (derived) *macdonnellii* (derived) *raricostatoides* (derived) *zieteni* *subplanicosta*
Oxynotum	*lymense*	*Gleviceras* (derived) *Oxynoticeras* (derived)
Obtusum	*stellare* *landrioti* *planicosta* *obtusum* *capricornoides* *turneri* *birchi*	*stellare* *planicosta* *obtusum* *turneri*
Semicostatum	*hartmanni* *brooki* *sulcifer* *Arnioceras* spp. *alcinoe* *Agassiceras* sp.	*hartmanni* *alcinoe*
Bucklandi	*striaries* *Arnioceras* sp. *pseudokridion* *scipionianum* *gmuendense* and *bucklandi* *rotiforme* *conybeari*	*sauzeanum* *scipionianum* (derived) *gmuendense* (derived) *?vercingetorix* (derived) *meridionalis* (derived) *Coroniceras* sp. (derived)
Angulata	*marmorea* *liasicus* *laqueus* *hagenowi* *portlocki*	*liasicus*
Planorbis	*johnstoni* *planorbis*	*johnstoni* *planorbis*

B

TABLE 9.4

Example of a numbering system used in a biostratigraphic zonation with planktonic foraminiferans (P = Paleogene, N = Neogene)

Geochronologic Units			Biostratigraphic Zones, Planktonic Foraminifera	
Standard Ages		Epochs		
Calabrian			Pleistocene	N23
				N22
				N21
				N20
Piacenzian	Rossellian		Pliocene	N19
Tabianian				N18
Messinian	Castellanian	Late	Miocene	N17
Tortonian				N16
				N15
Serravallian	Cessolian	Middle		N14
				N13
				N12
				N11
				N10
Langhian				N9
				N8
Burdigalian	Girondian	Early		N7
				N6
				N5
Aquitanian				N4
Chattian		Late	Oligocene	P22
				P21 B
				A

216

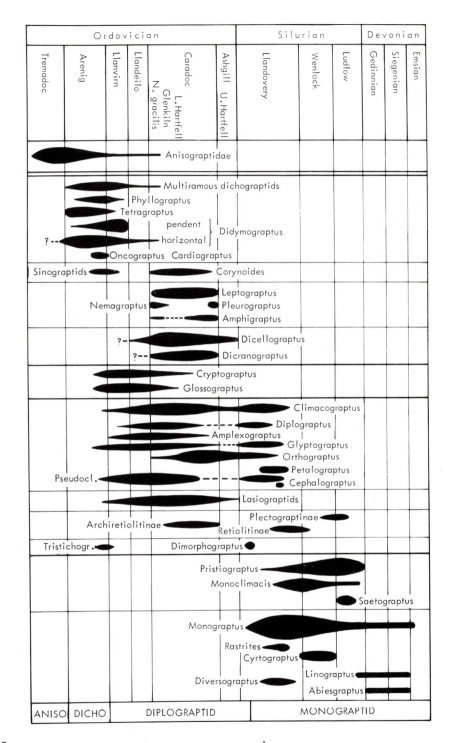

FIGURE 9.5

Example of a faunal range chart. This one is for graptolites. (From Bulman, O. M. B., 1970, Treatise on invertebrate paleontology, Pt. V, 2nd ed., Fig. 74, p. V98, courtesy of the Geological Society of America and University of Kansas.)

placed by other species. Inevitably, when traced over an extended distance, local environmental factors and regional geographic changes put their stamps upon the faunal assemblage. Facies-linked species are likely to be replaced first, but eventually nektonic and planktonic elements change in response to major environmental controls, usually climatic. Such changes are reflected in the provincialism seen in many faunas throughout the Phanerozoic record.

The correlation of biozones over long distances is sometimes enhanced by the expansion, contraction, and shifting of the geographic ranges of the constituent species through time. This means that there may be both geographic and temporal overlap and interfingering between the zonal indicators in one province with those in another (Fig. 9.6). Some fossil groups tend to be more widely distributed than others. The ammonites, for example, because of the nektonic mode of life of many, although not all, species and their frequent nekroplanktonic dispersal, are usually cited as being closest to the ideal index fossils. Some of the Jurassic ammonite zones are coming to be very widely known on almost a worldwide scale; this is particularly the case in the Lower Jurassic. At the beginning of the Middle Jurassic, however, the pandemism of the ammonites was ended by

changes in the distribution of land and sea and by climatic changes. From the Bajocian through the Lower Callovian stages, the zonal schemes of different regions have few taxa in common.

9.6 INDEX FOSSILS AND LITHOSTRATIGRAPHIC UNITS

The term **index fossil** has long been used, particularly in North America, in a sense that is different from its use in biostratigraphy, and this is potentially confusing. A guide to these fossils is Shimer and Shrock's (1944) *Index Fossils of North America*. Traditionally, some formations have been identified and mapped as much on the basis of their characteristic fossils as they have on distinctive lithologic features. Many such fossils have become well known to nonpaleontologists and even quite famous as guide fossils in determining certain formations. One example is seen in the Burlington Limestone of Iowa and Illinois, which invariably contains the brachiopod *Spirifer grimesi* as a characteristic fossil. This species has become so familiar that other formations with a completely different lithology also have been labeled Burlington, solely because they contain this "Burlington species." In this sense, the fossils are simply another of the lithologic constituents of the

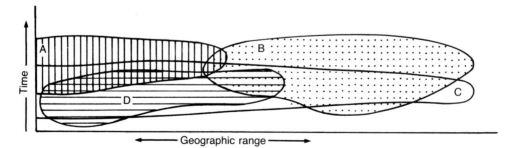

FIGURE 9.6
Ranges of four species as they overlap in space and time. Species C is a pandemic form, species A clearly is an endemic species.

rock formation and there is normally no attempt to see them as zonal index species. Despite the fact that several stratigraphic codes have sought to discourage the nonbiostratigraphic use of fossils, there are already many cases where lithostratigraphic units have been virtually defined and named on the basis of their fossil content alone. Keroher (1967) cited numerous examples of stratigraphic units that were originally named for their characteristic fossils. Thus, for example, the Mississippian Archimedes, Pentremital, and Encrinital (= Burlington) limestones gained widespread recognition through the midcontinent region during the nineteenth century. Lawson (1979) mentioned many familiar mapped units in Britain, such as the Cambrian Tetragraptus Shales, the Trigonia Grit of Jurassic age, and the Paludina Marble of the Cretaceous. He pointed out that this procedure was a common practice in Europe also. Perhaps the important point to stress is that such "index" fossils should be easily recognized in the field. Thus, McLaren (1959) cited cases where mapping, even at a reconnaissance level, was greatly expedited by the recognition of well-marked faunal changes. There are, on the other hand, cases where there has clearly been confusion over the difference between lithostratigraphic and biostratigraphic classifications. So, for example, the Upper Eocene Inglis and Williston formations of northern Florida are defined on the basis of their microfossils and are totally indistinguishable as practical mappable units in the field.

9.7 PROBLEMS AND PITFALLS

Although the basic concepts of biostratigraphic correlation would seem simple enough, there are numerous difficulties to be overcome before a reasonable degree of accuracy can be achieved. The problems are, for the most part, inherent in the science, and geologists have long been aware of them.

However, as stratigraphy becomes more sophisticated, and the divisions are ever more finely drawn, not only does the need for precise correlation grow in proportion, but so does the magnitude of the problems. The chief difficulties that arise are summarized below.

Taxonomy

It has been stated elsewhere that biostratigraphy begins with taxonomy, and this is probably the biggest problem of all. Taxonomic procedures are typically highly subjective; comparisons between two different faunas may become almost meaningless unless there is some assurance that approximately the same systematic criteria are being applied to both. As is well known, excessive zeal on the part of both "splitters" and "lumpers" has led to no end of trouble in all the biological sciences, and it is for this reason that various methods using a numerical approach to taxonomy have been developed. It must be admitted that numerical taxonomy has had, as yet, only a limited success and it certainly has received a mixed reception from biologists.

Missing Taxa

In biostratigraphy there is often a tendency to treat the absence of a given index fossil as being of equal importance as its presence. Indeed, the gaps in the range charts have been the obvious places to draw the boundaries between adjacent biozones. Assuming that the strata concerned lie within the time range of the taxon concerned, its absence in a sample may be due to a number of reasons:

1. Facies control (ecologic factors)
2. Not preserved (dissolution and other postdepositional factors)
3. Rare, so missed in the collecting process (human factor)
4. Misidentified (human factor)

Because there is no direct balance between the presence and absence of a species when comparing faunal lists between sections, some degree of unreliability is inevitable.

Range Charts

Hazel (1977) called attention to the fact that many range charts are constructed on the basis of unsound logic. A common procedure is to use a composite section as the basis for determining the range of specified taxa in a given local section. The fact that the composite section was itself constructed from data gathered at such local sections means that, as Hazel put it, "the data have been justified by the model, and the model by the data."

Repeatability of Zonal Schemes

The establishment of biozonal divisions within a stratigraphic succession is naturally dependent upon detailed taxonomic studies and is thus very much a subjective process. The repeatability of many zonal schemes is open to question; not only is there a lack of consistency in the way different sections are subdivided, but in any one section it is unlikely that two workers will reach a consensus on the way it should be zoned. That these are very real problems has been demonstrated in experiments (Zachariasse et al., 1978).

Depositional and Postdepositional Factors

Biostratigraphy is based upon data influenced by the interaction between evolution and sediment accumulation. The first is a continuous, although not uniform process; the second is markedly discontinuous and very variable in rate, particularly in shallow-marine environments. It is not surprising, therefore, that the actual distribution of fossils within those time slices recorded in rock strata are often highly distorted and fragmented facsimiles of their real distribution in

space and time. Figure 9.7, from Davaud (1982), illustrates how the variation in the rate of sedimentation both squeezes and stretches the distribution of a population as found in a given section. These same sedimentary variables also are the cause of a fragmentation and apparent reordering process that may so obscure the picture that correlation between sections becomes virtually impossible. As shown in Figure 9.8, again from Davaud (1982), the actual biostratigraphic record in the four sections contains no clue as to the real biochronological sequence. Thus, no proper correlation could be attempted without additional data or a prior knowledge of at least one chronologic sequence. Derivation of a biochronological scale from such apparently conflicting data is possible only by trial and error; at best, the result is only a synthetic scale that should be considered temporary and subject to constant modification. Strenuous efforts are being made to overcome some of these problems, particularly in the direction of developing a more objective (usually meaning quantitative) methodology.

9.8 QUANTITATIVE METHODS IN BIOSTRATIGRAPHY

Need for Quantitative Methods

The adoption of quantitative techniques in biostratigraphy is increasingly necessary for three reasons. First, there is a need for a more objective approach to the methodology of stratigraphic correlation. Second, better numerical models to explain certain geologic phenomena are required, and third, there is the need for techniques to exploit the massive data-handling capabilities of computers, particularly as they can be used to process the enormous amount of information produced from subsurface sampling. Quantitative techniques are now becoming more widely used in geology; to encourage this

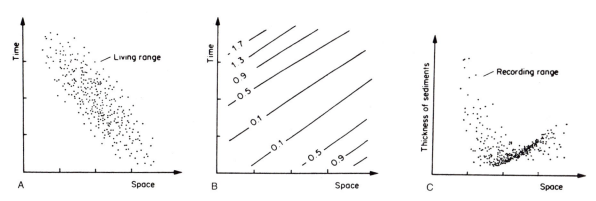

FIGURE 9.7

Theoretical example of distortion in the actual distribution of a fossil population within a sedimentary succession. Variations in the rate of sediment accumulation and the effects of penecontemporaneous erosion and reworking are mainly responsible. A. Actual space-time distribution of a species population. B. Graphic presentation of changing sedimentation rate in the same space-time domain. C. Actual distribution of species population in sedimentary succession. (From Davaud, E., 1982, The automation of biochronological correlation, *in* Cubitt, J. M., and Reyment, R. A., eds., Quantitative stratigraphic correlation Fig. 1, p. 87. Copyright © 1982. Reprinted by permission of John Wiley & Sons, Ltd.)

trend, the International Geological Correlation Program (IGCP) initiated in 1976 a project entitled "Evaluation and Development of Quantitative Stratigraphic Correlation Techniques," better known as "Project No. 148." It was initially intended to run for five years and concentrate on biostratigraphic problems, but it was later extended for an additional two years and eventually came to embrace a wide range of stratigraphic topics. For some of the papers emanating from Project No. 148, see Cubitt (1978) and Cubitt and Reyment (1982). Up to the present, the majority of quantitative biostratigraphic studies have been concerned with microfossils, many from boreholes.

Study of Faunal Similarities

Correlation between different stratigraphic sections on the basis of biostratigraphic data is largely a question of assessing the similarity of the assemblage in one succession with

that in another. The more species found to occur in common, the more likely it is that the two successions are of the same age. Examples of two quantitative techniques that have been used in similarity studies are contained in the next sections. One is a graphic method, and the other uses multivariate analysis.

Graphic Correlation Method The graphic correlation method was proposed by Shaw (1964), and later reviewed by Miller (1977), to provide a precise method for biostratigraphic correlation between two sections. It does this by displaying a graphic plot of the relationship between the two sections, which are set out along the X and Y axes of the graph. What is plotted are the positions of first and last appearance of selected taxa, as measured up from the base of the section. If the thickness of the units in the two sections being compared is the same, the regression line between them will plot as a straight line

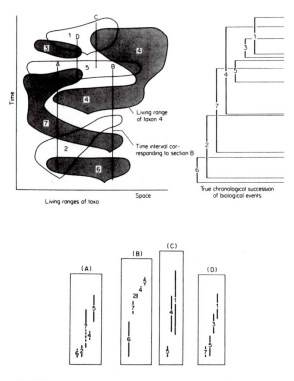

FIGURE 9.8
Theoretical reconstruction showing the difference
between the true distribution of taxa in space and
time and their actual distribution in stratigraphic
sections. Theoretical example (top) showing distribution in space and time of seven different taxa
and their true chronologic sequence. Four stratigraphic sections (A, B, C, and D) are indicated.
Sedimentary record (bottom) of biological events in
sections A, B, C, and D. Distortion is due to differential rates of sedimentation as shown in graphic
presentation in Figure 9.7. (From Davaud, E., The
automation of biochronological correlation, *in* Cubitt, J. M., and Reyment, R. A., eds., Quantitative
stratigraphic correlation Fig. 3, p. 88 and Fig. 4, p.
89. Copyright © 1982, John Wiley & Sons. Reprinted by permission of John Wiley & Sons, Ltd.)

The procedure begins with the selection of
a standard reference section (SRS), which in
practice is naturally the most complete and
fossiliferous section available and is usually
set out as the X (horizontal) axis. The ranges
of all species on the chronologic scale are established for the SRS as points of first (FAD
= first appearance datum) and last appearance (LAD) along the X axis, and all other
sections of similar age are compared with it
(Fig. 9.9). With the new section, as many as
possible of the species found in the SRS are
located and the positions of all FADs and
LADs are established along the Y axis. Projections of all bases and tops on both axes are
then plotted as points on the graph. The next
step is to determine the best-fit regression
line through the scatter of points; this is
termed the *line of correlation* (LOC). It is clear
that if the section being compared with the
SRS was one with a consistently slower or
faster rate of rock accumulation, the distances between the various FADs and LADs
will be scaled differently on the new section
than on the SRS, and the regression line will
then lie at some angle other than 45° from
the axis. Similarly, a change in the rate of accumulation will show up as a dogleg in the
regression line. The extreme case here would
be an erosional break, in which case the
missing part of the section would plot as a
horizontal segment of the LOC (Fig. 9.10). It
should be noted that missing section due to
faulting would plot in a similar way; there is
no direct way of determining this. It should
be remembered also that the straightness of
the LOC is almost certainly not a true representation of the situation; it is merely an average of the many small doglegs that reflect
the episodic nature of the depositional process. At the scale used in most studies, this
departure from reality can be ignored and the
straight line considered as showing the average rate of rock accumulation.

If the SRS has been carefully selected and
refined, the ranges of species plotted can be
considered, at least initially, as total ranges,

at 45° to each axis. Similarly, if the ranges of
the species are the same in the two sections,
the pairs of points marking species appearances and disappearances will plot also along
the regression line.

FIGURE 9.9

Graphic plots of first appearances (bases) and disappearances (tops) of species in a standard reference section (X) and a comparison section (Y₁). The slope of the line of correlation indicates that the average rate of rock accumulation was more rapid in section Y₁ than in section X. (After Miller, F. X., 1977).

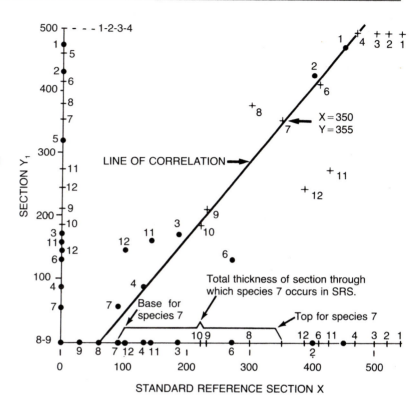

and the ranges of species in the section being compared as local ranges. Normally, the FADs from the comparison section will plot on the left side of the LOC and the LADs on the right. Occasionally, however, one or both will plot on the opposite side; in such cases, it is clear that the new section has provided useful new information and extended the previously known range (Fig. 9.11). The SRS should be then adjusted accordingly, at which time it becomes what Shaw (1964) termed a composite standard reference section (CSRS). Eventually, as the number of sections being compared grows, the CSRS becomes increasingly accurate, and the faunal list embraced by it will be considerably larger than that in the original SRS. It probably also will contain facies-controlled elements that were absent in the original section but were added later from comparison sections.

Despite the addition of new data and the expansion of the CSRS, it is obvious that the true time range of the taxon in question is probably unknowable. This is what Edwards (1978) called the "range chart problem." What is involved is the gap between the biologic event of a species first appearance/final disappearance in an area, for whatever reason (evolution, migration, extinction, etc.), and the biostratigraphic event of the lowest occurrence (LO) or highest occurrence (HO) observed in a given stratigraphic section. Plotted on what is known as a no-space graph, observed ranges of taxa are compared with an unscaled hypothetical sequence of biostratigraphic events (Fig. 9.12), which is revised and retested until a final hypothetical sequence is derived. The data can be readily coded for computer analysis.

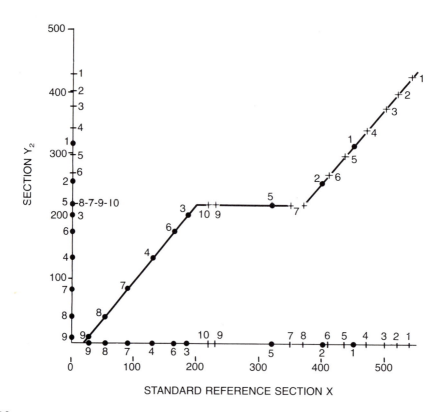

FIGURE 9.10
In this example, the line of correlation has a horizontal segment indicating that part of
the succession in section X is missing and unrepresented in section Y₂. This omission
may be due to faulting, erosion, or nondeposition. (After Miller, F. X., 1977.)

Multivariate Techniques Although there are some notable exceptions, such as Lyell's subdivision of the Tertiary on the basis of percentages of extant to extinct species (Table 9.5), biostratigraphic zonation using macrofossils does not typically involve large numbers or any special quantitative treatment. The situation is different in the case of microfossils such as foraminiferans, ostracodes, diatoms, radiolarians, coccoliths, and so on, where large data arrays are commonplace. It is here that the massive data-handling capabilities of computers can be used effectively. Even with microfossils, the traditional approach was to use only a few selected species in making biostratigraphic de-

terminations; as Hazel (1970) pointed out, much useful information was lost. Faunal associations having important evolutionary and ecologic implications could be identified only if species assemblages were evaluated and the significance of grouping within faunas was recognized.

Of the numerous multivariate analytical techniques available, cluster analysis (CA) has come to have wide usage in the handling of biostratigraphic data. Cluster analysis is particularly appropriate when dealing with compact groupings of data points; both between-group and within-group relationships can be shown with considerable fidelity. The method is less responsive in situations where

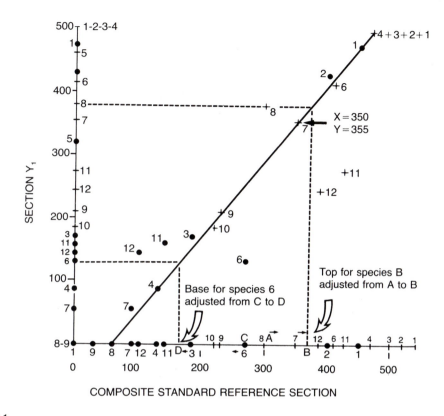

FIGURE 9.11
How new data are added to the standard reference section to expand species ranges. The standard reference section now becomes a composite reference section. Species 8 is found in younger rocks in the comparison section Y₁ than in the SRS, so the latter must be adjusted accordingly. Similarly, species 6 occurs lower in the comparison section than it does in the SRS. (After Miller, F. X., 1977.)

gradational relationships are present. Clustering is a procedure in which objects that are most similar are grouped together and separated from objects that are less similar. Numerous methods of cluster analysis are available, but it is the hierarchial procedure, in which the most similar objects are grouped first and the successively less similar groupings are merged at lower levels of similarity, that is most widely used in biostratigraphy. The measure of similarity between two samples is known as the *correlation coefficient;* several different ones have been proposed (Hazel, 1977). The selection of one or another

depends upon the kind of data available and the analysis proposed.

In biostratigraphic correlation, the normal type of problem is one in which it is necessary to define the faunal similarity between any two stratigraphic sections; that is, to compare them in terms of the number of species common to both. The data array comparing every section with every other can be set out using a binary code: 2 for presence and 1 for absence. Table 9.6A is a hypothetical data array in which the distribution of 12 selected species is compared in eight different stratigraphic samples. For the purpose of

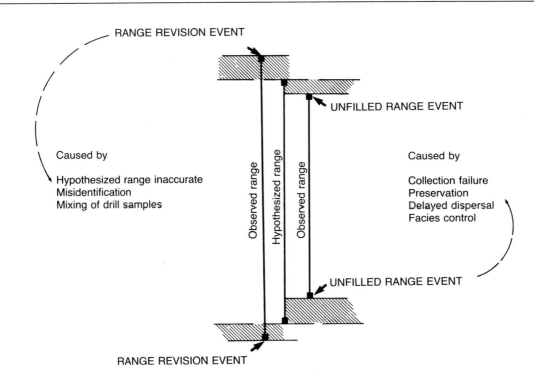

FIGURE 9.12
The range chart problem. The relationship between the hypothesized range of a species and its actual range. (Based on data from Edwards, 1978).

statistical treatment, a match of 2 with 2 is termed a positive match, 1 with 1 a negative match, and 2 with 1 a mismatch.

The simplest measure of similarity is the ratio of matches to the number of species compared and may be expressed in Sokal and Michener's simple matching coefficient (S_{sm}):

$$S_{sm} = \frac{C + n}{C + n + m}$$

where C = number of species common to both samples (positive matches), n = number of negative matches, and m = number of mismatches (items present in one sample but not in the other).

TABLE 9.5
Lyell's (1830–1833) subdivision of the Tertiary strata of the Paris Basin on the basis of ratios of extinct to extant species

Division	Extinct Species (%)	Extant Species (%)
Newer Pliocene	4	96
Older Pliocene	58	42
Miocene	83	17
Eocene	97	3

TABLE 9.6A
Data array for 12 species contained in 8 samples

Samples	Species											
	1	2	3	4	5	6	7	8	9	10	11	12
A	2	2	1	2	2	1	1	1	1	1	2	1
B	2	1	1	2	1	2	2	1	2	2	1	2
C	2	2	2	2	1	1	2	2	1	2	2	1
D	1	1	1	1	1	2	1	1	2	1	1	1
E	2	2	1	1	1	1	1	1	1	1	2	1
F	2	1	1	2	2	2	2	2	1	1	1	2
G	1	1	1	1	2	1	2	2	2	2	2	2
H	1	1	1	1	1	2	1	1	1	2	1	2

TABLE 9.6B
Half-matrix of similarity coefficients for data array in Table 9.6A using Sokal's and Michener's coefficient (S_{sm})

	A	B	C	D	E	F	G	H
A	—	0.4	0.5	0.4	0.8	0.6	0.3	0.4
B		—	0.5	0.6	0.4	0.6	0.5	0.6
C			—	0.1	0.5	0.5	0.4	0.3
D				—	0.6	0.4	0.5	0.8
E					—	0.4	0.3	0.6
F						—	0.5	0.4
G							—	0.5
H								—

TABLE 9.6C
Half-matrix of similarity coefficients for data array in Table 9.6A using Jaccard's coefficient (S_j)

	A	B	C	D	E	F	G	H
A	—	0.2	0.44	0	0.6	0.33	0.2	0
B		—	0.36	0.285	0.11	0.55	0.4	0.428
C			—	0	0.375	0.36	0.36	0.1
D				—	0	0.125	0.125	0.25
E					—	0.11	0.11	0
F						—	0.4	0.25
G							—	0.25
H								—

Another widely used similarity coefficient is the Jaccard coefficient (S_j), expressed as

$$S_j = \frac{C}{N_1 + N_2 - C}$$

where C = as above, N_1 = number of species in first sample, and N_2 = number of species in second sample. Alternatively, it may be expressed more simply as

$$S_j = \frac{C}{C + m}$$

It should be noted that because negative matches do not contribute toward similarity, they are ignored in the Jaccard coefficient.

The S_{sm} and S_j coefficients for the data in Table 9.6A are set out in Tables 9.6B and 9.6C as half-matrices. From these data, the most similar pair are selected and linked, then the next most similar pair, and so on until all have been clustered out. From the clusters a **dendrogram** can be constructed. Beginning with the mutually highest correlations, these are connected to form the initial clusters. In the next step of the hierarchy, already-clustered items are treated as a single item by arithmetic averaging for the purpose of clustering with the remaining items (see Figs. 9.13 and 9.14). A fuller account of procedures is contained in Davis (1986). It is obvious that the binary nature of the data used in these calculations, not to mention the large volume of data accumulated, particularly from boreholes, makes this procedure well suited to computer analysis. Numerous programs are available, and computer handling of biostratigraphic problems is now routine. One of the earliest studies was by Bonham-Carter (1967), who developed a FORTRAN program based on the computations of Sokal and Sneath (1963). After producing a similarity matrix, the data were then clustered and presented in a dendrogram (Fig. 9.13). A FORTRAN program was also developed by Millendorf and Heffner (1978) to produce similarity matrices that could incorporate the

various relative biostratigraphic value (RBV) procedures that are described in the next section. These were then used as input by Millendorf, Srivastava et al. (1978) for the DENDROGRAPH program developed by McCammon and Wenninger (1970) to produce dendrograms drawn on a Calcomp plotter.

The most basic numerical procedures are binary and use only present/absent attributes. They take no account of just why a species is absent, but neither do they treat in any way the relative abundance of species, which is often controlled by factors with little or no temporal significance. An environmental and/or preservational bias is also reduced by the "range through" procedure, advocated by many workers, in which the absence of a species is ignored, and it is counted as present if it occurs at stratigraphically lower and higher horizons.

A further refinement in quantitative procedures is seen in the recognition of the fact that the biostratigraphic utility (relative biostratigraphic value) of different taxa varies widely and that calculations should be weighted so that each species will contribute to the similarity coefficient in proportion to its RBV. Brower (1985) listed vertical range (V) as the most critical attribute in considering the biostratigraphic value of a species, followed by facies independence (F) and geographic range (G). These three attributes were assigned numerical values between 0.0 and 1.0, as set out in Table 9.7. From these numbers, an RBV can be assigned to the species. Three kinds of RBV have been proposed by different authors. McCammon (1970) suggested that, for a given species i, $RBV1_i = F_i(1 - V_i) + (1 - F_i)G_i$, in which all three attributes are weighted equally.

Brower et al. (1978) and Brower (1984) considered that the importance of vertical range should be recognized and suggested a modified formula:

$$\frac{F_i(1 - V) + G_i(1 - V_i)}{2}$$

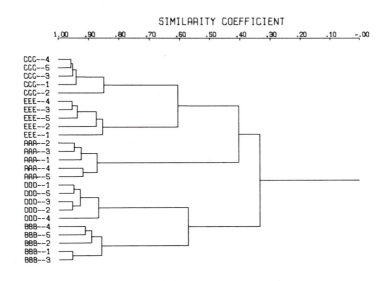

TEST DECK NUMBER 1.
```
    25    25    1    0    1    0    1    0    2    2
(2511,2X,A6)
11112122120121211211212102    DDD--1
21122222111211121112112212    BBB--5
22211111011211202221221021    CCC--3
12112122122121211211211212122 DDD--3
21121211211011121122221001    EEE--2
20222220112111211112111212    BBB--1
21122211122112211012112212    BBB--2
12211112121212221221111122111 AAA--4
02211111212212122221221121    CCC--1
21222221122101211121111212    BBB--3
21121212211121221C1221221     EEE--1
22221111210212222212221121    CCC--4
21121212211111121211221222    EEE--4
21121211211112112121221222    EEE--5
12110120221210111221212122    DDD--2
12211122212122211001112111    AAA--2
1C21112121222122221112110     AAA--1
12112122122221212121112122    DDD--4
21121011211011021212221222    EEE--3
1C112122022121211121110112    DDD--5
21122201112112112101212      BBB--4
12211202021220C122221110111   AAA--3
12222111212212221221221       CCC--2
22221111210212122212221121    CCC--5
12211222221222122211122111    AAA--5
```
A

TEST DECK NUMBER 2.
```
    25    25    1    0    1    0    0    0    2    2
(2511, 1X, A6)
2221121112122121211211212121221   1
2111211111112122220201212          2
1121212121111120221121211          3
2111212112121111112211121          4
1112221122121112121212111          5
2112122022211211111212212          6
12111112221111221110011111         7
22111122211111C11111211221         8
10211111222211122111111112         9
212121111121110120211111221        10
2210112221122111211211112          11
12110222211202111101111221         12
2111221111111221122211112          13
21121111110121121211111112         14
11122110112212221211102            15
11112121121212112111222212         16
11212211211111222112121212         17
11122022112122221212211211         18
2212221112121C021121221212221      19
21121112021111222212121222         20
11112211221111111221112111121      21
12211222112012122222121211         22
11222111021111210111111111         23
112221122122121211210222222        24
12121201211101222121122221         25
```
B

SIMILARITY COEFFICIENT

```
1.00  .90  .80  .70  .60  .50  .40  .30  .20  .10  -.00
CCC--4
CCC--5
CCC--3
CCC--1
CCC--2
EEE--4
EEE--3
EEE--5
EEE--2
EEE--1
AAA--2
AAA--3
AAA--1
AAA--4
AAA--5
DDD--1
DDD--5
DDD--3
DDD--2
DDD--4
BBB--4
BBB--5
BBB--2
BBB--1
BBB--3
```

FIGURE 9.13
Typical computer printout of a data array (2 = present, 1 = absent, 0 = no information), together with the resultant dendrogram plot using Sokal and Michener's coefficient. (From Bonham-Carter, G. F. 1967, FORTRAN IV program for Q-mode cluster analysis of non-quantitative data using IBM 7090/7094 computers: Kansas Geol. Survey Computer Contr. 17, Table 4, p. 4, and Fig. 1, p. 7.)

placing double weight on the vertical range as compared with the other attributes.

Finally, RBV3, also suggested by Brower et al. (1978) and Brower (1984) and defined as $RBV3_i = (1 - V_i)$, was specifically intended for the Deep Sea Drilling Project cores, in which the distribution of taxa is little influenced by facies changes. The taxa are also geographically wide-ranging, leaving only the vertical range as a measure of relative value.

FIGURE 9.14

Example of how cluster analyses can be used to differentiate zonal assemblages. A. Lithostratigraphy and locations of triloite samples in sections. B. Dendrogram based on comparison of 65 samples containing 24 trilobite species. C. Biostratigraphic interpretation of results. (From Hazel, J. E., 1977, Use of certain multivariate and other techniques in assemblage zonal biostratigraphy: Examples utilizing Cambrian, Cretaceous and Tertiary benthic invertebrates, *in* Kauffman, E. G., and Hazcl, J. E., eds., Concepts and methods of biotratigraphy, Figs. 2A, 2B, 2C, p. 194–196. Reprinted by permission of Dowden, Hutchinson and Ross, Stroudsburg, PA.)

DISTRIBUTION OF SAMPLES AND LITHOLOGIC UNITS, UPPER FRANCONIA SANDSTONE, LAKE PEPIN, MINNESOTA

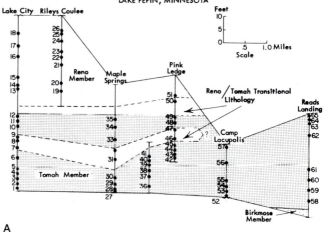

A

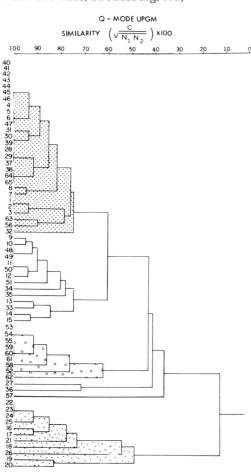

B

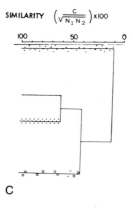

C

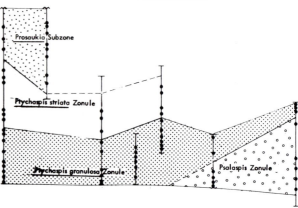

D

TABLE 9.7
Relative biostratigraphic values for various fossils (after Brower, 1985)

Types of Species	G	F	V	RBV1	RBV2	RBV3
Classic index fossil	1.00	1.00	0.01	0.990	0.990	0.990
Near index fossil	0.90	0.95	0.10	0.900	0.832	0.900
Short-ranged facies tracer	1.00	0.10	0.10	0.990	0.495	0.900
Long-ranged facies tracer	1.00	0.10	0.90	0.910	0.055	0.100
Index fossil in one stratigraphic section	0.10	1.00	0.10	0.900	0.495	0.900
Nonoccurring species	0.00	0.40	0.00	0.400	0.200	1.000
"Trash fossil"	0.05	0.10	0.90	0.055	0.008	0.100

In applying these modifications to multi-variant analysis techniques, the similarity coefficient formulae may be rewritten to accommodate the RBV weightings. Thus the Jaccard coefficient becomes

$$S_{ij} = \frac{RBV_{ij}}{RBV_i + RBV_j - RBV_{ij}}$$

As can be seen in Figure 9.15, in which the local ranges of species with high RBV values are contrasted with those having low RBV values, the accuracy of biostratigraphic correlation can be greatly improved by the careful selection of species and the use of weighting procedures.

The use of RBVs has been a matter of some controversy. As pointed out by Millendorf, Brower et al. (1978), if facies changes or paleoecological groupings are being considered, unweighted data are probably preferable. In biostratigraphic correlation, however, weighting can be used to emphasize those species with the highest RBV values. This can be done by weighting those species with high RBVs or by eliminating those species with low RBV values and treating the remaining reduced data sets either in weighted or in unweighted form. In general, it was concluded by Millendorf, Brower et al. (1978) that larger

data sets were preferable to small ones and that, for a given number of species, better results were likely from weighted than from unweighted data.

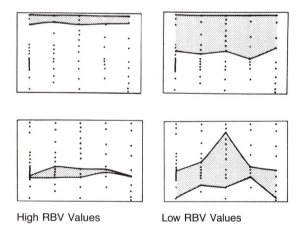

High RBV Values   Low RBV Values

FIGURE 9.15
Comparison of the stratigraphic range and correlative sensitivity of five stratigraphic sections by using species of low relative biostratigraphic values (left) and those with high values (right). The points on each section are sample locations. (From Brower, J. C., 1985, The index fossil concept and its application to quantitative stratigraphy, *in* Gradstein, F. M. et al., eds., Quantitative stratigraphy, Fig. 6, p. 57. Reprinted by permission of Kluwer Academic Publishers, Dordrecht, Holland.)

Similarity matrices

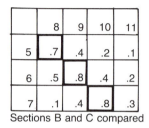

	5	6	7
1	.5	.2	.1
2	.5	.9	.5
3	.1	.2	.4
4	.1	.3	.7

Sections A and B compared

	8	9	10	11
5	.7	.4	.2	.1
6	.5	.8	.4	.2
7	.1	.4	.8	.3

Sections B and C compared

	12	13	14	15
8	.7	.3	.2	.3
9	.6	.9	.1	.1
10	.2	.4	.5	.4
11	.1	.1	.5	.7

Sections C and D compared

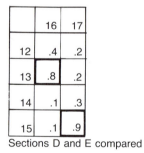

	16	17
12	.4	.2
13	.8	.2
14	.1	.3
15	.1	.9

Sections D and E compared

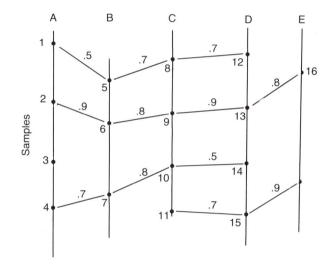

Matchings derived from similarity matrices
applied to cross section with similarity
coefficients given

FIGURE 9.16
Lateral tracing method used in correlating five different stratigraphic sections (A–E).
(From Millendorf and Millendorf, 1982, The conceptual basis for lateral tracing of biostratigraphic units, *in* Cubitt, J. N., and Reyment, R. A., eds., Quantitative stratigraphic
correlations, Figs. 3, 6, p. 103, 104. Copyright © 1982, John Wiley & Sons. Reprinted by
permission of John Wiley & Sons, Ltd.)

An extension of these quantitative procedures can be used in what is termed *lateral tracing,* in which faunal comparisons between adjacent sections can be carried step by step across considerable distances. It should be obvious that in regional studies involving wide areal extent, the samples from distant sections, even in an essentially isochronic unit, will not cluster significantly. This is because of progressive changes in the composition of the faunas. In lateral tracing, samples from two adjacent stratigraphic sections are matched A with B and this is repeated step by step—B with C, C with D and so on—along a line or network of sections. At each pairing of sections, all the samples in the section with the smaller number of samples are matched, the remaining samples in the more prolific section being discarded. The procedure in lateral tracing is described by Millendorf and Millendorf (1982) and is summarized in Figure 9.16. In addition to section-to-section comparisons, a study of similarities between vertically adjacent pairs of samples within individual sections is often productive. Thus, a low similarity coefficient

FIGURE 9.17
Similarity coefficients may demonstrate faunal discontinuities and, particularly in sections showing lithologic homogeneity, indicate subtle erosional breaks or periods of nondeposition.

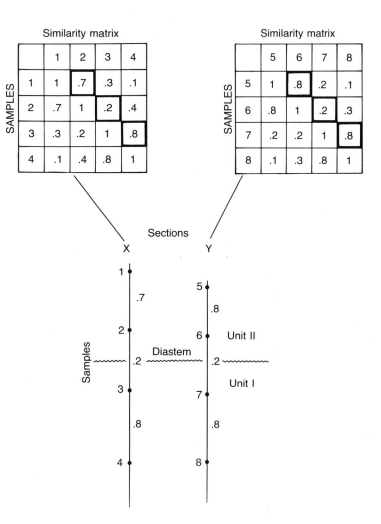

recorded between two adjacent samples is likely an indication of a marked change in the faunal assemblage. If there is no corresponding change in lithology, an erosional break should be suspected (Fig. 9.17). A FORTRAN program to perform lateral tracing was described by Millendorf and Heffner (1978).

Biostratigraphy based on microfossils from borehole samples involves some special problems because of the mixing of drill cuttings in the drilling process. The only way a meaningful stratigraphic succession can be established is by noting first appearances of species as successive formations are reached by the drill. Both stratigraphic and ecologic data can be obtained also by noting percentage shifts in species as drilling proceeds. Compared with the data from outcrop sections, the biostratigraphic picture derived from well logs is more or less blurred, a situation somewhat analogous to that caused by "noise" in electronically derived data.

9.9 CONCLUSION

The position of this chapter in the middle of the book is appropriate because its subject matter, in many ways, represents the keystone in the science of stratigraphy. It is worth emphasizing again that stratigraphy, as a study separate from that of sedimentology, is concerned particularly with the passage of time. It is especially important that time is measured in a way that can be linked directly with the sedimentary record itself. Only fossils can provide the means of doing this.

REFERENCES

Arkell, W. J., 1933, The Jurassic System in Great Britain: Oxford, Oxford Univ. Press, 681 p.

Bonham-Carter, G. F., 1967, FORTRAN IV program for Q-mode cluster analysis of non-quantitative data using IBM 7090/7094 computers: Kansas Geol. Survey Computer Contr. 17, 28 p.

Brower, J. C., 1984, The relative biostratigraphic values of fossils: Computers and Geosciences, v. 10, p. 111–132.

_____ 1985, The index fossil concept and its application to quantitative stratigraphy, in Gradstein, F. M. et al., ed., Quantitative stratigraphy: Hingham, MA, Reidel, p. 43–64.

Brower, J. C., Millendorf, S. A., and Dyman, T. S., 1978, Methods for the quantification of assemblage zones based on multivariate analysis of weighted and unweighted data: Computers and Geosciences, v. 4. p. 221–227.

Cubitt, J. M., 1978, Introduction to Proceedings of the 6th Geochautauqua on quantitative stratigraphic correlation: Computers and geosciences, v. 4, p. 215–216.

Cubitt, J. M., and Reyment, R. A., eds., 1982, Quantitative stratigraphic correlation: New York, John Wiley & Sons, 301 p.

Davaud, E., 1982, The automation of biochronological correlation, in Cubitt, J. M., and Reyment, R. A., eds., Quantitative stratigraphic correlation: New York, John Wiley & Sons, p. 85–99.

Davis, J. C., 1986, Statistics and data analysis in geology, 2nd ed.: New York, John Wiley & Sons, 646 p.

Edwards, L. E., 1978, Range charts and no-space graphs: Computers and Geosciences, v. 4 p. 247–255.

Hallam, A., 1975, Jurassic environments: Cambridge, Cambridge Univ. Press, ix + 269 p.

Hancock, J. M., 1977, The historic development of concepts of biostratigraphic correlation, in Kauffman, E. G., and Hazel, J. E., eds., Concepts and methods of biostratigraphy: Stroudsburg, PA, Dowden, Hutchinson and Ross, p. 3–22.

Hazel, J. E., 1970, Binary coefficients and clustering in biostratigraphy: Geol. Soc. America Bull., v. 81, p. 3237–3252.

_____ 1977, Use of certain multivariate and other techniques in assemblage zonal biostratigraphy: Examples utilizing Cambrian, Cretaceous and Tertiary benthic invertebrates, in Kauffman, E. G., and Hazel, J. E., eds., Concepts and methods of biostratigraphy: Stroudsburg, PA, Dowden, Hutchinson and Ross, p. 187–212.

Hedberg, H. D., ed., 1976, International stratigraphic guide: New York, John Wiley & Sons, 200 p.

Keroher, G. C., 1967, Some uses of fossil names in the evolution of stratigraphic nomenclature in the mid-continent, *in* Teichert, C., and Yochelson, E. L., Essays in paleontology and stratigraphy: Dept. of Geol., Univ. of Kansas Spec. Pub. 2, p. 21–48.

Lawson, J. D., 1979, Fossils and lithostratigraphy: Lethaia, v. 12, p. 189–191.

Ludvigsen, R., and Westrop, S. R., 1985, Three new Upper Cambrian stages for North America: Geology, v. 13, p. 139–143.

Lyell, C., 1830–1833, Principles of geology: London, John Murray, 3 vols.

Marr, J. E., 1898, The principles of stratigraphical geology: Cambridge, Cambridge Univ. Press, 304 p.

McCammon, R. B., 1970, On estimating the relative biostratigraphic value of fossils: Geol. Inst. Univ. Uppsala Bull. (n.s.), v. 2, p. 49–57.

McCammon, R. B., and Wenninger, G., 1970, The dendrograph: Kansas Geol. Survey Computer Contr., v. 48, 28 p.

McLaren, D. J., 1959, The role of fossils in defining rock units with examples from the Devonian of western and arctic Canada: Am. Jour. Sci., v. 257, p. 734–751.

Millendorf, S. A., Brower, J. C., and Dyman, T. S., 1978, A comparison of methods for the quantification of assemblage zones: Computers and Geosciences, v. 4, p. 229–242.

Millendorf, S. A., and Heffner, T., 1978, FORTRAN program for lateral tracing of time-stratigraphic units based on faunal assemblage zones: Computers and Geosciences, v. 4, p. 313–318.

Millendorf, S. A., and Millendorf, M. T., 1982, The conceptual basis for lateral tracing of biostratigraphic units, *in* Cubitt, J. N., and Reyment, R. A., eds., Quantitative stratigraphic correlation: New York, John Wiley & Sons, p. 101–106.

Millendorf, S. A., Srivastava, G. S., Dyman, T. A., and Brower, J. C., 1978, A FORTRAN program for calculating binary similarity coefficients: Computers and Geosciences, v. 4, p. 307–311.

Miller, F. X., 1977, The graphic correlation method in biostratigraphy, *in* Kauffman, E. G., and Hazel, J. E., eds., Concepts and methods of biostratigraphy: Stroudsburg, PA, Dowden, Hutchinson and Ross, p. 165–186.

Monty, C. L. V., 1968, D'Orbigny's concepts of stage and zone: Jour. Paleontology, v. 42. p. 689–701.

NACSN North American Commission on Stratigraphic Nomenclature, 1983, North American stratigraphic code: Am. Assoc. Petroleum Geologists Bull., v. 67, p. 841–875.

Palmer, A. R., 1963, Biomere—a new kind of biostratigraphic unit: Jour. Paleontology, v. 39, p. 149–153.

————— 1982, Biomere boundaries: A possible test for extraterrestrial perturbation of the biosphere: Geol. Soc. America Spec. Paper 190, p. 469–475.

Shaw, A. B., 1964, Time in stratigraphy: New York, McGraw-Hill, 365 p.

Shimer, H. W., and Shrock, R. R., 1944, Index fossils of North America: Cambridge, MA, M.I.T. Press, 837 p.

Sokal, R. R., and Sneath, P. H. A., 1963, Principles of numerical taxonomy: San Francisco, Freeman, 359 p.

Woodward, H. B., 1892, On geological zones: Geol. Assoc. London Proc., v. 12, p. 295–315.

Zachariasse, W. J., Riedel, W. R., Sanfilippo, A., Schmidt, R. R., Brolsma, M. J., Schroeder, H. J., Gersonde, R., Drooger, M. M., and Broekman, J. A., 1978, Micropaleontological counting methods and techniques in an exercise on an eight metres section of the Lower Pliocene of Cape Rossello, Sicily: Utrecht Micropal. Bull. 17, p. 1–265.

PART THREE
LITHOSTRATIGRAPHY

LITHOSTRATIGRAPHY IS NATURALLY AN ESSENTIAL PART OF ANY STRATIGRAPHIC AC-
count, but the intention in the present context is to assume that a general
background in sediments and sedimentary processes has been, or will be,
acquired elsewhere. This being said, and keeping in mind certain practical
aspects of stratigraphic data gathering used in industry, the topics selected
for this part are somewhat specialized and are intended to complement
conventional sedimentologically oriented texts. Chapter 14 is presented as
a "big picture" approach to sedimentary environments, and it is where the
present text makes contact with what might be termed the more orthodox
approach to stratigraphy.

10
STRATIGRAPHY FROM BOREHOLES

Some drill and bore the solid earth
and from the strata there extract a register.

William Cowper

10.1 INTRODUCTION

Borehole data have played such a vital role in stratigraphic studies of all kinds that some review of the methodology would seem appropriate in any stratigraphic text. The existing literature on borehole logging is already enormous, so this chapter will be no more than a summary. For more detailed accounts the reader is referred to Pirson (1977), Asquith and Gibson (1982), Visher (1984), North (1985), Doveton (1986), and Rider (1986) and to the notes and manuals prepared by the commercial logging companies such as Schlumberger, Dresser Atlas (now part of Western Atlas, International, Inc.), and Welex.

Well logs are used to determine the subsurface succession and for correlating subsurface formations. Subsurface data are integrated with surface sections in regional studies but in some areas boreholes may be the only source of stratigraphic information. In the petroleum industry, boreholes provide the geologist with information on productive zones for hydrocarbons and their thickness, structural settings, and reservoir characteristics. The value of the reservoir is naturally dependent upon how much oil and gas it contains and how easily they move through the pore system. Thus, two very important properties are the **porosity,** which is a measure of the proportion of pore space in a unit volume of rock, and the **permeability,** which is a measure of the relative ease with which fluids may pass through the pores. Evaluation of a reservoir also requires a measure of the hydrocarbon saturation (the proportions of gas, oil, and water), pore fluid pressure, and so on. All of these physical parameters can be measured by means of the various logs that can be run in a borehole. Two sources of information are available from boreholes: one set of data is obtained during the drilling process, and the other is derived from a number of logging tools of varying sophistication that measure electrical, radioactivity, and other properties of the rock formations and which produce a variety of so-called *wireline* logs.

10.2 SAMPLING FOR LITHOLOGY AND PALEONTOLOGY

Mudlogging

Direct sampling of the rocks in a borehole is done by means of drill cuttings (**ditch samples**) obtained by a **rotary drill** (Fig. 10.1), brought to the surface by circulation of **drilling mud** (Fig. 10.2) in the hole, and recorded in a sample log. Prior to the 1920s, when wireline logging was introduced, this was the sole source of information available. Although it might seem that actual chips of rock from the various formations encountered by the drill would be all that is required in establishing the subsurface succession, the situation is, in fact, more complex. One of the limitations on logging from drill cuttings stems from the delay between the time the drill penetrates the formation and the arrival of the cuttings at the surface. This *lag time,* as it is known, naturally increases with depth. For example, if the bottom of the hole is at 4000 m, it may take 1½ hours or longer

for the cuttings to reach the surface, depending on drilling speed and the rate of mud circulation. There is also the problem of mixing of cuttings from elsewhere in the section and caving from the sides of the hole above. This "stratigraphic leakage," as it is known, means that in any ditch sample, there is always a mixture of lithologies from higher parts of the section. It is important, therefore, to take careful note of the first appearance of a new lithology as drilling proceeds. Examination of the drill cuttings is done with a binocular microscope, and the different lithologies are noted as percentages, which are set out in a percentage strip log (Fig. 10.3).

From the strip log, a litholog is prepared in which the lithologies are plotted on a drilled-depth scale (Fig. 10.4), using standard symbols and sometimes colors. Additional information is contained in notations alongside the graphic plot. Because of lag time and stratigraphic leakage, the picking of formation tops (that is, the determination of the depth to a given formation) is not always easy. In practice, major changes in lithology are correlated with changes in drilling rate, but the end result is still a rather blurred picture of the stratigraphic succession. It was, in fact, the need for greater precision in measuring the depths to subsurface formations that provided the incentive for the development of the various wireline logs in the first place. Well cuttings are examined at intervals of anything from 1 to 10 m or more, depending on drilling speed. In near-surface or soft formations, drilling speed can be so high that samples are taken only at 25-m intervals. The cuttings may be further examined petrographically by thin section, and additional analysis for their microfossil content (foraminiferans, ostracodes, radiolarians, spores and pollen, etc.) is a common procedure.

Although drill cuttings provide the means of direct sampling of formations, they may not, in fact, always provide an accurate picture. This is because of mechanical effects

FIGURE 10.1
Rotary drill bit. (Photograph courtesy of the Hughes Tool Company.)

imposed by the drill bit and chemical effects of the drilling mud, as described by Graves (1986). In dense or well-indurated rocks, such as massive limestones or siliceous silt-stones, drill cuttings may retain their *in situ* texture. In many cases, however, the drill bit crushes or pulverizes the rock so that the material that finally reaches the surface as "bit flour" or "bit sand" bears little resemblance to the original rock formation or may be lost al-

FIGURE 10.2
Drill cuttings from a typical washed sample.

together. As drill cuttings are brought up in the drilling mud, they are passed through a shaker screen or shale shaker (Fig. 10.5). In practice, the smallest mesh capable of passing the mud is used, but this often allows sand-size and smaller material to pass through. Thus, a friable, poorly cemented sandstone might be completely disaggregated in the drilling process and not be recorded, whereas a shale that in outcrop might, in fact, be softer and more easily weathered would be adequately sampled from the drill cuttings. In practice, special precautions are taken if it is suspected that sampling is biased in this way. The normal procedure is to pass the mud through fine screens or to sample the sediment removed by centrifugal desanders and desilters. The effects of drilling muds must also be considered. The solution of evaporites and the effects of hydration on certain clays are obvious

examples where alternative procedures, such as the use of oil-based drilling muds, must be followed.

From ditch samples, the determination of depth to formation tops is rarely accurate to less than a meter or so, and thin units such as shale partings may be easily missed altogether. For these reasons the data from electric, radioactivity, sonic, and other logs are used to supplement the litholog because they can be considerably more accurate in the determination of formation tops. With the exception of the temperature log, which is run while running into the borehole, all these logs are run from the bottom of the hole to the top so that the cable may be kept taut for more accurate depth measurement. Depth measurements in borehole logs are made, not down from the ground surface, but from the level of the derrick floor, or rotary table (Fig. 10.6), and recorded with the notation

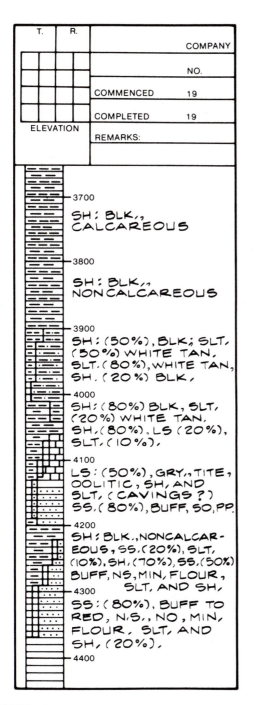

FIGURE 10.3
Percentage strip log. (From Silver, B. A., 1983, Techniques of using geologic data: Oklahoma City, Inst. for Energy Devel., Fig. 4.2, p. 80. Reprinted by permission.)

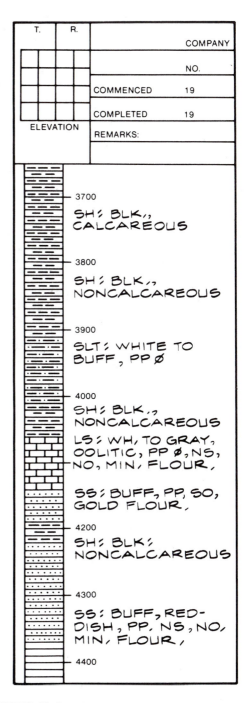

FIGURE 10.4
Typical litholog. (From Silver, B. A., 1983, Techniques of using geologic data: Oklahoma City, Inst. for Energy Devel., Fig. 4.3, p. 80. Reprinted by permission.)

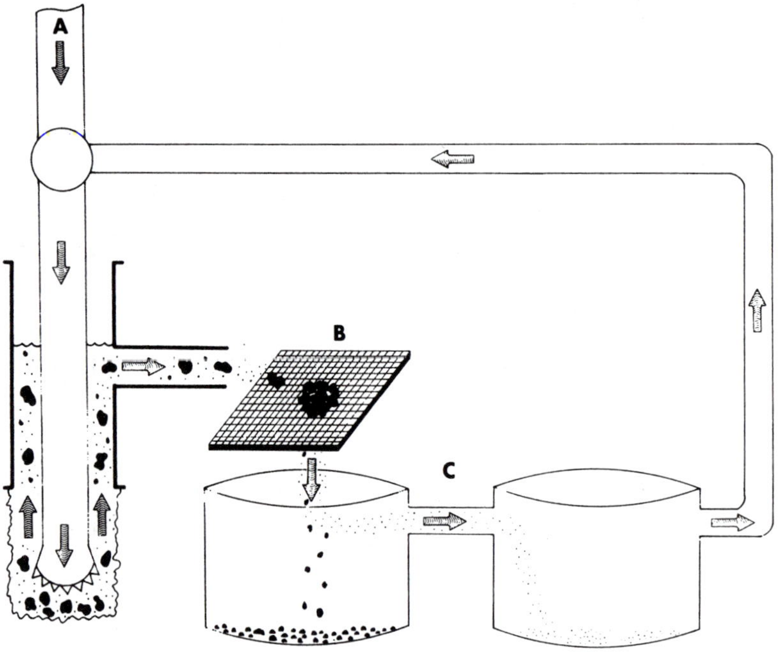

FIGURE 10.5
Diagram showing the flow of drilling mud through a rotary drilling system. Viscous mud is pumped down the drill pipe (A) and out through the drill bit. Drill cuttings are then carried up outside the drill pipe in the return flow to the surface. The cutting-laden mud is then passed through a series of sieves on the shale shaker (B), where the larger fragments are retained for sample study. Partially cleansed mud is then passed through several settling tanks to remove the fine sediment before the mud is pumped back down the well to begin the cycle again (C). (From Poag, C. W., 1977, Biostratigraphy in Gulf coastal petroleum exploration, *in* Kauffman, E. G., and Hazel, J. E., Concepts and methods of biostratigraphy, Fig. 1, p. 216. Reprinted by permission of Dowden, Hutchinson and Ross, Stroudsburg, PA.)

D.F. (alternatively, K.B. for **kelly bushing**—the collar at the top of the **kelly,** which is the square or hexagonal pipe that transmits torque to the **drill string**). Measurements are always given as depths to formation tops, and formation thicknesses, as such, are not inserted on subsurface logs as they are on surface sections.

Cores

From the above discussion, it is clear that there are some very obvious limitations to sampling by drill cuttings. When detailed and accurate information on lithology is required, such as in reservoir rock, cores are taken (Fig. 10.7). Coring is expensive in terms of rig time, so it is used only where necessary.

Cores are taken using a special tool known as a core barrel, which is simply a tube whose lower end is tipped with diamond teeth and which is rotated by the drill string. As the core is cut, it passes up inside the core barrel. The cylinders of rock removed in this way range from 5 to 15 cm in diameter and may be as long as 15–20 m. Core samples provide information that is as good as or even better than that obtained from surface outcrop. In addition, core samples play an important role in calibrating logs obtained from the various wireline tools.

If additional lithologic or paleontologic information is required, and no core is available, use is often made of a sidewall sampler. The type in wide use is the percussion sampler, in which a cylindrical bullet is fired by

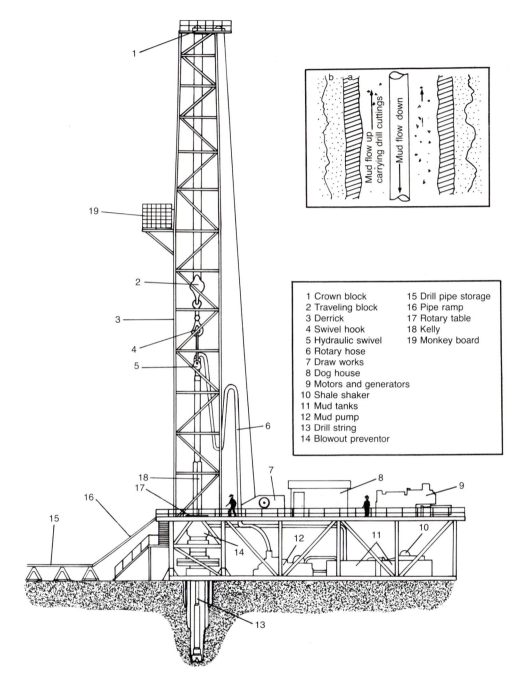

1 Crown block
2 Traveling block
3 Derrick
4 Swivel hook
5 Hydraulic swivel
6 Rotary hose
7 Draw works
8 Dog house
9 Motors and generators
10 Shale shaker
11 Mud tanks
12 Mud pump
13 Drill string
14 Blowout preventor
15 Drill pipe storage
16 Pipe ramp
17 Rotary table
18 Kelly
19 Monkey board

Mud flow up carrying drill cuttings

Mud flow down

FIGURE 10.6
Major features of a rotary drilling rig. On inset, a = mudcake b = invaded zone.

FIGURE 10.7
Typical drill core sample.

satisfactory as core samples, it is a case of making the best of it, because it is obviously too late to think of coring after the hole is drilled. Somewhat better sidewall samples can be obtained using another type of tool in which small electrically driven diamond drills are used.

Drilling-Time Logs

Variations in the hardness of the rock are typically reflected in the speed of progress in drilling so that drilling-speed logs are a useful source of information. They are constructed by noting the time taken to penetrate a given distance or, alternatively, the distance penetrated in a given time. In many cases, an automatic recorder (geolograph) is used in which the mud weight, weight of the bit, and rotational speed are all logged.

Paleologs

The microfossils identified within the formations drilled are plotted on a paleolog. The most important role of this log is biostratigraphic, to show the age of the formation, and it is an essential correlation log in wildcat wells and in exploration in general. In addition, other logs may be prepared to illustrate changes in paleoecologic parameters such as water depth, salinity, and so on, as interpreted from the fossil assemblages. With well-preserved and abundant microfaunas and floras, quite elaborate syntheses of changing paleoecologic conditions are sometimes attempted; these can prove invaluable supplements to the study of lithofacies changes.

As with drill cuttings, paleontologic samples often contain fossils from higher stratigraphic horizons because of mixing in the drilling process and caving from the walls of the hole. Such fossils are referred to as **infiltrated fossils,** and they may be the cause of erroneous biostratigraphic determinations if not recognized. It is clearly important to note

an electrically triggered powder charge laterally into the formation to take out a small core or plug (Fig. 10.8). The bullet is attached to the gun by two wires and is dragged back up the hole when the tool is withdrawn. The plugs measure up to about 5 cm in length, so they can usually penetrate the **mud cake** (see Fig. 10.6). Although they are never as

FIGURE 10.8
Sidewall core sampling. Schematic illustration of a sidewall sampler, sidewall core, and its record on a lithological log. (From Rider, M. H., Geological interpretation of well logs, Fig. 8–20, p. 90. © 1986, Blackie & Son Ltd.)

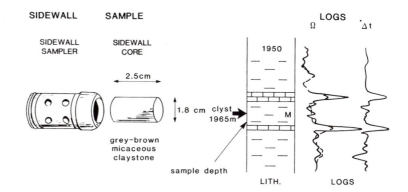

the first appearance of new faunal elements as drilling proceeds.

10.3 WIRELINE LOGS

Log Format

Wireline logs are run after the well is drilled and are obtained by means of sensing devices of various kinds contained in an insulated tube known as a **sonde** which is lowered to the bottom of the hole and then pulled to the surface. Continuous measurements are made by all the instruments and the results recorded, usually on film as galvanometer deflections keyed to a depth (or time) scale.

Throughout the petroleum industry, a standard format is used in recording well-logging data; this is known as the API (American Petroleum Institute) log grid. When printed out, this takes the form of a long strip of graph paper 8.25 in. (21.6 cm) wide, with a narrow depth column 0.75 in. (19 mm) wide between tracks 1 and 2 (Fig. 10.9). Vertically, the log (in the U.S.) is scaled in inches; in the most commonly used scales, each inch represents 100 ft, 50 ft, or 20 ft (known as the 1-in., 2-in., and 5-in. scales). The grid is subdivided further so that each division represents 10 feet of hole in the smaller scales and 2 ft of hole in the 5-in. scale. Horizontally, across each track, the scaling of the subdivisions varies depending upon which logs are being

recorded. Three types of grid are commonly used, as shown in Figure 10.9. Track 1 is always linear, but one or both of tracks 2 and 3 may be logarithmic. Sometimes tracks 2 and 3 are combined into a four-cycle logarithmic grid. A common practice is to run a number of logs (as many as 10 or more different measurements) simultaneously in various tool combination configurations, as described later. It is impossible to record all of this information on film, so the log data are recorded also on digital magnetic tape. These digital logs can be converted later to film recordings or, alternatively, can be transmitted immediately from the well site by microwave link or telephone line.

Caliper Logs

The caliper log records the changing diameter of the hole. In hard or dense rocks such as limestone, the hole is close to nominal diameter, but in soft shales or poorly cemented sandstone, fracturing by the drill or erosion by drilling mud causes caving of the walls, resulting in a larger hole (Fig. 10.10). In porous rocks the buildup of mud cake on the walls of the borehole often causes a decrease in hole size.

Temperature Logs

Because temperature is an important variable that affects resistivity (described below), a

FIGURE 10.9
The three common log grids.
(Courtesy Schlumberger Corpo-
ration.)

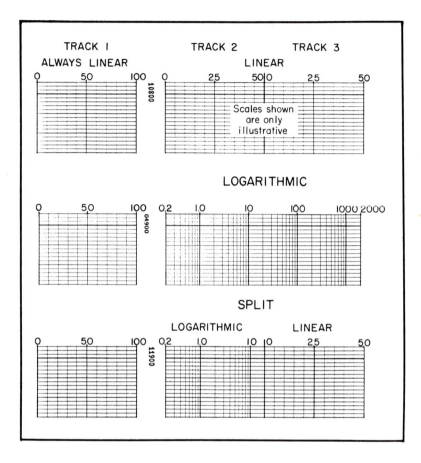

temperature log is essential for calibration purposes. The bottom hole temperature (BHT) is also important in evaluating the type of hydrocarbons present, particularly in terms of assessing the maturity of organic matter and its conversion into hydrocarbons. The temperature recorded is actually that of the drilling mud rather than that of the ambient rock or pore-space fluid temperature. Allowances must be made for the cooling effect of drilling mud and experiments have shown that it may take several months before equilibrium is reached. Normally, several temperature readings are taken with every log run after drilling stops. With several values available, an appropriate time/correction curve can be calculated.

Resistivity Logs

Resistivity logs, together with the SP (self potential) logs discussed later, are the basic **electrical logs.** They were introduced in the 1920s and so are among the oldest and most widely used of the subsurface tools. They are based upon the electrical properties of rocks with varying porosity and pore-space fluids. When rocks are dry they are essentially nonconductors of electricity, i.e., insulators, because the majority of the common rock-forming minerals are themselves insulators. The situation is different, however, with most of the fluids that are normally present in pore spaces in the rock. Pore-space water, whether it is recent ground water or **connate water,**

FIGURE 10.10

Typical response of caliper log curve. (From Rider, M. H., Geological interpretation of well logs, Fig. 3–1, p. 16. © 1986, Blackie & Son Ltd.)

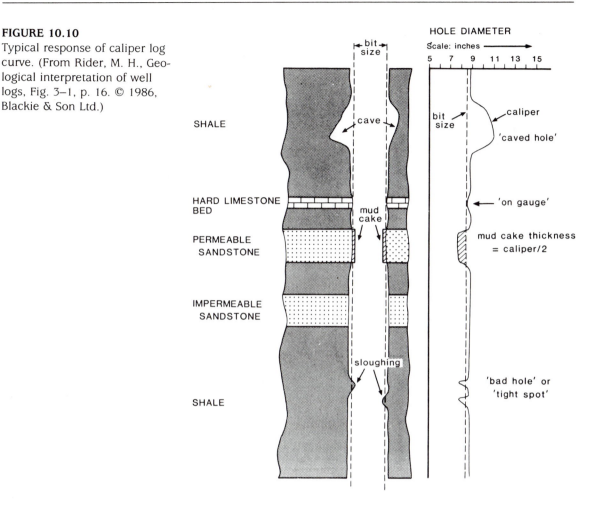

is more or less saline, and because such brines are electrolytes they conduct electricity. Roughly, the more saline the formation water, the better it is as a conductor. Generally speaking, salinity increases with depth.

Resistivity is the inverse of conductivity. In boreholes the change in resistivity from one formation to the next is what is measured as the sonde is pulled up the hole. The changes are a reflection of variations in the amount and kind of pore-space fluid contained in each formation. For a given fluid, the resistivity is inversely proportional to the porosity of the rock. Oil and gas are nonconductors, so

their presence in the pore space increases the resistivity. For a given oil- or gas-filled porosity, the smaller the water film over the mineral grains, the higher the resistivity. A further variable to be considered is the clay content of the rock; not only is the pore water in the clays a conductor but the clays themselves play a role. In general the presence of wet clay reduces the resistivity, but this varies among different clay minerals. It seems that clays with the greatest ability to exchange cations, expressed per unit weight of clay, are the more conductive. Montmorillonite is an example of a conductive clay; kaolinite is one of the least conductive.

Briefly then, the resistivity of a rock is mainly controlled by four parameters: (1) the nature of the pore-filling fluid; (2) the porosity of the rock, (3) the presence and type of clay minerals; and (4) the temperature of the fluid: the higher the temperature, the lower the resistivity, which translates into a measure of depth—the greater the depth, the lower the resistivity.

The drilling mud is another factor to be considered. Because it is under pressure, the liquid portion of the mud (mud filtrate) invades the formation to a certain distance depending on the porosity and permeability. The depth of mud filtrate penetration is known as the **invaded zone.** In the immediate vicinity of the borehole the original fluid filling the formation is almost completely replaced in what is called the **flushed zone.** As the filtrate passes into the rock it leaves the solid portion of the mud (mud cake) plastered to the wall of the borehole (see Fig. 10.6). Because the resistivity of the filtrate in the flushed zone is known, if its temperature also is known, the only variable is the rock itself, so its porosity and permeability characteristics should be seen as a distinctive signature on the log.

Resistivity is measured by passing a constant electric current down to an electrode mounted in a sonde. In the so-called normal system (Fig. 10.11) there are two electrodes—one power electrode (A) and one measuring electrode (M)—on the sonde. A second measuring electrode (N) may be on the surface or, in some instances, on the bridle at the lower end of the cable, provided it is a long distance from the first measuring electrode on the sonde. The potential difference between electrodes M and N is what is measured. By increasing the spacing between electrodes A and M, greater penetration into the formation can be achieved; on the other hand, if the spacing is too great, thin formations are likely to be missed. In practice, a

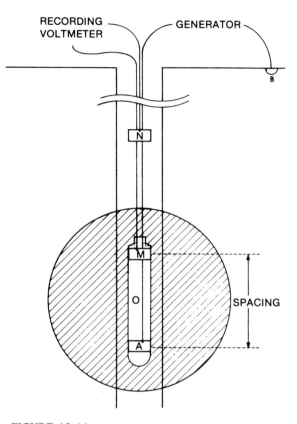

FIGURE 10.11
Arrangement of electrodes in the normal system.(From Well logging and interpretive techniques, Western Atlas International, Inc., Fig. 1, p. 2. Reprinted with permission.)

short normal (SN) spacing 16 in. (41 cm) apart is run, usually together with a long normal spacing of 64 in. (163 cm). As shown in Figure 10.12, the two logs are printed together in track 2 or tracks 2 and 3 combined, and the spacing notation appears at the top of the log. In certain cases, such as when thin beds of relatively high electrical resistance occur between thicker units, sensitivity is improved by using a modified arrangement of electrodes known as the lateral system (lateral log) in which both measuring electrodes

FIGURE 10.12
Electric logs as typically plotted together.(From Well logging and interpretive techniques, Western Atlas International, Inc., Fig. 6.1, p. 40. Reprinted with permission.)

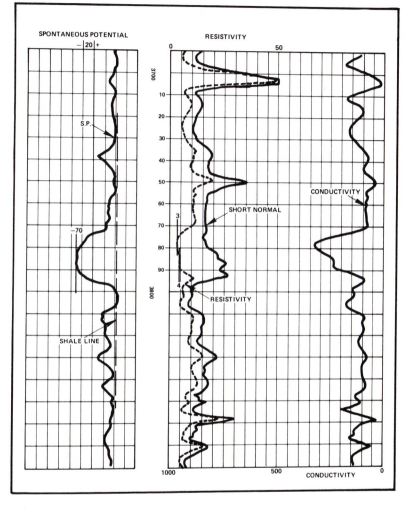

(M and N) are on the sonde in addition to the power electrode (A) (Fig. 10.13). The shapes of normal and lateral curves in several situations are compared in Figure 10.14.

The *microlog* is another means of measuring resistivity; this is done by means of closely spaced electrodes in contact with the wall of the borehole. In effect, the microlog measures both the resistivity of the mud cake, where it forms opposite porous formations, and the fluid in the rock immediately adjacent. From this can be obtained the

thickness of the mud cake, which is a good indicator of the porosity of the adjacent formation. The *laterolog* is a resistivity log in which the drilling mud is electrically charged so that as it enters the formation the current is focused laterally. The **induction log** measures the value of the true resistivity of rock formations directly and was developed for use in boreholes drilled with an oil-base mud or with air. The sonde beams an alternating current into the rock surrounding the borehole, and the eddy currents so induced gen-

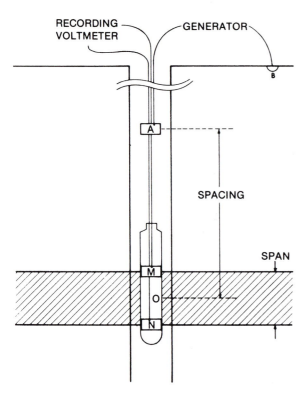

FIGURE 10.13
Arrangement of electrodes in the lateral system. (From Well logging and interpretive techniques, Western Atlas International, Inc., Fig. 2, p. 2. Reprinted with permission.)

erate a secondary magnetic field. Pickup coils in the induction sonde measure the strength of the field, which is proportional to the conductivity of the rock formation. The induction log is a particularly sensitive log for the determination of the oil–water contact in reservoir rock.

Self Potential (SP) Logs

The self potential log, which measures natural electric currents in the borehole, is generated partly by electrochemical and partly by electrokinetic effects. If an electrode is suspended in a borehole and connected to a reference electrode on the ground surface, a natural potential difference can be measured. No external electric currents are applied, and it is for this reason that the phenomenon is known as *self potential,* or *spontaneous potential.* This effect was first discovered in France by Conrad Schlumberger and his son-in-law H. G. Doll, and SP logs became widely used by the Schlumberger Company during the 1920s.

The electrochemical potential is a consequence of the fact that the drilling mud and the formation water have different salinities (concentrations of NaCl). These differences create spontaneous currents. In porous formations the mud filtrate invades the rock and moves across the interface; from the more concentrated saline formation water to the less saline filtrate, there is a net flow of Cl^- ions, with an electrical current flow in the opposite direction. This electromotive force is known as the liquid-junction potential, or *diffusion potential.* In shale formations, a spontaneous current with opposite polarity, known as the *shale potential,* is generated. Clays are relatively more permeable to Na^+ than to Cl^- ions and therefore function as a semipermeable membrane. This means that there is a net movement of Na^+ into the less saline filtrate, with the current flowing in the same direction. In other words, the mud filtrate opposite a porous sandstone will be negatively charged because of the diffusion-potential effect, and opposite a shale it will be positively charged because of the shale potential. As a consequence, around the boundary between the shale and the porous bed, an electrical current circulates. It is focused at the shale contact (Fig. 10.15) so that the log record provides an indicator of the formation top, although, in general, SP logs give rather poor bed definition.

The electrokinetic (streaming potential) effect is probably minor and is due to the passing of the mud filtrate through a negatively charged mud cake as it invades a porous for-

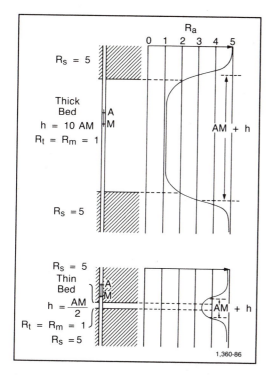

A   Normal curves—bed less resistive than adjacent formations.

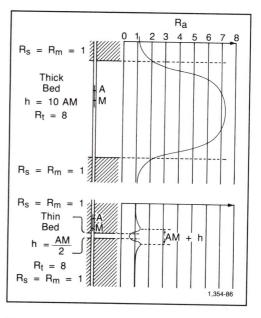

B   Normal curves—bed more resistive than adjacent formations.

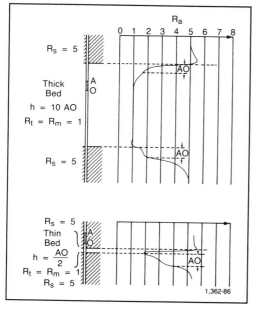

C   Lateral curves—bed less resistive than adjacent formations.

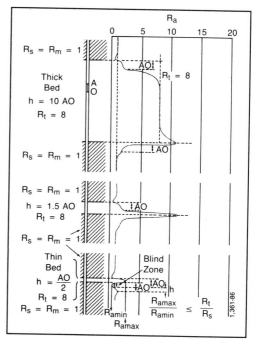

D   Lateral curves—bed more resistive than adjacent formations.

FIGURE 10.14

Response of normal and lateral spacing of electrodes in various stratigraphic successions. (Courtesy Schlumberger Corporation.)

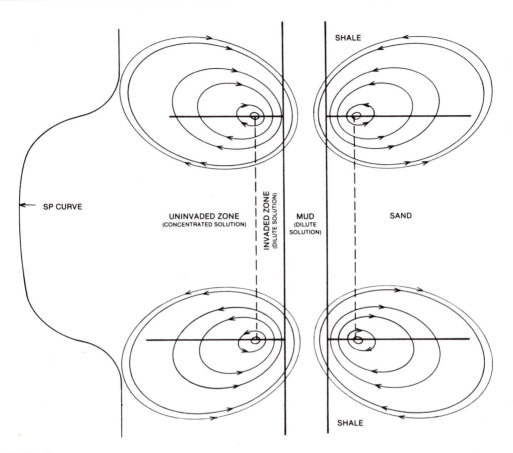

FIGURE 10.15
Diagram showing potential and current distribution in and around a permeable bed.
(From Well logging and interpretive techniques, Western Atlas International, Inc., Fig. 5,
p. 7. Reprinted with permission.)

mation, with a positive charge generated on the formation side and a negative charge on the borehole side. Normally, this effect is disregarded in electrical logging. The SP is measured in millivolts and the log is seen as a deflection from a base line (Fig. 10.16) arbitrarily selected by reference to the line recorded opposite shales (the **shale line**). No absolute values are measured, only changes in potential as the tool passes up the borehole. SP logs are useful in determining shales and also distinguishing fresh water from brines. The SP log is run at the same time as

the resistivity log and the two are usually plotted side by side, with the SP curve generally recorded in track 1.

Sonic Logs (Velocity Logs)

The acoustic velocity log measures the speed of sound through rock. This is done by recording the travel time of a sound pulse from a signal generator to three or more detectors all mounted in a sonde pulled up the borehole. The log is used in determining porosity, pore-space fluids, and interstitial fluid pres-

FIGURE 10.16
Example of SP Log in a sand-shale series. (From Well logging and interpretive techniques, Western Atlas International, Inc., Example 112A. Reprinted with permission.)

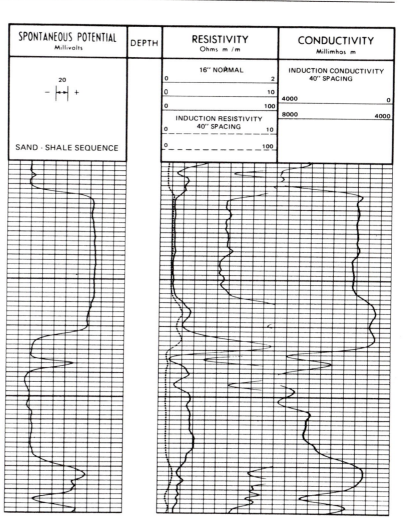

sure (Fig. 10.17). Together with the density log, it is also used in the construction of synthetic seismograms (Fig. 10.18) and in the interpretation of seismic records. The sonic log can be used to determine interval velocities (described in Chapter 11) and so is a useful means of calibrating seismic sections. This, indeed, was the reason it was first invented by Schlumberger in 1934. The sound pulse used in sonic logging has a much higher frequency than that used in seismic surveys (20–40 kHz as compared with 5–50 Hz). This means that beds as thin as 50 cm

can be detected in sonic logs, whereas, even at shallow depths, a stratigraphic interval of 10 m is about the limit of resolution of the seismic tool (Fig. 10.19). At greater depths, resolution is even poorer, rising to 50 m or more, depending on wavelength and velocity. Further discussion on this topic is presented in Chapter 11.

Gamma Ray Logs

The **gamma ray log** records the natural gamma radiation from the formation. Most

FIGURE 10.17
Some typical sonic log re-
sponses. (From Rider, M. H.,
Geological interpretation of well
logs, Fig. 8–1, p. 77. © 1986,
Blackie & Son Ltd.)

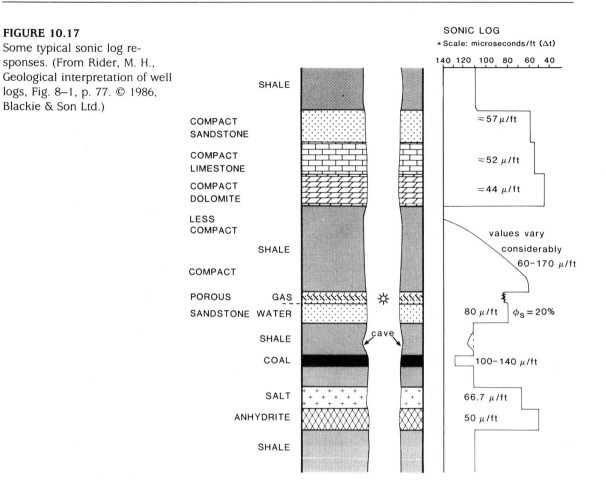

gamma ray activity comes from minerals containing nuclides of the uranium-radium series, the thorium series, and potassium-40. Of these, potassium is by far the most widely distributed in sedimentary rocks, being a common constituent of clays, many of which are the weathering products of feldspars and micas in granites. Many clays contain up to 0.3 percent potassium and of this, 0.02 per-cent is the unstable isotope potassium-40. Inevitably, therefore, shales are the most im-portant gamma ray producers. Organically rich shales may be particularly radioactive because they also often have an appreciable uranium content (up to 0.01 percent). The gamma ray log is useful in correlating hori-zons within shale successions and is a sensi-tive indicator of shale breaks. Sandstones, unless they are arkosic, typically give low readings, but there is considerable response to any clay or organic material present in a sand, so, used in conjunction with other logs, gamma ray logs are useful tools in reservoir evaluations (Fig. 10.20).

The gamma ray emission is measured in API gamma ray units, the official unit, based on standards established in a test pit in Hous-ton, Texas. Gamma ray logs must be field cal-ibrated at the wellhead before and after each run, and the calibration data are printed on the log; the gamma ray curve usually is in track 1. The measuring tool used is normally

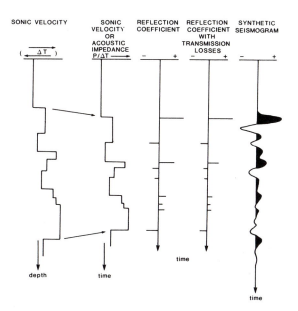

SONIC VELOCITY   SONIC VELOCITY OR ACOUSTIC IMPEDANCE P/ΔT →   REFLECTION COEFFICIENT   REFLECTION COEFFICIENT WITH TRANSMISSION LOSSES   SYNTHETIC SEISMOGRAM

FIGURE 10.18
Construction of a synthetic seismic trace from the sonic log. (From Rider, M. H., Geological interpretation of well logs, Fig. 11–4, p. 126. © 1986, Blackie & Son Ltd.)

a scintillation counter, and because it is relatively small, quite thin formations can be accurately measured. A useful feature of the gamma ray log is that it can be run in holes that have been cased, although at depths less than about 20–25 m the cosmic ray background obscures the record.

In the spectral gamma ray logging technique a spectrometer analysis is made, and it is now possible to distinguish between the three gamma ray sources: potassium, uranium, and thorium (Fig. 10.20). This has greatly improved the sensitivity of gamma ray logging in indicating shales. It has been shown, for example, that uranium as a gamma ray source is a poor shale indicator because it is widely disseminated in secondary minerals. Thorium, on the other hand, is a sensitive shale indicator, with potassium almost as reliable. Carbonates are typically not

radioactive and show gamma activity only when organic matter is present and the radioactivity is derived from uranium. The gamma ray count increases with increased shale, with potassium and thorium as the sources. Gamma ray logs have come to be among the most widely used logs and are frequently an important tool in well-to-well correlation. Too much reliance should not be placed on the smallest variations in the log because these are merely statistical artifacts and are not repeatable.

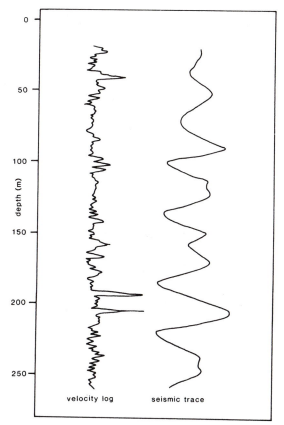

FIGURE 10.19
Contrasting frequency content of the sonic log and a seismic trace. (Redrawn from Sheriff, R. E., 1980, Seismic stratigraphy: Boston, Internat. Human Resources Devel. Corp.)

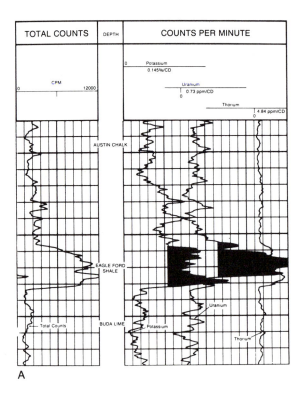

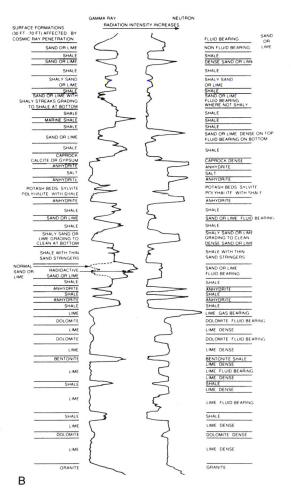

FIGURE 10.20
A. Spectral gamma ray logs. B. Typical responses in gamma ray and neutron logs. (From Well logging and interpretive techniques, Western Atlas International, Inc., (A) Fig. 11.1, p. 85, and (B) Fig. 3, p. 4. Reprinted with permission.)

Neutron Logs

The **neutron log** measures the response of formations to induced radiation provided by a radium-beryllium neutron source in a capsule passed up the hole (Fig. 10.21). The source bombards the rock formation with fast neutrons, and their degradation in energy as they pass through the formation is measured by two detectors on the sonde. The intensity of secondary gamma radiation from the formation bombarded is proportional to the hydrogen atom concentration of the pore-filling fluid, and so the log is a good measure of porosity. The presence of gas is also detect-able because it has a lower hydrogen atom concentration than does oil or water. The gamma ray and neutron logs are run at the same time, with the natural gamma ray counter separated from the neutron log device by a suitable distance in the sonde. By incorporating a delaying circuit for storage of the gamma ray readout, it can be matched with the neutron log on the depth scale (Fig. 10.22).

Density Logs

One way of measuring the density of any material is to measure its electron density, and

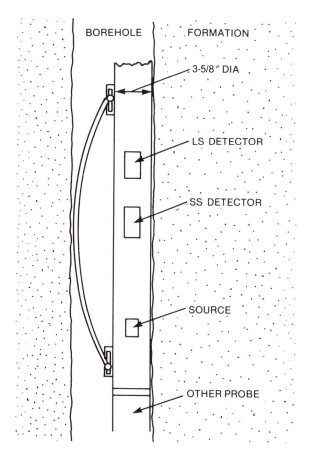

FIGURE 10.21
Schematic diagram of a compensated neutron tool. The source and detectors are held pressed against the borehole wall by spring-loaded arms. (Redrawn from Well logging and interpretive techniques, Western Atlas International, Inc., Fig. 10, p. 8. Reprinted with permission.)

the **density log** does this with the orbital electrons of the matter making up the rock. It works on the principle that the probability of a gamma ray, emitted from a gamma ray source, colliding with an orbital electron is proportional to the electron density. The log curve follows roughly the same pattern as that of velocity and neutron logs (Fig. 10.22). A density logging tool contains a gamma ray source (cobalt-60 and cesium-137 are com-

monly used) and a scintillation counter. Penetration depth is only about 15 cm into the formation so that the hole diameter, mud cake, and drilling mud are all factors that influence the readout. It is important that the tool stays in contact with the borehole wall; this is achieved by a spring-loaded backup arm and belly springs. For the same reason, the logging speed is kept down to about 8–10 m per minute to avoid the tendency for the tool to bounce off the irregularities of the wall of the borehole. In practice the best results are achieved when the sonde can be calibrated with rocks of known density in the borehole.

Dipmeter Logs

The dipmeter is a sensing device to determine the attitude of bedding planes that intersect the borehole. It is clear that over the short distance between a bedding plane trace on one side of the borehole to that on the other side, the actual displacement is very small (Fig. 10.23), so the device must be sensitive. The modern dipmeter tool has three or four arms that project radially from a central housing and touch the walls of the hole. Small electrodes on each arm produce a microresistivity trace as they pass up the borehole. The device incorporates a compass and a pendulum so that the geographic orientation of the tool, and its deviation from the vertical, can be monitored. The microresistivity traces are all very similar in shape, but correlate horizontally only if the bedding is horizontal. With increasing dip of the bedding, the vertical displacement from one trace to the next becomes greater. Computation of the displacements results in a graphic presentation known as a "tadpole plot" (Fig. 10.24). The log is calibrated in degrees of dip angle, and in most arrangements the tadpoles plot farther to the right with increasing dip angle. The tadpole tails point in the direction of dip.

FIGURE 10.22
Neutron and density logs compared. (From Doveton, J. H., Log analysis of subsurface geology, concepts and computer methods, Fig. 4, p. 39. Copyright © 1986, John Wiley & Sons. Reprinted by permission of John Wiley & Sons, Inc.)

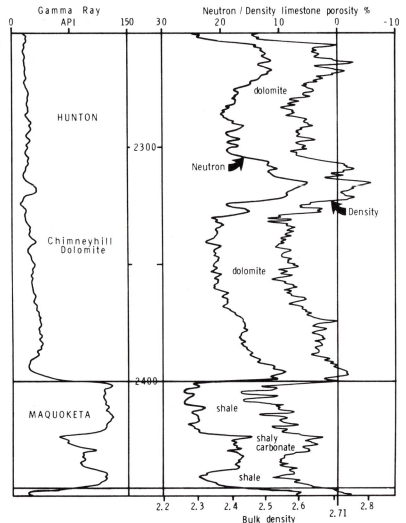

Individually, tadpoles are of little interest; it is only clusters of tadpoles that seem to have common dip or azimuth trends that are significant. To aid interpretation, a color code is widely used:

Green   Dips with a common azimuth and essentially constant dip

Blue   Dips with a common azimuth and decreasing dip with depth

Red   Dips with a common azimuth and increasing dip with depth

Although dipmeters have obvious applications in the interpretation of regional structures, they are increasingly used in stratigraphic analyses. For example, compactional drapes of shales over the relatively incompressible sandstone lenses of channel deposits or over biohermal carbonate reefs can be detected. On a much smaller scale, **cross-bedding** in many sandstones can often be determined and, used with other logs, it is possible to recognize diagnostic signatures for

FIGURE 10.23

Top: Borehole wall registration of dipping planar geologic feature as an inclined elliptical trace. Bottom: Four-arm dipmeter orientation traces and microresistivity curves. (From Doveton, J. H., Log analysis of subsurface geology, concepts and computer methods, Fig. 2, p. 125. Copyright © 1986, John Wiley & Sons. Reprinted by permission of John Wiley & Sons, Inc.)

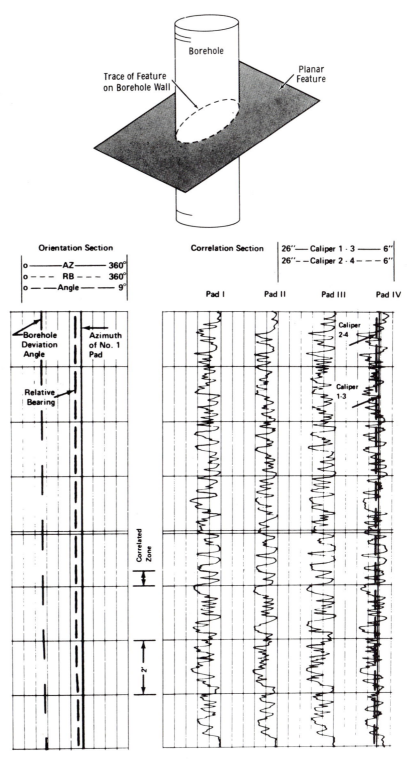

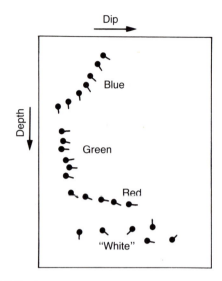

FIGURE 10.24
Tadpole, or vector, plot of computed dips showing color codes. "White" designates no trend.

particular environments, such as aeolian dunes, delta distributaries, alluvial channels, barrier beaches, and so on (Figs. 10.25 and 10.26).

Tool Combinations

It takes time to run a wireline log, so there is an obvious advantage in running as many logs as feasible at the same time. This is made possible by combining three, or sometimes more, tools into a long string that may exceed 20 m in length. One problem with this technique is that the various sensing devices are at different places in the tool string, and so cannot all measure a given rock formation at the same time. This is overcome by storing the information as it is logged in memory devices for playback as the appropriate depth is logged by the bottom tool in the string. Obviously, there always will be only a partial record for the formations at the bottom of the hole, and this requires a decision as to the sequence of tools in the string. An example

of a combination log is the commonly used IES log (for induction electrical survey) that usually includes SP, 16-in. normal, and deep induction logs, often in combination with a gamma ray log.

10.4 INTERPRETATION OF WELL LOGS

It is obvious that with all the variables possible within a mixed succession of sedimentary rocks, no single log provides a unique signature for any one lithologic type. When several logs are examined together, however, the task of identifying a rock formation and its pore-filling fluid becomes considerably easier. In other words, if one kind of log does not provide a definitive answer as to one or more characteristics of a rock unit, another log probably will.

The most widely used logs have been the SP and the resistivity (SN) curves, and in published reports they are usually plotted side by side to give a characteristic pinching and swelling pattern. These are so familiar to subsurface geologists that in many cases well-to-well correlations are made entirely on the basis of these logs. In published cross sections, if there is no notation as to the kinds of logs illustrated (and it is frequently omitted), one can assume they are the SP and SN curves (Fig. 10.27). In recent years the gamma ray log also has become widely used (Fig. 10.28). The response of the gamma ray log is roughly the converse of the SP, showing high values opposite shales and low at sandstones, the coarsest and best sorted sands having the lowest rate of gamma ray emission. Dense carbonates, evaporites, and coals all have typically low SP readings, although porous limestones and dolomites can show values similar to those of many sandstones. It should be remembered that in the case of porous sandstones and carbonates the nature of the pore-filling medium, whether it is water, oil, or gas, will alter the shape of the

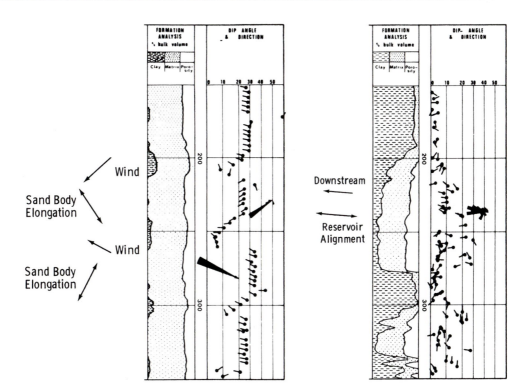

FIGURE 10.25
Dip patterns in (left) eolian dunes and (right) a braided stream alluvial deposit. (From Doveton, J. H., Log analysis of subsurface geology, concepts and computer methods. Copyright © 1986, John Wiley & Sons. Reprinted by permission of John Wiley & Sons, Inc.)

curve. As was seen in discussing the principles involved in well logging, what is actually being measured in many cases, either directly or indirectly, is the porosity and permeability of the formations. Porosity and permeability together are, in the sandstones, linked with grain size and sorting and to a lesser extent with rounding and shape of the grains. Studies by Beard and Weyl (1973) showed that porosity varies with the sorting of the sand, being considerably better (42.4 percent) in very well-sorted sand than in very poorly sorted sand (27.9 percent). Although porosity is apparently independent of grain size for sands of the same sorting, permeability improves with increasing grain size and improved sorting. The shape of the sand grains also plays a role, and with increasing angularity of grains, there is a proportional increase in both porosity and permeability.

In ancient sandstones, the effects of compaction and cementation have to be considered if those sandstones are to be compared with unindurated sands in laboratory studies. Again, cementation has an important effect, as does the presence of clay. Many sandstones that theoretically should be good reservoirs have proved to be "tight" because of occlusion of pore throats by clay which, in some cases, is authigenic and not detrital. If core samples are available, the SP values can be correlated directly with the grain size; if

FIGURE 10.26
Gamma ray and dipmeter logs used together in a turbidite sequence. (From Selley, R. C., 1978, Ancient sedimentary environments, Fig. 10–8, p. 254. Reprinted by permission of Chapman and Hall.)

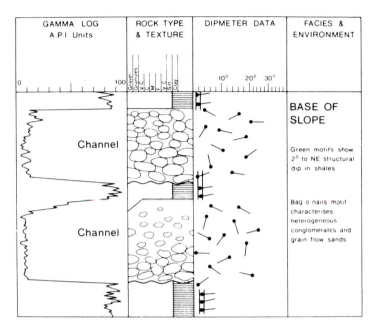

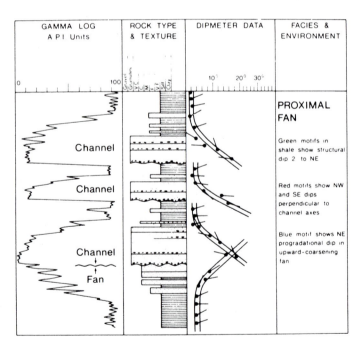

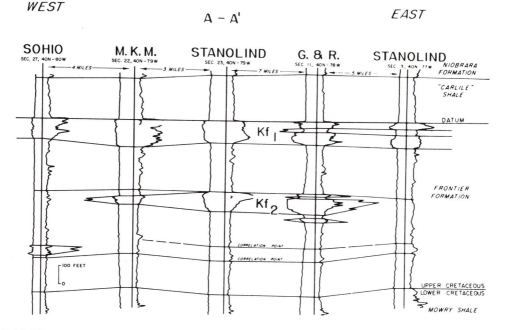

FIGURE 10.27
Example of stratigraphic correlation using electric logs. This is a west-east section A-A′
in the Frontier Formation in the Upper Cretaceous of the Salt Creek Field, Wyoming.
(From Barlow, J. A., Jr., and Hann, J. D., 1970, Regional stratigraphy of Frontier Formation and relation to Salt Creek Field, Wyo., *in* Halbouty, M. T., Geology of giant petroleum fields, AAPG Mem. 14, Fig. 7, p. 153, reprinted by permission of American Association of Petroleum Geologists.)

not, the highest SP value (at the sand line)
can be calibrated as indicating the coarsest
sand sampled in well cuttings. With decreasing sand-grain size there is usually a progressive increase in the proportion of clay present; the gamma ray log typically provides a
check of this. If the SP, resistivity, and
gamma ray logs are read in terms of variations in clastic grain size, progressive
changes through the section can be detected
and characteristic log motifs recognized as
diagnostic facies indicators.

The different log shapes seen in clastic sequences can be grouped into four major
types (Figs. 10.29 and 10.30):

1. Barrel shaped—abrupt lower and upper
   boundaries of sand underlain and overlain
   by shales
2. Funnel shaped—gradational lower boundary with shale, gradual upward reduction
   in clay content, and increase in proportion
   of sand and in sand-grain size; abrupt upper contact with shale or siltstone
3. Cone shaped—abrupt basal contact with
   shale, then a gradual decrease in sand
   and sand-grain size and concomitant increase in clay; upper contact with shale or
   siltstone gradational
4. Zigzag shaped—small-scale and often regular alternations of shale/siltstone and

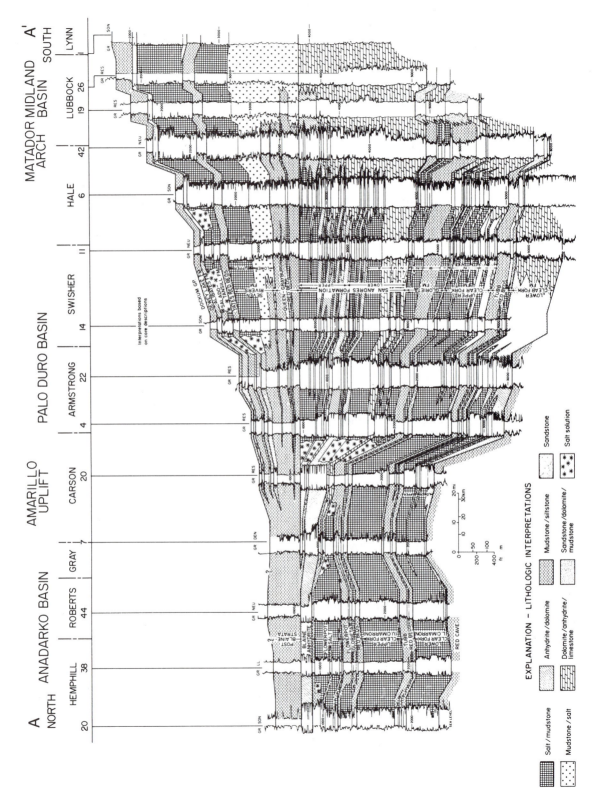

EXPLANATION – LITHOLOGIC INTERPRETATIONS

Salt / mudstone		Sandstone
	Anhydrite / dolomite	Salt solution
	Mudstone / siltstone	
Mudstone / salt	Sandstone / dolomite / mudstone	
	Dolomite / anhydrite / limestone	

FIGURE 10.28 (Opposite)
Cross section of Permian salt-bearing units in the subsurface of the Texas Panhandle, based on gamma ray, neutron, and density logs. (From Presley, M. W., 1987, Evolution of Permian evaporite basin in Texas Panhandle, AAPG Bull., v. 71, Fig. 5, p. 173, reprinted by permission of American Association of Petroleum Geologists.)

sandstone; within each sandstone the lower boundary may be gradational and the upper one abrupt (coarsening upward) or vice versa (fining upward)

It is important to note the scale of the log, because the overall shape of the small divisions in the type 4 log will often be similar to those in types 2 and 3. Each of these log signatures is suggestive of certain types of sedimentary environment, but may not be entirely diagnostic without further data. Perhaps the most useful additional input is provided by information on certain mineral-grain indicators. In particular, the relative abundance of glauconite, calcareous **bioclastic** grains, mica flakes, and carbonaceous material is a good measure of marine influ-

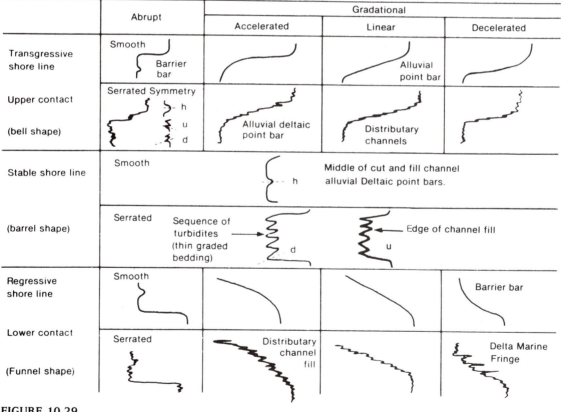

FIGURE 10.29
Classification of SP curve shapes and interpretation in terms of sedimentary environments. (Redrawn from Well logging and interpretive techniques, Western Atlas International, Inc., Fig. 34, p. 21. Reprinted with permission.)

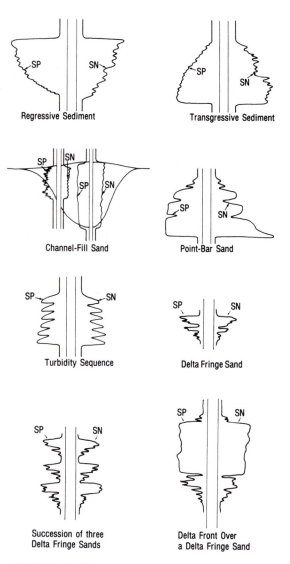

FIGURE 10.30

Sedimentary environment as expressed in SP and short normal (SN) curves. (From *Geologic well log analysis,* Third Edition, by Silvain J. Pirson. Copyright © 1983 by Gulf Publishing Company, Houston, TX. Used with permission. All rights reserved.)

ence. Subdivision into marine and nonmarine facies on the basis of only two indicators, glauconite and carbonaceous material, was suggested by Selley (1976, 1978), who also demonstrated very effectively the usefulness

of dipmeter logs in the environmental analysis of sand bodies (Fig. 10.25).

For more detailed discussion of well-log interpretation see Lynch (1962), Jageler and Matuszak (1972), Pirson (1977), and Selley (1978).

10.5 CORRELATION

Lithologic correlation of subsurface formations from well to well within a single field or other relatively small area is normally possible using any one or more of the different types of logs run. Unless there are rapid facies changes or structural complications, there is usually little room for ambiguity. With good well control, structure, isopach, reservoir, and other maps can be drawn on or between any number of horizons marked by distinctive "kicks" on the logs. Lithofacies maps such as sand/shale ratio maps can also be plotted from electric logs and gamma ray logs. With increasing distance between wells the correlations become more speculative, and in practice an increasing number of different logs together are brought to bear on the problem, usually supplemented by paleologs, if available. Correlation may be by direct comparison of certain distinctive wiggles and shapes on the logs (Fig. 10.31), or it may be based on interpretation of sedimentary environments, in which case the emphasis is on facies patterns within a regional framework. Thus, correlations would be influenced by the knowledge that, for example, a marine slope or a deltaic or marine transgressive sequence was being studied. In this latter aspect, subsurface correlation differs little from correlation of formations seen in outcrops. Correlation between subsurface formations and surface sections is not always simple and cannot necessarily be done directly. Many of the well-to-well correlations are based upon distinctive "kicks," "shoulders," and other features of the curves constructed by the various mechanical, electrical, and radioactivity logging devices. It should be remembered that

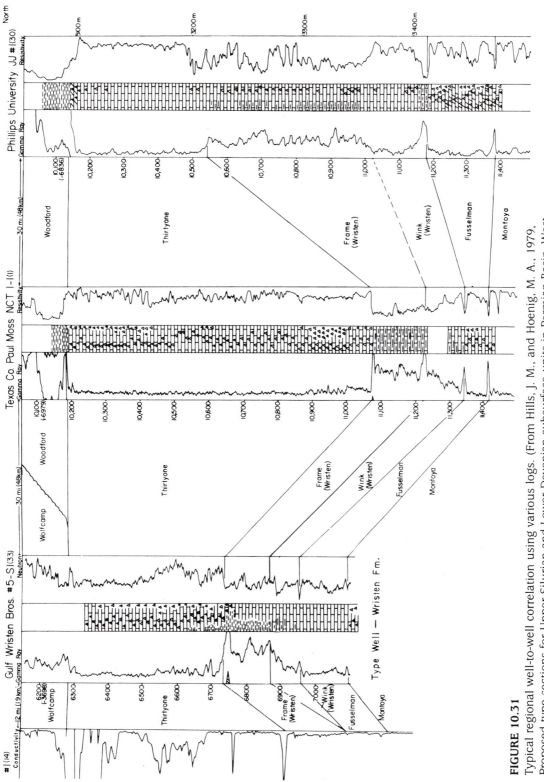

FIGURE 10.31

Typical regional well-to-well correlation using various logs. (From Hills, J. M., and Hoenig, M. A., 1979, Proposed type sections for Upper Silurian and Lower Devonian subsurface units in Permian Basin, West Texas, AAPG Bull., v. 63, Fig. 4, p. 1510, reprinted by permission of American Association of Petroleum Geologists.)

these logs plot changing values in features of the strata which are usually completely undetectable in outcrop. In one study by Ettensohn et al. (1979) in eastern Tennessee, subdivisions of Devonian-Mississippian black shales, based on subsurface gamma ray logs, were successfully located in surface outcrop sections by using a scintillometer. More commonly, the subdivision of the subsurface section into mappable units may not correlate at all with the units selected for surface mapping. Conversely, marked differences between formations and members in surface sections may not be so distinctive in well logs and may be virtually undetectable.

In the graphic presentation of well logs, a common practice is to superimpose lithologic symbols onto a wireline log curve (Fig. 10.32) or sometimes onto a pair of curves (Fig. 10.33). When one of these curves is a caliper log, the line of the log curve often gives (not surprisingly) a reasonable facsimile of how the outcrop profile would appear as a response to weathering. Note, for example, the relatively resistant dolomite in Figure 10.33. Gamma ray logs, because of their sensitive response to shales, also tend to have a natural "outcrop look" about them, as Figure 10.32 shows.

10.6 CONCLUSION

In some cases, stratigraphic data from the subsurface far exceed those from surface outcrops and even may be the only kind of information available. The acquisition, compilation, and use of these data are procedures that make subsurface stratigraphy almost a subdiscipline in its own right. Outside the petroleum industry, there is often little awareness of just how large a body of stratigraphic information is acquired from boreholes and also just how specialized much of that information is. The interpretation of wireline logs, in particular, has become an extremely sophisticated art, and a surprising

amount of geological interpretation is based on such logs.

One of the best examples of geologic exploration of a very large area that was done entirely from subsurface data is seen in the North Sea and adjacent northwest European continental shelf. Despite very detailed knowledge of the geology of the surrounding onshore areas, surprisingly little was known of the offshore situation beyond what could be projected from the land and gleaned from scattered bottom samples. Prior to the late 1950s, onshore oil exploration had proved disappointing, and the few developed fields in Britain, Holland, and Germany were small. Real exploration interest can be dated from 1959 with the discovery of the giant Groningen gas field in northern Holland, producing from the Lower Permian sandstones. After that, subsurface investigation began in earnest and, during the 1960s, several offshore gas fields were brought in. The first major North Sea oil discovery was made in 1970, when Phillips drilled the discovery well of the Ekofisk Field (Fig. 10.34).

Quite apart from the obvious economic benefits derived from the exploration of the North Sea and western European continental shelf, our understanding of the origin and early history of the North Atlantic region has been greatly enhanced by data that would have been entirely unavailable without boreholes.

REFERENCES

Asquith, G., and Gibson, C., 1982, Basic well log analysis for geologists: Am. Assoc. Petroleum Geologists Methods in Exploration Ser., No. 216.

Beard, D. C., and Weyl, P. K., 1973, Influence of texture on porosity and permeability of unconsolidated sand: Am. Assoc. Petroleum Geologists Bull., v. 57, p. 349–369.

Doveton, J. H., 1986, Log analysis of subsurface geology, concepts and computer methods: New York, John Wiley & Sons, 273 p.

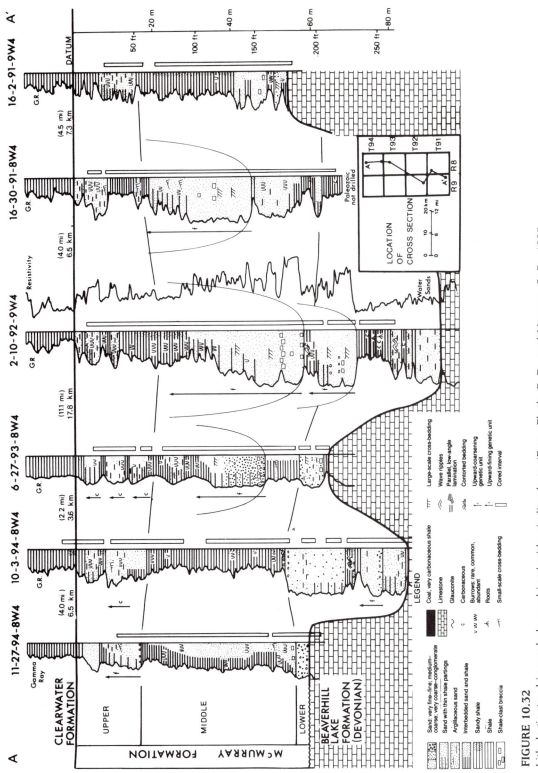

FIGURE 10.32

Lithologic graphic symbols combined with wireline logs. (From Flach, P. D., and Mossop, G. D., 1985, Depositional environments of Lower Cretaceous McMurray Formation, Athabasca oil sands, Alberta, AAPG Bull., v. 69, Fig. 3, p. 1198, reprinted by permission of American Association of Petroleum Geologists.)

271

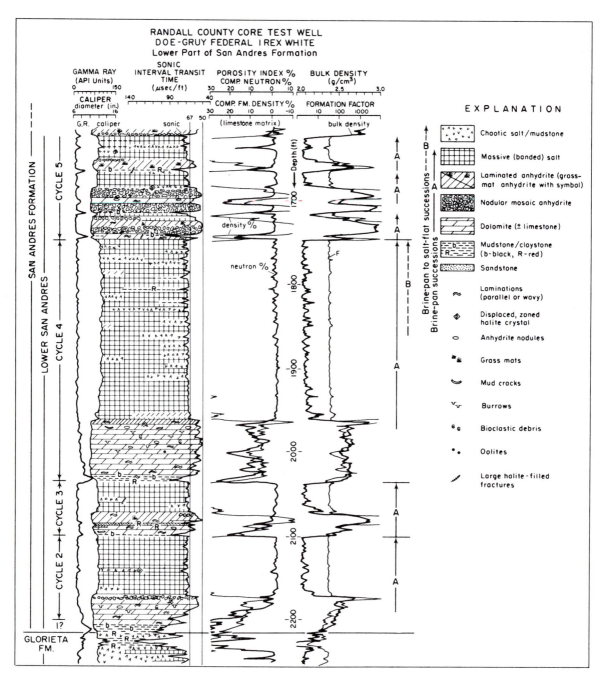

FIGURE 10.33

Lithologic graphic symbols combined with a pair of wireline logs. (From Presley, M. W., 1987, Evolution of Permian evaporite basin in Texas Panhandle, AAPG Bull., v 71, Fig. 7, p. 175, reprinted by permission of American Association of Petroleum Geologists.)

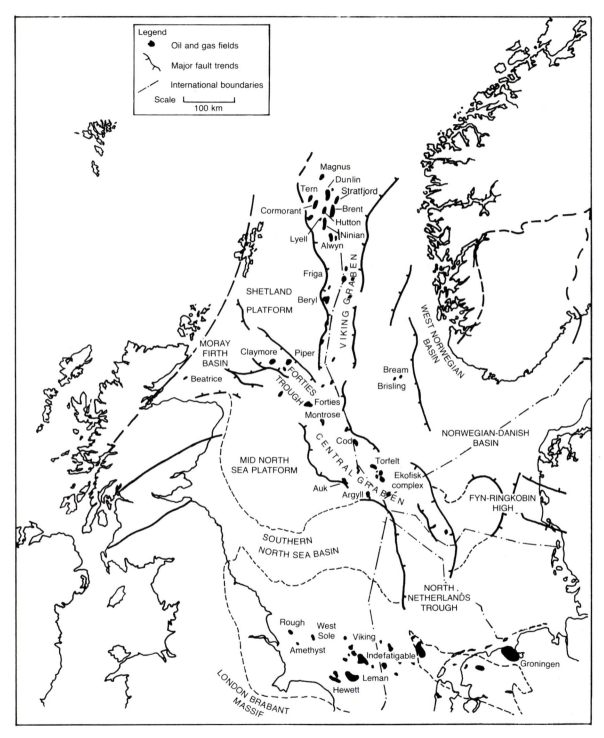

FIGURE 10.34
Chief structural features of the North Sea Basin, showing major oil and gas fields.

Ettensohn, F. R., Fulton, L. P., and Kepferle, R. C., 1979, Use of scintillometer and gamma-ray logs for correlation and stratigraphy in homogeneous black shales: Geol. Soc. America Bull., v. 90, pt. 1, p. 421–423.

Graves, W., 1986, Bit-generated rock textures and their effect on evaluation of lithology, porosity and shows in drill-cutting samples: Am. Assoc. Petroleum Geologists Bull., v. 70, p. 1129–1135.

Jageler, A. H., and Matuszak, D. R., 1972, Use of well logs and dipmeters in stratigraphic trap exploration, *in* King, R. E., ed., Stratigraphic oil and gas fields: Am. Assoc. Petroleum Geologists Spec. Pub. 10, p. 107–135.

Lynch, E. J., 1962, Formation evaluation: New York, Harper and Row, 422 p.

North, F. K., 1985, Petroleum geology: London, Allen and Unwin, 607 p.

Pirson, S. J., 1977, Geologic well log analysis, 2nd ed.: Houston, Gulf Pub. Corp., 370 p.

Rider, M. H., 1986, Geological interpretation of well logs: New York, John Wiley & Sons, 175 p.

Selley, R. C., 1976, Subsurface environmental analysis of North Sea sediments: Am. Assoc. Petroleum Geologists Bull., v. 60, p. 184–195.

———— 1978, Ancient sedimentary environments, 2nd ed.: Ithaca, NY, Cornell Univ. Press, 287 p.

Visher, G. S., 1984, Exploration stratigraphy: Tulsa, PennWell, 334 p.

11
SEISMIC STRATIGRAPHY

Pierce the dark soil, and as they pierce and pass,
Make bare the secrets of the earth's deep heart, . . .

Percy B. Shelley

11.1 INTRODUCTION

Seismic studies are concerned with vibrations that travel through the earth. These vibrations may be natural, as with earthquakes, or artificially induced by means of explosive or other sound sources. The sound waves that are transmitted through rocks are in the frequency range of 10–100 cycles per second (humans hear in the range of 20–20,000 cycles per second). Seismic surveys, as exploration tools, first came into use in the 1920s as an aid to oil exploration, the main objective being to find buried structures, such as anticlines, that might prove to be traps for oil and gas. Seismologists are concerned with earthquakes, and their contribution to our understanding of deep earth structures has come from their interpretation of earthquake waves; the majority of all other studies has been concerned with the search for hydrocarbons. As with drilling, seismic exploration is carried out almost exclusively by specialist companies in contract to the oil companies.

It is in the offshore region that seismic surveys in recent years have made their biggest contribution to geological knowledge. Here they have provided a wealth of information

275

on major bodies of crustal rocks that otherwise would be largely inaccessible. Taking advantage of increasingly sophisticated technology applied to signal enhancement and signal interpretation, seismic surveys have now become valuable stratigraphic tools and are having an increasing impact on the development of stratigraphy as a whole. This chapter is intended as a brief introduction to the principles of seismic surveying and the interpretation of seismic data; excellent introductory texts are Coffeen (1978), Sheriff (1978), Kearey and Brooks (1984), and Dobrin and Savit (1988). There are several further references to seismic data in later chapters.

11.2 PRINCIPLES OF SEISMIC SURVEYING

The basic principle in seismic surveying involves the transmitting of sound waves into the earth and the recording and analyzing of the reflected echoes from stratal surfaces underground. This is a type of echo-ranging system, roughly similar to that used by bats, whereby they can avoid obstacles in total darkness. Sound waves transmitted into the ground are both reflected back from subsurface strata and deflected, or refracted, as their velocity changes in traveling through strata of differing density. In the early days of seismic exploration, the track of sound waves refracted along certain high-velocity layers was studied and the configuration of the strata mapped. Except for certain specialized uses, this so-called refraction shooting has been now almost entirely supplanted by reflection shooting, in which the travel time of sound waves moving in near-vertical directions from subsurface reflecting horizons are recorded.

The most common, and for many years almost the only, method of generating vibrations was to use dynamite set in **shotholes** in the ground. Among nonexplosive methods is the thumper, in which a 2700 kg (3 tons)

weight is dropped on the ground, or the Dinoseis method developed by Sinclair Oil, using a gas explosion in a closed chamber set on the ground. In the Vibroseis (trade name of Continental Oil Company) method an oscillatory sound source is used rather than an impulsive one. In marine surveys, explosives are now generally discouraged because of their effect on marine life, and other methods, such as the airgun, in which a bubble of compressed air is injected into the water, are used. Reflection signals, or echoes, are picked up by sensitive sound detectors, known as **geophones** (hydrophones in marine surveys), arranged in groups and spaced along an electrical cable laid along the ground surface (Fig. 11.1). The geophones in a group are arranged with various configurations designed to reduce unwanted interference and are wired together so that the entire group acts as one geophone. Along the seismic survey line there are usually 24, 48, or 96 groups on a cable and, after amplification and filtering to remove excess "noise," signals are transmitted to magnetic tape recording equipment housed in a recording truck. Data are then computer-processed and, after enhancement, the final data tape can be used in the printing on paper of a seismic section.

It is vibrations of the ground that are received as electrical signals and transmitted to the recording instrument. Here the data from each geophone group are displayed as a wiggly line, or *trace* (Fig. 11.2). In most recording arrangements a wiggle to the right is termed a peak, representing an upward motion of the ground surface, and one to the left is called a trough. In what is known as a wiggle trace section, the traces are seen as wiggly lines alone (Fig. 11.3). Although the fine details of the wiggles can be seen, an overall view of major structures or bedding fracture is less easy to comprehend. A common practice is to use the variable-area type of display in which only peaks above a chosen amplitude are filled in to improve read-

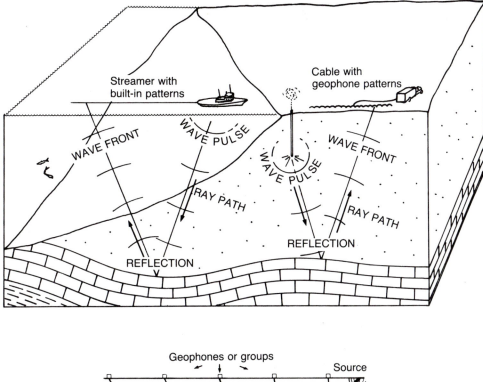

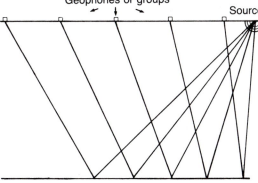

FIGURE 11.1
How seismic reflections are generated.

ability of the trace (Fig. 11.4). The record as printed typically has 24, 48, or more traces side by side, depending on the number of geophone groups on the cable. The traces are close enough that the higher-amplitude peaks and troughs of adjacent traces often overlap. At the top of the record are the shot numbers, and down the side are the timing intervals, with timelines for every 10 milliseconds running across the record (Fig. 11.4). The peaks and troughs of a trace are grouped into what are termed *wavelets*. Each wavelet comprises one or two peaks and troughs that extend over a time interval of about 50 or 100

FIGURE 11.2
Origin of seismic traces. A. Seismic profile. B. Time section. (From Silver, B. A., 1983, Techniques of using geologic data: Oklahoma City, Inst. for Energy Devel., Fig. 2.16, p. 59.)

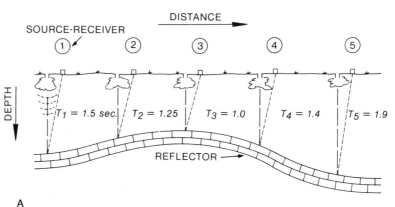

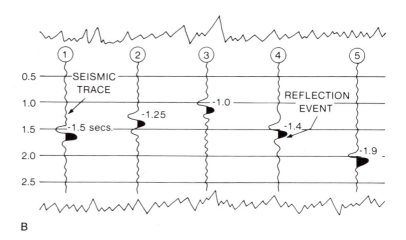

milliseconds. Wavelet configuration—the amplitude of peaks and troughs and the number and clustering of peaks and troughs—impart what is referred to as "character" (Fig. 11.5). Once recognized and if persistent over a distance on the record section, distinctive characters are often a powerful aid in correlation, particularly if faulting and other structures, poor records, or gaps in the record are causing problems. Character interpretation in terms of lithofacies, bedding thickness and type, and so on is something of an art, typically requiring long experience and a thorough knowledge of the geologic setting. In recent years, improved signal quality and

signal enhancement, together with computer processing, is making character interpretation easier and more objective. Of particular value are colored displays that help to show the significance of particular features, such as reflection strength, frequency, and velocity (Fig. 11.6).

Primary seismic reflections are a response to density/velocity contrasts along bedding surfaces or unconformities. Oil–water and gas–water contacts also generate reflections for the same reason. The rock strata vary considerably in the way they respond to seismic energy, depending on the density of the rock and the velocity of sound waves passing

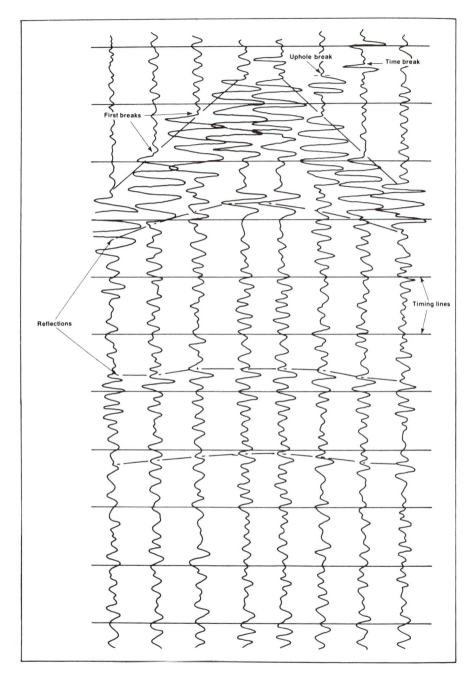

FIGURE 11.3
Typical seismic record section of the wiggle trace type. (From Coffeen, J. A., 1978, Seismic exploration fundamentals, Fig. 1–35, p. 24. Reprinted by permission of PennWell Books, Tulsa.)

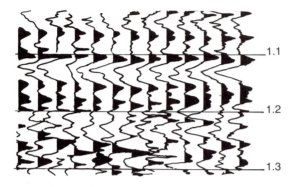

FIGURE 11.4
Portion of a variable-area type seismic record.

through. What is known as the acoustic impedance (the product of velocity and density) can be calculated for a given rock layer, and from this a reflection coefficient can be determined. As a seismic wave travels down through the earth, it will be reflected back by many interfaces. In general, the greater the reflection coefficient across the interface, the stronger the signal that is returned. It is important to remember that, because most reflections are composite in nature and controlled by interference effects, there is not necessarily any direct correlation between seismic signals and geologic stratal surfaces.

Neither is there a simple depth/travel time relationship for seismic waves traveling down into the earth. Not only does velocity increase with increasing depth, but wave paths invariably include a horizontal component.

In an attempt to overcome the problems of interpretation of seismic surveys in terms of real stratal surfaces, use is made of two types of well logs. Reflectivity is controlled by velocity and density, and these are measured on sonic and density logs, respectively. Hence, it is possible to calculate reflection coefficients from such logs and to construct a synthetic seismogram to show the seismic response of the various horizons making up the section (Fig. 11.7). Conversely, using seismic data, an equivalent of a well log known as a seismic log (Chapter 10) can be constructed. In effect, this is a kind of synthetic sonic log. Although the relationships between these two data sources are obvious, there are some fundamental differences to be considered. For one thing, seismic traces are plotted in two-way travel time, whereas logs are plotted in depth. Another difference lies in the fact that in seismic surveys it is the impedance (velocity × density) that is measured, whereas, in logs, it is the velocity alone and the density alone that are recorded.

FIGURE 11.5
Character in seismic traces. (From Coffeen, J. A., 1978, Seismic exploration fundamentals, Fig. 8–4, p. 179. Reprinted by permission of PennWell Books, Tulsa.)

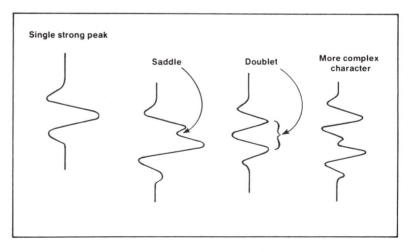

FIGURE 11.6

From a color print, showing the result of image processing and amplitude codification. The lightest bands are green and yellow for highest amplitude; in this case, indicative of limestones. (From Fontaine, J. M., Curray, R., Lacaze, J., Lanaud, R., and Yapaudjian, L., 1987, Seismic interpretation of carbonate depositional environments: AAPG Bull., v. 71, Fig. 2, p. 284, reprinted by permission of American Association of Petroleum Geologists.)

AMPLITUDE

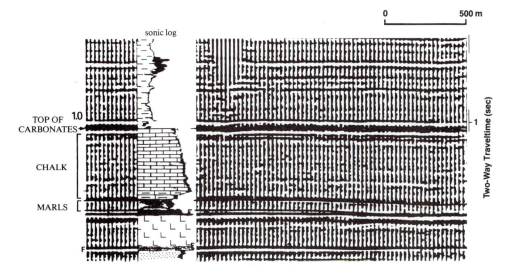

FIGURE 11.7

Synthetic seismic trace derived from a sonic log. (From Fontaine, J. M., Curray, R., Lacaze, J., Lanaud, R., and Yapaudjian, L., 1987, Seismic interpretation of carbonate depositional environments: AAPG Bull., v. 71, Fig. 5, p. 286, reprinted by permission of American Association of Petroleum Geologists.)

11.3 STRATIGRAPHIC INTERPRETATION OF SEISMIC DATA

Until relatively recently the chief role of seismic surveying in oil exploration was in its application to the interpretation of structural problems and the finding of structural traps. The main exception to this was seen in the case of reef limestones. The velocity contrasts between the limestones and their enclosing shales, and the presence of drape folds over the reef structures (Fig. 11.8), meant that reefs could often be located with a fair measure of certainty. One of the first big successes was the discovery in 1947 of the Leduc field in Alberta, Canada. In other types of sedimentary succession, vertical changes in lithology and facies changes are rarely amenable to mapping by the seismic method because of the limited resolution of the seismic pulse. This is why, in its application to oil exploration, seismic methods have not been notably successful in locating other types of stratigraphic traps such as pinchouts, facies changes, **shoestring sands,** and the like.

Nature of Reflections

Limitations of the Seismic Tool Critical lithologic changes normally take place through vertical distances that are much shorter than the wavelength of a seismic wave. Therefore, a stratigraphic succession that would be readily subdivided and described in outcrop section or in a well log is beyond the resolving power of the seismic tool and is not discernible as such in the seismic section. Unconformities and stratal surfaces can be detected by seismic surveys provided there is sufficient velocity/density contrast across them. An important aspect of seismic stratigraphy is concerned with detecting angular relationships between sets of these stratal surfaces. In one sense, therefore, what are being seen in the seismic section are still structures, but they are, for the most part, original and contemporaneous structures internal to the particular sedimentary succession or related to it, not later deformational features like folds, faults, and **diapirs.**

Conformable sequences of sediments consist of individual beds laid down one on top of the other, although not necessarily horizontally or sequentially vertically upward. Each bedding surface represents, as we have seen, a pause in deposition or a change in the character of the sediment. The seismic record of such successions will show conformable reflections, of which some will undoubtedly originate from individual bedding surfaces, but the majority actually will be interference composites generated from several bedding interfaces and representing an average of groups of bedding surfaces. In other words, although all bedding surfaces are potential reflectors, their spacing is normally too close for them to be discerned singly. This is because the frequencies used in seismic surveys are low, with wavelengths measured in tens of meters at shallow depths, increasing to hundreds of meters at greater depths. The actual resolution possible generally ranges between $\frac{1}{4}$ and $\frac{1}{8}$ wavelength but this varies, depending on such factors as noise, the quality of the record, and to some extent on an intuitive factor dependent on the interpreter. The basic relationship of wavelength to other attributes is expressed as wavelength (λ) = velocity × period (velocity/frequency). At shallow depths velocities usually range from 1500 to 2000 m/sec, with a dominant frequency of about 50 Hz (50 oscillations per sec). A simple calculation shows that wavelengths of 30 to 40 m are common. In general, at greater depths rock densities increase and so seismic velocities increase also, rising to 5000 or 6000 m/sec. Because deep reflections usually are of lower frequency (around 20 Hz is an average value), wavelengths of 250–300 m are typical of deeper sections. It follows from these relationships that the resolving power of the seis-

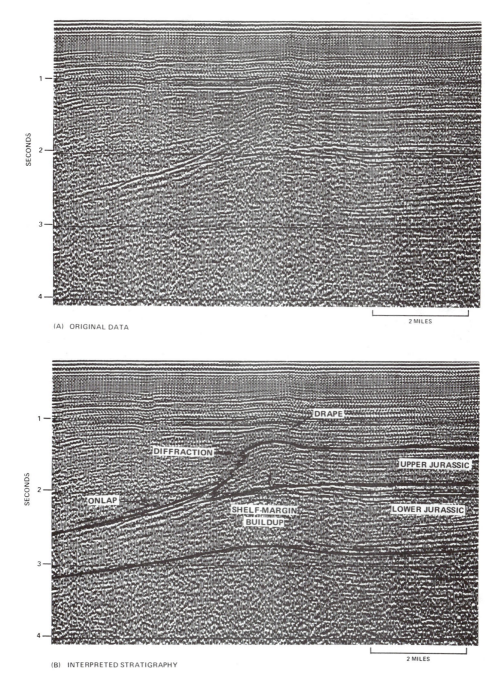

(A) ORIGINAL DATA

2 MILES

(B) INTERPRETED STRATIGRAPHY

2 MILES

FIGURE 11.8

Shelf-margin carbonate buildups detected in seismic section. The abrupt change in the dip of the reflectors and the draping over the buildup can be seen, as can the onlap of cycles onto the buildup, as indicated. A. Original data. B. Interpretive stratigraphy. (From Bubb, J. N., and Hatlelid, W. A., 1977, Seismic stratigraphy and global changes of sea level, Part 10: Seismic recognition of carbonate buildups, *in* Payton, C. E., ed., Seismic stratigraphy—Applications to hydrocarbon exploration: AAPG Mem. 26, Fig. 6, p. 191, reprinted by permission of American Association of Petroleum Geologists.)

mic tool becomes progressively poorer with increasing depth. For further discussion of problems of seismic resolution see Sheriff (1977, 1985).

Seismic Parameters Individually or collectively, seismic reflections can be described in terms of such characters as configuration, continuity, amplitude, frequency, internal velocity, and external form (Table 11.1). Probably the most important of the reflection parameters is *configuration* because it is from the configuration pattern that the general depositional setting may be determined. The more common configuration patterns are illustrated in Figure 11.9.

Of particular importance are clinoform patterns. As described in Chapter 1 (Fig. 1.5), clinothem deposits are laid down on a sloping surface separating a shallow-water environment (above wave base) from a relatively deep-water environment (below wave base). Although the basic pattern is simple, numer-

ous variations are possible, depending on such factors as differences in sediment supply, rate of deposition, or overall water depth. In major deltas, for example, progradation of the delta front (foreset beds) into a basin would be seen in the oblique pattern of reflections shown in Figures 11.9C, D, and E and 11.10. With stable or falling sea level and abundant sediment supply, the tangential pattern (Figs. 11.9D and 11.10) would be likely. Here the clinoform surfaces of the delta front pass down asymptotically into **fondoform** surfaces of **prodelta (bottomset)** deposition. The parallel oblique pattern (Figs. 11.9E and 11.10) would likely indicate little or no bottomset deposition, although the abrupt downward termination of clinoform reflectors also would be expected if detection of the bottomset beds was beyond the resolving power of the seismic tool. A similar situation might exist at the upward termination of clinoform reflections at the **undaform** surface and the absence of reflectors there might

TABLE 11.1
Seismic reflection parameters used in seismic stratigraphy and their geologic significance (from Mitchum, P. R. Jr., et al., 1977, Seismic stratigraphy and global changes of sea level, part 6: Stratigraphic interpretation of seismic reflection patterns in depositional sequences, *in* Payton, C. E., ed., Seismic stratigraphy—Applications to hydrocarbon exploration: AAPG Mem. 26, Table 2, p. 122, reprinted by permission of American Association of Petroleum Geologists)

Seismic Facies Parameters	Geologic Interpretation
Reflection configuration	Bedding patterns Depositional processes Erosion and paleotopography Fluid contacts
Reflection continuity	Bedding continuity Depositional processes
Reflection amplitude	Velocity-density contrast Bed spacing Fluid content
Reflection frequency	Bed thickness Fluid content
Interval velocity	Estimation of lithology Estimation of porosity Fluid content
External form (areal association) of seismic facies units	Gross depositional environment Sediment source Geologic setting

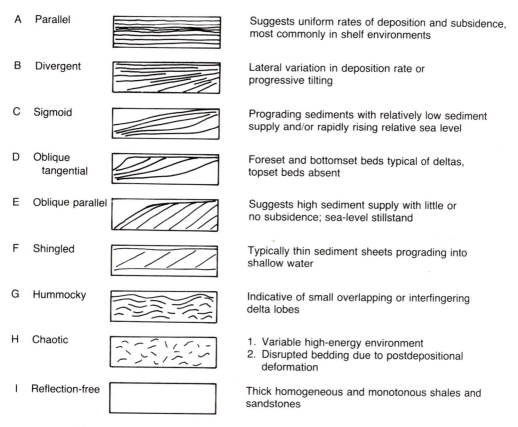

A	Parallel	Suggests uniform rates of deposition and subsidence, most commonly in shelf environments
B	Divergent	Lateral variation in deposition rate or progressive tilting
C	Sigmoid	Prograding sediments with relatively low sediment supply and/or rapidly rising relative sea level
D	Oblique tangential	Foreset and bottomset beds typical of deltas, topset beds absent
E	Oblique parallel	Suggests high sediment supply with little or no subsidence; sea-level stillstand
F	Shingled	Typically thin sediment sheets prograding into shallow water
G	Hummocky	Indicative of small overlapping or interfingering delta lobes
H	Chaotic	1. Variable high-energy environment 2. Disrupted bedding due to postdepositional deformation
I	Reflection-free	Thick homogeneous and monotonous shales and sandstones

FIGURE 11.9
Common seismic configurations.

be due either to extremely thin **topset** beds or their absence through erosion. Thin or absent topset beds are an indication of stillstand or falling sea level. With rising sea level and/or basin subsidence, topset beds become increasingly important and therefore discernible in the seismic record, with the result being the sigmoid pattern seen in Figures 11.9C and 11.10. According to Vail et al. (1977), this configuration is interpreted as implying also a relatively low-energy environment with low sediment supply.

Variations in the *continuity* of reflectors are amenable to interpretation in terms of bedding continuity. Although, as stated earlier, seismic reflection horizons do not represent bedding surfaces as such, continuity and uniformity of bedding is seen in good continuity of reflectors. Seismic sections showing good continuity are usually interpreted as indicating a widespread and uniform depositional environment. When an individual reflection has good continuity, but shows lateral changes in other characteristics, such as amplitude, it is likely that it is a bedding plane facsimile with chronostratigraphic significance. As was discussed in Chapter 1, many bedding surfaces are traceable over considerable distances and can be considered as isochronous surfaces. Vail et al. (1977) demonstrated the time-stratigraphic implications of primary seismic reflectors and there is lit-

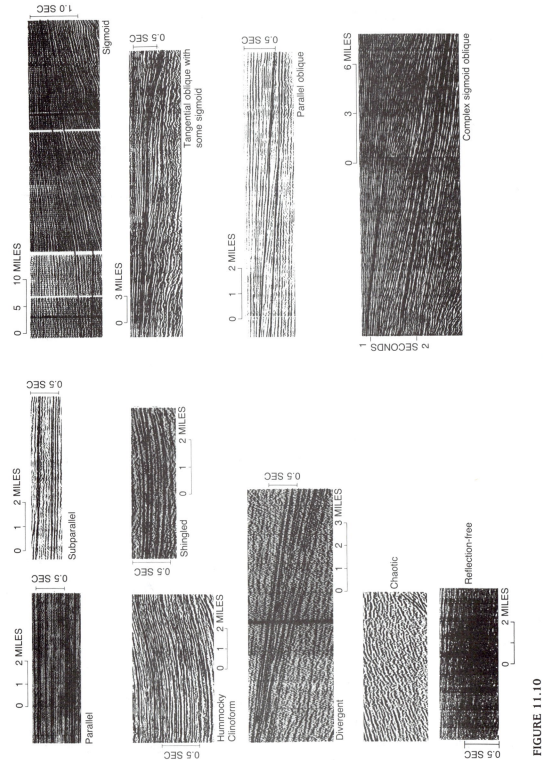

FIGURE 11.10

Portions of seismic record sections, showing typical configuration types. (From Mitchum, Vail, and San-gree, 1977.)

tle doubt that continuous reflections are in many cases isochronous surfaces (Fig. 11.11).

Variations in the *amplitude* of reflected seismic waves are largely a function of the acoustic impedance (the product of velocity and density) across the reflecting interface (Sheriff, 1977). Lateral changes in amplitude indicate changes in the character of the bedding, notably bedding thickness. As shown in Figure 11.12, the relationship between amplitude and bedding thickness is not, however, a direct one because of interference effects. As a layer (in practice, a group of beds) becomes progressively thinner, wavelets generated by velocity contrasts at the base and top of the layer become progressively closer. This telescoping effect sometimes results in enhancement as peak coincides with peak and trough with trough (increasing the amplitude). However, with different spacing, peak coincides with trough so the two tend to cancel each other out, thus reducing the amplitude. The progressive pinching out of a layer and its eventual disappearance will occur earlier on the seismic section than it does in

actuality. This is because once the thickness drops below the limit of seismic resolution, it is no longer discernible. This is known as internal convergence (Fig. 11.13).

The seismic parameter that comes closest to providing a direct indication of the lithology of the reflector is the *interval velocity* (the average velocity of seismic waves between reflectors). There is, however, no unique seismic signature to be matched against each lithologic type because mineralogy and texture are often less important than are porosity, density, and pore-fluid pressure. All of these are influenced by external controls, notably depth. As described earlier, velocity and density logs from wells that are tied into the seismic section are invaluable in interpreting some of these variables.

As with all waves, seismic waves oscillate, or vibrate, at a given *frequency,* measured in oscillations per second and expressed in hertz (Hz) or kilohertz (kHz) This value is derived by dividing the wave velocity by wavelength. As stated earlier, the sensitivity of the seismic tool in discerning thinner stratal units is controlled by the wavelength. *Wavelength,*

FIGURE 11.11

A seismic record section showing good continuity. (From Lindseth, R. O., 1979, Synthetic sonic logs—A process for stratigraphic interpretation: Geophysics, v. 44, Fig. 21, p. 24, copyright by the American Geophysical Union.)

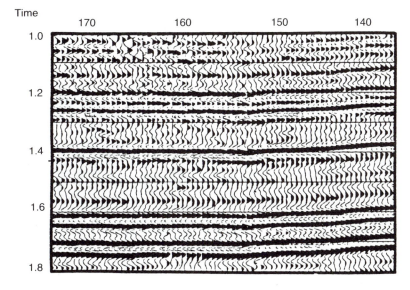

FIGURE 11.12
The effect of changing stratal thickness on wavelet form. (From Coffeen, J. A., 1978, Seismic exploration fundamentals, Fig. 8–6, p. 183. Reprinted by permission of PennWell Books, Tulsa.)

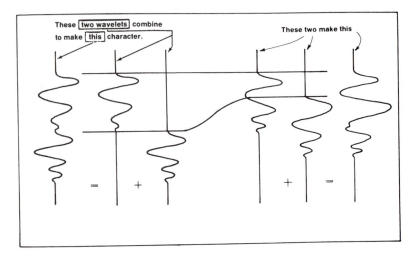

in turn, is controlled by frequency and velocity (λ = velocity/frequency) and because velocity is an inherent feature of the rock, the only controllable variable is the frequency. High-frequency sound sources give good resolution, but their ability to penetrate rock is considerably less than with low-frequency sound. The selection of a suitable sound source is important and involves a trade-off. To achieve maximum penetration, a low frequency is preferred, but to obtain better resolution, a higher-frequency sound source is required (Fig. 11.14). Bed thickness is one of the attributes that can be discerned from frequency variations. Frequency changes are also useful in determining the presence of various pore-filling fluids.

Geologic Sections and Seismic Sections

Although there is a general resemblance between seismic sections and geological cross sections, it should not be forgotten that there is a fundamental difference in the measurements used in the vertical dimension. A geological section is drawn to scale on the basis of depths measured down from the surface, or thicknesses of superimposed beds or formations, usually with some vertical exaggeration. Seismic sections, on the other hand, although naturally having a linear horizontal scale, are scaled in a vertical direction on the basis of travel time of seismic waves. No simple conversion from time to distance is possible because as rock density increases with

FIGURE 11.13
Internal convergence of reflectors.

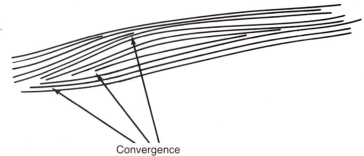

Convergence

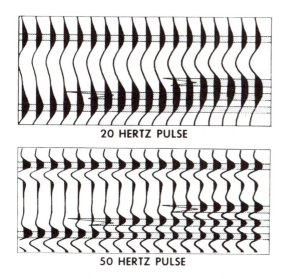

20 HERTZ PULSE

50 HERTZ PULSE

FIGURE 11.14

Higher frequencies may produce better resolution of bed thickness. (From Vail, P. R. et al., 1977, Seismic stratigraphy and global changes of sea level. Part 5, Chronostratigraphic significance of seismic reflections: in Payton, C. E., ed., Seismic stratigraphy—Applications to hydrocarbon exploration, AAPG Mem. 26, Fig. 17, p. 115, reprinted by permission of American Association of Petroleum Geologists.)

depth, so does seismic velocity. Thus, at shallow depth a 100-millisecond interval might represent a vertical distance of 100 or 150 m, whereas at greater depth the same time interval might represent twice or three times that distance. Computer manipulation of data can remove these variables, and it is now possible to produce seismic sections displayed in "real" terms of formation depth and thickness (Fig. 11.15). This is sometimes done for special purposes, but it is expensive, and so most published papers and reports in industry still largely use seismic sections based on travel time.

Vertical distance on seismic sections is, to some extent, influenced also by certain characteristics inherent in the seismic tool. For example, the depth to a given reflective interface can usually be determined only with some margin of error. This is because the reflection from what appears to be a particular interface (a known geological formation top or unconformity surface) is, in fact, not from that interface alone, but also from many smaller interfaces below and above it, all of which reflect some energy. In other words, the reflection recorded is a composite; for this reason the greatest amplitude of the wavelet is achieved only after some delay as signal strength is built up by interaction between the components of the composite reflection. What this means is that the pick of a particular reflecting horizon may occur some distance above the actual horizon.

Isochronous Surfaces

As described in Chapter 1, bedding surfaces are probably isochronous surfaces and often cut across facies boundaries. A similar point was made by Sheriff (1980), who suggested that sediments might be deposited over long intervals of time, but that every so often they were "rearranged" by some relatively sudden event, such as a storm, to form an isochronous surface. A repetition of these "rearrangement" events would impose a pattern on the sedimentary succession that would, in effect, divide it into a series of isochronous horizons. Seismic reflections are, as we have seen, randomly derived as interference composites from bedding surfaces, and so these horizons will inevitably contain a component of time parallelism. It is interesting to note that this is a form of event stratigraphy to which the seismic tool seems ideally suited. For all its shortcomings in terms of signal quality, resolution, and a legion of unknown variables, in this particular context, seismic surveys seem able to discern a subtle pattern as no other method apparently can.

11.4 SEISMIC FACIES

In discussing the lithofacies of a rock, what is referred to is the overall aspect of the rock

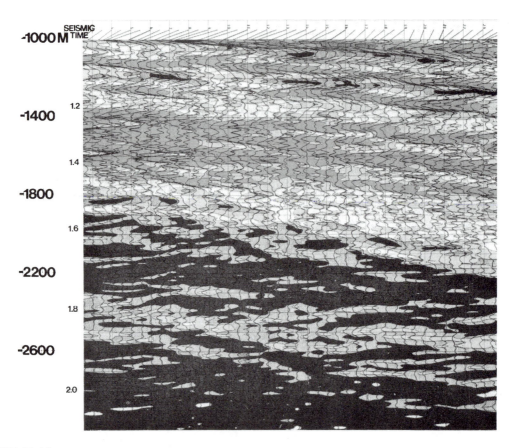

FIGURE 11.15
Example of a seismic section with true scale depth (from a colored print). (From Lindseth, R. O., and Beraldo, V. L., 1985, A late Cretaceous submarine canyon in Brazil, *in*
Berg, O. R., and Woolverton, D. G., eds., Seismic stratigraphy II, AAPG Mem. 39, Fig. 12,
p. 179, reprinted by permission of American Association of Petroleum Geologists.)

body which sets it apart from adjacent rocks.
According to Brown and Fisher (1979), a seismic facies is the sonic response to a lithofacies, and they defined it as follows: "A
seismic facies is an areally definable 3-
dimensional unit composed of seismic reflections whose elements, such as reflection configuration, amplitude, continuity, frequency
and interval velocity, differ from the elements
of adjacent facies units. A seismic facies unit
is interpreted to express certain lithologic,

stratification, and depositional features of the
deposits that generate the reflections in the
unit."

In tracing an individual reflection over any
distance, it may undergo changes in character which can often be interpreted as due to
lithofacies changes. As described earlier, a
single bedding surface may pass from sand
to silt or mud in response to changing bottom
environment or increasing water depth. In the
same way, a seismic reflection may pass

from one lithofacies to another; this will probably be indicated by changes in wave form, frequency, or amplitude.

Of all the reflection attributes discussed, the different kinds of configuration provide the most obvious basis for a first-order classification of seismic facies patterns. Sangree and Widmier (1974) suggested a classification with six different types: parallel, divergent, oblique, sigmoidal, chaotic, and reflection-free (Fig. 11.16). The interpretation of these types is summarized in Table 11.2. In another study, Brown and Fisher (1980) recognized four basic types of seismic facies: parallel/divergent, progradational, mounded/draped, and onlap/fills. Continuity and amplitude were suggested by Sheriff (1977) as diagnostic attributes. Thus, seismic facies with good continuity and high amplitude are indicative of a uniform depositional environment over a wide area. More specifically, such reflection characteristics suggest shales interbedded with thick sands, silts, or carbon-

ates, laid down in a neritic environment. Further combinations of continuity and amplitude characteristics are summarized in Table 11.3.

11.5 SEISMIC SEQUENCES

Reflection Packages

In seismic surveying, it has been found that within the overall section reflection "packages" can be recognized. These are portions of the total record containing groupings of reflections that are conformable and/or similar in their general characteristics and that are separated from the surrounding section by discontinuities of one kind or another. Unconformity-bounded units as depositional sequences can be recognized in the seismic section as seismic sequences. Internally, a seismic sequence is defined as a package of concordant reflections that have a certain homogeneity in terms of configuration type

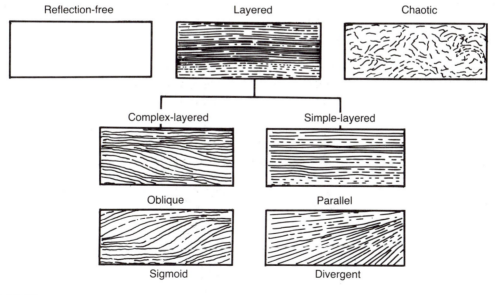

FIGURE 11.16
A classification of seismic facies. (Redrawn after Sangree and Widmier, 1974.)

TABLE 11.2
Classification and interpretation of seismic facies patterns (after Sangree and Widmier, 1974)

Parallel	Uniform conditions on a stable or uniformly subsiding surface
Divergent	Areal variation in rate of deposition, progressive tilting, or a combination of both
Oblique	Progressive outbuilding on a surface sloping down into deeper water; top of oblique pattern indicates shallow-water and high-energy environment (e.g., delta plain)
Sigmoidal	Continuing subsidence and low depositional energy, sometimes deposition in fairly deep water; often sand-poor
Chaotic	High energy; variability or disruption after deposition
Reflection-free	1. Clastic environment; uniform lithology, such as homogeneous marine shales 2. Nonclastic environment; massive carbonate or salt

and that can be interpreted as a conformable succession of genetically related strata. The bounding unconformities can best be recognized in the seismic section as surfaces of discontinuity marking reflection terminations (Fig. 11.17). If no angular discordance exists, the unconformity may still be recognized provided there is a marked density/velocity contrast across the break, although confirmation of this is usually possible only if the reflection can be traced to a place where it is marked as a surface of discontinuity by reflection terminations. Whereas the gross internal features of the depositional sequence can be discerned in seismic sections, details at the bounding unconformities may be lost because of limitations of the seismic tool. For example, as shown in Figure 11.18, if the beds adjacent to the unconformity surface get thinner as they approach asymptotically they eventually are no longer resolvable on the seismic record. As a result, the reflections appear to have abrupt terminations.

Unlike stratal surfaces, unconformity surfaces are not isochronous, but they do have a measure of chronostratigraphic significance.

TABLE 11.3
Interpretation of seismic facies (after Sheriff, 1977)

High continuity and high amplitude	
Marine	Continuous strata deposited in relatively widespread, uniform environment; neritic sands and shales.
Nonmarine	Fluvial clays and coals
Low amplitude	
Sand-prone	Grade landward into nonmarine, low-continuity, variable-amplitude seismic facies, often fluvial; grade basinward into high-continuity, high-amplitude marine facies
Shale-prone	Grade landward into silt- or sand-prone facies with high continuity and amplitude; grade basinward into prograded slope facies

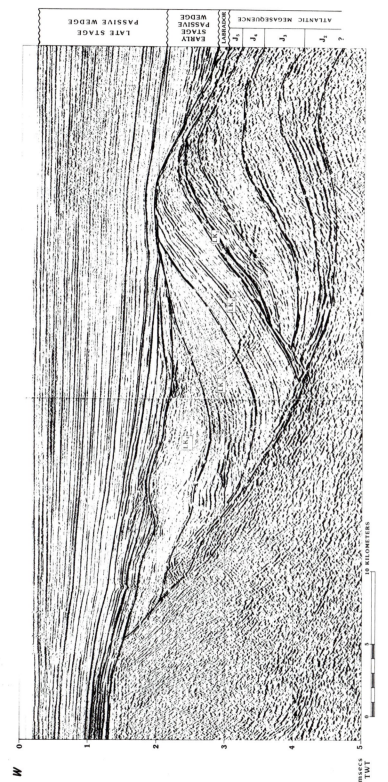

FIGURE 11.17

Seismic record section from the western flank of Avalon basin graben on Grand Banks, Newfoundland, showing reflection packages delineated in the Jurassic and early Cretaceous. The solid lines show sequence (second-order cycle) boundaries. Sequence boundaries are base J2 (220), base J3 (196), base J4 (173), base J5 (14). (From Hubbard, R. J., 1988, Age and significance of sequence boundaries on Jurassic and Early Cretaceous rifted continental margins: AAPG Bull., v. 72, Fig. 6, p. 58–59, reprinted by permission of American Association of Petroleum Geologists.)

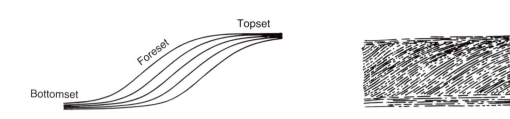

FIGURE 11.18
Delta topset, foreset, and bottomset sediments and their likely appearance on seismic
section.

Clearly, all the strata below the hiatus are older than those above. In one direction or another, when traced laterally, the hiatus becomes progressively smaller and may eventually disappear within a conformable succession. This zero hiatus horizon has been termed by Vail et al. (1977) a *conformity,* defined as a surface separating younger from older strata, but which exhibits no sign of erosion or nondeposition (Fig. 11.19). It might simply be considered as a bedding plane, although, strictly speaking, it is probably no more than a seismic entity and may not even exist as a single physical surface. There will be further reference to this aspect of seismic stratigraphy in later chapters.

Seismic Sequences and Sea-Level Changes

A depositional sequence, bounded below and above by unconformities, represents the record of a single cycle of sediment accumulation. The basal unconformity marks the surface over which the sediments were deposited. The upper unconformity is the basal unconformity of the next cycle above and is an indication that an unknown portion of the sedimentary cycle is missing as a result of erosion. The seismic record of such sedimentary cycles shows a series of reflections that are, or were initially, horizontal. Traced laterally, each reflector terminates against an inclined surface of discontinuity that marks the basal unconformity. Two examples of these termination relationships are shown in Figure 11.20. Most seismic sequences are unconformity-bounded units of one kind or another; along continental margins they usually represent the deposits laid down during a marine transgression, thus this type of discontinuity relationship is termed *coastal onlap*. Although a seismic reflection package above a coastal onlap discontinuity can be interpreted as the sediments laid down during a relative rise in sea

FIGURE 11.19
Location of hiatuses in seismic sections.

Angular discordance between reflectors indicates unconformity

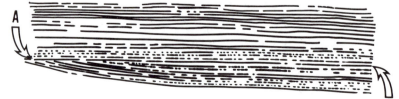

Traced laterally it passes into conformable succession

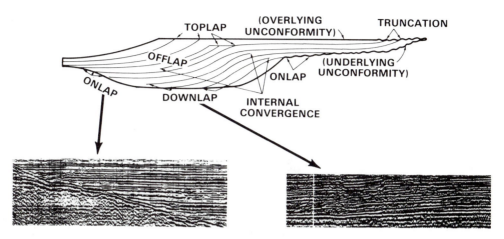

FIGURE 11.20

Several types of discontinuity surface recognized in seismic sections. (From Mitchum, R. M., Jr. et al., 1977, Seismic stratigraphy and global changes of sea level, part 6: Stratigraphic interpretation of seismic reflection patterns in depositional sequences, *in* Payton, C. E., ed., Seismic stratigraphy—Applications to hydrocarbon exploration, AAPG Mem. 26, Fig. 1, p. 118, and Figs. 3b and 3c, p. 120, reprinted by permission of American Association of Petroleum Geologists.)

level, the record of a marine regression is not necessarily due to a fall in sea level. There is no symmetrical arrangement of seismic reflectors that can be interpreted in terms of rising and falling sea level. The top of a reflection package may show reflection terminations in what is known as a *toplap* configuration (Fig. 11.21). Although this may indicate a stillstand or fall in sea level, it may also be due to depositional **progradation** resulting from an increase in the supply of terrigenous clastics. If the supply of clastics is sufficient, a net marine regression may occur even during a time when sea level is rising. It should be clear that depositional and seismic records of marine transgression and regression cannot be directly interpreted in terms of sea-level fluctuation and least of all of eustatic sea-level change. Work by Vail and associates at Exxon Corporation (Vail et al., 1977) has done much to sort out the seismic record of these complex relationships. Dis-

cussion of the Vail sea-level curve is contained in the next two chapters.

11.6 CONCLUSION

Even if all the numerous reflection parameters are considered and the many combinations of signal characteristics are carefully noted, it should be clear that there is no unique seismic signature for each lithofacies and sedimentary environment. Correct interpretation of seismic records must involve familiarity with the overall geologic setting, age relationships, and structural style of the area being studied. It cannot be emphasized too strongly that a great deal of seismic interpretation is subjective and depends upon the skill and experience of the interpreter. As pointed out by Brown and Fisher (1979), stratigraphic interpretation of seismic facies depends heavily upon a process of elimination. A review of raw data and a careful analysis of

FIGURE 11.21

Toplap relationship in seismic section. (Top: From Vail, P. R. et al., 1977, Seismic stratigraphy and global changes of sea level, part 3: Relative changes of sea level from coastal onlap, *in* Payton, C. E., ed., Seismic stratigraphy—Applications to hydrocarbon exploration, AAPG Mem. 26, Fig. 6, p. 70. Bottom: From Mitchum, R. M., Jr. et al,, 1977, Seismic stratigraphy and global changes of sea level, part 6: Stratigraphic interpretation of seismic reflection patterns in depositional sequences, *in* Payton, C. E., ed., Seismic stratigraphy—Applications to hydrocarbon exploration, AAPG Mem. 26, Fig. 2, p. 119. Both reprinted by permission of American Association of Petroleum Geologists.)

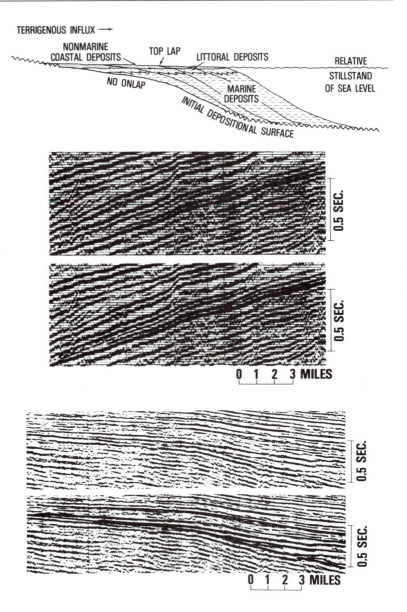

all the reflection characteristics mentioned above, together with many other data not dealt with here, will probably produce several apparently viable alternative depositional interpretations. The next step is largely intuitive and is controlled by the individual interpreter and his or her familiarity with the region and knowledge of the geologic setting.

REFERENCES

Brown, L. F., Jr., and Fisher, W. L., 1979, Seismic stratigraphic interpretation of depositional systems and its role in petroleum exploration: Am. Assoc. Petroleum Geologists, Stratigraphic interpretation of seismic data school.

_____ 1980, Seismic stratigraphic interpretations and petroleum exploration: Geophysical principles and techniques: Am. Assoc. Petroleum Geologists Continuing Educ. Course Notes Ser. 16, 56 p.

Coffeen, J. A., 1978, Seismic exploration fundamentals: Tulsa, PennWell, 277 p.

Dobrin, M. B., and Savit, C. H., 1988, Introduction to geophysical prospecting, 4th ed.: New York, McGraw-Hill, 867 p.

Kearey, P., and Brooks, M., 1984, An introduction to geophysical exploration: Oxford, Blackwell, 296 p.

Sangree, J. B., and Widmier, J. N., 1974, Interpretation of depositional facies from seismic data: paper presented at Symposium on Contemporary Geophysical Interpretation, Geophys. Soc. of Houston, 4–5 Dec., 1974.

Sheriff, R. E., 1977, Limitations on resolution of seismic reflections and geologic detail derivable from them, *in* Payton, C. E., ed., Seismic stratigraphy—Applications to hydrocarbon exploration: Am. Assoc. Petroleum Geologists Mem. 26, p. 3–14.

———— 1978, A first course in geophysical exploration and interpretation: Boston, Internat. Human Resources Dev. Corp., 313 p.

———— 1980, Seismic stratigraphy: Boston, Internat. Human Resources Devel. Corp., 227 p.

———— 1985, Aspects of seismic resolution, *in* Berg, O. R., and Woolverton, D. G., eds., Seismic stratigraphy II: An integrated approach to hydrocarbon exploration: Am. Assoc. Petroleum Geologists Mem. 39, p. 1–10.

Vail, P. R., Todd, R. G., and Sangree, J. R., 1977, Stratigraphic interpretation of seismic reflection patterns in depositional sequences, *in* Payton, C. E., ed., Seismic stratigraphy—Applications to hydrocarbon exploration: Am. Assoc. Petroleum Geologists Mem. 26, p. 99–116.

12

CHANGING SEA LEVELS

So in all lands we may sense the former presence of the sea.

Rachel Carson

12.1 INTRODUCTION

From the earliest days of stratigraphic investigation, evidence for the former presence of the sea in areas that are now dry land had been noted. Periodic incursions of marine waters onto the continents have been widespread and clearly are important events of earth history. On a regional or local scale, detailed studies of facies changes reveal that throughout the stratigraphic column much of the sedimentary succession is cyclical and contains a record of constantly changing sea level. Prior to the revelations of seismic stratigraphy, what was not known in anything but the broadest outline was whether these marine transgressions and regressions were worldwide and due to eustatic effects, were the consequences of epeirogenic or tectonic movements of the continents, or resulted from other causes. There long had been a general awareness of the apparent synchroneity of some of the changes, such as the widespread transgressions in the early Cambrian and in the Ordovician, the regression of late Permian age, the Cenomanian transgression in the Cretaceous, and so on, but the mechanisms were unknown. It was at one time supposed that the earth had undergone periods of mountain building on a global scale, alternating with periods of crustal quiescence, and fluctuations of sea level were generally thought to be linked in some way to such episodes.

Along any given coastal margin, it is obvious that the shifts in relative sea level recorded in the sediments may sometimes be due to an actual (eustatic) rise or fall in world sea level and sometimes to other, more local, mechanisms of coastal uplift or downwarping, sediment compaction and subsidence, or outbuilding, and so on. Most typically, more than one mechanism was operating at the same time so that the net rise or fall of sea level was likely the result of quite a complex interplay of several factors. At any one place, the sedimentary record is usually ambiguous as to the nature and relative importance of the different mechanisms involved. It is only when a sea-level curve constructed for one place is compared with those of similar age from elsewhere, preferably on other continents, that it is possible to discern a eustatic sea-level fluctuation "signal" and separate it from the "noise" of relative sea-level changes caused by local or regional effects.

12.2 EVIDENCE FOR SEA-LEVEL CHANGE

Within the stratigraphic succession, there are potentially three methods that can be used to determine changes in the former level of land and sea. These are to use the evidence of lithologic changes, fluctuating between deeper- and shallower-water facies; to calculate the areas of former inundations over continental shelves and continental interiors; and finally, to use seismic stratigraphic methods

to determine the number, magnitude, and type of unconformities within continental shelf successions as indicators of marine advance and retreat.

Facies changes in both clastic and nonclastic successions are usually not difficult to decipher in terms of which direction sea level was moving; that is, rising or falling. What is less clear is just how much and how rapidly sea level was changing. It is also often difficult to determine precisely when a particular marine transgression reached its climax and sea level began to fall again (Fig. 12.1). Opinions vary as to what facies most accurately reflect these changes, and this will be the subject of later discussion. Just as important as the study of facies changes is the accurate determination of the age of transgressions. When dating is based on biostratigraphic evidence, the selection of suitable zone fossils may be influenced by the nature of the transgression or regression, as was described in Chapter 8. Biostratigraphic control is so essential that it is only in those parts of the stratigraphic column with adequate faunas and floras that it has been possible to construct anything other than a very generalized sea-level curve.

In the Jurassic, as a result of work by Hallam (1977, 1978, 1981, 1984), the sea-level curve is now becoming very well-documented. Only slightly less detailed is the curve for the Cretaceous, based on the works of, among others, Kauffman (1977). For the Devonian, work by House (1975a, b) can be cited, and for the Carboniferous, where the detailed studies of coal-bearing cyclothems early revealed the relationship between facies changes and sea-level fluctuations, the work of Ramsbottom (1977) has been of prime importance.

In the second method, the areas covered, or formerly covered, by marine sediments of a given age or age range (Fig. 12.2) are measured on equal-area projection maps. Because continentwide and intercontinental cor-

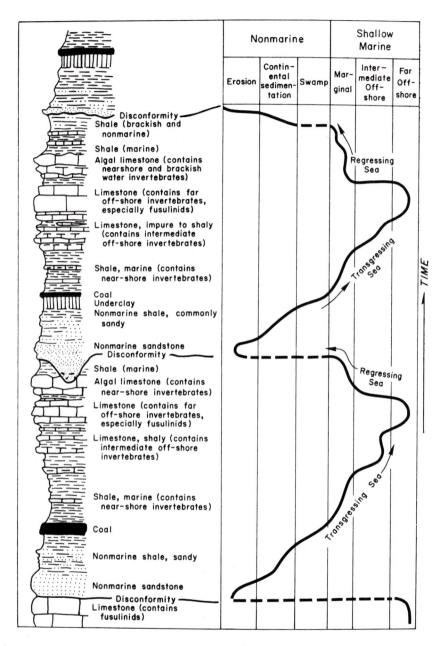

FIGURE 12.1
Carboniferous cyclothem showing the successional facies changes related to sea-level fluctuations. (From Crowell, J. C., 1978, Gondwanan glaciation, cyclothems, continental positioning and climate change: Am. Jour. Sci., v. 278, Fig. 1, p. 1346. Reprinted by permission of American Journal of Science.)

FIGURE 12.2
Cratonic areas covered by marine incursions during the Phanerozoic. (Data from Hallam, 1977, and Wise, 1974.)

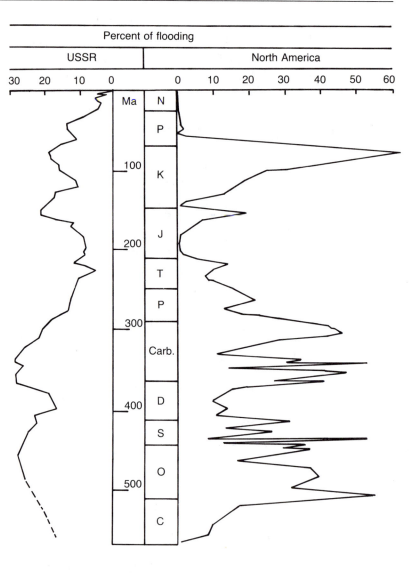

relations are involved, accurate long-range biostratigraphic correlation is again essential. Paleogeographic maps for different geologic periods are the basis for the calculations involved (Fig. 12.3). These vary considerably in their accuracy from continent to continent, with North America and the USSR perhaps having the best control. As shown in Figure 12.2, the graphic plots of percentages of areas inundated on the two platforms match reasonably well and are considered quite accurate. Naturally, the older the succession, the more likely it is to have been removed by erosion, buried, or metamorphosed, so the accuracy of the areal method tends to decrease with increasing age. The sedimentary envelope deposited during a given marine incursion inevitably thins to a featheredge at some distant strandline. Removal by erosion during the subsequent period of emergence

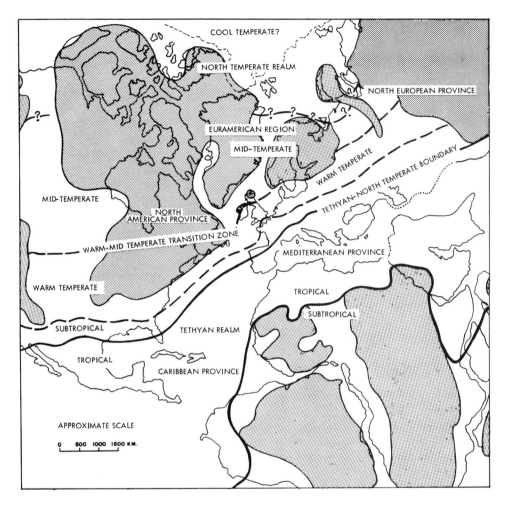

FIGURE 12.3
Generalized paleogeographic map showing the distribution of land and sea during the
maximum of a Cretaceous transgression. (From Kauffman, E. G., 1975, Dispersal and
biostratigraphic potential of Cretaceous benthonic bivalves in the western interior, *in*
Caldwell, W. G. E., ed. The Cretaceous system in western North America: Geol. Assoc.
Canada Spec. Paper 13, p. 163–194. Reprinted by permission of Geological Association
of Canada.)

following marine regression means that the
location of the actual former shoreline often
can be only estimated by extrapolation. Even
if they are present, the basal transgressive
sands and strandline deposits are not notably
fossiliferous and the accurate dating or iden-

tification of given transgressions in their further reaches may not be easy. Despite such
problems, the areal-plot method is considered quite objective and has been used with
considerable success. An additional benefit is
that some measure of the actual rise in me-

ters of sea level can be calculated by combining the areal data with what can be estimated about the former **hypsometric curve** (the cumulative-frequency profile of average elevations) of the continents at times in the past. This topic will be discussed further in a later section.

The seismic stratigraphic method has made the biggest contribution to our understanding of sea-level fluctuations because it has provided details of the stratigraphy where sea-level changes make the most impression—along the continental margins. The most complete and undisturbed successions lie beneath the continental shelves of passive margins, but they, naturally, are largely inaccessible to conventional stratigraphic methods. It was only with the growing interest of the oil industry and under the impetus of exploration for offshore oil that detailed seismic information, supplemented by increasing numbers of borehole logs, has become available. The method, pioneered by Vail and his associates at Exxon Corporation (Vail et al., 1977), used large numbers of seismic sections from continental shelves around the world and was based on the recognition of numerous erosional unconformities and (as discussed in Chapter 11) coastal onlap sequences that could be precisely dated by borehole tie-ins. Many of them could be correlated worldwide and were assumed to be caused by eustatic sea-level changes. The steps involved in the construction of a sea-level curve are shown in Figure 12.4. After analyzing the depositional environments, a stratigraphic section is drawn showing the distribution of marine and coastal facies and the presence of onlap and offlap contacts. From this section, a chronostratigraphic chart is prepared in which the data are plotted against geologic time. It is then possible to construct a chart that identifies the cycles of relative rise and fall of sea level and shows the vertical component and the magnitude of the changes.

12.3 MECHANISMS

Volume Changes

Major eustatic sea-level fluctuations are caused either by changes in the volume of the ocean basins or by changes in the volume of ocean waters. In the former case, the only plausible explanation in view of the magnitude of the changes is to assume that they are linked with plate tectonic movements. In the second case, the only mechanism known to account for the high amplitude of the changes is periodic glaciation. Table 12.1 lists other causes, but they are very minor in their effect. The major changes involving plate tectonic movements occurred over long periods measured in tens to hundreds of millions of years. At the other extreme, the glacio-eustatic sea-level changes occurred over relatively short cycles, typically of less than a million years. Sea-level changes that were apparently globally synchronous and that occurred over cycles of intermediate length are more difficult to explain.

Long-Term Sea-Level Fluctuations

Perhaps the most widely held view is that variations in the rate of seafloor spreading are the major cause of sea-level change. When the rate of seafloor spreading slows down, the rate of cooling of oceanic crust relative to its movement increases and thermal contraction occurs closer to the mid-ocean ridge (Fig. 12.5). The resultant seafloor subsidence and increase in ocean basin volume is marked by eustatic sea-level fall. Conversely, an increase in spreading rate causes an increase in the volume of mid-ocean ridges and adjacent oceanic crust, with resulting decrease in the cubic holding capacity of the ocean basin.

In the Cretaceous, for example, Hays and Pitman (1973) suggested that a sudden increase in the seafloor spreading rate from 110 to 85 m.y. ago was responsible for the major marine transgression at that time.

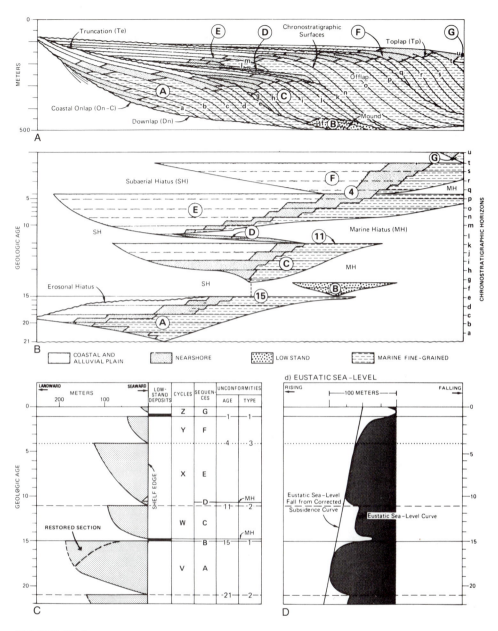

FIGURE 12.4

Procedure used by Vail and Todd (1981) in preparing a sea-level curve from seismic data. A. Stratigraphic cross section is prepared from seismic record sections and key horizons are dated by extrapolation into well sections and by other means. B. Chronostratigraphic diagram is prepared showing the relationships of the lithofacies within the various reflection packages to the time dimension. C. Reflection packages are interpreted in terms of cycles of coastal onlap. D. Using a correction factor for crustal subsidence, the coastal onlap curve is modified to become a eustatic sea-level curve. (From Vail, P. R., and Todd, R. G., Northern North Sea Jurassic unconformities, chronostratigraphy and sea level changes from seismic stratigraphy, *in* Illing, L. V., and Hobson, G. D., eds., Petroleum geology of the continental shelf of northwest Europe: London Inst. Petroleum, Fig. 2, p. 220. Reprinted by permission of Elsevier Science Publishers B. V.)

TABLE 12.1
Causes of relative change of sea level

Eustatic changes
Changes in volume of ocean water
    Changes in volume of land ice (up to 150 m at 1 cm/yr)
    Desiccation of basins (15 m+ at 1 cm/yr)
    Changes in mean oceanic temperatures (1m/°C)
Changes in volume of ocean basins
    Variations in volume of oceanic ridge (300 m at 1 cm/1000 yrs)
    Flooding of continental crust and formation of ocean trenches
    Sedimentation (few m at 1 cm/1000 yrs)
Noneustatic changes (local and regional)
Regression
    Epeirogenic upwarping
    Postglacial rebound
    Uplift and/or faulting in tectonically active areas
    Depositional progradation
Transgression
    Epeirogenic downwarping
    Sinking along hinge line of postglacial uplift
    Thermal subsidence of crust
    Downwarping and/or faulting in tectonically active areas
    Sinking of coastal tract due to compaction of underlying sediment
    Coastal retreat due to active erosion

Rather than changes in the rate of spreading, Hallam (1977) preferred an increase in the length of the spreading ocean-ridge systems. In yet another proposal by Schlanger et al. (1981), it was regional uplift of large areas of the central Pacific as a result of lithospheric heating that was responsible. These authors cited the widespread evidence for large-scale vulcanicity between 110 and 70 m.y. B.P. in support of their hypothesis.

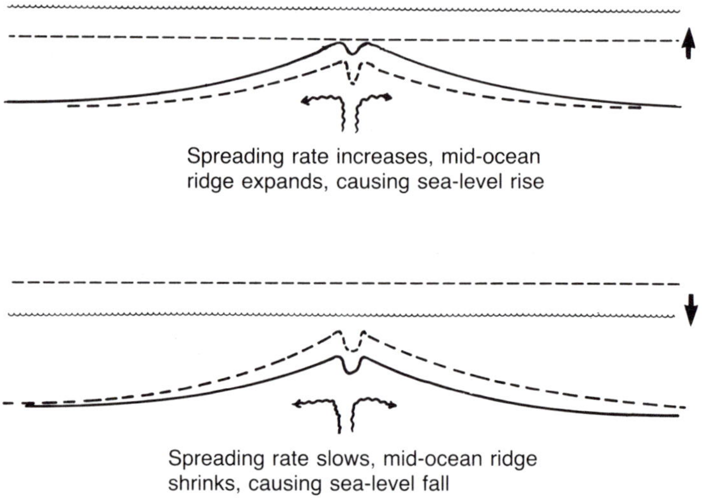

FIGURE 12.5
Seafloor spreading and sea-level fluctuations.

Spreading rate increases, mid-ocean ridge expands, causing sea-level rise

Spreading rate slows, mid-ocean ridge shrinks, causing sea-level fall

According to Worsley et al. (1984), seafloor spreading is only one manifestation of plate tectonic movements that control sea-level fluctuations. They suggested that, during rifting, continental crust is stretched and continental areas are increased, favoring marine transgression. Conversely, the subduction along continental margins of Andean type, or continental collision of Himalayan type, decreases continental area and leads to marine regression. A further control of ocean-basin volume involves thermally induced uplift of continental blocks, and this obviously results in marine regression. This effect is seen when large continental masses of Pangaea-like dimensions have accreted. At such times, it is postulated that the thermal effect of the large masses of thick and poorly conductive crust retards heat flow from the mantle.

These various controls can be viewed within the long cycle of continental accretion, rifting, and the opening and closing of major ocean basins that was first described by Wilson (1966), and since referred to as the **Wilson cycle** (discussed in more detail in Chapter 14). The interaction of eustatic changes, induced by seafloor changes on the one hand and movements of the continents on the other, were plotted by Worsley et al. (1984) to produce a master curve of global sea-level change extending over some 400 m.y. (Fig. 12.6). The curve is broadly similar to those derived by Vail et al. (1977), Pitman (1978), and Hallam (1984).

Certain aspects of the Worsley et al. (1984) model are reminiscent of that proposed by Sloss and Speed (1974), who suggested that emergence and submergence of the cratons is caused by variations in the movement of subcrustal melt produced by heat flow in the mantle. According to this model, solid–melt transformation in the asthenosphere undergoes episodic change. Variations in the proportion of melt produced in the asthenosphere below the continents and the rate of its migration are reflected in changes in the thickness of the subcrustal asthenosphere. As heat input to the asthenosphere increases, the melt fraction thickens and the continental crust above rises. Eventually a critical concentration is reached, and there is outflow of melt from subcontinental to suboceanic asthenosphere. This causes contraction or deflation of the asthenosphere below the craton and it will sink. Therefore, transfer of subcontinental melt causes not only the craton to sink but the seafloor spreading rate to increase. The effect of both of these processes is seen in a marine transgression, which is worldwide by virtue of the synchroneity of cratonic movements and obviously because of its eustatic component.

12.4 VAIL SEA-LEVEL CURVE

As mentioned earlier, the global sea-level curve (Figs. 12.7 and 12.8) proposed by Vail et al. (1977) was based upon seismic evidence of episodes of coastal onlap. From seismic record sections from around the world, sedimentary sequences between bounding unconformities and their correlative conformities were interpreted as responses to eustatic sea-level change. The sea-level curve as compiled from these data showed that variations in sea level were of three kinds: cycles with a wavelength of 200–400 m.y. (so-called first-order cycles); cycles with a wavelength of 10–80 m.y. (second-order cycles); and cycles of 1–10 m.y. (third-order cycles). In the original version of the curve (Vail et al., 1977), the shape of the cycles showed a markedly "sawtooth" configuration, depicting relatively slow sea-level rises and extremely rapid (virtually instantaneous) falls marked by the flat parts of the sawteeth.

The marked asymmetry of the cycles in the original Exxon curve was rather surprising because many of the mechanisms that have been proposed as responsible for eustatic fluctuations, possibly with the exception of glaciation, presumably operate in

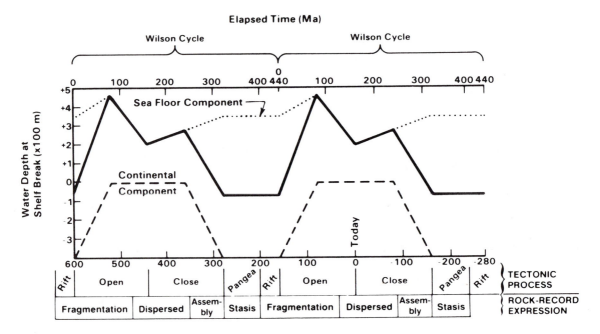

FIGURE 12.6
Curve proposed by Worsley et al. (1984) to demonstrate the relationship between long-term eustatic sea-level fluctuations and the Wilson cycle of continental rifting and oceanic evolution. At times of continental accretion (Pangea phases) the average age of oceanic floor is at a maximum and oceanic floor heat flow at a minimum, yielding deeper ocean basins and lower eustatic sea levels. As a new Wilson cycle is initiated at continental breakup the average age of oceanic floor decreases, heat flow increases, and mean ocean depth decreases, causing a eustatic rise of sea level. (From Worsley, T. R. et al., 1984. Global tectonics and eustasy for the past 2 billion years: Marine Geology, v. 58, Fig. 13, p. 391.) Reprinted by permission of Elsevier Science Publishers, B. V.

such a way that marine transgressions proceed at about the same rate as do the regressions. The cycles may vary in amplitude, but there was, to many workers, no obvious reason why the cycle curve should not be a symmetrical one. Considerable controversy surrounded the Vail curve when it was first introduced, and, because a good deal of the supporting data remained in company files, it was not easily resolved.

In later revisions of that part of the Vail curve from Triassic to Pleistocene (Vail and Todd, 1981; Vail et al., 1984; Haq et al., 1987) new ideas on interpretation were incorpo-

rated, and the marked sawtooth shape has been modified (Figs. 12.9 and 12.10). What the original study had apparently not done was to differentiate between alluvial-coastal-plain and shallow-marine facies, largely because of limitations of the seismic tool. This led to an erroneous interpretation of the data, which might show sea level still rising when, in fact, it had begun to fall. Inevitably, this reduced the time between maximum marine advance and subsequent sea-level fall and produced the "instantaneous" regression (the flat parts on the sawtooth curve) that had been unacceptable to many workers.

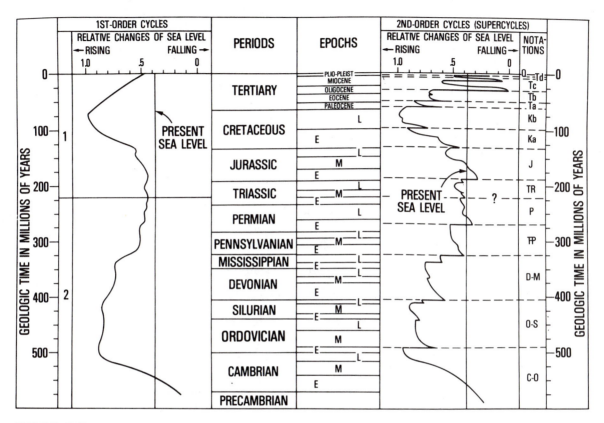

FIGURE 12.7
The original Vail sea-level curve for the Phanerozoic. (From Vail, P. R. et al., 1977, Seismic stratigraphy and global changes of sea level, part 4: Global cycles of relative changes of sea level, *in* Payton, C. E., Seismic stratigraphy—Applications to hydrocarbon exploration: AAPG Mem. 26, Fig. 1, p. 84, reprinted by permission of American Association of Petroleum Geologists.)

Vail et al. (1984) claim that there is adequate stratigraphic evidence that the coastal onlap cycles can be matched between continental margins around the world (Fig. 12.11) and that this global synchroneity can be explained only by eustatic sea-level fluctuations. There is little argument with the first-order cycles because one or other of the plate tectonic mechanisms described earlier would seem to provide an adequate explanation. Similarly, few would disagree with the glacio-eustatic origin of third-order cycles at times

of glaciation. Unfortunately, certain third-order cyclical changes occur at times when there is evidence of a nonglacial earth. The same problem arises with many of the second-order cyclical changes. Although Vail et al. (1984) insist on a eustatic cause, the apparent lack of a mechanism throws considerable doubt on their interpretation of the curve.

It was suggested by Pitman (1978) that the curve was not, in fact, a measure of eustatic change, but rather one compounded of an in-

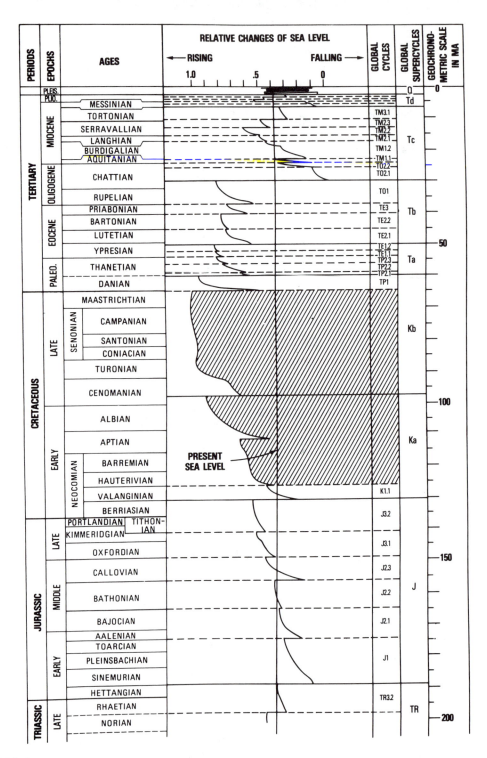

FIGURE 12.8
Global cycles of relative change of sea level during Jurassic-Tertiary time, as depicted in the original Vail et al. model. The shaded portion of the chart involves confidential proprietary data. (From Vail, P. R. et al., 1977, Seismic stratigraphy and global changes of sea level, part 4: Global cycles of relative changes of sea level, *in* Payton, C. E., Seismic stratigraphy—Applications to hydrocarbon exploration: AAPG Mem. 26, Fig. 2, p. 85, reprinted by permission of American Association of Petroleum Geologists.)

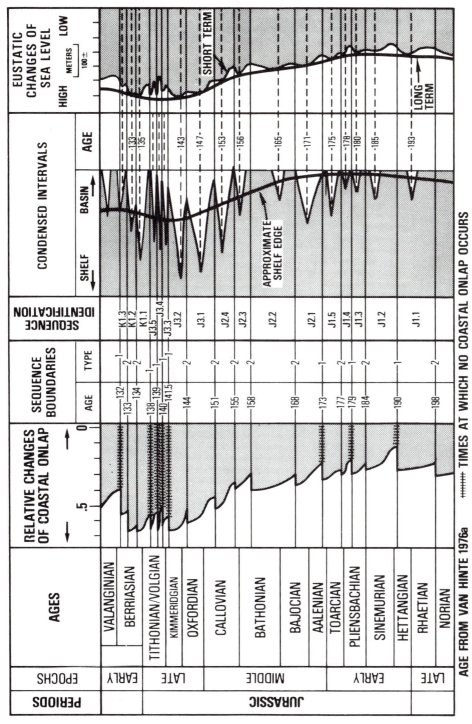

FIGURE 12.9

A late version of the Vail curve for Jurassic supercycle. (From Vail, P. R. et al., 1984, Jurassic unconformities, chronostratigraphy and sea level changes from seismic stratigraphy and biostratigraphy, *in* Schlee, J. R., ed., Interregional unconformities and hydrocarbon accumulation: AAPG Mem. 36, Fig. 2, p. 132, reprinted by permission of American Association of Petroleum Geologists.)

311

FIGURE 12.10

Sequence stratigraphy and the eustatic sea-level curve for the past 65 million years (From Haq, B. U., et al., Chronology of fluctuating sea levels since the Triassic: Science, v. 235, Fig. 2, p. 1159. Copyright 1987 by the AAAS.)

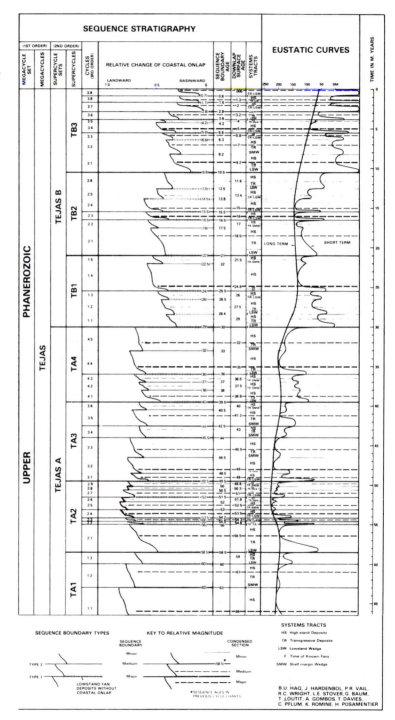

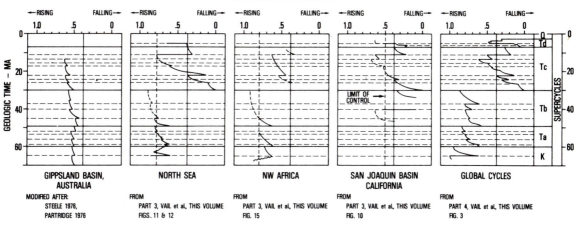

FIGURE 12.11
Correlation of regional cycles of coastal onlap from four continents. (From Vail, P. R. et al., 1977, Seismic stratigraphy and global changes of sea level, part 4: Cycle of relative changes of sea level, *in* Payton, C. E., ed., Seismic stratigraphy—Applications to hydrocarbon exploration: AAPG Mem. 26, Fig. 5, p. 90, reprinted by permission of American Association of Petroleum Geologists.)

teraction between a slow but varying eustatic component and crustal subsidence along Atlantic-type (passive) continental margins, caused by a combination of subcrustal movements. In collecting the original stratigraphic data for a global sea-level curve, only passive-type margins could be used because in active margins the record is obscured by too much local tectonic "noise." If all passive margins are progressively subsiding, and at approximately similar rates, such regional noneustatic effects do not obscure the effects of eustatic sea-level change, but only modify them. What this means is that the global sea-level curve might record changes in the *rate* of eustatic change rather than sea-level changes themselves. Thus, a marine transgression occurred when the *rate of eustatic sea-level rise increased* or the *rate of fall decreased.* Conversely, a regression was the result of a declining rate of sea-level rise or an increasing rate of sea-level fall (Fig. 12.12).

It also has been pointed out (Miall, 1986) that because the Vail curve is, in fact, no more than a modal composite of numerous regional curves, the global synchroneity may be more apparent than real. It is doubtful that biostratigraphic determinations always are accurate enough to differentiate the shorter sea-level events. This, and the possibility that certain small sequences might be missed, could lead to miscorrelation of the unconformities involved. It also is suggested that correlations between sea-level curves from widely separated locations, even on different continents, might be essentially synchronous without the necessity for a eustatic cause. Instead, the matching simply may indicate that the continental margins share a common basin history of subsidence or uplift.

This might be the interpretation of data compiled by Poag and Ward (1987) from COST (continental offshore stratigraphic test)

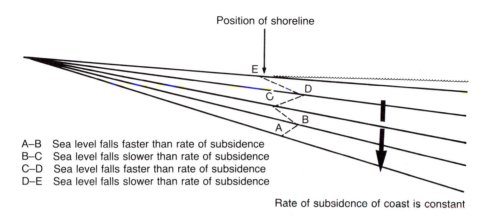

FIGURE 12.12
Relationship between changes in rate of sea-level fall and shoreline position.

and other wells in the United States and Irish margins of the North Atlantic Basin. The depositional sequences and unconformities show good agreement between all the wells and, in turn, with the major Cenozoic unconformities of the Vail model (Fig. 12.13). In other data compiled by Heller et al. (1982) from COST wells off the eastern United States, the evidence indicated that, in addition to an ongoing relatively slow subsidence due to thermal cooling, water and sediment loading, and compaction, other tectonic movements could be detected. These movements were responsible for episodes of relatively rapid downwarping and coastal onlap, the timing of which seemed to be synchronous with seafloor-spreading changes in the North Atlantic. That these movements were more than local effects is indicated by an apparent synchroneity of subsidence events between wells 1200 km apart.

That eustacy is not the prime mechanism in many cases is also indicated in work by Hubbard (1988), based on a detailed study of Jurassic and Early Cretaceous sequence boundaries in the Beaufort Sea, the Grand Banks of Newfoundland (see Fig. 11.17), and the Santos Basin, Brazil. As with the Vail curve, this study used the same impressive array of data from seismic surveys and boreholes, available only to major oil companies (in this case British Petroleum). The results of this work suggested that the bounding unconformities marking onlap events were principally caused by changes in the rate of basin subsidence, sediment supply, and cyclical long-term tectono-eustatic sea-level changes. Strongly supported also was the premise indicated by earlier studies that apparent intercontinental correlations might simply reflect similar tectonic histories of basin margins following rifting. It would be hardly surprising, for example, to find that the margins facing each other across the South Atlantic or the southern Australian and opposing Antarctic margins exhibited matching histories of marine onlap. As a result of this study, it was concluded that the majority of the first- and second-order cycles of the Vail curve are not, in fact, of global extent but rather are regional phenomena largely controlled by coeval tectonics.

In summary, it seems that, although the Vail curve has come to be widely accepted as a useful stratigraphic tool, there remain numerous questions to be answered. It would seem that some, at least, of the first-order changes might be of eustatic origin and

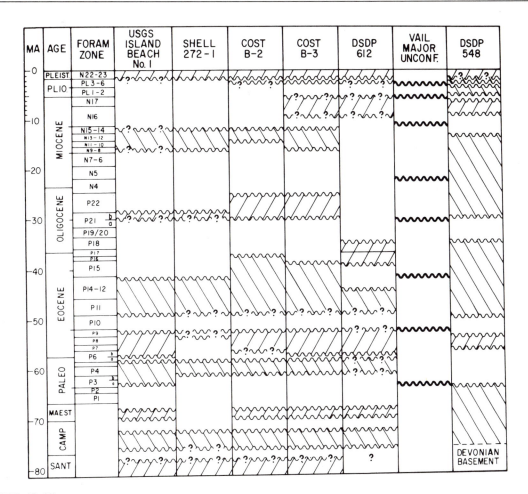

FIGURE 12.13
Summary of the stratigraphy from COST and other wells on the outer shelf of New Jersey and southwest of Ireland (DSDP 548). Shaded parts of the columns are depositional sequences, blanks are hiatuses. (From Poag, W., ed., 1987, Geologic evolution of the United States Atlantic margin: New York, Van Nostrand Reinhold, Fig. 9–5, p. 345.)

caused by long-term plate tectonic movements. There is also no real problem in the case of third-order cycles during glacial periods. It is the third-order cycles during nonglacial times, and most of the second-order cycles, that probably cannot be satisfactorily explained in terms of eustatic sea-level changes. They must, it is widely held, be due to an interaction between basin subsidence of varying rate and the long-term ocean-basin volumetric changes. In the view of Summerhayes (1986), the significance of the Vail curve as a chronostratigraphic tool is probably exaggerated, and to use the global chart as a primary method of correlation is, in Miall's (1986) view, a case of circular thinking. This is also the view of Hubbard (1988), who recommended that the global approach to basin analysis should be discarded and that each sedimentary basin be treated as a unique entity. Further, the Vail curve ignores the fact that subcrustal movements are probably occurring on a scale large enough to be of global significance. As Sloss (1988) put it "cratons, their margins, and their interior basins 'do not just lie there' passively waiting to be encroached upon by rising sea levels or laid bare to erosion as sea level falls."

An interaction between eustatic sea-level changes and regional tectonic changes is, in fact, accepted by Vail et al. (1984). However, it would seem that an important area of disagreement between Vail et al. (1984) and their numerous critics lies in how they view the relative speed and magnitude of the various mechanisms. Whereas the work of Pitman (1978), Watts (1982), Parkinson and Summerhayes (1985), and Hubbard (1988) suggest that it is regional tectonic movements that are superimposed upon the long-term ocean-basin volumetric (that is, eustatic) changes, Vail et al. (1984) take what is in some ways an opposite position. This states that their curve reflects "eustatic sea level changes superimposed on regional tectonic and sedimentary regimes that change at much slower rates." It is the resultant necessity for short-term eustatic sea-level changes that remains the chief bone of contention in the continuing debate.

12.5 SYMMETRY AND ASYMMETRY IN SEA-LEVEL CYCLES

In the debate over the shape of sea-level curves, there are two areas of controversy. One is concerned with the nature of the mechanisms that control sea-level changes (or more particularly with their possible interactions) and the other with the interpretations of the facies changes that supposedly record the sea-level shifts. Although a particular causative mechanism might operate in such a way as to produce a symmetrical sea-level curve, its interaction with other mechanisms might cause a rapid rise and slow fall or vice

versa. On the other hand, asymmetry of the curve is not necessarily proof of this interaction because some eustatic mechanisms themselves may produce markedly asymmetrical curves over both long and short spans of time. One study by Rice and Fairbridge (1975), for example, did suggest that rapid pulses of ocean-ridge thermal expansion might be followed by long periods of cooling and contraction to give rapid eustatic sea-level rises and slow retreats.

Vail et al. (1984), even though their original sawtooth curve was "smoothed out," still insisted that many transgressive-regressive cycles were asymmetric. In the Jurassic to early Cretaceous interval, they recognized 17 global unconformities; eight of these are termed Type 1 unconformities, marking rapid sea-level fall, and nine are Type 2 unconformities, caused by slow falls of sea level (Fig. 12.14). In addition, 16 marine condensed sections, marking rapid eustatic rises of sea level, have been identified.

In the case of glacio-eustatic fluctuations, the evidence for asymmetry is conflicting. According to Ramsbottom (1977), Carboniferous cyclothems indicate slow transgression and comparatively rapid regression; both the cyclothems and groupings of cyclothems known as mesothems exhibit this asymmetry (Fig. 12.15). The curves of Ross and Ross (1985) for Carboniferous mesothems show the same shape, as do the third-order cycles of the Vail Cenozoic curve. On the other hand, Heckel's (1986) work on Pennsylvanian glacio-eustatic marine transgressive-regressive cycles in the mid-continent region of North America shows asymmetry in the opposite direction, namely with slow, interrupted regression, indicating gradual ice cap buildup, followed by rapid marine transgression resulting from rapid ice cap melting. Although the dominant period of the major cyclothems ranges between 250,000 and 400,000 years, the shape of the curve is similar to that of the shorter 100,000-year period

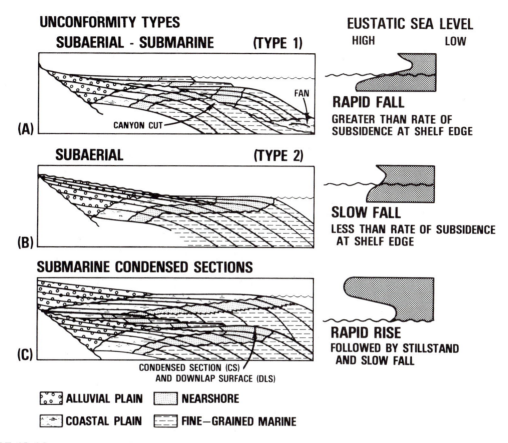

FIGURE 12.14
Diagrammatic charts showing the relation between Type 1 and Type 2 unconformities
(A, B) and submarine condensed sections (C) to eustatic sea-level changes. (From Vail,
P. R. et al., 1984, Jurassic unconformities, chronostratigraphy and sea level changes
from seismic stratigraphy and biostratigraphy, *in* Schlee, J. R., ed., Interregional unconformities and hydrocarbon accumulation: AAPG Mem. 36, Fig. 4, p. 134, reprinted by
permission of American Association of Petroleum Geologists.)

of Pleistocene fluctuations. According to
Broecker and Van Donk (1970), the typical
Pleistocene curve reflects gradually falling
sea level and ice sheet expansion in a series
of subcycles over a period of 90,000 years,
followed by rapid ice melting and sea-level
rise in less than one-tenth of that time (Fig.
12.16). This discrepancy between the results
of the different studies may be due to differences in scale because few of the third-order
cycles of Vail et al. (1977) are less than 2

m.y., whereas the longest of Heckel's (1986)
cycles covers 400,000 years. It is possible,
therefore, that a larger, "slow-transgression"
cycle is made up of a series of short "fast-
transgression" subcycles (Fig. 12.17), although, it must be admitted, there is little direct evidence for this. Clearly, this whole
topic of sea-level curve shape is still very
much an area of active investigation.

It is possible, also, that an apparent asymmetrical shape of individual transgressive-re-

FIGURE 12.15
Mesothemic cycles in the Carboniferous of northwest Europe.
(Reproduced by permission of
the Geological Society from
Rates of transgression and
regression in the Carboniferous
of N.W. Europe by W. M. C.
Ramsbottom in Geol. Soc. London Jour., v. 136, 1979.)

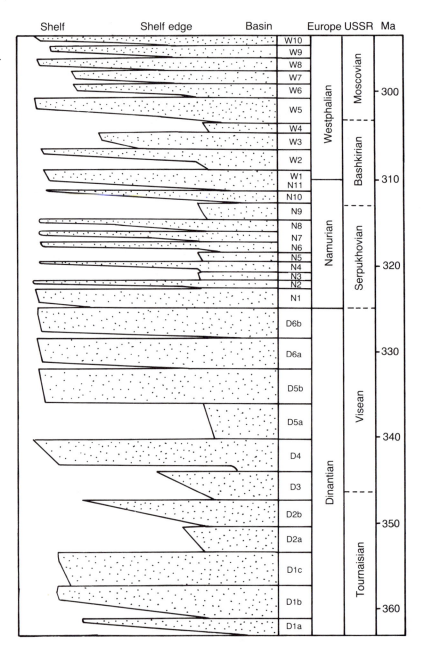

gressive cycles does not reflect the actual rate of sea-level change, but rather is due to the nature of the sedimentary record of the cy-cles. It is important to note that the sedi-ments laid down during the transgressive part of the cycle differ from those deposited as the sea level falls, so that truly symmetri-cal sedimentary cycles in which the regres-sive phase is a mirror image of the transgres-sive portion are rare or possibly nonexistent.

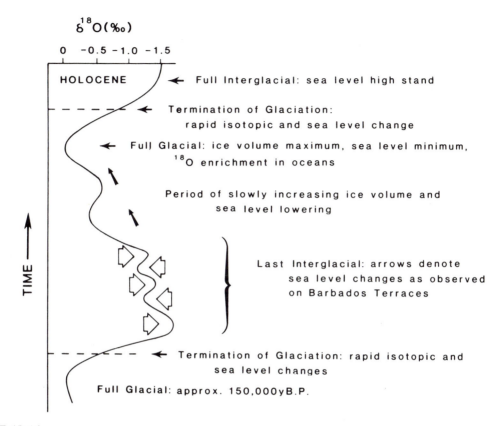

FIGURE 12.16
Sea-level curve during the latest 150,000 year glacial-interglacial episode, as interpreted from the oxygen isotope record. (From Williams, D. F. 1984, Correlation of Pleistocene marine sediments of the Gulf of Mexico and other basins using oxygen isotope stratigraphy, *in* Principles of Pleistocene stratigraphy applied to the Gulf of Mexico: Internat. Human Resources Devel. Corp., Boston, Fig. 3–1, p. 68.)

Transgressive sediments are typically thin, particularly if the transgression is a rapid one. Curray (1964), for example, described the sediments marking the very pronounced post-Wisconsin rise of sea level as comprising only a thin veneer of basal transgressive sands. Regressive progradation, on the other hand, often is typified by a relatively thick sedimentary succession, whether of clastics or carbonates, so that as far as the sedimentary record of the transgressive-regressive cycle is concerned, it will inevitably be asymmetrical in thickness. As Wilson (1975) pointed out, marine transgressions are apt to *appear* more rapid than they actually are.

The sedimentary record does not apparently contain any obvious time/thickness or time-linked facies relationship that might guide us in establishing the *actual* timetable of the sea-level cycle. Fossils are no help either, because in many of the shorter cycles in which asymmetry is most marked, sedimentation in the regressive phase is far too rapid for details of accumulation rates to be discerned by biostratigraphic indicators. The illusion of rapid transgression may be due also

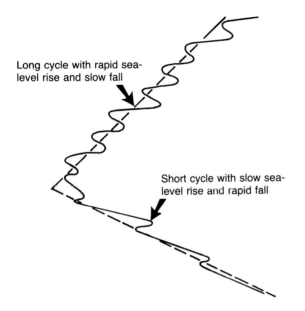

Long cycle with rapid sea-level rise and slow fall

Short cycle with slow sea-level rise and rapid fall

FIGURE 12.17
How two different asymmetrical sea level curves might be superimposed.

to the "kickback" effect as described by Matthews (1974). This is seen in carbonates, which may continue to build up and maintain shallow-water facies in the face of progressive submergence. The end of such conditions would come quickly once a critical depth was exceeded and the carbonate buildup was "drowned." This event would be marked by a sudden change upward from shallow-water to deeper-water facies.

The construction of a sea-level curve is not entirely objective; interpretation of the sedimentary succession is another factor to be considered. Factors discussed above, such as the "kickback" effect and the appearance of regressive facies, resulting from depositional progradation while a transgression is still in progress, introduce variables that are difficult to assess. It is not surprising, therefore, that there is often a difference of opinion as to which facies mark the beginning, the peak, and the end of a particular transgressive-regressive cycle. In the view of some, the

transgression peaked in the middle of the facies, indicating the deepest water, whereas in other interpretations, the top of the deeper-water succession is said to mark the maximum transgression. The interpretations by four different authors of an Upper Cretaceous cyclothem of the Western U.S. is shown in Figure 12.18. When comparing studies by workers in rocks of various ages, it is seen that there is no unanimity as to the shape of eustatic sea-level curves (Fig. 12.19). Hallam (1978), for example, suggested that Jurassic eustatic cycles began, for the most part, with a relatively rapid sea-level rise followed by a protracted stillstand, and then terminated with a rapid sea-level drop. In a few cases, a rapid rise was followed by a slow regression. In the Cretaceous and Tertiary of Western Australia, Quilty (1980) found symmetrical cycles similar to those of Hallam. Cooper (1977), on the other hand, noted a marked asymmetry in Cretaceous cycles, with rapid advances and slow retreats.

12.6 AMPLITUDE OF EUSTATIC SEA-LEVEL CHANGE

Continental Freeboard

The periodic flooding of the cratonic interiors through the Phanerozoic has been shown to involve sea-level fluctuations of approximately the same order of magnitude. What Kuenen (1939) called the "freeboard" of the continents (that is, their relative elevation above sea level) has apparently remained more or less constant. For most of Phanerozoic time, according to Wise (1972), the freeboard has been within ±60 m of a normal value 20 m above present-day sea level. This reflects an equilibrium state in which the various geologic agents of erosion and sedimentation and tectonic processes in the mantle are more or less in a state of balance, regardless of the changing configuration of continents and ocean basins through time.

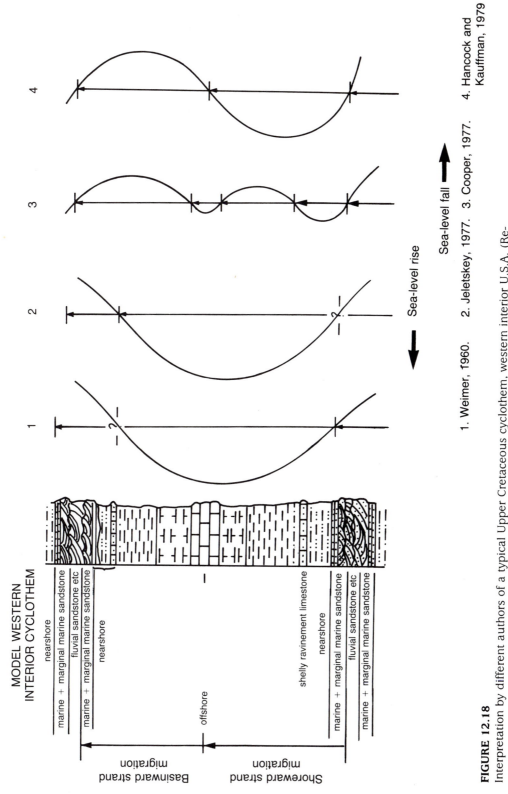

FIGURE 12.18

Interpretation by different authors of a typical Upper Cretaceous cyclothem, western interior U.S.A. (Reproduced by permission of the Geological Society from The great transgressions of the Late Cretaceous, Hancock, J. M., and Kauffman, E. G., in Geol. Soc. London Jour., v. 136, 1979, Fig. 3, p. 177.)

321

FIGURE 12.19
Possible eustatic models for the
European Jurassic. A. Short
phases of rapid sea-level rise interrupted by longer phases of
stillstand. B. Moderate rise followed by moderate fall without
intervening phase of stillstand.
C. Slow rise followed immediately by rapid fall. D. Rapid rise
followed immediately by slow
fall. E. Rapid rise and fall interrupted by longer phase of stillstand. (From Hallam, A., 1978,
Eustatic cycles in the Jurassic:
Palaeogeog., Palaeoclim., Palaeoecol., v. 23, Fig. 6.4, p. 124.
Reprinted by permission of Elsevier Science Publishers B. V.)

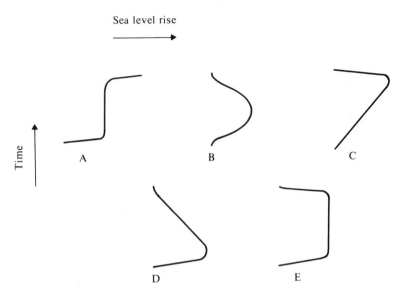

Estimates of the actual amount of nonglacial eustatic sea-level change have been
made using numerous techniques. Sleep
(1976), for example, concluded, on the basis
of evidence of paleotopographic relief and
present-day elevations of marine sediments
in tectonically stable areas, that a vertical
amplitude of eustatic sea-level change on the
order of 300 m was indicated. Using a synthesis of data from many sources, Vail et al.
(1977) suggested, for the Late Triassic to
Present, a range from 350 m above to 200 m
below present sea level. Estimates based on
calculations of changes in ocean-basin volume, consequent on variations in seafloor-
spreading rates, are more difficult to assess
because there is a strong possibility that
other mechanisms are involved as well. Flemming and Roberts (1973), however, suggested a maximum rise of sea level of 300 m
and a fall of 800–1000 m. The above figures
are, of course, for maxima and minima on
the sea-level curve; stillstands at these elevations were probably brief. As mentioned earlier, Wise (1972, 1974) suggested that for 80
percent of the Phanerozoic the amplitude of
sea-level changes was considerably less,

ranging from 80 m above to 40 m below present sea level. It also should be remembered
that, because of the implications of epeirogenic mechanisms, such as that proposed by
Sloss and Speed (1974), the evidence assembled in the above reports may not, in fact, indicate eustatic sea-level changes alone.

Geoidal Changes

The **geoid** refers to the sea-level surface that
generally is assumed to reflect the oblate
spheroidal shape of the earth. It is the datum
used for all measurement of topographic elevations and relative sea-level changes. Unfortunately, and as has been known since the
earliest studies of artificial satellite orbits, the
earth has quite an irregular shape, and deep-
seated gravity anomalies cause the mean sea
surface to have broad undulations. Across the
Indian Ocean, for example, the sea surface is
180 m lower in the Maldive Islands than it is
in New Guinea. Morner (1981) suggested that
irregularities of such magnitude introduce a
random factor that may reduce any systematic attempt to analyze sea-level changes to a
pointless exercise! This is, perhaps, an un-

duly pessimistic view, but it is clear that possible variations of this nature cannot be ignored.

12.7 SHORT-TERM SEA-LEVEL CHANGES

Variation in the rate of seafloor spreading, with the concomitant changes in mid-ocean ridge volumes, is generally held to be the basic cause of long-term changes in eustatic sea level, and it is thought that ridge expansion can cause sea-level changes of as much as 300–500 m, but only at a rate of, at the most, 10–12 m per million years. Superimposed on this comparatively slow and long-term change are fluctuations that are several orders of magnitude faster and that are difficult to account for by any seafloor spreading mechanism. At times when there is independent evidence of glaciation, an explanation for rapid sea-level change is not difficult to find. Such glacio-eustatic sea-level changes will be discussed separately.

Basin Isolation

At other, i.e., nonglacial, times, the only known mechanism that might cause rapid sea-level change is the desiccation or flooding of isolated marine basins, such as that of the Mediterranean during the so-called Messinian salinity crisis in the Miocene; this crisis, it is estimated, caused sea-level rise of some 12–15 m. This small figure, and an apparent lack of convincing evidence of other desiccation basins in the past, have tended to discount this mechanism as playing any significant role in past sea-level change. Berger and Winterer (1974) and Hsü and Winterer (1980), on the other hand, claimed that desiccation of basins has been an important factor and cite, as an example, evidence for an early Cretaceous basin, isolated during the early rifting of the South Atlantic, that was five times as large as the Mediterranean. There undoubtedly have been episodes of basin iso-

lation during both oceanic opening (Red Sea Stage) and closure. Extensive salt deposits, such as the Jurassic Louann Salt in the subsurface of the Gulf Coast region, can be interpreted as supporting evidence for basin isolation.

Glacio-Eustatic Sea-Level Changes

Until comparatively recently, glacio-eustatic sea-level changes were discussed virtually only in terms of Pleistocene glacial and interglacial episodes. From being a sort of aberrant phenomenon tacked on to Phanerozoic sea-level curves and hardly discernible at most scales, it is now apparent that glacio-eustatic sea-level changes have, in fact, played a significant role over much longer spans of time. It now is realized, for example, that the late Cenozoic Ice Age began much earlier than had previously been suspected. Direct evidence of glaciation in the form of tills has been found in sediments as old as Miocene, whereas, in marine successions in the southern ocean, ice-rafted material dated as Oligocene or older has been described. According to Matthews (1984), there is a high probability that an ice cap appeared on the Antarctic continent as early as the Eocene or even before that, as the continent moved into polar regions and as separation from Australia and the initiation of the circum-Antarctic wind/ocean current pattern caused thermal isolation. Although direct evidence of an ice cap, in the form of tills and moraines, has been destroyed by, or is buried under, the modern ice cap, the ice-rafted material does set a minimum age of Oligocene. On the other hand, the absence of yet older ice-rafted debris does not necessarily preclude an earlier ice cap because it was only when the ice margin reached tidewater and began to calve icebergs that morainic material would be rafted out to sea. As Matthews (1984) pointed out, Antarctica is a large continent with plenty of room for a considerable ice cap in the interior. This argument would also seem a reasonable answer to the botanical evidence from the Ross Sea region for a mild climate in the Eocene (Kemp and Barrett, 1975). It is possible that areas at or near sea level might have enjoyed a fairly mild climate, due to maritime influences, even though an ice cap existed in the interior. Pleistocene sea-level fluctuations are discussed further, later in this chapter and in Chapter 15.

Of the pre-Pleistocene glaciations, the Permo-Carboniferous glaciation is best known. Tillites, glacially striated pavements, and other indicators of continental glaciation are widespread in the southern continents that formerly comprised Gondwana and testify to an ice age that lasted for some 90 m.y., from early Carboniferous to mid-Permian time. This is not to say that during this time Gondwana was covered by a single huge ice cap; rather, ice centers expanded and contracted and progressively shifted as the supercontinent moved across the south pole. As can be seen on the apparent polar wandering curve (Fig. 12.20), the earliest glacial centers were in central Africa and South America and some of the latest were in Australia, where glacial conditions began much later and persisted long after they had disappeared in South America. There is no reason to doubt that the same Milankovitch mechanism that controlled the well-documented waxing and waning of the Pleistocene ice sheets also played a similar role in the Permo-Carboniferous, although direct evidence of this periodicity in the areas formerly glaciated is scanty. There is, however, good evidence of glacial cycles to be seen in the glacio-eustatic sea-level fluctuations demonstrated by the cyclothems that figure so prominently in many Carboniferous successions of the northern hemisphere. In Kansas, for example, Moore (1964) recognized over 100 repetitive cycles, and although in many places the cyclothems are modified by local effects, there

is little doubt that they are glacio-eustatic in origin. In European and North African successions, ranging from Visean or early Namurian (Early Carboniferous) to early Kazanian (Late Permian) age, over 100 cycles also have been documented. More recently, Heckel (1986) described 55 cycles of marine advance and retreat (Fig. 12.21) in strata of mid-Desmoinesian (mid-Pennsylvanian) to mid-Virgilian age (latest Pennsylvanian) and, as mentioned earlier, was able to discern minor cycles of 40,000–120,000 years, comparable in length with those in the Pleistocene. More prominent are larger cycles of 235,000–400,000 years; these compare with a similar periodicity seen in the earlier Cenozoic glaciations, thought to be associated with long-term fluctuations in the Antarctic ice cap. In a model described by Moore et al. (1982), the presence of a single ice cap over Antarctica resulted in a different response to the Milankovitch insolation variations than when both northern and southern ice caps had appeared by Pleistocene time. As discussed in Chapter 15, the volume of the early Cenozoic ice sheets fluctuated with a 41,000-year rhythm; it is only after the beginning of the Pleistocene that the 100,000-year ice-volume cycle becomes discernible.

The late Ordovician–early Silurian glaciation, the evidence for which is seen in glacial deposits in the Sahara region of Africa and in South America (Beuf et al., 1971; Berry and Boucot, 1973), began in the late Caradoc, or possibly earlier, and lasted until late Llandovery or early Wenlock time, a period of some 25 m.y. At least three advance-retreat sequences were described by Allen (1975), and studies of facies and faunal changes through this interval by several workers (McKerrow, 1979; Brenchley and Newall, 1980; M. E. Johnson et al., 1981, 1985) showed that rapid eustatic sea-level changes occurred, and it is likely that many of them were of glacio-eustatic origin. A brief glacial period in the Late Devonian (Famennian) has been documented

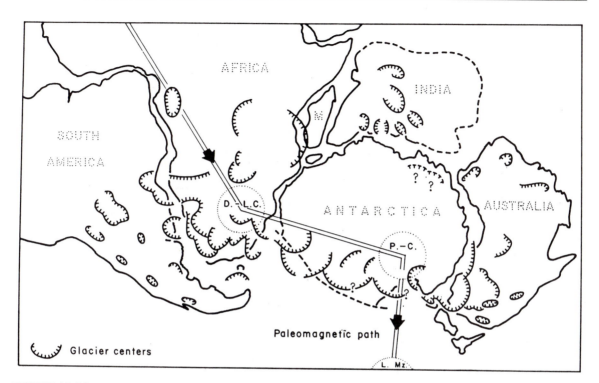

FIGURE 12.20
APW curve for the south pole from the Devonian through the Permian. D.-L.C. = Devonian-Lower Carboniferous, P.-C. = Permian-Cenozoic, L. Mz. = Lower Mesozoic.) (From Crowell, J. C., 1978, Gondwanan glaciation, cyclothems, continental positioning and climate change: Am. Jour. Sci., v. 278, Fig. 3, p. 1356. Reprinted by permission of American Journal of Science.)

in South America (Caputo, 1985), and of the 14 eustatic transgressive-regressive cycles described by J. G. Johnson et al. (1985), two in the Famennian were thought to be glacio-eustatic in origin.

It is clear from this brief summary that ice ages have been quite important features in earth history. As can be seen in Figure 12.22, the three major glacial periods of the Phanerozoic together total more than 160 m.y. In addition, there were probably other minor glaciations, such as that in the Late Devonian, so that there is the possibility that nearly one-third of Phanerozoic time has been marked by ice ages. It should not be forgotten that the Precambrian also contains a con-

siderable record of glaciations. These are discussed briefly in Chapter 16.

12.8 SEA-LEVEL FLUCTUATIONS DURING THE QUATERNARY

Quaternary glacio-eustatic sea-level changes have profoundly influenced sedimentary processes on the continental shelves. If the sediments of earlier marine transgressions and regressions along passive margins are preserved at all, they are typically very thin. This is because all the more recent marine advances and recessions have been roughly of the same order of magnitude, so that old shorelines were often repeatedly reoccupied.

FIGURE 12.21

Sea-level curve for part of the middle Upper Pennsylvanian sequence in the mid-continent region. (From Heckel, P. H., 1986, Sea-level curve for Pennsylvanian eustatic marine transgressive-regressive depositional cycles along midcontinent outcrop belt, North America: Geology, v. 14, Fig. 2, p. 332.)

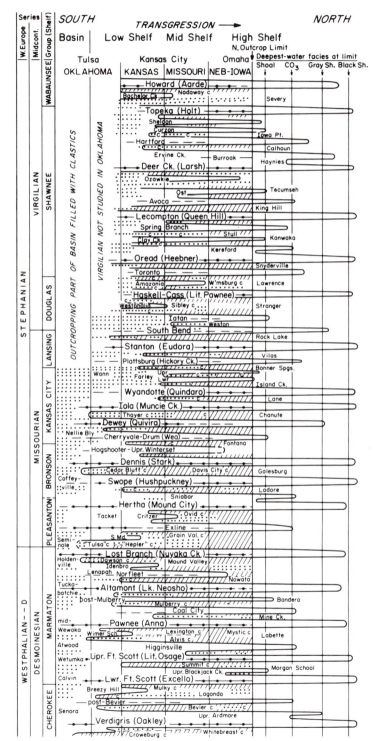

FIGURE 12.22

Ice ages in the late Proterozoic and Phanerozoic. A summary of data on the worldwide distribution of diamictites with a glacigenic component. (Compiled from data by Hambrey, M. J., and Harland, W. B., 1981, Earth's pre-Pleistocene glacial record: Cambridge Univ. Press, Cambridge, England.) According to Matthews (1984), the so-called Neogene glaciation began in the Paleogene. Burke and Waterhouse (1973) suggest that there was intermittent glaciation through the Cambrian and Ordovician.

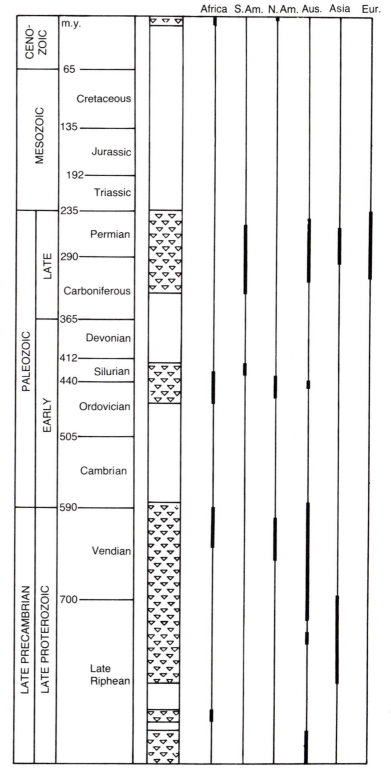

327

The result is that each new suite of transgressive sediments tended to form at the expense of the sedimentary record left during previous interglacial episodes. In South Florida, for example, a succession recording five marine units totals only 15 m in thickness (Enos and Perkins, 1977). The situation is very different in regions of tectonic instability where, in some cases, the erosional and depositional pattern of fluctuating sea levels may be preserved either as submerged offshore terraces or in a staircase-like series of uplifted beaches or reef tracts superimposed on a rising coast. On the Huon Peninsula of New Guinea more than 20 reef tracts are preserved, the highest and oldest being more than 250,000 years old (Figs. 12.23 and 12.24). Studies of the lowest seven reef complexes by Bloom et al. (1974) revealed that each was formed during a eustatic highstand of sea level. With radiometric dates from each of the seven reef tracts, and assuming a constant rate of uplift of the coastline, it was possible to construct a detailed sea-level curve dating back to 140,000 years B.P. This interval is through the whole time span of the

Wisconsin glaciation. It showed that there had been a regular recurrence of marine transgressions marking interstadial periods at intervals of about 20,000 years. Although less complete, the sea-level fluctuations derived from studies of elevated reef tracts in Barbados (Mesolella et al., 1969; Steinen et al., 1973) is closely similar (Fig. 12.25).

The record of glacial and glacially related events becomes considerably more detailed once we reach the latest glacial episode, the Wisconsin (or Würm in the alpine succession), and it becomes even better in post-Wisconsin time. Second- and third-order fluctuations of climate can be detected, and the broader climatic and sea-level curves established for major cycles within the late Cenozoic are seen to involve many smaller subcycles. A late Pleistocene to Recent eustatic sea-level curve has been established on the basis of numerous observations from tectonically stable areas of the world; it shows a relatively rapid rise from around 90 m below present sea level about 18,000 years B.P. to approximately 10 m below sea level about 6,000 years B.P. From 6,000 B.P. to the pres-

FIGURE 12.23
Emergent reef terraces, Huon Peninsula, New Guinea. (Photograph courtesy of A. L. Bloom.)

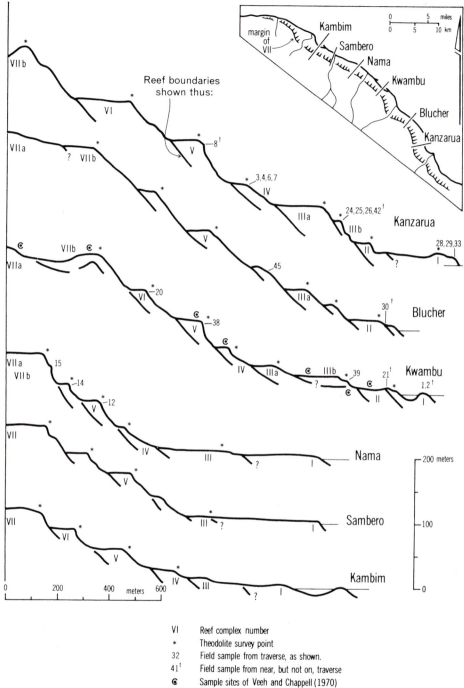

FIGURE 12.24

Profiles of emergent reef terraces on the Huon Peninsula, New Guinea. Terraces are numbered from the youngest (I) to the oldest (VII). (From Bloom, A. L. et al., 1974, Quaternary sea level fluctuations on a tectonic coast: New ^{230}Th/^{234}U dates from the Huon Peninsula, New Guinea: Quaternary Research, v. 4, Fig. 4, p. 190.)

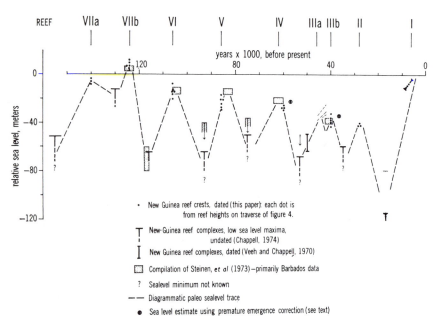

FIGURE 12.25

Late Pleistocene sea-level fluctuations based on New Guinea and Barbados elevated terraces. (From Bloom, A. L. et al., 1974, Quaternary sea level fluctuations on a tectonic
coast: New ^{230}Th/^{234}U dates from the Huon Peninsula, New Guinea: Quaternary Research, v. 4, Fig. 5, p. 203.)

ent, the rate of sea-level rise has slowed (Fig.
12.26).

12.9 SEA LEVELS AND SEDIMENTS ON PASSIVE MARGINS

Fluctuating sea level affects the way in which
sediment is transferred to the shelf; this may
be the result of direct or indirect processes.
Where there is a strong tidal regime, a large
proportion of river-borne sediment is carried
directly out to sea, but without effective removal by tidal currents and in the absence of
vigorous wave action, alluvial sediment accumulation results in the formation of a
delta. Indirect transfer occurs as a consequence of the erosion and removal of sediment that has been in storage within the
coastal boundary. Swift (1974) described the
coastal boundary in terms of two types of

"valve" or bypass mechanism. Shelf sediment
contributed directly by rivers is said to be derived by *river-mouth bypassing,* whereas erosion of the shoreline to release sediment to
the shelf is termed *shoreface bypassing.* The
latter is an indirect process of sediment transfer through the coastal boundary because the
sediment first undergoes a period of storage
in the shoreface.

The relative proportions of river-mouth
and shoreface-derived sediment carried to
the shelf depends very largely on whether sea
level is rising or falling and at what rate. A
rapid rise of sea level consumes the source
areas of terrigenous sediments and results in
the drowning of river estuaries, which then
become sediment sinks, so that the amount
of sediment reaching the shelf is reduced. At
the same time, the shoreface retrogrades, becoming an important source of sediment.

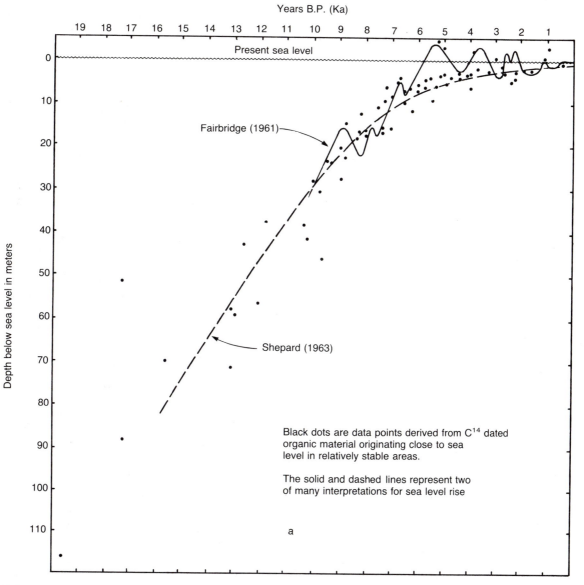

FIGURE 12.26
Post-Wisconsin sea-level rise. (After Fairbridge, R. W., 1961, Eustatic changes in sea level, *in* Physics and chemistry of the Earth, v. 4, p. 99–185 and Shepard, F. P., 1963, Thirty five thousand years of sea level, *in* Clements, T., ed., Essays in marine geology in honor of K. O. Emery, Univ. of S. Calif., p. 1–10.)

When the rate of sea-level rise slows down, estuaries return to an equilibrium state as they become silted up and depositional progradation back on to the shelf begins. It should be noted that in areas suitable for carbonate deposition, outbuilding will begin at this stage also. It is under these conditions that a swing back from retrogradational to progradational facies may commence even before the actual sea-level rise has peaked. With falling sea level the shoreface progrades and, in effect, becomes a sediment sink; then river-mouth bypassing becomes dominant. Swift (1974) used the terms **autochthonous** for shelf sediment derived from the shoreface

and **allochthonous** for that transported by river-mouth bypassing.

Because of the rapid rise of sea level since the end of the Wisconsin glaciation, much of the sedimentary cover of the modern shelves was laid down under a previously existing regime and is only now undergoing reworking and modification. The term **relict** has been used to describe these older sediments but one must distinguish between the sediments themselves and the sedimentary structures and topography. Thus, much of what Swift (1974) termed autochthonous sediment is actually relict in the sense that it was derived from an earlier shoreface and coastal bound-

TABLE 12.2
Classification of marine transgressions and regressions (based on Curray, 1964)

I. Erosional regression
Due to emergence consequent on rapidly falling sea level or tectonic uplift, with little or no deposition

II. Mixed erosional and depositional regression
Similar to I but with some deposition, most likely a thin cover of strandline deposits left behind by the retreating shore

III. Discontinuous depositional regression
Balance shifting from falling sea level as primary cause, toward sedimentary progradation; rate of sea-level fall slower and/or rate of sediment supply higher

IV. Depositional regression
Primary cause of shoreline advance due to rapid sediment accumulation, as in a delta; if sea level is falling, it is more than balanced by sediment accumulation

V. Depositional Transgression
Sea level now rising, due to subsidence, eustatic rise, or both

VI. Discontinuous depositional transgression
Due to increasing rate of transgression or decreasing rate of deposition; discontinuous sedimentary units due to inability of deposition to continually keep pace with rising sea level

VII. Rapid erosional transgression
Submergence too rapid for transitional and littoral deposits, so marine sediments lie directly on subaerially eroded surface in onlap relationship

VIII. Erosional transgression
Sea level relatively stable with negligible sediment supply; wave action intense enough to cause net erosion and shoreline retreat; often aided by compaction of underlying sediments, e.g., as in "destructional phase" in deltas after distributary switching

ary dating from a former sea-level lowstand. Insofar as the sedimentary cover of modern passive margins contains some record of the most recent sea-level changes, we may expect to find three types of sediment: (1) a basal suite of sediment of the drowned coastal boundary and shallow offshore dating from the last glacial epoch; (2) the autochthonous sand sheet derived from wave attack on the shoreface during the initial period of rapid sea-level rise from around 18,000 B.P. to around 6,000 B.P.; and (3) a Recent veneer of finer allochthonous sediments. These reflect the increasing importance of river-mouth bypassing as the rate of marine advance slowed down since 6,000 B.P. and be-

cause, as the estuaries and bays became filled with sediment, they began to return to an equilibrium state.

Variations in the rate at which the various controls operate on all types of shelves, both passive and active, lead to different kinds of transgression and regression. Curray (1964) recognized eight categories (Table 12.2), which were represented graphically as arbitrarily defined areas on a sea-level chart (Fig. 12.27). On the basis of shelf profiles at numerous localities, Curray then plotted sedimentary histories as lines on the master chart and made some interesting comparisons (Fig. 12.28). In each location, the period of rapidly rising sea level from approximately 18,000

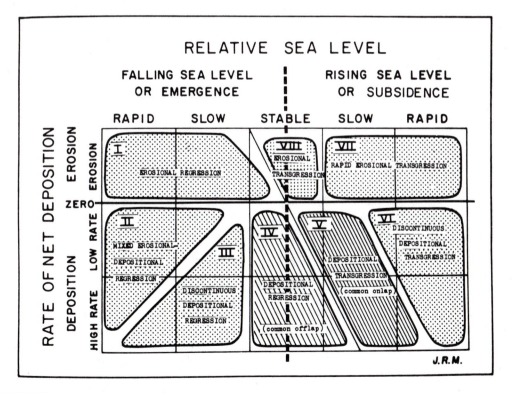

FIGURE 12.27
Classification of marine transgressions and regressions. (From Curray, J. R., 1964, Transgressions and regressions, *in* Miller, R. L., ed., Papers in marine geology (Shepard commemorative vol.), Fig. 10–10, p. 191.

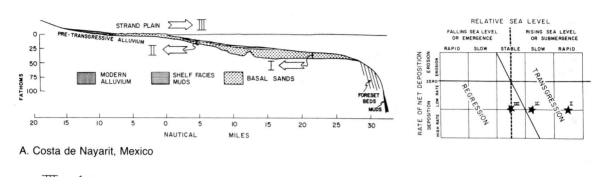

A. Costa de Nayarit, Mexico

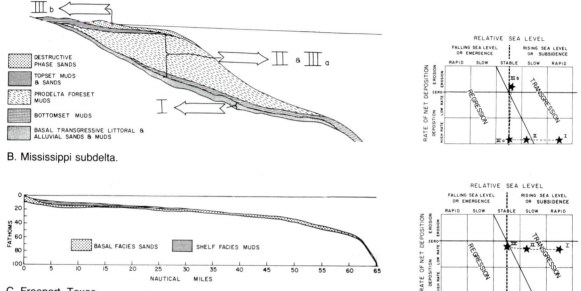

B. Mississippi subdelta.

C. Freeport, Texas.

FIGURE 12.28
Post-Wisconsinan history of three typical coastal sections compared in diagrammatic
section and when plotted on a sea-level chart. (From Curray, J. R., 1964, Transgressions
and regressions, *in* Miller, R. L., ed., Papers in marine geology (Shepard commemorative vol.), Figs. 10–4, 10–5, and 10–9.)

years B.P. to about 7,000 years B.P. was shown
to have resulted in a layer of sand, the "basal
transgressive sand" of Curray's (1964) description, which is the autochthonous sand
sheet of Swift's (1974) terminology. It is overlain by "shelf facies muds"—allochthonous
silts and muds—that mark the slowing down
of sea-level rise after about 7,000 B.P., the
equilibration of estuaries and bays, and the
shift from shoreface bypassing to the river-
mouth bypassing mode.

To reiterate the main points of the above
discussion, during marine regression, stillstand, and/or slow transgression, the sediment contribution to the shelf is allochthonous and dominantly composed of muds and
silts. Rapid transgression, on the other hand,
results in shoreface bypassing and an autochthonous contribution, largely of sands. Because there is only a limited amount of sand
available for reworking, such basal transgressive sands are typically thin. Except in re-

gions of tectonism, sea-level fluctuations, through much of geologic time, have been relatively slow when compared with the rapid glacio-eustatic fluctuations experienced during ice ages. According to this model, most marine shelf deposits on passive margins should be allochthonous muds and silts, whereas sandy intercalations presumably represent periods of rapid transgression. It follows also that the type of sediment that is transferred to the continental shelf is controlled not by sea-level changes *per se* but rather by the rate of sea-level change.

12.10 CONCLUSION

Of all the cyclical phenomena operating on the earth's surface, it is certainly sea-level change that, either directly or indirectly, plays the dominant role in stratigraphy. The discussion in this chapter has touched upon a range of topics, and it is clear that many long-term geologic cycles, as well as those of shorter duration, are in some way linked to sea-level changes. Only in the distant hinterlands of large continents are sedimentary cycles perhaps isolated from what is happening in the oceans.

Cyclical sea-level change plays an important role in stratigraphic classification, although not in the sense of the degree envisioned by Vail et al. (1984). Instead, it is the link with the biostratigraphic boundaries that is most meaningful. In such sea-level curves as those of Vail et al. (1984) and Haq et al. (1987) and, for the Jurassic, of Hallam (1978) and Vail and Todd (1981), it is obvious that in some cases stage boundaries coincide with sea-level events. The reason is not hard to find. Many transgressions and regressions on and off the continental shelves lead to episodes of environmental stress for a host of shallow-marine organisms, many of which are the zone fossils used in biostratigraphic division. That the turnover of species, stage

boundaries, and sea-level change are linked is, in this light, explicable, and a clear indication of the way sea-level shifts put their stamp on the stratigraphic record.

REFERENCES

Allen, P., 1975, Ordovician glacials in the central Sahara, *in* Wright, A. E. and Moseley, F., Ice ages: ancient and modern: London, Seel House Press, p. 275–286.

Berger, W. H., and Winterer, E. L., 1974, Plate stratigraphy and the fluctuating carbonate line, *in* Hsü, K. J., and Jenkyns, H. C., eds., Pelagic sediments: On land and under the sea: Internat. Assoc. Sedimentologists Spec. Pub. 1, p. 11–48.

Berry, W. B. N., and Boucot, A. J., 1973, Glacioeustatic control of Late Ordovician early Silurian sedimentation and faunal changes: Geol. Soc. America Bull., v. 84, p. 273–284.

Beuf, S., Biju-Duval, B., DeChardel, O., Gariel, O., and Bennacef, A., 1971, Les grès du Paléozoique inférieur au Sahara—sédimentation et discontinuities, évolution structurale d'un craton: Inst. Française Pétrole-Science et Technique du Pétrole, No. 18, 464 p.

Bloom, A. L., Broecker, W. S., Chapell, J. M. A., Matthews, R. K., and Mesolella, K. J., 1974, Quaternary sea level fluctuations on a tectonic coast: New dates from the Huon Peninsula, New Guinea: Quaternary Research, v. 4, p. 185–205.

Brenchley, P. J., and Newall, G., 1980, A facies analysis of Upper Ordovician regressive sequences in the Oslo region, Norway—A record of glacio-eustatic changes: Palaeogeog., Palaeoclim., Palaeoecol., v. 31, p. 1–38.

Broecker, W. S., and Van Donk, J., 1970, Insolation changes, ice volumes and the O^{18} record in deep sea cores: Rev. Geophysics and Space Physics, v. 8, p. 169–198.

Burke, K., and Waterhouse, J. B., 1973, Saharan glaciation dated in North America: Nature, v. 241, p. 267–268.

Caputo, M. V., 1985, Late Devonian glaciation in South America: Palaeogeog. Palaeoclim., Palaeoecol., v. 51, p. 291–317.

Cooper, M. R., 1977, Eustacy during the Cretaceous: Its implications and importance: Palaeogeog., Palaeoclim., Palaeoecol., v. 22, p. 1–60.

Curray, J. R., 1964, Transgressions and regressions, in Miller, R. L., ed.: Papers in marine geology, p. 175–203.

Enos, P., and Perkins, R. D., 1977, Quaternary sedimentation in south Florida: Geol. Soc. America Mem. 147, 198 p.

Flemming, N. C., and Roberts, D. G., 1973, Tectono-eustatic changes in sea level and seafloor spreading: Nature, v. 243, p. 19–22.

Hallam, A., 1977, Secular changes in marine inundation of USSR and North America through the Phanerozoic: Nature, v. 269, p. 769–772.

_____ 1978, Eustatic cycles in the Jurassic: Palaeogeog., Palaeoclim., Palaeoecol., v. 23, p. 1–32.

_____ 1981, A revised sea-level curve for the early Jurassic: Geol. Soc. London Jour., v. 138, p. 735–743.

_____ 1984, Pre-Quaternary sea-level changes: Ann. Rev. Earth Planet Sci., v. 12, p. 205–243.

Haq, B. U., Hardenbol, J., and Vail, P. R., 1987, Chronology of fluctuating sea levels since the Triassic: Science, v. 235, p. 1156–1166.

Hays, J. D., and Pitman, W. C., III, 1973, Lithospheric plate motion, sea-level changes and climatic and ecological consequences: Nature, v. 246, p. 18–22.

Heckel, P. H., 1986, Sea-level curve for Pennsylvanian eustatic marine transgressive-regressive depositional cycles along midcontinent outcrop belt: North America, Geology, v. 14, p. 330–334.

Heller, P. L., Wentworth, C. M., and Poag, C. W., 1982, Episodic post-rift subsidence of the United States Atlantic continental margins: Geol. Soc. America Bull., v. 93, p. 379–390.

House, M. R., 1975a, Facies and time in Devonian tropical areas, Yorks: Geol. Soc. Proc., v. 40, p. 233–287.

_____ 1975b, Faunas and time in the marine Devonian, Yorks: Geol. Soc. Proc., v. 40, p. 459–490.

Hsü, K. J., and Winterer, E. L., 1980, Discussion of causes of worldwide changes in sea level: Geol. Soc. London Jour., v. 137, p. 509–510.

Hubbard, R. J., 1988, Age and significance of sequence boundaries on Jurassic and Early Cretaceous rifted continental margins: Am. Assoc. Petroleum Geologists Bull., v. 72, p. 49–72.

Johnson, J. G., Klapper, G., and Sandberg, C. A., 1985, Devonian eustatic fluctuations in Euramerica: Geol. Soc. America Bull., v. 96, p. 567–587.

Johnson, M. E., Cocks, L. R. M., and Copper, P., 1981, Late Ordovician–Early Silurian fluctuations in sea level from eastern Anticosti Island, Quebec: Lethaia, v. 14, p. 73–82.

_____ Rong, J. Y., and Yang, X-C., 1985, Intercontinental correlation by sea level events in the Early Silurian of North America and China (Yangtze Platform): Geol. Soc. America Bull., v. 96, p. 1384–1397.

Kauffman, E. G., 1977, Upper Cretaceous cyclothems, biotas, and environments, Rock Canyon anticline, Pueblo, Colorado: The Mountain Geologist, v. 14, p. 129–152.

Kemp, E. M., and Barrett, P. J., 1975, Antarctic glaciation and early Tertiary vegetation: Nature, v. 58, p. 507–508.

Kuenen, P. H., 1939, Quantitative estimations relating to eustatic movements: Geol. Mijnbouw, v. 18, p. 194–201.

Matthews, R. K., 1974, Dynamic stratigraphy: Englewood Cliffs, NJ, Prentice Hall, 370 p.

_____ 1984, Oxygen isotope record of ice-volume history: 100 million years of glacio-eustatic sea-level fluctuation, in Schlee, T. S., ed., Interregional unconformities and hydrocarbon accumulation: Am. Assoc. Petroleum Geologists Mem. 36, p. 97–107.

McKerrow, W. S., 1979, Ordovician and Silurian changes in sea level: Geol. Soc. London Jour., v. 136, p. 137–145.

Mesolella, K. J., Matthews, R. K., Broecker, W. S., and Thurber, D. L., 1969, The astronomical theory of climatic change: Barbados data: Jour. Geology, v. 77, p. 250–274.

Miall, A. D., 1986, Eustatic sea level changes interpreted from seismic stratigraphy: A critique of the methodology with particular reference to the North Sea Jurassic record: Am. Assoc. Petroleum Geologists Bull., v. 70, p. 131–137.

Moore, R. C., 1964, Paleoecological aspects of Kansas Pennsylvanian and Permian cyclothems, *in* Merriam, D. F., ed., Symposium on cyclic sedimentation: Kansas Geol. Soc. Bull., v. 169, p. 287–380.

Moore, T. C., Jr., Pisias, N. G., and Dunn, D. A., 1982, Carbonate time series of the Quaternary and Late Miocene sediments in the Pacific Ocean: A spectral comparison: Marine Geology, v. 46, p. 217–233.

Morner, N. A., 1981, Revolution in Cretaceous sea level analysis: Geology, v. 9, p. 344–346.

Parkinson, D., and Summerhayes, C. P., 1985, Synchronous global sequence boundaries: Am. Assoc. Petroleum Geologists Bull., v. 69, p. 685–687.

Pitman, W. C., III, 1978, Relationship between eustacy and stratigraphic sequences of passive margins: Geol. Soc. America Bull., v. 89, p. 1389–1403.

Poag, C. W., and Ward, L. W., 1987, Cenozoic unconformities and depositional supersequences of North Atlantic continental margins: Testing the Vail model: Geology, v. 15, p. 159–162.

Quilty, P. A., 1980, Sedimentation cycles in the Cretaceous and Cenozoic, Western Australia: Tectonophysics, v. 63, p. 349–366.

Ramsbottom, W. H. C., 1977, Major cycles of transgression and regression (mesothems) in the Namurian: Proc. Yorkshire Geol. Soc., v. 41, p. 261–291.

Rice, A., and Fairbridge, R. W., 1975, Thermal runaway in the mantle and neotectonics: Tectonophysics, v. 29, p. 59–72.

Ross, C. A., and Ross, J. R. P., 1985, Late Paleozoic depositional sequences are synchronous and worldwide: Geology, v. 13, p. 194–197.

Schlanger, S. O., Jenkyns, H. C., and Premoli-Silva, I., 1981, Volcanism and vertical tectonics in the Pacific Basin related to global Cretaceous transgressions: Earth and Planetary Sci. Letters, v. 52, p. 435–449.

Sleep, N. H., 1976, Platform subsidence mechanisms and "eustatic" sea-level changes: Tectonophysics, v. 36, p. 45–56.

Sloss, L. L., 1988, Forty years of sequence stratigraphy: Geol. Soc. America Bull., v. 100, p. 1661–1665.

Sloss, L. L., and Speed, R. C., 1974, Relationships of cratonic and continental margin tectonic episodes, *in* Dickinson, W. R., ed., Tectonics and sedimentation: Soc. Econ. Paleontologists and Mineralogists Spec. Pub. 22, p. 98–119.

Steinen, R. P., Harrison, R. S., and Matthews, R. K., 1973, Eustatic lowstand of sea level between 125,000 and 105,000 B.P. Evidence from the subsurface of Barbados, West Indies: Geol. Soc. America Bull., v. 84, p. 63–70.

Summerhayes, C. P., 1986, Sea level curves based on seismic stratigraphy: Their chronostratigraphic significance: Palaeogeog., Palaeoclim., Palaeoecol., v. 57, p. 27–42.

Swift, D. J. P., 1974, Continental shelf sedimentation, *in* Burk, C. A., and Drake, C. L., eds., The geology of continental margins: New York, Springer-Verlag, p. 117–135.

Vail, P. R., Hardenbol, J., and Todd, R. G., 1984, Jurassic unconformities, chronostratigraphy, and sea-level changes from seismic stratigraphy and biostratigraphy, *in* Schlee, J. R., ed., Interregional unconformities and hydrocarbon accumulation: Am. Assoc. Petroleum Geologists Mem. 36, p. 129–144.

Vail, P. R., Mitchum, R. M., Jr., and Thompson, S., III, 1977, Seismic stratigraphy and global changes of sea level, part 4: Global cycles of relative changes of sea level, *in* Payton, C. E., Seismic stratigraphy—Applications to hydrocarbon exploration: Am. Assoc. Petroleum Geologists Mem. 26.

Vail, P. R., and Todd, R. G., 1981, Northern North Sea Jurassic unconformities, chronostratigraphy and sea level changes from seismic stratigraphy, *in* Illing, L. V., and Hobson, G. D., eds., Petroleum geology of the continental shelf of north-west Europe: London, Inst. of Petrol., p. 216–235.

Watts, A. B., 1982, Tectonic subsidence, flexure and global changes of sea level: Nature, v. 297, p. 469–474.

Wilson, J. L., 1975, Carbonate facies in geologic history: New York, Springer-Verlag, 471 p.

Wilson, J. T., 1966, Did the Atlantic close and then re-open?: Nature, v. 211, p. 676–681.

Wise, D. U., 1972, Freeboard of continents through time: Geol. Soc. America Mem. 132, p. 87–99.

_____ 1974, Continental margins, freeboard and the volumes of continents and oceans through time, *in* Burk, C. A., and Drake, C. L., eds., The geology of continental margins. New York, Springer-Verlag, p. 45–58.

Worsley, T. R., Nance, D., and Moody, J. B., 1984, Global tectonics and eustasy for the past two billion years: Marine Geology, v. 58, p. 373–400.

13

CYCLES AND
SEQUENCES

The more things change, the more they are the same.

Alphonse Karr

13.1 INTRODUCTION

From what has been discussed in earlier chapters, it should be clear that a major preoccupation of stratigraphers is to find ways of dividing and measuring stratigraphic successions and geologic time. As with all classification systems, the idea that there might be "natural" divisions that are more logical than the ones in current use has always been intriguing. Such is the case with classifications used at present in division of the stratigraphic succession. Of all the controls on sediment accumulation, it is the fluctuation of sea level and the resulting ebb and flow of the seas onto the continents that is dominant. It is not surprising, therefore, that there have been numerous suggestions that the unconformities that mark periods of emergence and erosion would provide the most logical boundaries within the stratigraphic succession.

13.2 UNCONFORMITY-BOUNDED
UNITS

From the very beginnings of stratigraphy as a science, unconformities have played an important role in dividing the geologic column into convenient packages, and many of the original geologic systems, now recognized on a worldwide scale, were established as unconformity-bounded units. The original type sections provided biostratigraphic yardsticks for the delineation of the systems elsewhere, but from the earliest days it quickly became apparent that systemic boundaries were not natural divisions marked everywhere by erosional breaks with worldwide significance. It has been only by liberal interpretation of the original definitions of the systems that worldwide application has been possible. The evolution of the modern geologic time scale has been largely a matter of compromise. The criteria used in the original definition of the geologic systems were largely paleontological, but with many local lithostratigraphic influences. The progressive extension of the systemic boundaries to regions beyond the type areas has led to all kinds of problems. One instance is seen in the Cambrian System where, at its type area in Wales, the Cambrian-Ordovician boundary was established at a marked erosional break at the top of the shales of the Tremadoc Series. On the other hand, Tremadocian graptolites and other fossils are considered more akin to species typical of the Ordovician, and so, in other parts of the world, beds with a Tremadocian fauna are usually considered a natural part of the Ordovician, rather than of the Cambrian.

The episodes of major eustatic recession are usually marked somewhere by erosional breaks within the sedimentary veneers of the cratonic interiors. These are interregional unconformities that are often almost of continentwide extent. The sedimentary envelopes between them are the major unconformity-bounded units that come as close as we will ever get to "natural" chronostratigraphic divisions of major rank equivalent to the systems as originally defined. The difficulties encountered in recognizing the standard systemic boundaries, and in reconciling them with local boundaries, have led numerous authors

to propose new schemes for lithostratigraphic and chronostratigraphic subdivisions. Chang (1975) suggested that unconformity-bounded stratigraphic successions should be considered as an independent category of stratigraphic unit and proposed the term **synthem** for such units (see also Hedberg, 1976, p. 92). Although there has been acceptance of the idea of synthems, the term itself seems, until very recently, to have taken second place to the term **sequence,** which is perhaps best known in the sense used by Sloss (1963), who defined it as follows: "Stratigraphic sequences are rock stratigraphic units of higher rank than group, megagroup or supergroup, traceable over major areas of a continent and bounded by unconformities of interregional scope."

13.3 SEQUENCES

Sequences are the largest unconformity-bounded units and contain the record of major marine incursions onto the cratons; sometimes, as epeiric seas, right across them. The six sequences of the North American continent described and named by Sloss (1963) have their counterparts on other cratons, as shown in Figure 13.1. The interregional unconformities bounding the sequences are surfaces of erosional truncation (Fig. 13.2). They can be demonstrated most clearly when there is angular discordance between the basal bed of a sequence and the strata below. Regional dips in the sedimentary veneers of cratonic interiors are, however, typically small, and even major hiatuses involving long time periods may not be very obvious. On a local scale, the strata above and below the unconformity may be essentially parallel, but traced over many kilometers, the parallelism between lower and upper successions is rarely maintained, and the lower beds are seen to pinch out successively against the unconformity above. In surface geology, this can be demonstrated by regional mapping; in seismic sections, the angularity of successive

FIGURE 13.1

Correlation of sequences on three cratons. (From Petters, S. W., 1979, West African cratonic stratigraphic sequences: Geology, v. 7, Fig. 3, p. 530. By permission of The Geological Society of America.)

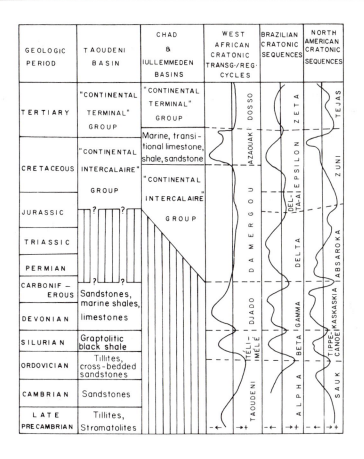

reflection terminations can often be seen directly.

The transgressions into the continental interiors, exemplified by sequences, occurred with numerous second- and higher-order sedimentary cycles within them. During the Sauk transgression onto the North American craton, for example, Aitken (1966) recognized eight so-called Grand Cycles in the Cambrian to Lower Ordovician of Alberta, Canada (Fig. 13.3). No major unconformities occur at the boundaries of the cycles, but each one begins with shaly and silty beds, which pass up into limestones in the upper part of the cycle. The contact at the top of the limestone is characteristically abrupt and overlain by the silty and shaly part of the overlying cycle. Individual Grand Cycles range in thickness from 200 to 800 m. The earliest of the Grand Cycles be-

gan in the early Albertan (approximately 540 m.y. ago), and the cycles came to an end in the Middle Llanvirnian (approximately 470 m.y. ago), so each cycle averaged about nine million years. The same Grand Cycles have been detected in the Cambrian of the Great Basin (Palmer, 1971) and in Newfoundland (Chow and James, 1987). There is little doubt that the cycles are of eustatic origin, possibly controlled by variations in the rate of sea-level rise.

13.4 SURFACE OF TRANSGRESSION

There has been much discussion as to the nature of interregional unconformities that formed the craton surface over which epicontinental seas of the past transgressed. The marked uniformity of many of the epeiric

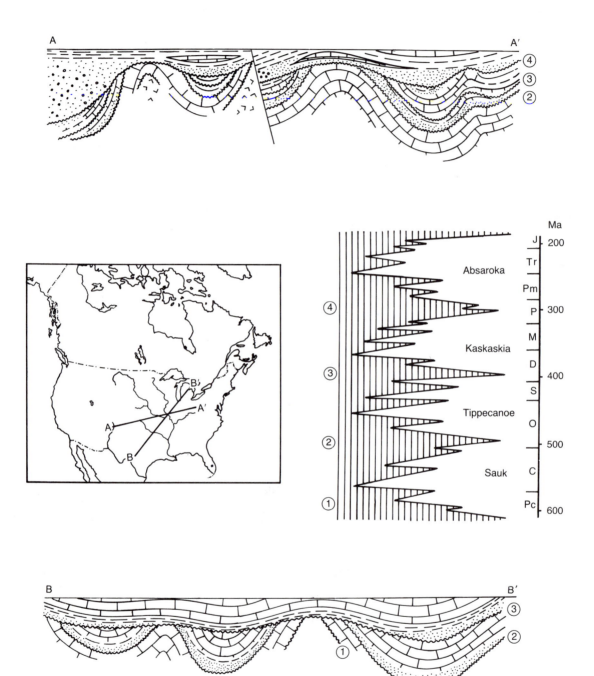

FIGURE 13.2

Sequences on the North American continent. (Redrawn from Sloss, L. L., 1963, Sequences in the cratonic interior of North America: Geol. Soc. America Bull., v. 74, Fig. 1, p. 97 and Fig. 2, p. 100. By permission of The Geological Society of America.)

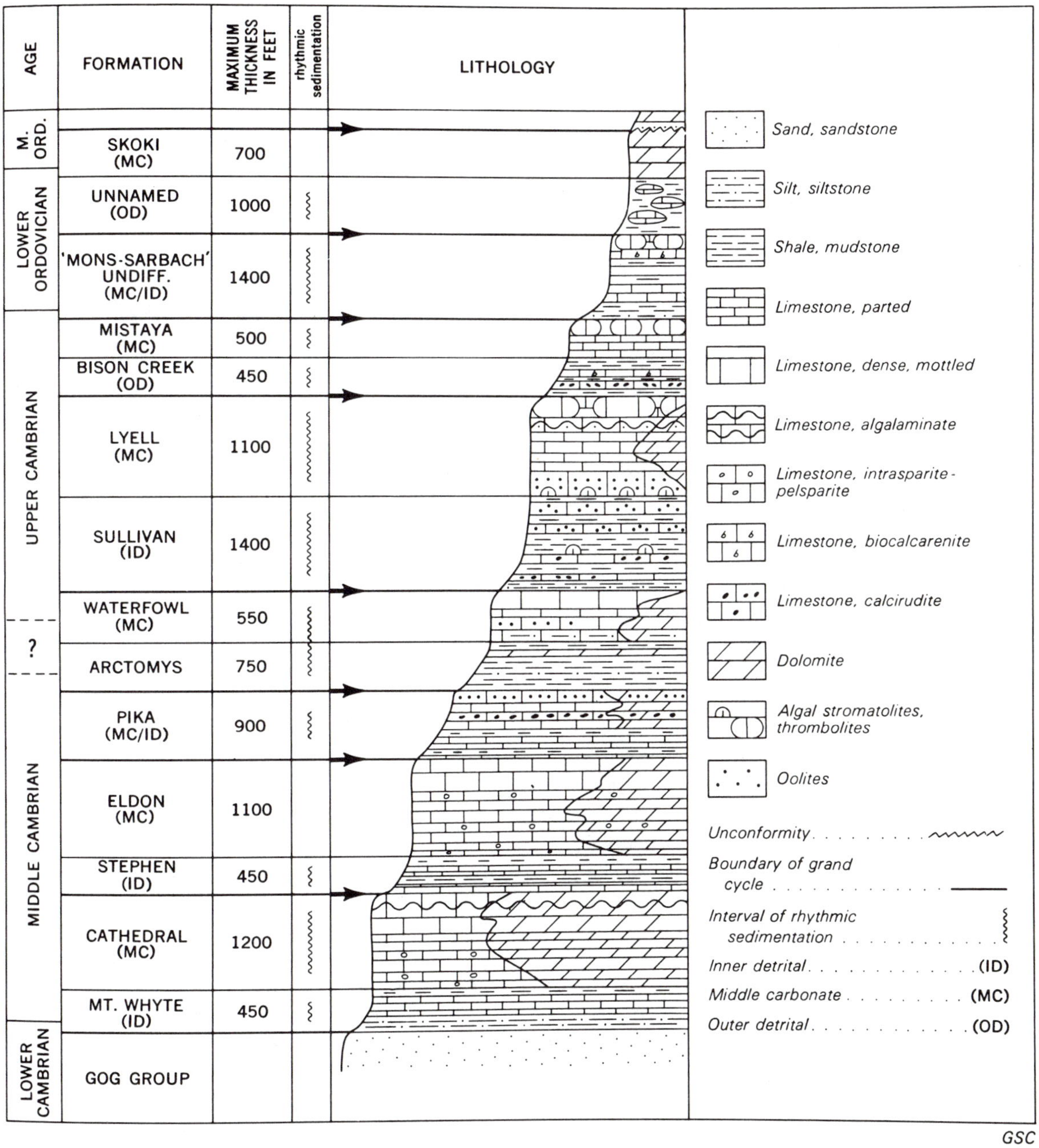

FIGURE 13.3
Composite stratigraphic column of Middle Cambrian to Middle Ordovician rocks of the
southern Rocky Mountains showing boundaries of eight Grand Cycles. (From Aitken,
J. D., 1966, Middle Cambrian to Middle Ordovician cyclic sedimentation, southern Rocky
Mountains of Alberta: Can. Petroleum Geology Rev., v. 14, p. 405–441.)

sedimentary formations suggests a surface of low relief. Assuming that structural warping of the craton hinterland was minimal, this is probably a reasonable supposition, and it is possible that during periods of marine regression an emergent craton surface might eventually come to form what the nineteenth-century geomorphologist W. M. Davis would have described as a **peneplain.** During non-glacial intervals of earth history, sea-level fluctuations, whether due to eustatic or epeirogenic causes or a mixture of both, were, in the context of geomorphic cycles, often very slow. It could be argued, therefore, that even if varied basement rocks were exposed, craton land surfaces would eventually be reduced to near peneplains by subaerial processes. It has been suggested that planation could have been accomplished during the marine transgression by coastal erosional processes and that the sea advanced over a beveled surface of its own making; however, this is unlikely in view of the extreme paucity of clastics in many typical epeiric successions. Even though the craton at the time of its submergence might be devoid of significant relief, it was certainly not flat, and there was a regional gradient from interior to margin. Shaw (1964) postulated gradients across the craton as low as "0.1′ to about 0.5′ per mile" (approximately 1 in 50,000 to 1 in 10,000), compared with average bottom slopes of modern marginal seas of "2′ to 10′ per mile" (approximately 1 in 3000 to 1 in 500). Such low gradients were, it was argued, the cause of the lack of tidal circulation in epicontinental seas, an essential feature of the Shaw-Irwin model of "epeiric clear water sedimentation" described in Chapter 14. Such very gentle slopes over continental distances may seem difficult to imagine, but it should be remembered that even on modern continents, regional gradients are often extremely small. For example, from the headwaters of the Mississippi to its mouth, the river flows across a surface with average gradient of

about 1 in 20,000. In the Amazon basin, the gradient is even less.

13.5 UNCONFORMITY-BOUNDED UNITS ALONG CONTINENTAL MARGINS

Within the unconformity-bounded units of sequence rank as defined by Sloss (1963) are many smaller units that also are bounded by unconformities, often of only local significance. In continental interiors, the alternations of sediment accumulation and erosion are typically controlled by changes in basin subsidence, by movement along intracratonic structural trends, or by changes in the rate of uplift in adjacent orogenic belts. Eustatic sea-level changes normally have little influence. In approaching the margins of a continent, the hiatuses represented by the major interregional unconformities become progressively smaller and the sedimentary record more complete. At the same time, the lesser unconformities are increasingly likely to be due to sea-level changes as well as more local effects due to marginal warping. This leaves a record that is seen in both facies changes and erosional intervals. However, the record is clearly modified by many local structural effects and variations in sediment supply and in climate.

It is the stratigraphic succession beneath the coastal belt and continental shelf, at least on passive margins, that contains the most complete record of sea-level fluctuations. This is where seismic surveys have, in recent years, provided access to information that had previously been unavailable. As described earlier, the sea-level curve established by Vail et al. (1977) was based largely on an analysis of unconformity-bounded units, the bounding discontinuities of which are usually visible in seismic sections because of the discordant relationships between "packages" of reflections. The actual surfaces of discontinuity likely contain no diagnostic features unless they are themselves marked by a large velocity contrast. In comparing the reflection patterns across an unconformity, there are four basic relationships. Thus, as an upper surface, the unconformity may be underlain by parallel strata or it may truncate inclined strata. As a basal surface, the unconformity may underlie parallel strata or it may be a surface against which inclined strata successively terminate (Fig. 13.4). Because, obviously, the upper boundary of a sedimentary package is the base of the one above it, all combinations of the four basic patterns are possible. Only when the sequences below and above the unconformity are both parallel and the surface itself contains no distinctive reflection characteristics does the seismic record sometimes fail. Normally, such "conformable" surfaces can be traced laterally to a place where some angular discordance between lower and upper successions is apparent.

Mitchum et al. (1977) suggested a further subdivision of the basic four patterns and thus recognized the influence of sediment source and basin configuration. At the lower

Reflections parallel

Reflections inclined and truncated

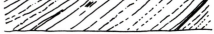

Erosion surface as top of sedimentary sequence

Reflections parallel

Reflections truncated downward

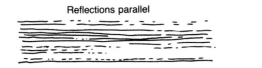

Erosion surface as base of sedimentary sequence

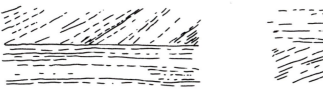

Combinations across unconformity surface

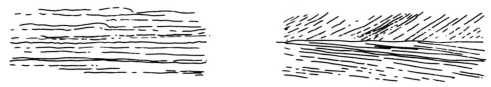

FIGURE 13.4
Possible relationships of seismic reflection packages across an unconformity.

boundary of a sedimentary unit, the strata are said to have a **baselap** relationship to the underlying surface if the junction is one of angular discordance. One type of baselap is seen to result from a marine transgression when shallow-water sediment laps up over a land surface. The bedding surfaces in the transgressing sediments are approximately horizontal, but landward they progressively pinch out, and their relationship to the underlying surface is said to be one of **onlap** (Fig. 13.5). In **coastal onlap,** the onlapping sediments are derived from the land surface over which the sea is encroaching, and the beds become progressively younger in a landward direction. Such onlapping marine sediments may sometimes lie above an erosion surface, but more commonly are found to transgress across an earlier series of deltaic or coastal plain and alluvial deposits. The normal situation is one in which the sediments are derived from the land surface over which the transgression is taking place; in small basins, however, onlapping sediments may, in fact, have been derived from the opposite shore. This happens when the supply of sediment is great enough to fill the basin. To recognize these sediment source differences, the term **proximal onlap** was suggested for the first situation and **distal onlap** for the second.

Where sediment supply is abundant there may be net progradation basinward, regardless of whether sea level is rising or falling. These prograding, **clinothem** beds will have a relationship of **downlap** to the surface—the fondoform—over which they are are encroaching. They have a small angle of original dip, which typically tends to flatten asymptotically into the basin so that they eventually pinch out. The beds become progressively younger basinward away from the sediment source; this is termed **distal downlap** (Fig. 13.5). In small basins, it is the encroaching downlap sediments that may reach completely across to the opposite shore to result in distal onlap there. A combination of distal downlap and distal onlap would eventually fill the basin with sediment, and this relationship would seem to be similar to what Brown and Fisher (1977) called *uplap* in describing sediment fill behind fault blocks or salt diapir domes on continental shelves (Fig. 13.6). Some of the best examples of downlap are seen associated with an advancing delta front.

In a complete cycle of transgression/regression, onlapping beds pass upward into sediments of progressively shallower-water aspect in an **offlap** relationship to the erosion surface that marks the end of the cycle. This type of offlap is probably not common on

FIGURE 13.5
Relations of strata to the boundaries of depositional sequences. (From Mitchum, R. M., Jr., et al., 1977, Seismic stratigraphy and global changes of sea level, part 2, The depositional sequence as a basic unit for stratigraphic analysis, *in* Payton, C. E., ed., Seismic stratigraphy—Applications to hydrocarbon exploration: AAPG Mem. 26, Fig. 1, p. 54, reprinted by permission of American Association of Petroleum Geologists.)

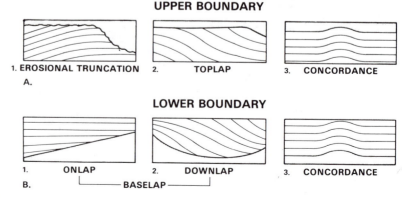

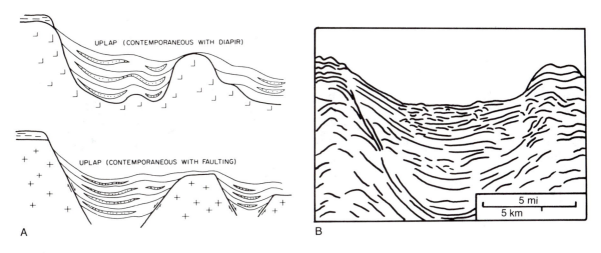

FIGURE 13.6
Sections showing uplap relationships within a sedimentary succession: A. Diagrammatic cross sections of two settings for uplap facies. (From Brown, J. R. L., and Fisher, W. L., 1977, Seismic-stratigraphic interpretation of depositional systems: Examples from Brazilian rift and pull-apart basins, *in* Payton, C. E., ed., Seismic stratigraphy—Applications to hydrocarbon exploration: AAPG Mem. 26, Fig. 16, p. 238, reprinted by permission of American Association of Petroleum Geologists.) B. Typical appearance of a seismic record of uplap behind a diapir dam.

anything but a small scale. Sea-level fluctuations large enough to leave a significant sedimentary record are not usually marked by a thin progradational succession that signals the marine regression and is simply a mirror image of the thin retrogradational succession of the marine transgression. Instead, as described in a previous section, regression is likely marked by sedimentary progradation whether sea level is actually rising or falling. The best examples of offlap, in seismic sections at least, are seen in areas of active delta outbuilding or where there is vigorous biogenic reef construction. Variations in the rate of sediment supply during rising sea level, change in the rate of sea-level change, or changes in the rate of subsidence can all result in an alternation of onlap/offlap units (Fig. 13.7).

The various unconformity relationships described above can usually be detected in the seismic record with little difficulty; however,

the precise horizon of the unconformity surface may sometimes be in doubt because of signal problems. As described by Vail and Todd (1981), many erosional surfaces generate such a strong reflection that a second peak appears on the wave form below the principal reflection in what is termed a *follow cycle*. If this is not removed by processing, this follow cycle may be mistaken for the unconformity, which is then picked too low. The correct horizon can usually be located by noting reflections above the unconformity, such as downlapping reflectors, because these terminate against the true unconformity, whereas truncated reflectors below the unconformity often terminate against the false unconformity.

An unconformity as part of a stratigraphic section is sometimes given a name. The normal practice has been to name it after the episode of uplift and erosion responsible, if it is part of a named orogenic disturbance. Vail

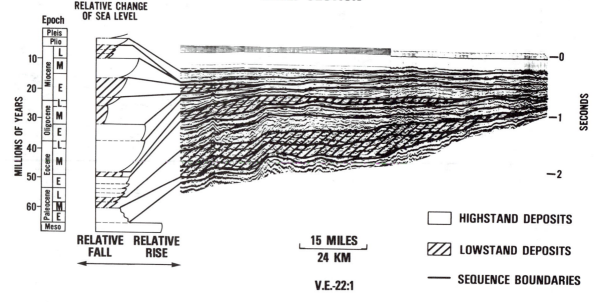

SHELF SECTION

FIGURE 13.7

North Sea Tertiary shelf seismic example of depositional patterns during highstand and lowstands of sea level. (From Vail, P. R., et al., 1977, Seismic stratigraphy and global changes of sea level, part 4, Global cycles of relative changes of sea level, *in* Payton, C. E., ed., Seismic stratigraphy—Applications to hydrocarbon exploration: AAPG Mem. 26, Fig. 11, p. 75, reprinted by permission of American Association of Petroleum Geologists.)

and Todd (1981), however, recommend that the name of the chronostratigraphic unit immediately above the unconformity be used instead (Fig. 13.8); thus, for example, the unconformity below the Jurassic Kimmeridgian Stage should be called the basal Kimmeridgian unconformity.

13.6 CYCLES AND CYCLOTHEMS

Virtually all the major unconformity-bounded units consist of marine or dominantly marine sediments. Because they represent deposition during marine transgressions and regressions, which are cyclical phenomena, unconformity-bounded units should be considered as cyclical units of one kind or another. Sedimentary successions have long been de-

scribed in terms of cycles and rhythms and the two words are more or less interchangeable. Some workers have suggested that the term *rhythm* be confined to the ABAB type of repetition, as in glacial varves, but this is certainly not a generally held view. Symmetrical cycles are those expressed as ABCDCBA, whereas asymmetrical cycles are of the ABCDABCD type. The terms cycle and rhythm can obviously refer not only to the sediments themselves, but also to associated phenomena or the time involved. The term *cyclothem* is strictly a lithostratigraphic term and refers to the sediments laid down during a single transgressive/regressive cycle. As originally defined by Weller (Wanless and Weller, 1932), cyclothem was used in reference to the well-marked and numerous repetitive

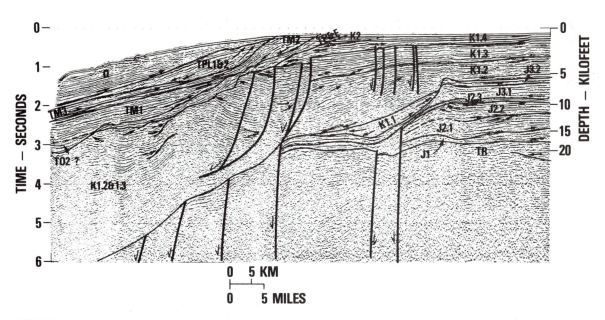

FIGURE 13.8
Seismic line, offshore West Africa, showing the major sequence boundaries. (From Todd, R. G. and Mitchum, R. M., Jr., 1977, Seismic stratigraphy and global changes of sea level, part 8: Identification of Upper Triassic and Lower Cretaceous seismic sequences in Gulf of Mexico and offshore West Africa, *in* Payton, C. E., ed., Seismic stratigraphy—Applications to hydrocarbon exploration: AAPG Mem. 26, Fig. 9, p. 157, reprinted by permission of American Association of Petroleum Geologists.)

successions of strata of Pennsylvanian age in the northeastern U.S.; implicit in its definition was an association with coals, although a coal bed was not necessarily present in all cyclothems because some cyclothems were considered incomplete.

The "ideal" cyclothem, as seen in the Pennsylvanian of Illinois, was said to contain, according to Weller (1958), ten lithologic units or members (Fig. 13.9), but this succession is not expressed everywhere, and after allowing for local variations and missing members, the simplest and basic cyclothem contains five members:

E Shale   } Marine hemicyclothem
D Limestone

C Coal
B Underclay } Nonmarine hemicyclothem
A Sandstone

The cyclothemic sequence supposedly represents sedimentation in a coastal-plain environment undergoing progressive subsidence and/or sea-level rise. The lower half of each cyclothem contains nonmarine beds (the nonmarine hemicyclothem) culminating in a coal. Coal-forming conditions were terminated by marine transgression, and the cyclothem was completed by marine sediments (the marine hemicyclothem) culminating in shales laid down during the regressional phase of the cycle. Emergence and erosion followed in many cases—the basal sandstone of the subsequent cyclothem is usually seen to be lying on an erosion surface. The period of emergence and intensity of erosion varied, and in some cases, a large proportion of the previous cyclothem was removed before the onset of the next cyclothemic episode. The

FIGURE 13.9
Stratigraphic section of a completely developed cyclothem, typical of the Pennsylvanian of Illinois and adjacent region. (Redrawn after Weller, 1958.)

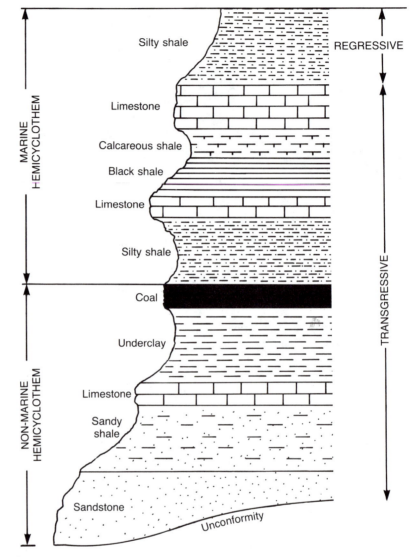

many deviations from the ideal cyclothem may be due to this erosion factor, but incomplete cyclothems might also have resulted because the relative changes in sea level were too small to bring about the full development of the sequence and the cyclothem was aborted. Splitting of many members in cyclothems, and lateral facies changes, indicate that uplift or subsidence was often an important influence. On the other hand, the long-distance correlation in many cyclothems suggests that eustatic sea-level fluctuations were also important, as was discussed in Chapter 12.

Studies of Pennsylvanian coals in recent years have suggested that the concept of the classical coal-bearing cyclothemic sequence is oversimplified and misleading. There are, apparently, many cases where the observed relationship of coals, clays, sandstones, and limestones do not seem to fit any standard pattern similar to the Weller model. These observations suggest that regional and even local structural effects play a more important role than had previously been suspected. According to Ferm (1975), many of the observed coal-bearing cyclothems seem to owe their origin to pulsating influxes of clastic sediments, alternating with periods of nonclastic deposition characterized by coals or limestones. This would indicate deposition in a deltaic environment in which regional subsidence, deltaic progradation, and distributary switching may have been local controls superimposed on a glacio-eustatic control and which may in some areas even have played the dominant role.

Cyclothems of the type originally described in the Pennsylvanian have been described from coal-bearing successions in other areas and of various ages. In other studies, the term cyclothem has come to be used to describe the deposits of any transgressive/regressive cycle, including evaporite/carbonate cycles (Winston, 1972) and even fluvial cycles (e.g., Allen, 1964). Cycles of cycles have been recognized by some workers; Moore (1936) coined the terms **megacyclothem** for a series of cyclothems and **hypercyclothem** for a series of megacyclothems (Table 13.1). In one sense, and particularly for mapping purposes, cyclothems and other units of higher rank could be considered lithostratigraphic units and this ranking of units as approximately equivalent to conventional lithostratigraphic units. Cyclothems are, however, unconformity-bounded units and, as such, have come to be incorporated within a different hierarchy of terms. Ramsbottom (1977), for example, grouped cyclothems within **mesothems** (Table 13.1 and Fig. 13.10), which he defined as unconformity-bounded units of middle rank, having a time duration roughly similar to that of a stage. In the Namurian (Carboniferous) of northwest Europe, the seven stages recognized contain about 60 cyclothems grouped into 11 mesothems deposited over a time span of something less than 15 m.y. (Ramsbottom, 1979). For the unit of highest rank, the term *synthem* (Chang, 1975) was proposed (Table 13.1). It is interesting to note that the original cyclothem concept, as defined by Weller, was intended to apply to cycles believed to be eustatic in origin. More recently, Ramsbottom (1979) suggested that for cyclical units of noneustatic origin a term other than cyclothem should be used.

Mitchum et al. (1977), in their study of eustatic sea-level cycles, relied heavily on the evidence of unconformity-bounded units and demonstrated the cyclical nature of such units. They preferred, however, to use the term *sequence* in describing their basic unconformity-bounded units. Although these units were an order of magnitude smaller than the sequences of Sloss (1963), Mitchum et al. considered them to have the same general characteristics and so the same term was adopted. The original sequence in the sense of Sloss (1963) was, in effect, elevated in rank to become a **supersequence.** From the description of Mitchum et al. (1977), it is clear that the term *sequence* used in their sense is synonymous with the mesothem of Ramsbottom (1977) (see Table 13.1).

13.7 UNCONFORMITY-BOUNDED UNITS IN TIME

Strictly speaking, a sequence, mesothem, or other unconformity-bounded unit is a lithostratigraphic unit. It does, however, have considerable chronostratigraphic significance because the sediments contained within it were all laid down after the basal unconformity and before the upper unconformity. In other words, a mesothem or sequence represents a particular sedimentary episode involving a transgression and regression over a given interval of time, and so the basal unconformity is time-transgressive. Because the bounding

TABLE 13.1
Terminology used to describe unconformity-bounded units

Approximate Time Span	Approximate Chronostrat. Equivalent	Unconformity-Bounded Units				Vail et al (1977)		Moore (1936)
		Chang (1975)	Ramsbottom (1977)	Salvador* (1987)	Hubbard (1988)		Sea-Level Fluctuations	
200–400 m.y.	Erathem			Supersynthem	Megasequence		1st-order cycle	Hypercyclothem
40–60 m.y.	System	Synthem	Synthem	Synthem	Sequence	Supersequence (= sequence of Sloss, 1963)	2nd-order cycle (global supercycle)	—
								Megacyclothem
5 m.y.	Stage	Interthem	Mesothem	Subsynthem	Subsequence	Sequence†	3rd-order cycle (global cycle)	—
250,000–1 m.y. (Ramsbottom) 200,000–250,000 yrs.	Chronozone		Cyclothem (common usage)	Miosynthem			Paracycle	Cyclothem

*No time connotation is implied in the case of divisions by Salvador (1987).

†Approx. = holostrome. The holostrome of Wheeler (1958, p. 1055) is a time-stratigraphic unit defined as a restored transgressive-regressive depositional sequence, it includes the strata that may later have been removed by erosion. (Wheeler, H. E., 1958, Time-stratigraphy: Am. Assoc. Petroleum Geologists Bull., v. 42, p. 1047–1063.)

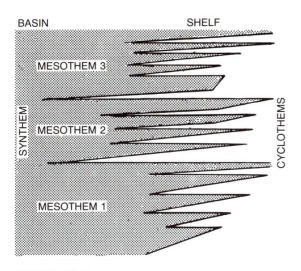

FIGURE 13.10
Nomenclature of eustatic cycles. (After Ramsbottom, 1979.)

unconformities are diachronous, the interval of time actually represented by the strata in a sequence must vary from place to place. It is obviously shortest in those areas at the extreme limit of the particular marine incursion and grows larger as the sequence is traced oceanward. Conversely, the time intervals represented by the hiatuses below and above grow smaller oceanward until they disappear entirely in a conformable succession representing, in theory at least, continuous sedimentation. The time unit embracing the time span during which an entire sequence was deposited is known as a **sechron,** defined as the time interval represented by the sediments contained between the lower and upper boundaries of the sequence, where the bounding unconformities have disappeared into a conformable succession. The place where the lower boundary is conformable may not necessarily be the same place as where the upper unconformity disappears (Fig. 13.11).

Mitchum et al. (1977) and Vail et al. (1984) discussed the chronostratigraphic signifi-

cance of unconformities and pointed out that, although the duration of the hiatus between the strata below and above the unconformity varies from place to place, the unconformity itself should be considered a chronostratigraphic boundary because it separates rocks of different ages and no chronostratigraphic surface crosses it. Like other workers before them, Vail et al. (1984) see unconformity-bounded units as natural divisions. They suggest that cycles of coastal onlap, as seen in the asymmetrical shape of the cycles in their sea-level curve, mark the times of global unconformities. Hence, these can be used to divide the stratigraphic section chronostratigraphically into depositional sequences.

13.8 UNCONFORMITY-BOUNDED UNITS IN STRATIGRAPHIC CODES

Unconformity-bounded units only recently have come to be recognized by the International Subcommission on Stratigraphic Classification (ISSC), and the term synthem originally proposed by Chang (1975) has been proposed as the basic unconformity-bounded unit (Salvador, 1987). In naming such units, it is suggested that an appropriate geographic name, derived from a location where the unit is well developed, be assigned (Fig. 13.12). If necessary, synthems may be subdivided into two or more **subsynthems,** and two or more synthems may be combined into a **supersynthem.** For a further category of small or minor synthems within a larger synthem, the term **miosynthem** is proposed (Table 13.1).

The North American Stratigraphic code (NACSN, 1983) does not recognize major unconformity-bounded units as separate entities, and they have no place in the formal classification. Mention is made of "sequences" and "synthems," but the recommendation is that they be treated only as informal units, at least for the time being (Appendix 3). Although the code has not in-

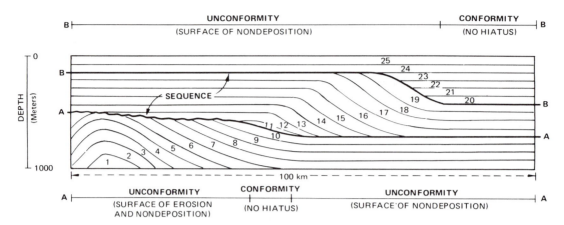

FIGURE 13.11
Unconformities and correlative conformities (A–A and B–B). (From Mitchum, R. M., Jr., et al., 1977, Seismic stratigraphy and global changes of sea level, part 2, The depositional sequence as a basic unit for stratigraphic analysis, *in* Payton, C. E., ed., Seismic stratigraphy—Applications to hydrocarbon exploration: AAPG Mem. 26, Fig. 2, p. 58, reprinted by permission of American Association of Petroleum Geologists.)

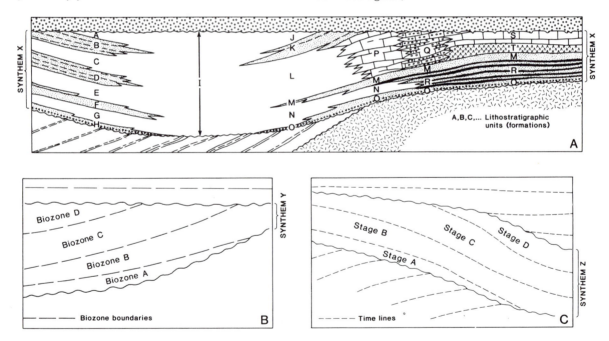

FIGURE 13.12
Relationship of unconformity-bounded units to other kinds of stratigraphic units included within them. Unconformity-bounded units may include any number of other kinds of stratigraphic units: lithostratigraphic (A), biostratigraphic (B), chronostratigraphic (C), and so on. The boundaries of unconformity-bounded units can be parallel or, most commonly, intersect at an angle with the boundaries of the stratigraphic units they include. (From Salvador, A., 1987, Unconformity-bounded stratigraphic units: Geol. Soc. America Bull., v. 98, Fig. 1, p. 235.)

cluded major unconformity-bounded units as physical entities, it does, rather curiously, recognize the temporal parameters involved and establishes so-called **diachronic units.** These comprise the unequal spans of time represented in the various physical stratigraphic units that are not bounded by isochronous surfaces.

As pointed out earlier, unconformity-bounded units do have considerable chronostratigraphic significance, although they are clearly not bounded by isochronous surfaces. The time span represented by the sediments between the bounding unconformities varies from place to place, and it is this that is described in terms of diachronic units. The North American Code recognizes such units so as to provide a basis for measuring the variability in age of the beginning and the end of deposition within diachronous stratigraphic units as they are traced over any distance. The fundamental diachronic unit is the **diachron,** and a hierarchical ranking of divisions within a diachron is proposed, using the terms **episode, phase, span,** and **cline** in order of decreasing rank. Another stratigraphic category that should be mentioned in the present context, because of the possibility of confusion, is the **allostratigraphic** unit. This is described as a "mappable stratiform body of sedimentary rock that is defined, and identified on the basis of its bounding discontinuities" (NASCN, 1983). Although, out of context, this definition seems to fit the unconformity-bounded units under discussion, the additional remarks of the code seem to preclude them. What seems meant by the code is that the deposits of such unconformity-bounded units exhibit some degree of internal genetic homogeneity, even if there is some lithic heterogeneity. From the examples illustrated in the code, it would appear that it is relatively small-scale units such as alluvial terrace gravels, glacial morainic deposits, lake deposits, and the like that it is intended

to classify, although this is not very clear from the description.

There is little doubt that not all the recommendations set out in the 1983 code will stand the test of time, and there will be criticism of some of them. This is, however, not only expected, as the authors of the code point out, but desirable. Concepts that prove to be unacceptable, or of no use, will simply fade away, as has happened on numerous occasions in the past. This is a normal and healthy process.

13.9 CONCLUSION

The material in this chapter is, in many ways, a reiteration of the theme of Chapter 1, namely, that sediment accumulation at whatever scale is markedly repetitive and cyclical. It should be emphasized that there is nothing very new about this—some general awareness of underlying rhythms was evident many years ago in Grabau's (1936) "pulsation" theory and in such works as Umbgrove's (1946) classic book *The Pulse of the Earth.* Thus, the current preoccupation with cyclicity, "catastrophic uniformitarianism," punctuated equilibrium, and other evidence of what Byers (1982) called "the fall of continuity" might be said to strike a familiar chord that has been heard before.

REFERENCES

Aitken, J. D., 1966, Middle Cambrian to Middle Ordovician cyclic sedimentation, southern Rocky Mountains of Alberta: Can. Petroleum Geology Rev., v. 14, p. 405–441.

Allen, J. R. L., 1964, Studies in fluviatile sedimentation: Six cyclothems from the Lower Old Red Sandstone, Anglo-Welsh basin: Sedimentology, v. 3, p. 163–198.

Brown, L. F., Jr., and Fisher, W. L., 1977, Seismic-stratigraphic interpretation of depositional systems: Examples from Brazilian rift and pull-apart basins, *in* Payton, C. E., ed., Seismic stratigraphy—Applications to hydrocarbon exploration: Am. Assoc. Petroleum Geologists Mem. 26, p. 185–204.

Byers, C. W., 1982, Stratigraphy—The fall of continuity: Jour. Geol. Education, v. 30, p. 215–221.

Chang, K. H., 1975, Unconformity-bounded stratigraphic units: Geol. Soc. America Bull., v. 86, p. 1544–1552.

Chow, N., and James, N. P., 1987, Cambrian Grand Cycles: A northern Appalachian perspective: Geol. Soc. America Bull., v. 98, p. 418–429.

Ferm, J. C., 1975, Pennsylvanian cyclothems of the Appalachian Plateau, a retrospective view, *in* McKee, E. D., Crosby, E. J., and others, eds., Paleotectonic investigations of the Pennsylvanian System in the United States, part II, Interpretative summary and special features of the Pennsylvanian System: U.S. Geol. Survey Prof. Paper 853, p. 57–64.

Grabau, A. W., 1936, Oscillation or pulsation: XVI Internat. Geol. Congress, Washington, DC, p. 539–553.

Hedberg, H. H., ed., 1976, International stratigraphic guide: New York, John Wiley & Sons, 200 p.

Mitchum, R. M., Jr., Vail, P. R., and Thompson, S., III, 1977, Seismic stratigraphy and global changes of sea level, part 2: The depositional sequence as a basic unit for stratigraphic analysis, *in* Payton, C. E., ed., Seismic stratigraphy—Applications to hydrocarbon exploration: Am. Assoc. Petroleum Geologists Mem. 26, p. 53–62.

Moore, R. C., 1936, Stratigraphic classification of the Pennsylvanian rocks of Kansas: Geol. Survey Kansas Bull. 22.

N.A.C.S.N. (North American Commission on Stratigraphic Nomenclature), 1983, North American stratigraphic code, 1983: Am. Assoc. Petroleum Geologists Bull., v. 67, p. 841–875.

Palmer, A. R., 1971, The Cambrian of the Appalachian and eastern New England regions, eastern United States, *in* Holland, C. H., ed., Cambrian of the New World: New York, Wiley Interscience, p. 169–218.

Ramsbottom, W. H. C., 1977, Major cycles of transgression and regression (mesothems) in the Namurian: Proc. Yorkshire Geol. Soc., v. 41, p. 261–291.

———— 1979, Rates of transgression and regression in the Carboniferous of N.W. Europe: Geol. Soc. London Jour., v. 136, p. 147–153.

Salvador, A., 1987, Unconformity-bounded stratigraphic units: Geol. Soc. America Bull., v. 98, p. 232–237.

Shaw, A. B., 1964, Time in stratigraphy: New York, McGraw-Hill, 365 p.

Sloss, L. L., 1963, Sequences in the cratonic interior of North America: Geol. Soc. America Bull., v. 74, p. 93–114.

Umbgrove, J. H. F., 1946, The pulse of the earth: The Hague, Martinus Nijhoff, 358 p.

Vail, P. R., Hardenbol, J., and Todd, R. G., 1984, Jurassic unconformities, chronostratigraphy, and sea-level changes from seismic stratigraphy and biostratigraphy, *in* Schlee, J. R., ed., Interregional unconformities and hydrocarbon accumulation: Am. Assoc. Petroleum Geologists Mem. 36, p. 129–144.

Vail, P. R., Mitchum, R. M., Jr., and Thompson, S., III, 1977, Seismic stratigraphy and global changes of sea level, part 4: Global cycles of relative changes of sea level, *in* Payton, C. E., ed., Seismic stratigraphy—Applications to hydrocarbon exploration: Am. Assoc. Petroleum Geologists Mem. 26, p. 83–97.

Vail, P. R., and Todd, R. G., 1981, Northern North Sea Jurassic unconformities, chronostratigraphy and sea level changes from seismic stratigraphy, *in* Petroleum geology of the continental shelf of northwest Europe: London, Heyden, p. 216–235.

Wanless, H. R., and Weller, J. M., 1932, Correlation and extent of Pennsylvanian cyclothems: Geol. Soc. America Bull., v. 43, p. 1003–1016.

Weller, J. M., 1958, Cyclothems and larger sedimentary cycles of the Pennsylvanian: Jour. Geology, v. 66, p. 195–207.

Winston, G. O., 1972, Oil occurrences and Lower Cretaceous carbonate-evaporite cyclothems in South Florida: Am. Assoc. Petroleum Geologists Bull., v. 56, p. 158–160.

14

SEDIMENTARY SYSTEMS

The solid parts of the present land appear, in general, to have been composed of the productions of the sea.

James Hutton

14.1 INTRODUCTION

Traditionally, sediments and sedimentology are subjects that have formed a large part of stratigraphic studies, and this is reflected in the format of most stratigraphic texts. The purpose of this chapter is not to emulate this approach or, even less, to duplicate material that can, and often does, fill whole books. Rather, its purpose is to provide a link with these other sources. A summary as brief as this must use the most fundamental classification, and it is consequently based upon the largest tectonic elements on earth. The rationale for the classification is indicated by plate tectonic divisions and plate boundary relationships. These suggest that five broad sedimentary associations can be recognized. These divisions, which might be termed *tectonic settings,* are stable continental platforms during submergence, stable continental platforms emergent, passive (trailing-edge) continental margins, ocean basins, and active continental margins. Each of these settings is, to a large extent, self-contained, both in the context of modern sedimentary environments and in the context of geologic history, although over time sediments originating in one setting may end up in another. At any one time, the differences between the settings are very apparent and are of the kind and degree that suggest that they are quite "natural" divisions.

14.2 CRATONS SUBMERGED—EPEIRIC SETTINGS

The interior regions of all the continental blocks of the world consist of one or more continental nuclei that comprise the ancient stable **cratons** around which are wrapped the mobile belts represented by fold/thrust mountains.

One of the most striking features of the sedimentary rocks that cover large areas of the continents is that they bear witness to periodic and widespread inundations by **epeiric,** or **epicontinental,** seas. Over many thousands of square kilometers on all the continents are relatively thin, sheetlike sedimentary formations that were laid down under shallow-marine conditions of remarkable uniformity, over a surface of low relief. At times of maximum inundation of the North American continent, it has been calculated that as much as 90 percent of the present craton surface was under water. The unmetamorphosed sedimentary covers on the cratons range in age from Proterozoic through Phanerozoic and typically are veneers nonconformably overlying metamorphosed and crystalline basement rocks. Traced laterally,

individual formations of the cratonic cover become thicker and change in character as they pass into marginal belts. There they invariably have become involved in structural deformation and, in places, metamorphism during one or more orogenic/collision episodes. In the heartland of the cratons are large areas which for long periods have been more or less emergent. These are the **shield** areas. It is on the shields that the most extensive exposures of Precambrian rocks can be seen and from them come much of the world's supply of important economic minerals. Those portions of the craton which, from time to time, were submerged by epeiric seas and became covered by younger sediments, have been termed **platforms.**

The sedimentary successions of the platforms provide us with the best stratigraphic record in terms of completeness, state of preservation, and lack of folding and faulting. This is in marked contrast to the continental margins, where the great sedimentary wedges have been involved in orogenic movements and the record, if not rendered indecipherable by metamorphism, is often difficult to interpret because of structural complexity. Despite the prominence of epeiric sea sediments in stratigraphic successions, they have tended to be neglected in studies of sedimentary environments. In many cases discussions of shallow shelf seas are taken to include epeiric seas also, and the implication is that epeiric sea environments differed from shelf environments only in their size. This is probably not true. Epeiric seas remain something of a mystery because, unlike most other major sedimentary environments, and we except Hudson Bay as a doubtful example, no modern analogues are to be found on earth today. This is not to say that conditions in the interior of such epicontinental seas were completely alien. The close resemblance, in many cases, between the lithology of epeiric sea formations and that of modern sediments provides some measure of assurance in deciphering ancient environments, but there remain many puzzling features. For example, the paraconformities, mentioned in Chapter 1, have yet to be explained adequately. Also intriguing are the widespread blanket sands and black shale formations, neither of which seem to have been formed under conditions that have convincing modern analogues. Whether clastic or carbonate deposition was dominant in an epeiric sea depended upon the proximity of sources of clastics. During the Early Paleozoic marine transgressions onto the North American and Siberian platforms, carbonates were very extensive because clastic sources were far removed (Fig. 14.1). Even modest amounts of suspended particulate matter in the water column will have a marked effect on light penetration and the suspension feeding activities of lime-secreting organisms. It follows that the clastic and carbonate sedimentary environments are almost mutually exclusive, as described by K. R. Walker et al. (1983).

Not all shallow interior seas were the sites of carbonate deposition. The great interior seaway that covered much of western North America in the Cretaceous (Fig. 14.2), for example, was dominated by clastic sediments, and frequently brackish-water rather than marine conditions prevailed. The seaway was, in fact, a partially submerged foreland basin undergoing active subsidence and receiving floods of terrigenous clastics from the rising Sevier orogenic belt to the west. The tectonic setting was, therefore, quite different from the stable cratonic interiors of the classic Early Paleozoic epeiric seas. This difference would seem, in the view of some workers at least, to preclude the Cretaceous seaway as a true "epeiric" sea. On the other hand, Shaw (1964) did define as epeiric any sea that was "upon the main mass of the continent," and this definition is supported by the *Glossary of Geology* (Bates and Jackson, 1987), which also considers the term *epicontinental* a synonym. Perhaps this problem of

semantics could be resolved by recognizing epeiric and epicontinental seas as being different, reserving the term epeiric only for situations involving flooding of a stable craton and with no orogenic downwarping involved.

Carbonate Sedimentation in Epeiric Seas

There is no question that the unique characteristic of many epeiric seas of the past was the importance of limestone deposition. The wide extent of some individual formations attests to the very uniform conditions that must at times have prevailed over thousands of square kilometers. On a regional scale, there is often a roughly concentric arrangement of facies, with limestones toward the periphery of a craton and dolomites in the interior (Fig. 14.1). In some cases, the dolomites, in turn, are concentric to an innermost zone characterized by evaporites. This distribution of facies has been noted in Cambrian, Ordovician, and Silurian successions on the North American craton and also in the Silurian of the Russian platform. A similar pattern is seen in the distribution of Mississippian limestones in western North America, although at this time clastic sediments were encroaching from the west. An explanation for this arrangement was suggested by Irwin (1965) and Shaw (1964) in an "epeiric clear-water sedimentation" model (Fig. 14.3). According to this hypothesis, ancient epeiric seas were so shallow that their interiors lacked normal astronomical tides because of the effect of frictional drag. Shaw (1964) cited the northeastern part of Florida Bay as a modern example. Here water movement is due solely to wind-driven waves and currents. In this model, only around the periphery of the craton margin was normal tidal influence effective. Toward the interior of the epeiric sea, the progressive diminution of tidal circulation was reflected in the succession of concentric lithofacies belts mentioned, belts which Shaw

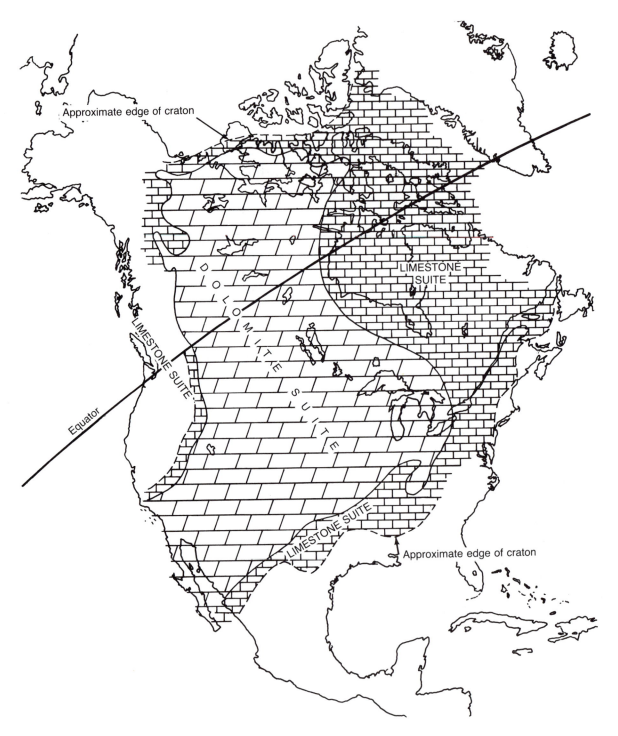

FIGURE 14.1
Carbonate facies in the epeiric sea that covered North America during the Silurian. (After Berry, W. B. N., and Boucot, A. J., 1970, Correlation of the North American Silurian rocks: Geol. Soc. America Spec. Paper 102. Redrawn from Figs 11 and 12, p. 104.

FIGURE 14.2
Interior seaway of North America during the late Cretaceous.

(1964) described as the "basic autochthonous pattern." Superimposed on this pattern and accentuating it was the effect of the biogenic carbonate reefs that preferentially accumulated in the outer zone of vigorous water movement and further restricted tidal circulation. That the apparent lack of tidal circulation did not lead to stagnant or anoxic conditions is demonstrated by the fact that the interiors of epeiric seas were able to support a varied marine fauna. Berry and Boucot (1970), in their study of the Silurian of the North American craton, could see no difference between the invertebrate fauna of the limestone and dolomite suites, although, not surprisingly, they reported a more restricted fauna in the evaporite suite.

Although plausible, the Irwin/Shaw model is by no means proven. It is claimed by some workers that epeiric seas were, in fact, influ-

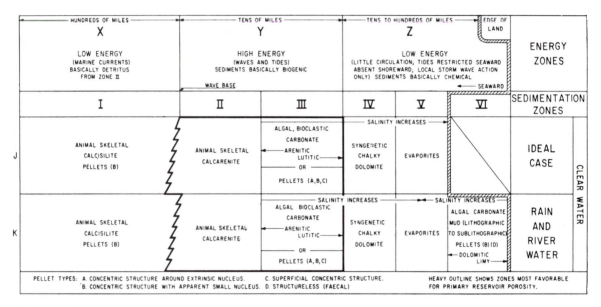

FIGURE 14.3

Energy zones and associated facies in epeiric seas. (From Irwin, M. L., 1965, General theory of epeiric clear water sedimentation: AAPG Bull., v. 49, Fig. 5, p. 452, reprinted by permission of American Association of Petroleum Geologists.)

enced by tides. For example, Klein and Ryer (1978) pointed out that tidal circulation patterns in certain extensive Holocene shelf seas suggested a general correlation between tide range and the width of shelf seas. Far from frictional drag damping the tidal influence, it seems that the greater the distance onto the shelf, the greater is the tidal range. Klein and Ryer (1978) and Driese et al. (1981) also cited numerous examples of sedimentological evidence of tidal influence in rocks from epeiric

sea successions in several places and of various ages, from Precambrian to Cretaceous. One of the biggest problems in resolving such debates is the absence of modern analogues. Nowhere today are shallow-water carbonates accumulating on the vast scale they did in the past. The Bahama Banks, Campeche Bank, and the Persian Gulf are modern carbonate banks, but the resemblance to ancient epeiric carbonate environments is only superficial. Apart from the considerable contrast in scale, all late Cenozoic carbonate successions have been influenced by Pleistocene glacio-eustatic sea-level fluctuations.

Although the interiors of cratonic platforms are stable, they are not devoid of structure, and epeiric-sea sedimentary successions were, in places, influenced by **intracratonic** upwarping of arches and downsagging of basins. It seems likely that sedimentation kept basins more or less filled, and, although they are marked by greater than average thicknesses of sediment, there is little to indicate that they formed major topographic lows on the epeiric seafloor. They

did, however, apparently influence the location of reef trends, as for example, in the Silurian of the Michigan Basin area (Fig. 14.4).

Modern and Ancient Reefs

Extensive carbonate reefs were a feature of many epeiric seas, and again comparisons with modern reefs have been made. In modern examples there are very clear-cut differences between wave-resistant reefs, built by active framebuilders, and carbonate banks, passively accumulated through hydrodynamic agencies or by sediment-trapping

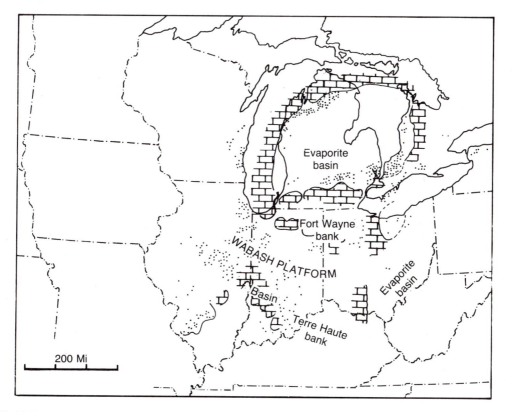

FIGURE 14.4
Composite paleogeographic map for the Silurian of the Michigan Basin, showing carbonate banks, barrier reefs, and pinnacle reefs. (Redrawn from Shaver, R. H., 1974, Silurian reefs of northern Indiana: Reef and interreef macrofaunas: Am. Assoc. Petroleum Geologists Bull., v. 58, p. 934–956)

organisms. In ancient reef structures (**bio-herms**), however, the distinction is often obscure. It is likely that many of the so-called "reefs" of the geologic past were formed under a variety of conditions and few, in fact, can be considered directly analogous to modern coralgal reefs. The recognition of fragments of corals, stromatoporoids, and other organisms, generally considered framebuilders, among the fossils in many ancient reefs perhaps led early workers to the erroneous assumption that they were dealing with reefs essentially similar to the coralgal structures of the present day (Braithwaite, 1973). More detailed studies have, in many cases, resulted in new interpretations.

Dunham (1970) made an important distinction between what he called *stratigraphic* and *ecologic* reefs. A stratigraphic reef, although a three-dimensional mass of limestone, may or may not have had much topographic relief during its formation, and it was not formed by framebuilding organisms. An ecologic reef, on the other hand, was formed as a wave-resistant feature by framebuilders, or organic binders, such as the corals and encrusting algae characteristic of modern reefs. The essential structural difference depends upon whether the binding of the reef mass was organic or inorganic. An organically bound reef is essentially an edifice built up by the colonial reef-building organisms themselves. Lime sand and mud in the interstices of the reef framework may be important, even volumetrically the dominant component, but it will play no role in the structure of the reef itself. In contrast, in an inorganically bound reef, organic skeletal elements, as well as the clastic infilling material, will be cemented together. Within an *in situ* reef, the difference between organic and inorganic binding is probably impossible to discern and is revealed only in the broken talus blocks derived from the reef. Only if the reef has been inorganically bound will lime **clasts** be present in the fracture pieces. Clearly, in an

organically bound structure, the clastic component will have been loose and will have trickled out of the talus block at the time of its dislodgement.

Carbonate reef buildups are usually discernible in seismic surveys. The shelf, or **back-reef,** facies is characterized by reflections with good continuity and high amplitude. These pass into discontinuous or chaotic reflections in the reef core, the abrupt change or termination of reflectors being quite characteristic. In the basinal facies beyond the reef tract, reflection continuity improves again, but with moderate amplitude. There is usually a pronounced velocity contrast between the reef carbonates and the nonreef sediments surrounding the buildup, so the gross outline of the reef usually can be accurately delineated. A further diagnostic feature is the presence of **drape folding** in nonreef sediments that lie above the relatively incompressible reef core. These folds typically die out stratigraphically upward. Bubb and Hatlelid (1977) listed four categories of carbonate buildups that are most easily recognized in seismic sections: **shelf-margin reefs, barrier reefs, patch reefs,** and **pinnacle reefs** (Fig. 14.5).

Even at the climax of a major marine transgression some topographic features in the cratonic interior remained emergent, and numerous small areas of more-resistant basement rocks formed islands. During the Sauk (Cambrian–Early Ordovician) sea transgression onto the North American continent, for example, the sea apparently lapped around exposures of Precambrian rocks such as the Keweenawan mountains in Wisconsin, the Tulsa mountains in Oklahoma, and the St. Francis Mountains in Missouri. Other small knobs of Precambrian rock sticking up out of the epeiric sea have been reported from Kingston, Ontario, and near El Paso, Texas. In the last case, the highlands apparently persisted from the late Cambrian into the early Ordovician before being drowned in

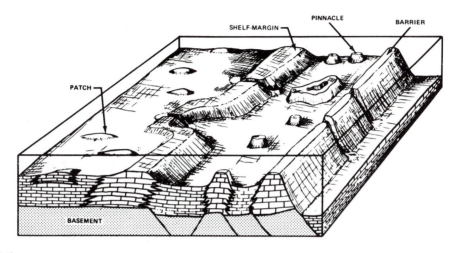

FIGURE 14.5
Types of carbonate buildups most easily recognized in seismic sections. (From Bubb,
J. N., and Hatlelid, W. G., 1977, Seismic stratigraphy and global changes of sea level,
part 10: Seismic recognition of carbonate buildups, *in* Payton, C. E., ed., Seismic stratigraphy—Applications to hydrocarbon exploration: AAPG Mem. 26, Fig. 6, p. 191, reprinted by permission of American Association of Petroleum Geologists.)

shallow-marine carbonates. According to Kottlowski et al. (1973), the extremely long life
for such a small topographic feature (relief of
only 230 m) was due to its isolation from
high-energy wave action by surrounding extensive supratidal to shallow-water mud flats
and biohermal reefs, facies considered typical of the Sauk epeiric sea in that region. The
generally arid climate of that area (Chafetz,
1980) also may have been a factor.

Clastic Sediments in Epeiric Seas

Not all epicontinental seas were dominated
by carbonates. When the seas were less extensive and marginal land areas were available as a source, clastic sediment became an
important component. If old stable land
masses were present, these were likely of low
relief; this would mean that fluctuations of
sea level were reflected in quite rapid shoreline fluctuations. It was under these conditions that on the North American continent

many of the extensive mid-continent coal
swamps formed during the Pennsylvanian.
Many of the characteristic cyclothems, with
alternations of marine and nonmarine facies,
were, as described in Chapter 12, of glacio-
eustatic origin. The fact that single cyclothems have been traced for over 1000 km
suggests that the craton surface was one of
low relief passing from a flat shallow-marine
platform onto an equally flat swampy land
surface.

Sedimentary facies were influenced by
other factors also, such as climate and tectonic setting. At times, rising highland areas
within the craton were sources of clastics, as
during the Carboniferous on the European
platform and during the Permian in North
America. Collision tectonics at the craton
margin also influenced facies patterns, as
during the Cretaceous in North America. The
Cretaceous interior seaway was the latest of
the widespread floodings of the cratons—at
least eight major pulses of transgression and

regression have been recognized not only in western North America, but on a global scale. In the western United States, the sea-level fluctuations are well-documented in a series of sandstone and shale formations typified by the widespread Mesa Verde Group and the Mancos Shale (see Fig. 14.24).

Carbonate platforms were not entirely devoid of clastics, although the general inundation of the craton left few potential source areas emergent. Intracratonic land areas, such as exposed ancient shield terranes, were likely of low relief. If tectonically active source regions were present, they were at the craton margin and thus far removed. Perhaps the most typical clastics associated with carbonate epeiric sequences are the so-called basal, or blanket, sands which, although thin, often have very wide distribution. Basal transgressive sands are a common feature of sedimentary successions in marginal seas and are usually pictured as paralic sands, and possibly coastal dunes, reworked by the encroaching strandline during an episode of rising sea level. The well-known Ordovican St. Peter sandstone of the North American midcontinent region is one of the best examples. Although usually less than 130 m thick, it, or correlative sands, were described by Krumbein and Sloss (1963) as extending from Indiana to Colorado. The source of the sand was apparently to the north from emergent areas of the Canadian Shield and, according to Dapples (1955), originated as shoreline sands, repeatedly reworked during the course of the Sauk marine transgression. This accounts for its remarkable purity and the well-rounded shape of the grains. On the North American craton, in the absence of incursions from rising tectonic sources or uplifted hinterland areas, the total amount of sand was limited and became progressively less during successive transgressions during the Paleozoic. During each cycle some sand was disinterred and reworked, but much of it was eventually retained within the dominantly carbonate succession.

In the short term it is unlikely that continuous sand sheets are spread at the front of an ongoing transgression. As Ryer (1977) pointed out, variations in rates of transgression and rates of sediment supply result in a situation in which erosion by the advancing strandline is, from time to time, interrupted by periods of progradation. The preserved record of such fluctuating conditions is seen in a series of discontinuous sand bodies representing buried barrier islands or beaches. Franks (1980), in describing the Early Cretaceous transgressions in Kansas, suggested that the encroachment of marine facies landward occurred in a stepwise fashion because the net result of progradation during transgression would, for a time at least, be an upward building of barrier-island sands (Fig. 14.6). This would continue until the sand supply diminished, at which time the barrier would be overwhelmed and the whole process would begin again some distance landward. The fact that so many of the major basal sands of epeiric-sea sequences are apparently more or less continuous is because of the pulsatory nature of transgressions. Many sands originated as **fluvial,** tidal channel, or foreshore sands (Driese et al., 1981), and it was presumably only by repeated reworking over the long term that they became smoothed out and merged to form a single sand sheet.

Isolated occurrences of sands elsewhere within epeiric successions are usually ascribed to wind transportation. In certain localities irregular masses of sand have been interpreted as sediments that infiltrated downward into solution features on ancient karstic surfaces. A modern example is seen in central Kentucky, where sands from the caprock Big Clifty sandstone are found as cave and solution-hollow infillings well down into the underlying St. Genevieve and St. Louis limestones.

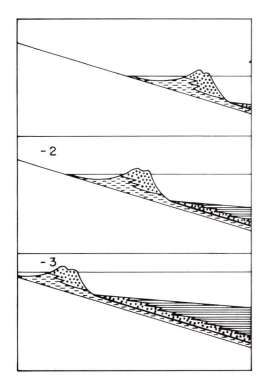

OPEN-SEA
DEPOSITS

BARRIER-ISLAND
AND TRANSGRESSIVE
SHEET SANDS

LAGOON OR
BAY DEPOSITS

FIGURE 14.6
Transgression and shore erosion resulting in partial destruction of lagoon sediments and the development of transgressive disconformities. Nearshore deposition results in a transgressive sand sheet. (From Franks, P. C., 1980, Models of marine transgression—Example from Lower Cretaceous fluvial and paralic deposits, north-central Kansas: Geology, v. 8, Fig. 3, p. 59.)

14.3 PASSIVE (MID-PLATE) MARGINS

Modern Shelf Sedimentation

The bulk of the clastic sediment forming the continental shelf and continental rise is derived originally from the denudation of the continental interior. The sediment is not all immediately delivered to and distributed over the continental shelf; a portion of it remains for a time in what Swift (1974) termed the *coastal boundary,* which includes the inner shelf and shoreface, the shoreline, the intracoastal zone of **lagoons,** estuaries, coastal swamps, and so on. These collectively form a natural unit that was most affected by the short-term glacio-eustatic sea-level fluctuations of the Late Cenozoic. On broad mid-plate margins, such as those of eastern North and South America, the regional slope is gentle, and so very large areas of the modern shelf are within the 130-m bathymetric contour and were emergent during glacial periods. Much of the veneer of sediment that covers the modern shelf may be termed *relict* sediment, implying that textural and morphological features were inherited from an earlier regime of shallower-water or emergent conditions. There are, for example, alluvial deposits and even forest peats with tree stumps in the position of growth now submerged beneath shallow shelf seas. These sedimentary associations are obviously not in equilibrium with or characteristic of the present-day marine environment. On exposed shelves the "modern" sediment is dominantly a thin and ephemeral veneer of shifting sand, with sand waves and ridges often in more or less continuous movement.

A certain proportion of the total sediment transported from the continental interior, particularly the finer-grained material of the suspended load, passes through the coastal boundary in the discharge plume of rivers and out into deeper water in the ebb-tide jet. The coarser traction load generally moves more slowly but tends to be swept out during times of flood discharge. The processes involved in moving and storing sediment within the coastal boundary continue the size sorting that has been going on throughout the journey downriver from the interior. The distribution of various size fractions is closely linked with different types of coastal morphology; the coarser material moves to the beaches, whereas the fines accumulate in tidal flats, lagoons, and flood-plain deposits. Sediment enters storage differently in differing geomorphological environments, as follows.

The Coastal Plain Deposits of sand, silt, and mud are spread across the coastal plain as a result of laterally migrating river channels and overtopping of banks when the rivers are in flood.

Beaches Sands at river mouths, particularly in locations with weak tidal scour, but with relatively vigorous wave action, are entrained by longshore processes and accumulate to build sandy **beaches.** During periods of normal weather, beaches tend to accrete and become sediment sinks. During storms, erosion causes a mass transfer of material into the offshore zone, but this is more or less a seasonal to-and-fro movement with little net transfer of sediment into deeper water (Fig. 14.7).

Tidal Flats and Lagoons These are found behind barrier coasts and in sheltered bays. A mixed load of sand-, silt-, and clay-size material is entrained by the incoming tide, is carried through tidal channels, and settles during the period of slack water. Because of

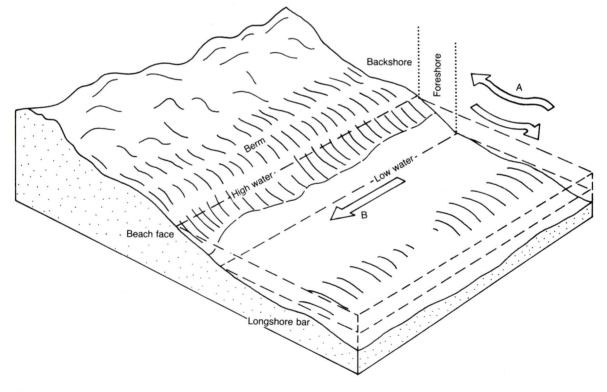

FIGURE 14.7
Sand movement in the beach environment. A. Back-and-forth movement between beach face/berm in the foreshore zone and longshore bars in the nearshore (offshore) zone.
B. Longshore currents.

the greater cohesiveness and depositional stability of the fines, they tend to remain behind; the sand-size fraction is picked up and returned by the ebb tide.

Deltas Major deltas form important constructional elements in continental shelves, and much of the large volume of sediment accumulated stays in permanent storage as part of the shelf wedge. Consequently, only relatively small amounts of sediment are involved in further transfer onto the surrounding shelf, at least in the short term. Over longer periods, falling sea level may result in reentrainment of distributary sands, whereas rising sea level increases wave attack on the delta shoreface beaches. The regional setting of major deltas varies, and several classifications have been proposed. Most of these recognize a basic grouping into three fundamental types: fluvial-, wave-, and-tide-dominated deltas (Fig. 14.8). The varied sediments of the delta platform are collectively known as topset beds (those sediments on the undaform surface), and at their outer edge they pass, with a marked change in slope, into sediments of the delta slope, or prodelta, known as foreset (clinoform) beds. Foreset deposits are typically silty clays, and much of the material settles from suspension. Local oversteepening of the delta slope may, however, result in slumping and **turbidity currents.** Each successive foreset bed is laid down on an inclined surface and encroaches farther out into the basin. The gradual advance of the delta-front slope across the basin floor (the fondoform) causes the advancing foreset beds to encroach over the flat-lying deeper-water (bottomset) deposits of the prodelta. This occurs out beyond the front of the delta slope (Fig. 14.9).

Stratigraphy of Passive Margins

The sediments and sedimentary processes of a typical passive marginal shelf that have been discussed thus far are obviously representative of only a surface veneer and of a very brief span of time. In considering the earlier history of the shelf we are dealing with an accumulation of sediments 12 or more kilometers thick and are immediately faced with two problems. The first is the relative inaccessibility of these deeply buried older rocks. The second stems from the fact that the Holocene and Pleistocene sediments are not typical of the succession as a whole, so that extrapolation into the pre-Pleistocene must be done with some reservations. The overall aspect of the most typical mid-plate margin shelf is that of a great wedge, or prism, of sediments, in which individual units thicken oceanward and have outer edges emerging at the continental slope. Although sometimes the strata are exposed in outcrop, in many places they are covered by a thin layer of younger sediment or partially buried under slumps and slides. The outer edges of older and deeper formations are often buried beneath continental-rise deposits, but may pass into reef buildups or diapiric structures. Seismic sections over the shelf and slope off the eastern U.S. show that the basement rocks, consisting of continental crust, are block-faulted. The faults trend roughly parallel to the continental margin; they are of similar structural style and are similar in age to the Triassic rift valleys, which extend from the Maritime Provinces of Canada through New Jersey to the Carolinas and which are known to occur as far south as Florida in the subsurface. The rifts date from initial continental breakup and ocean opening; one or more faults farther east were destined to open up to become the Atlantic Ocean (Fig. 14.10). According to Burke (1976), the graben associated with continental rupture, and which immediately preceded seafloor spreading, developed between 210 and 170 m.y. ago in the northern Atlantic region. Between what is now west Africa and the northern coast of Brazil, the date of gra-

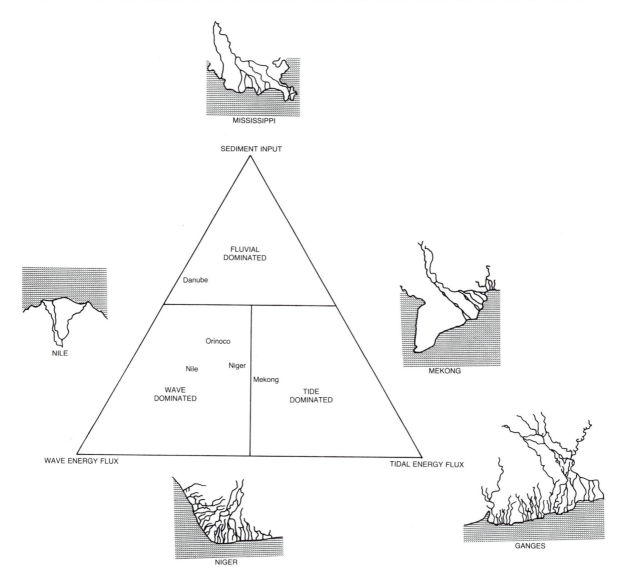

FIGURE 14.8
Classification of deltas, with five modern examples.

ben formation was between about 145 and 125 m.y. ago. The sequence above the basement rocks may be divided into distinctive suites that document changes in the tectonic and sedimentary environment during the evolution of a rifted margin. From the base up these can be summarized as the rift valley suite, evaporite suite, marine carbonate suite, and clastic offlap suite.

Rift Valley Suite During the earliest stages of North Atlantic rifting, in the Jurassic, the basins were, in many cases, isolated from the early Atlantic ocean and became depocenters

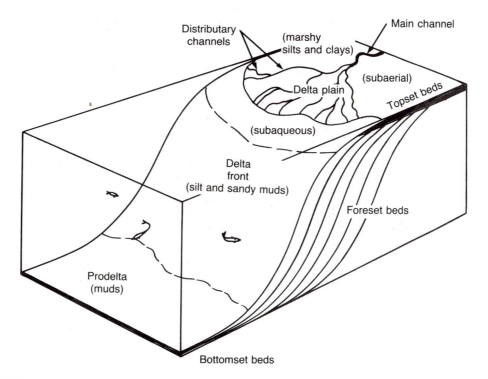

FIGURE 14.9
Schematic cross section through a delta, showing the relationships between topset, foreset, and bottomset beds.

for large volumes of coarse clastics. These were delivered by short-fetch streams from adjacent block-faulted highlands and deposited as alluvial fans and braided fluvial sediments along the boundary faults. With active basin subsidence, sediment accumulation was rapid, and drainage and sediment-dispersal systems within a basin probably developed in the form of lacustrine **fan deltas** prograding along the basin axis. Lacustrine and swamp deposits were also common. In the case of the North American Atlantic margin, basins of this type have been detected in the basement rocks, and there is no reason to suppose that the tectonic and topographic setting and the depositional environments in these deeply buried basins were any different from those farther west, which have now be-

come disinterred by erosional stripping since the Tertiary. In New Jersey, for example, the rift basins are seen to be filled by a variety of sediments, comprising the Late Triassic Newark Supergroup, consisting of conglomerates, arkoses, and red and black shales, in places reaching 6000 m in thickness (Fig. 14.11).

Evaporite Suite Sooner or later many of the early rift basins became connected with the ocean. If the connection was relatively restricted, the basins became evaporite basins and the sites of thick salt deposits (Fig. 14.12). This incipient ocean-basin phase has been termed the Red Sea Stage. Extensive salt deposits at the base of the marginal succession of the Gulf Coast region are typified by the Louann Salt of Jurassic age. Mo-

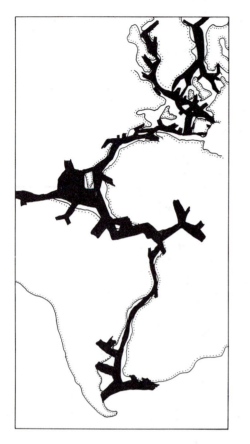

FIGURE 14.10
Distribution of the major graben around the Atlantic Ocean, formed during continental rupture. (Redrawn from Burke, K., 1976, Development of graben associated with the initial ruptures of the Atlantic Ocean: Tectonophysics, v. 36, Fig. 1, p. 95. Reprinted by permission of Elsevier Science Publishers B. V., Amsterdam.)

bilization and upward movement of this salt to form diapiric structures (salt domes) in overlying Mesozoic and Cenozoic sediments has resulted in many major sites for oil and gas accumulation. Of the same age and with the same stratigraphic setting is the Argo Salt of the subsurface in the offshore east of Nova Scotia. Diapirs associated with this salt have also been extensively drilled in the search for oil.

Marine Carbonate Suite With progressive opening of the Atlantic ocean basin and the establishment of more effective circulation, evaporites gave way to limestones, which were often extensive on the shelf. Even though fan deltas were still delivering clastics, much of this material was retained by inshore rift basins still active during the waning stages of block faulting. One of the most striking features of seismic depth sections across the shelf is the presence at many locations of a buried topographic high, lying beneath or close to the present shelf break. These features have been known for some time but their true nature was not fully understood at first. In places, there are indications that these topographic dams are buried volcanic arcs; others consist of salt diapirs (see Fig. 13.6B) or igneous intrusions. Very commonly they are carbonate buildups (Figs. 14.13, 14.14, and 14.15). In all cases, their stratigraphic relationship to the mass of contemporary sediments on the continent side appears to be a discordant one and the sediments seem to be "ponded" behind them. The shelf break, facing onto a progressively enlarging ocean, with upwelling deep water supplying abundant nutrients, was clearly a favorable locale for carbonate growth. Given suitable water temperatures, extensive reef buildups persisted over long periods as the margin subsided. Such barrier reefs were effective dams, behind which accumulated much of the land-derived sediment and which resulted in starved-basin conditions on the oceanward side. Shelf-edge carbonates are apparently a common feature within the lower sequences of mid-plate margin sedimentary wedges. Similar reef growths have been noted along the margin of West Africa (Lehner and DeRuiter, 1977) and Brazil (Brown and Fisher, 1977). Beneath the North American Atlantic shelf, reefs can be traced from the vicinity of Georges Bank southward to the Blake Plateau/Bahamas region and the Turks and Caicos Islands. Along the north

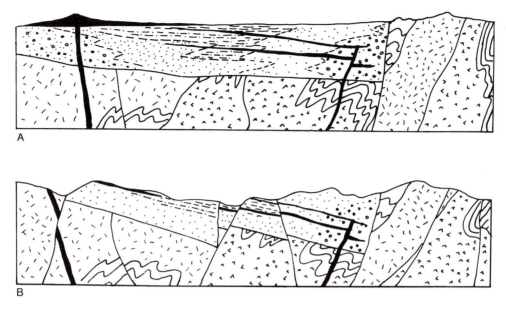

FIGURE 14.11
Two stages in the history of the Triassic rift valleys of the northeastern United States and Canada.

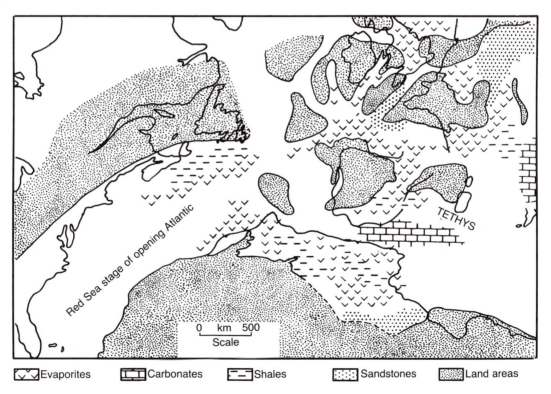

Evaporites   Carbonates   Shales   Sandstones   Land areas

FIGURE 14.12
Evaporite basin or basins that formed during the Red Sea stage of the evolving North Atlantic. (Based on reconstruction by Jansa, L. F., Bujak, J. P., and Williams, G. L., 1980, Upper Triassic salt deposits of the western North Atlantic: Can. Jour. Earth Sci., v. 17, p. 547–559.)

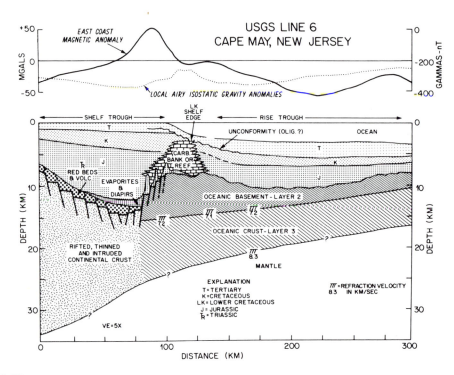

FIGURE 14.13
Schematic cross section, based on reflection data, through a passive-margin continental
shelf. The line of section crosses the eastern end of the Long Island platform. (From
Grow, J. A., Mattick, R. E., and Schlee, J. S., 1978, Multichannel seismic depth sections
and interval velocities over outer continental shelf and upper continental slope between
Cape Hatteras and Cape Cod: AAPG Mem. 29, Fig. 13, p. 80, reprinted by permission of
American Association of Petroleum Geologists.)

coast of Cuba, reefs are known also. These
presumably are continuous with those under
the western Florida platform, around the Gulf
of Mexico to the Edwards–Stuart City reef
trend of the Louisiana and East Texas subsurface, and thence southward to the Golden
Lane reef of the Tampico area, Mexico (Fig.
14.16). Shelf and platform carbonates and
marls are typically uniformly bedded and in
seismic sections are seen to have reflection
signatures characterized by good continuity,
uniform spacing, and high amplitude. There
is often good reflection continuity with clinoform limestones on the slope. This is in
marked contrast to slope sandstone and shale

clinoforms, typical of the clastic offlap suite,
which terminate abruptly in a toplap relationship with the base of delta front sandstones
prograded to the shelf margin (Galloway and
Brown, 1973).

Clastic Offlap Suite The termination of
widespread shelf-carbonate deposition was
caused, in many cases, by an increasing influx of terrigenous clastic sediment. The cessation of block faulting along the margins
also meant that rift basins no longer served
as sediment traps so that active progradation
of deltas across the shelf and onto the slope
shifted the main depocenters oceanward.

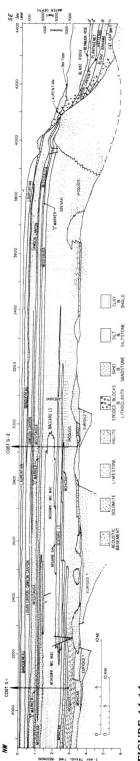

FIGURE 14.14

Interpretation of a seismic section across the continental shelf and rise, eastern seaboard, United States. Note tie-in with COST well G-2. (From Poag, C. W., 1985, Depositional history and stratigraphic reference section for central Baltimore Canyon Trough, *in* Poag, C. W., ed., Geologic evolution of the United States Atlantic margin. Fig. 7.6. Reprinted by permission of Van Nostrand Reinhold.)

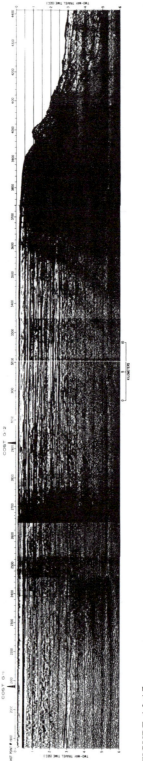

FIGURE 14.15
Seismic section used in the drafting of Figure 14.14. (From Poag, C. W., 1985, Depositional history and stratigraphic reference section for central Baltimore Canyon Trough, *in* Poag, C. W., ed., Geologic evolution of the United States Atlantic margin, Fig. 7.6. Reprinted by permission of Van Nostrand Reinhold.)

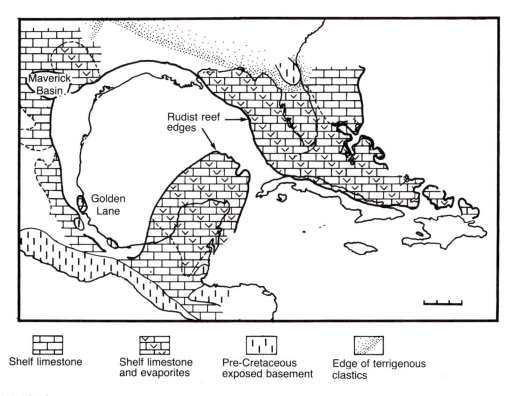

Shelf limestone

Shelf limestone and evaporites

Pre-Cretaceous exposed basement

Edge of terrigenous clastics

FIGURE 14.16
Shelf-edge carbonate reefs of Lower and Middle Cretaceous time, eastern North America and Gulf of Mexico. (Adapted from Bryant, W. R., Meyerhoff, A. A., Brown, N. K., Jr., Furrer, M. A., Pyle, T. E., and Antoine, J. W., 1969, Escarpments, reef trends and diapiric structures, eastern Gulf of Mexico: Am. Assoc. Petroleum Geologists Bull., v. 53, p. 2506–2542.)

This trend was accelerated by the subsidence of the continental margin as it moved away from the spreading center. Isostatic depression under sediment loading added a further component.

In the North Atlantic region, the end of major shelf-edge reefal development and the initiation of the clastic offlap suite occurred, according to Poag (1985), in the Hauterivian (Lower Cretaceous). This is the age of the oldest unit that can be traced in seismic sections across the top of the shelf-edge carbonate reefs, which are seen as very high-amplitude reflectors with good continuity. COST well logs show this sequence to consist of shaly limestones and thin sands. Prograding of delta systems across the shelf was rapid at first but slowed down beyond the shelf edge. The seismic record of these sediments is dominated by pronounced clinoform reflectors, and the frequent interruption of these reflectors by **growth faults** is indicative of instability consequent on rapid accumulation. Beyond the shelf edge, slope systems comprise turbidites, exhibiting offlap relationships during times of maximum sediment supply, but onlap at times of active canyon cutting. Only in areas remote from clastics did carbonates continue to accumulate; in some cases they transgressed delta lobes.

From the Late Cretaceous, further evolution of marginal sediment systems was increasingly influenced by changes in both the remote continental hinterland and the ocean basin. The gradual development of a mature continental drainage system, with larger rivers, led to the concentration of major terrigenous clastic depocenters in deltas. Elsewhere on the shelf, sediment delivery was from river-mouth or shoreface sources and distribution patterns were determined by tide- or storm-dominated transport, essentially as in the model that is operating at the present day. According to Brown and Fisher (1979), the seismic record shows that two phases of clastic accumulation can be recognized. An early phase of rapid progradation by deltaic sediments is marked by oblique clinoform seismic reflectors, whereas a change to sigmoid style clinoforms is indicative of the slowing down of progradation and the growing importance of carbonate platform deposits, marking the second phase. In the case of the Brazilian margin, on which these conclusions were based, this change occurred in the Middle Tertiary. In areas free of terrigenous clastics, biogenic carbonate production increased, reflecting more efficient recycling of nutrients as **thermohaline circulation** became established in the expanding ocean. Upwelling of deep-ocean waters was also marked in some outer shelf areas by phosphorite deposits; for example, in the Miocene section of North Carolina and central Florida. Along the Atlantic margin of North America, only in the area of the Florida-Bahamas platform did carbonate accumulation continue uninterrupted on a major scale into the Tertiary. This was due to a combination of remoteness from terrigenous clastic sources and the influence of the Gulf Stream. The most recent veneer of shelf sediments, discussed in Chapter 12, has been influenced by glacio-eustatic sea-level fluctuations, probably since the Oligocene or early Miocene, and represents a third phase in the accumulation

of the clastic offlap suite. During periods of **lowstand** not only were sediments near the shelf break more susceptible to disturbance by wave action, but many major rivers discharged at or near the shelf edge. This resulted in an increase in slump and slide structures on the continental slope and in offlapping turbidite deposition across the continental rise.

14.4 DEEP-SEA SETTINGS

Deep-sea depositional environments can be divided into two broad categories, those dominated by siliciclastic sediment and those where such sediment is absent and where **pelagic** sediments are laid down. The siliciclastic sediment is derived from the continents, so it is that part of the ocean basin extending from the shelf break, across the continental rise, and out onto the abyssal plains that comprises the first environmental category. The second area is simply that part of the ocean floor beyond the reach of turbidity currents.

Continental Slope and Rise

The continental slope is mainly a region of slope instability, with sediment accumulation interrupted by intermittent gravity-driven events, such as submarine slides, slumps, and debris flows. On seismic records they can be seen as interrupted and chaotic reflectors. Cutting back into the slope, and, in places, extending back almost to the strandline, are submarine canyons. These function as conduits down which large volumes of sediment, derived from the continental shelf, are carried as sediment flows. The mass movements are typically periodic and may be triggered by storms or earth tremors in the upper canyon. The sediment flows emerge onto the continental rise and continue as turbidity currents, sometimes traveling far out into the abyssal plains beyond. The lower end of a

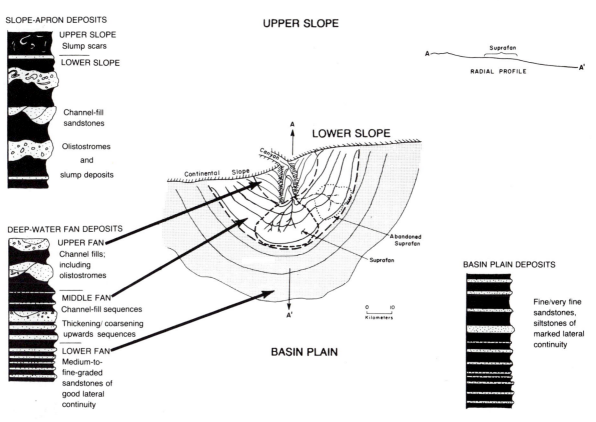

SLOPE-APRON DEPOSITS

UPPER SLOPE
Slump scars

LOWER SLOPE

Channel-fill
sandstones

Olistostromes
and
slump deposits

DEEP-WATER FAN DEPOSITS

UPPER FAN
Channel fills;
including
olistostromes

MIDDLE FAN
Channel-fill sequences
Thickening/ coarsening
upwards sequences

LOWER FAN
Medium-to-
fine-graded
sandstones of
good lateral
continuity

UPPER SLOPE

Suprafan

A — A'

RADIAL PROFILE

LOWER SLOPE

Canyon

Continental Slope

Abandoned
Suprafan

Suprafan

A'

0   10
Kilometers

BASIN PLAIN

BASIN PLAIN DEPOSITS

Fine/very fine
sandstones,
siltstones of
marked lateral
continuity

FIGURE 14.17
Morphology of a deep-sea fan, showing the typical facies associations. (From Normark, W. R., 1978, Fan valleys, channels and depositional lobes on modern submarine fans: Characters for recognition of sandy turbidite environments: AAPG Bull., v. 62, Fig. 1, p. 914, reprinted by permission of American Association of Petroleum Geologists.)

typical submarine canyon is marked by a fan built up by material transported down the canyon, and the continental rise is formed by the coalescing of many such fans. The various features of a typical deep-sea fan are seen in Figure 14.17. Turbidity current activity is markedly episodic and, in areas of active fan accumulation, occurs perhaps once every ten years according to Nelson (1976). Only the largest flows extend across the abyssal plains, with depositional pulses occurring perhaps only once every 1000 to 3000 years. In the modern oceans, turbidity currents are less frequent than they were in the geologi-

cally recent past. During the late Cenozoic glaciations, when sea levels were often glacio-eustatically lowered almost to the shelf break, turbidity current activity was apparently much more frequent and widespread. Ancient turbidites are common, particularly in what used to be called "eugeosynclinal" sequences, and many thick successions contain a record of thousands of turbidity current events (Fig. 14.18).

Pelagic Environment

True pelagic sediments are dominated by the calcareous and siliceous tests of planktonic

FIGURE 14.18
Turbidites in the Aberystwyth Grits of the Ordovician (Upper Llandoverian) of Wales.

organisms, settling from the zone of high productivity in surface waters, together with the clay particles that are ubiquitous in the water column. Biologic productivity in surface waters is largely controlled by water temperature and the supply of nutrients, as influenced by upwelling (Fig. 14.19). The distribution of biogenic bottom sediments, on the other hand, is controlled more especially by water depth and the position of the **carbonate compensation depth (CCD)**. Generally speaking, calcareous **oozes** are found only on those parts of the ocean floor above the CCD (Fig. 14.20); in deeper areas, siliceous oozes are to be found. In regions of low surface productivity, only red clays accumulate in deeper areas. These are derived largely from aeolian dust, stratospheric volcanic ash contributions, and, possibly, meteoritic particles. The depth to the CCD varies and is greatest below regions of high biologic productivity, notably in equatorial regions. This results in a greater thickness of oozes, the so-called equatorial bulge described by Berger and Winterer (1974) and Lancelot (1978).

It was for many years believed that the abyssal depths of the oceans were places where literally nothing happened except for the slow rain of sediment from above. The evidence accumulated from deep-sea drilling programs reveals a very different picture. It is clear, for one thing, that the stratigraphic

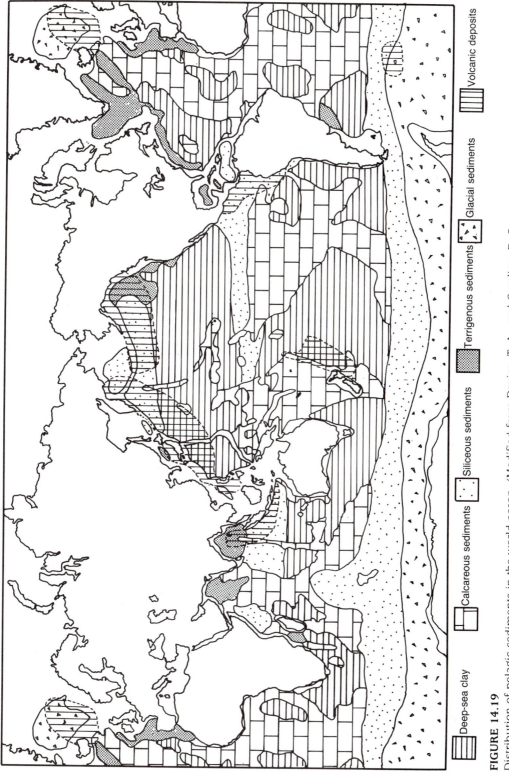

FIGURE 14.19

Distribution of pelagic sediments in the world ocean. (Modified from Davies, T. A., and Gorsline, D. S., 1976, Oceanic sediments and sedimentary processes, *in* Chemical oceanography, Riley, J. P., and Chester, R., eds., Fig. 24.7. Reprinted by permission of Academic Press, Orlando, FL.)

Deep-sea clay   Calcareous sediments   Siliceous sediments   Terrigenous sediments   Glacial sediments   Volcanic deposits

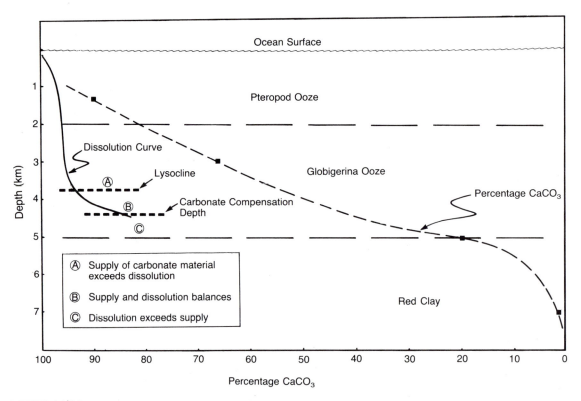

FIGURE 14.20
CaCO₃ percentage trend with depth and the relationship between the skeletal carbonate content of pelagic sediments and the dissolution rate. The lysocline marks the depth zone where a rapid increase in the solution rate occurs. The compensation depth is the level where the rate of supply of plankton carbonate skeletal material is balanced by its rate of solution. (Redrawn from Murray, J., and Hjort, J., 1912, The depths of the ocean: London Macmillan.)

succession of the seafloor is by no means complete. Far from being the repository for an unbroken record of passing time, there are many gaps in the succession. In the Cenozoic, for example, eight episodes during which hiatuses were abundant have been documented by Moore et al. (1978), who attributed them to changes in continent-ocean configurations and the concomitant effect on thermohaline circulation patterns in the oceans. There is evidence also of periodic anoxic events, marked by black muds. It is likely that these and other rapid changes in

the chemistry and temperature distribution of the oceans were the cause of major biological catastrophes. Plate movements have had profound long-term influences on the stratigraphic record of ocean basins. The movement of ocean floor away from a spreading center and its gradual sinking, consequent on thermal contraction, is reflected in the accumulation of a progressively more complete stratigraphic succession, at least back to the Late Jurassic, the age of the oldest pelagic sediments. Plate movements are clearly documented also by tracing the shift through

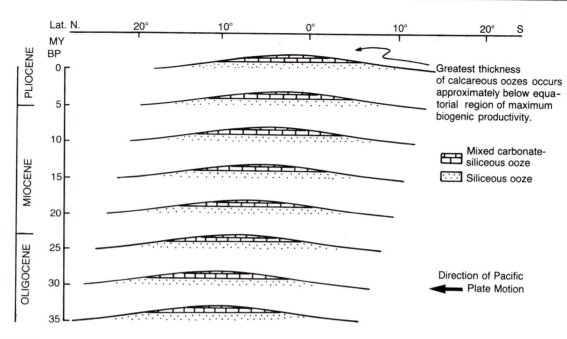

FIGURE 14.21

Diagram illustrating the northward movement of the Pacific plate and the resultant displacement, since the Oligocene, of the equatorial bulge in pelagic oozes. This reflects the zone of maximum biologic productivity in the surface waters. (After Berger and Winterer, 1974.)

time of deep-sea facies boundaries, such as the expanded sequence of the calcareous oozes in the equatorial bulge, mentioned above (Fig. 14.21). Numerous tools are available in studying the stratigraphy of deep-sea sediments, including biostratigraphic methods, isotope studies, tephrochronology, and magnetic stratigraphy. Because all of these are mentioned elsewhere in this book, they will not be discussed further at this point.

14.5 ACTIVE MARGINS

Under this heading is included all those plate marginal sedimentary settings directly affected by active tectonism. In contrast to a passive margin, whose history, as described earlier, usually is contained in a decipherable stratigraphic record, active margins are of several types and include both continental and oceanic plate relationships. The sedimentary settings are typically complex, and they rarely show anything but a general superposition of facies that are diagnostic of certain stages in an orogenic cycle. Active and passive margins have little in common in terms of sedimentary features, but they are in one sense linked through time by being at opposite ends of a Wilson cycle, as discussed later. At convergent plate margins, three different relationships are possible: collision may be intraoceanic, ocean to continent, or continent to continent (Fig. 14.22).

Intraoceanic (Volcanic Arc-Trench) Collison Margins

Intraoceanic margins are characterized by deep-sea trenches that mark the zone of subduction, and they may or may not have some

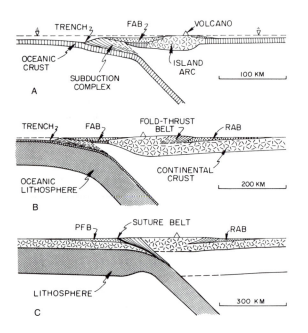

FIGURE 14.22

Comparison between the three types of collision margins: intraoceanic (A), continent–ocean (B), and continent–continent (C). (After Dickinson, W. R., 1977, Tectono-stratigraphic evolution of subduction-controlled sedimentary assemblages, *in* Talwani, M., and Pitman, W. C. eds., Island arcs, deep sea trenches and back arc basins: Maurice Ewing Series 1, Am. Geophys. Union, Fig. 1, p. 34.)

sedimentary fill, depending upon the regional setting. Parallel with such trenches are chains of volcanoes, the vulcanicity being triggered by descent of the lithospheric plate into the asthenosphere. As oceanic or trench sediments are scraped off the subducting plate, an **outer arc ridge** may form and even become emergent in places. From this ridge, sediment flow into the trench is seen in a series of submarine fans that feed so-called axial channel flows that move along the length of the trench. It is on the overriding plate that the volcanic arc, or magmatic arc (Dickinson and Seely, 1979), forms and this contributes sediment to **forearc** and **backarc basins** that flank the volcanic chain (Fig. 14.23). Accord-

ing to Sigurdsson et al. (1980), in a study of the Lesser Antillean volcanic arc, only about 20 percent of the total ejecta remains in the vicinity of the arc, the bulk of it being dispersed through ash falls and pyroclastic debris flows onto the ocean floor.

Ocean-to-Continent Collisions

This is the so-called Andean-type margin, where continental crust lies on the overriding plate. The major zone of thrusting and folding forms an orogenic mountain belt that lies along the oceanic edge of the continental plate, which, in this context is referred to as the **foreland.** As subduction progresses, the belt of folded and thrust rocks tends to migrate toward the foreland; in so doing, it flexes the crust downward in front of it to form a depression, known as a *foreland basin.* Typically, such basins are asymmetrical in cross section, being deepest near the fold-mountain belt, hence the term **foredeep** used by some authors. Lying, as they do, adjacent to an orogenic belt, foreland basins become major depocenters for sediment eroded from the rising mountains. The sediment is composed dominantly of gravels, sands, and silts, forming alluvial fans and fan deltas, spreading across the basin axis. A modern foreland basin occurs on the eastern side of the Andes; similar basins formed during the Jurassic and Cretaceous along the eastern flank of the present-day Rocky Mountains. In parts of Utah and Colorado, the sedimentary fill derived from uplift during the Sevier orogeny is over 10,000 m thick. At times of continental emergence, the sedimentary fill of foreland basins is largely nonmarine, as in the basins east of the modern Andes. The term *molasse* has been applied to sediments in such a setting; there will be further discussion of this term later.

At times of high eustatic sea level, a good deal of the fill may be of marine or brackish-water sediments. In the Cretaceous interior

FIGURE 14.23
Schematic diagram showing the relationship between forearc and backarc basins associated with a volcanic arc. This example is in the Lesser Antilles. (From Sigurdsson, H., et al. 1980, Volcanogenic sedimentation in the Lesser Antilles arc: Jour. Geology, v. 88, Fig. 4, p. 531. Reprinted by permission of University of Chicago Press.)

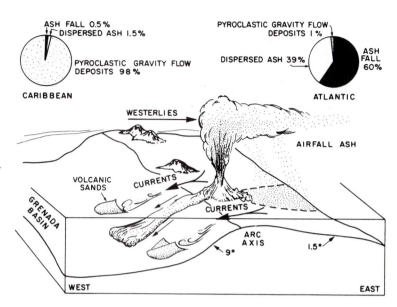

seaway east of the Sevier orogenic belt, for example, enormous clastic wedges, derived from the tectonic front, interfingered eastward into marine shales (Fig. 14.24). The migration of the fold-thrust belt toward the foreland inevitably begins to involve the foreland basins themselves, and old sedimentary fills are often difficult to discern beneath thick overthrust sheets. As described earlier, on the oceanic or subduction side of the orogenic belt, forearc and backarc basins also occur. Some of these, according to Dickinson and Seely (1979), may be residual basins in the sense that they represent portions of ocean floor caught between projections of continental crust onto the ocean plate at the time of collision. Such basins seem to be similar in concept to those visualized by Graham et al. (1975) and which they termed *remnant* basins.

Continent-to-Continent Collisions

The chief difference between this tectonic setting, the so-called Himalayan type, and that discussed previously lies in the presence of two forelands rather than one. One of the classic areas is in the Alpine-Mediterranean region, with foreland basins both north and south of the Alps. The well-known "Queenston Delta" of the Ordovician of New York forms part of a thick succession that accumulated in the foreland basin lying northwest of the Taconic fold belt that arose after the collision of the Laurentia and Baltica plates.

Under this heading also would be included the collisions of small blocks of continental crust (microcontinents), oceanic plateaus, and island arcs when they accrete onto larger continental blocks. These so-called **suspect** (or **exotic**) **terranes** (Fig. 14.25) are in some cases believed to have traveled long distances on plate surfaces from where they originally formed. Along some continental collision margins, including ancient ones of earlier continental configurations, it is thought that seafloor spreading and subduction have resulted in repetitive collisions of this type over considerable periods of time.

Strike-Slip Basins

Translational plate boundaries also provide a setting for sediment accumulation in basins

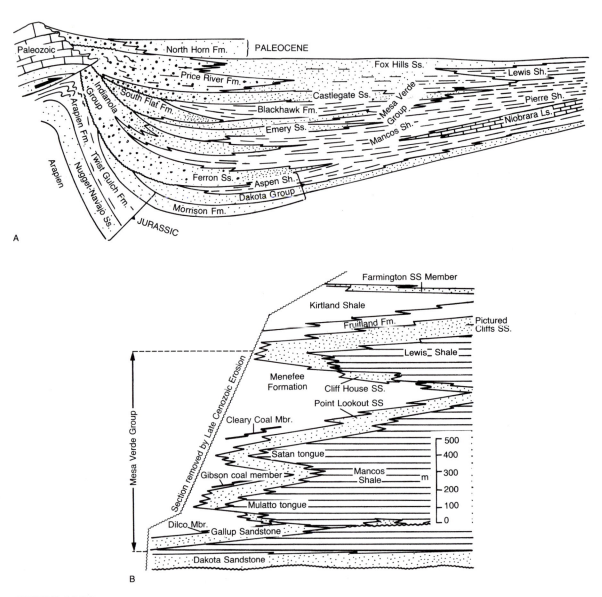

FIGURE 14.24
A. Diagrammatic cross section of Upper Cretaceous sediments in the foredeep east of the Sevier orogenic belt. B. Sea-level changes reflected in interfingering sands and shales of Late Cretaceous age in the San Juan Basin, Colorado.

of various types. These are formed usually in extensional areas, so they bear a superficial resemblance to the more familiar rift basins formed during the initial stage of continental breakup. Traced over any distance, the lateral

(strike-slip) faults, which are the surface manifestation of translational plate margin movement, are typically sinuous. As the adjacent plates move past one another, the fault plane is in some places under compression

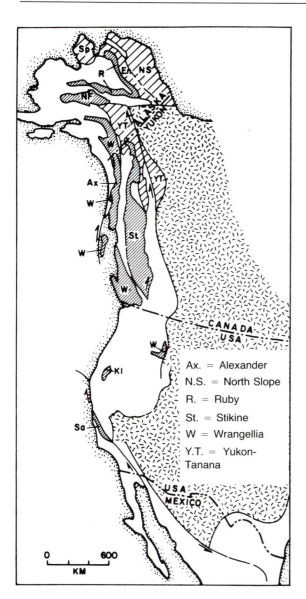

FIGURE 14.25
Exotic terranes in the northwest region. (After Coney, P. J., Jones, D. L. and Monger, J. W. H., 1980, Cordilleran suspect terranes: Nature, v. 288, Fig. 1, p. 330.)

Ax. = Alexander
N.S. = North Slope
R. = Ruby
St. = Stikine
W = Wrangellia
Y.T. = Yukon-Tanana

and in others under tension. These stresses and strains are taken up in numerous smaller subsidiary faults, often with an *en echelon* arrangement (Fig. 14.26). Among

these various stress patterns, so-called *pull-apart* basins develop; these become filled with considerable thicknesses of sediment, typically showing rapid lateral facies changes and indications of variable sediment supply. Examples of numerous strike-slip faults, associated with the San Andreas transform fault system, occur along the California coastline, and many of the associated basins, including the Ventura and Los Angeles basins, are oil producers. The Los Angeles basin, for example, with some 10,000 m of Neogene sediments, is estimated to have an ultimate recovery of 8 billion barrels of oil. Certain basins in the Gulf of California are also pull-apart basins.

Geosynclines

In the historical sense, the many and varied sedimentary successions formed in areas of active tectonism can be properly understood

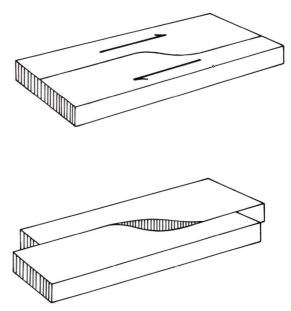

FIGURE 14.26
Strike-slip fault pattern and the origin of pull-apart basins.

only by reference to the **geosynclinal** concept. From the mid-nineteenth century on, much attention had been paid to the genesis and evolution of fold-mountain belts. Even quite early attempts at reconstructing the sedimentary environments prior to folding had demonstrated that the fold mountains of the world had been formed from great thicknesses of sediment accumulated in elongated downwarps in the earth's crust, usually situated along continental margins and subsequently compressed to form fold belts. The work of Stille (1936) and Kay (1947, 1951) marked important steps toward understanding mountain belts; they showed that a geosyncline actually consisted of two distinct down-sagging belts: an inner one (on the continent side), termed a **miogeosyncline,** and an outer one, known as a **eugeosyncline** (Fig. 14.27). Miogeosynclinal deposits were viewed as being dominated by shallow-water clastics and carbonates; eugeosynclinal deposits were typically dark shale and **graywacke** sequences, including many with repeated graded bedding, that were later to be explained as **turbidites.** Volcanic deposits and lavas were also features. In the years before the revelations of plate tectonics, the distri-

bution of sediments within geosynclines could be accounted for only by paleogeographic reconstructions with "borderlands" and island chains that presumably lay somewhere oceanward of the continental margins. In a nondrifting world, the problem of explaining what happened to such lands was, to put it mildly, difficult (Fig. 14.28).

With an understanding of plate tectonics, it finally became possible to construct a model that not only described the miogeosyncline/eugeosyncline relationship, but also explained how geosynclines turned into fold mountains. The orogenic phase has, in fact, been shown as only the last part of a continuum that extends back much further in time to embrace the very beginning of the ocean in which the sediments involved in the orogeny were laid down. This is the concept of the *Wilson cycle* (Wilson, 1966), which begins with rifting and the initiation of an ocean basin and continues through to reclosing of the basin. During this time, the great sedimentary wedges of the continental shelf, slope, and rise are emplaced along the trailing-edge margins of the continent. Later, as the ocean closes above one or more subduction zones, the sedimentary wedges are

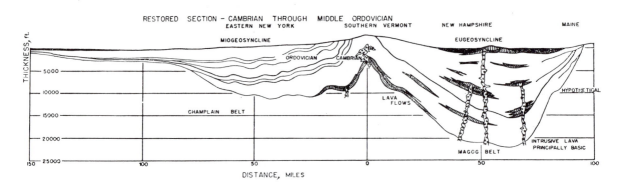

FIGURE 14.27

Palinspastic restoration of the Cambrian and Ordovician strata and volcanic rocks as they were before the Taconic orogeny in New England. Length of section is 425 km. (From Kay, G. M., 1951, North American geosynclines: Geol. Soc. America Mem. 48., pl. g. p. 26.)

FIGURE 14.28
Borderlands and geosynclines surrounding the North American continent during the Paleozoic. (From Schuchert, C., 1923, Sites and nature of the North American geosynclines: Geol. Soc. America Bull., v. 34, pp. 151–229.)

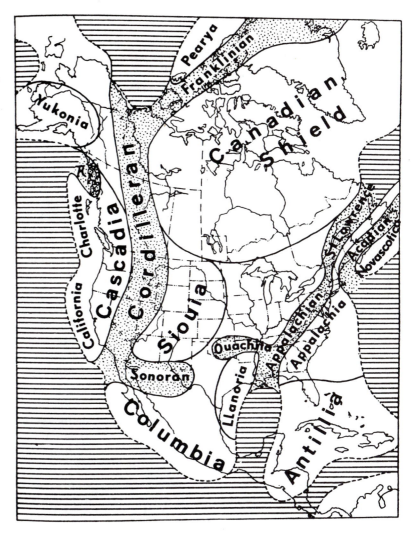

caught up in a plate boundary collision, and they are seen as the miogeosynclinal/eugeosynclinal couplet, in the historical sense. In this context, the eugeosynclinal sedimentary association includes not only the continental rise prism, but also the pelagic sediments beyond on the ocean floor, together with volcanic suites associated with subduction.

Although today the miogeosyncline/eugeosyncline concept of Stille and Kay has been considerably modified, the distinction between the two sedimentary associations is very obvious. In any fold belt there is, in fact, a third distinct suite of sediments to be considered. The three suites, labeled simply in a genetic sense, are the continental-shelf prism (the miogeosynclinal suite), the continental-rise prism (the eugeosynclinal suite), and finally, what can be termed syntectonic sediments, laid down in basins (e.g., foreland basins) that appeared as a consequence of plate collision and orogeny. Within the Wilson cycle, the eugeosynclinal suite includes ocean-floor sediments as well as volcanic arc successions, and these three sedimentary suites have a roughly sequential ordering. As early as 1897, Bertrand described sedimentary associations with sequential significance in what he called a geosynclinal cycle, containing four facies. The cycle began with a *pre-orogenic* facies, consisting of terrigenous clastics and platform carbonates. Stratigraphically above this facies was a *pre-flysch* facies, typified by fine-grained sediments, black shales, cherts, and dark limestones, interpreted as of deep-water origin. Above that came the *flysch* facies, consisting of alternating thin-bedded sandstones, mudstones, marls, and turbidites. Finally, came the *molasse* facies, synorogenic to postorogenic clastic wedges of sandstones and gravels, often of fluvial origin.

The term **flysch** has come to have a wide range of meanings, some of them misleading and at variance with the original concept. As originally used by alpine geologists, there was usually the implication that flysch deposits were laid down during, or possibly in the early stages of, an orogenic cycle. The sediment was derived from rising mountains and deposited in actively subsiding basins, invariably in a marine environment. **Molasse** is also a term long familiar to alpine workers, who usually considered such deposits as characteristic of the waning stages of an orogenic cycle. The term is usually applied to sandstones, often cross-bedded, and gravels, commonly of fluviatile origin. Included, in some situations, are coals, marls, and even shallow-marine sediments. They are typically found as clastic wedges on the flanks of orogenic belts in foreland basins, as described earlier (Van Houten, 1973).

14.6 CRATONS EMERGENT— TERRESTRIAL SETTINGS

The sedimentary associations that are found in cratonic interiors at times when they are

largely emergent, as at the present day, are, generally speaking, of nonmarine origin. Modern terrestrial sediments include alluvial, lacustrine, and swamp deposits, desert deposits, and glacial deposits. From the present-day perspective, such sediments seem to comprise a relatively thin veneer on the continental surface and, compared with the sediments accumulating at the continental margins and in ocean basins, are often localized and sporadic in their distribution. When long spans of geologic time are considered, however, such sedimentary associations take on a very different perspective. Emergent land surfaces are, on the whole, places where erosion rather than deposition is dominant. However, the active subsidence of intracratonic basins of various types and over long periods of time has ensured that a considerable record of nonmarine environments is preserved. In many places great accumulations of alluvial sands and gravels, conglomerates, **pyroclastics,** and so on reach thousands of meters in thickness and bear witness to the importance of tectonic controls in sediment accumulation.

Among the largest depocenters of nonmarine sediments are the foreland basins that develop in front (inboard or cratonward) of rising orogenic belts. In them, thousands of meters of sands, gravels, and muds eroded from the rising mountains comprise the molasse deposits described earlier in this chapter. Depocenters of this magnitude are found also in basins of extensional as well as compressional tectonic style. In the Basin and Range Province west of the Rockies and the Colorado Plateau, sediment fill in some intermontane basins is of enormous thickness; at Jackson Hole, Wyoming, for example, almost 12 km of Neogene sediments have been logged. Such sedimentary piles rival even those of the continental margins and are entirely nonmarine in origin.

Alluvial Sediments

Alluvial deposits can be divided into three broad groups: those deposited by braided streams, those of meandering streams, and those in deltas. Today, major deltas are found mainly at continental margins and were discussed earlier in this chapter. Along the course of a typical large river, in a setting uncomplicated by structure or climatic zonation, the river system is usually a braided stream in its upper portion and a meandering stream in its lower reaches. Braided streams normally flow over relatively steep gradients and transport excessive loads of sediment, mostly as bedload. At any one time much of this material lies as gravel-and-sand bars that divide the stream into many small, intertwining channels. The flow of such streams is variable, sometimes markedly so, if there are seasonal climatic changes (wet and dry seasons, spring melting, and so on). Particularly in the case of the coarser gravels, downstream transportation is intermittent and the deposits are highly variable. For a comprehensive account of braided streams, see Miall (1977). All streams tend toward a sinuous course as a consequence of their turbulent flow, and this tendency increases with progress downstream. Meandering streams flow over relatively low gradients and their discharge is much less variable than that of braided streams, although they usually flood one or more times a year.

Changes in the shape and location of the stream channel, and of associated deposits caused by the interplay of erosional and depositional processes, are relatively systematic and predictable. Two types of deposit are formed by meandering streams: channel deposits, laid down laterally by the progressive accretion of **point bars** (Fig. 14.29) on inside (convex) bends during times of normal flow, and **overbank deposits,** laid down during times of flood. During periods of peak flood-ing, deposition of silts and muds is at a maximum, and the grain size is coarsest immediately adjacent to the channel. The deposits form a **levee** that slopes gently away from the river. Across the surface of the **flood plain,** sediment accumulation is relatively slow and is dominantly of muds. In rivers close to grade, as in many coastal plains, the flood-plain surface may be permanently or intermittently near or below the water table and so is swampy, leading to the accumulation of peats. At moderate flood stages, when the levees may be still emergent, breaching of the levee results in the formation of crevasse splays that spread silts and muds in lobate accumulations across the flood plain.

Lacustrine Sediments

Lakes are typically ephemeral features, so that only the largest lakes are likely to leave any significant sedimentary record. Ancient lake deposits, particularly of pre-Cenozoic age, are rare. Among modern large lakes, the majority owe their origin to structural sags (for example, Lake Chad in north-central Africa, Lake Eyre in Australia) or are found in areas of active tectonism, in fault valleys. Many, such as Great Bear Lake, Lake Winnipeg, and the Laurentian Great Lakes, occupy areas of glacial scour.

Lakes are classified under two headings: hydrologically open lakes, with outlet streams, and those lakes in areas of inland drainage, which are termed hydrologically closed lakes, commonly characterized by **evaporite** deposits. Within the recent geologic past, many lakes have fluctuated between these states, normally in response to glacial-interglacial climatic changes. In the largest lakes, the physical conditions and the sediment may be quite similar to those in the ocean, except for the water chemistry and the absence of tidal influences. As in oceans, most of the coarser clastics are deposited

FIGURE 14.29
Diagrammatic section through a typical point bar on a meander bend. (From Allen, J. R. L., 1970, Physical processes of sedimentation, Fig. 45, p. 133: Winchester, MA, George Allen and Unwin Ltd.)

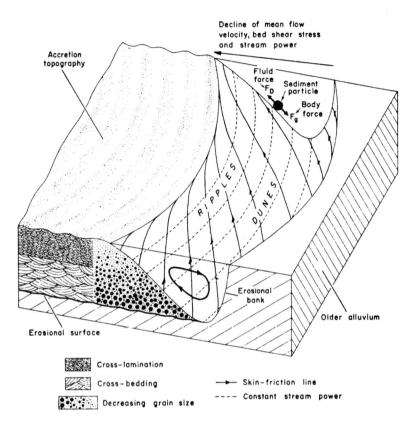

Cross-lamination

Cross-bedding

Decreasing grain size

→ Skin-friction line

---- Constant stream power

close to shore, in the form of beaches, beach ridges, and spits. In the deeper parts of a lake, below wave base, clays and silts accumulate from suspension at very slow rates. Many of these fine-grained sediments are well stratified and may show annual varves. Many lakes have a varied fauna and flora, including lime-secreting forms such as bivalves, gastropods, **ostrocodes**, and calcareous algae. Thin limestones and marls are to be found particularly in the less-turbid offshore region, where clastics accumulation is reduced. In lakes with closed drainage, the most characteristic sediments are evaporite deposits of various kinds. Terrigenous clastic sediment is usually widely distributed as a consequence of reworking as lake levels rise and fall, often seasonally.

Among examples of ancient lacustrine successions, one of the best known is the Green River Formation, of Middle Eocene age. According to Frazier and Schwimmer (1987), these lake sediments cover about 78,000 km^2 of southwestern Wyoming, northwestern Colorado, and northeastern Utah and in places are 1500 m thick. In some facies of the formation, remarkably preserved fossil fishes are common; this unusual feature, together with varves undisturbed by bioturbation, suggested to Bradley (1964) that lake-bottom conditions were from time to time anoxic and inimical to bottom-dwelling scavengers.

Desert Sediments

The sediments most generally considered characteristic of deserts are aeolian sands in the form of dunes. This is despite the fact that of the 20 percent of the modern land surface that is described as desert, only one-fifth is classified as sandy desert. Not only are rocky and stony deserts far more extensive, but it should be remembered that desert landforms in general are sculpted more by the erosive and transportational power of running water during infrequent flash floods than by the wind. Nevertheless, analyses of desert deposits have tended to emphasize the role of the wind, and numerous studies, especially the classic work of Bagnold (1941), have described the origin, internal structure, and texture of desert dunes. The morphology of aeolian sands can be described (Wilson, 1971, 1972) at three scales, the smallest being ripples having amplitudes and wavelengths measured in centimeters. Larger deposits, having a wide range of shapes (Fig. 14.30), are classified into various types of dunes, some of which are migratory under the influence of unidirectional winds (Fig. 14.31). Finally, some workers recognize a third category of major sand accumulations known as **draas,** or dune complexes, frequently measuring many kilometers across. Particularly where sand is in short supply, the areas of desert between the dunes are often more extensive than those covered by the dunes. These interdune areas are typically swept clear of sand and retain only pebbly **lag deposits.** Such topographic lows are characterized by ephemeral streams and **playa lakes.** In low-lying areas where the water table is close enough to the surface, so-called inland **sabkhas** occur where evaporative conditions are reminiscent of those in coastal sabkhas, with similar evaporite deposits. Under stable water-table conditions, vegetation may be present, and in shallow lakes algal mats are not uncommon.

Numerous examples of ancient desert deposits have been described in formations of many ages. The good sorting and rounding of the sand grains, together with medium- to large-scale cross-bedding are considered diagnostic features. One such formation is the Jurassic Entrada Sandstone of northern Utah and Colorado, described by Kocurek (1981) as deposited in an **erg** (sand sea) and compared with modern ergs in the eastern Rub'al Khali, Saudi Arabia. Large-scale cross-bedding was interpreted as forming in large migrating dunes (Fig. 14.32). The Triassic Wingate Formation, of Utah and Arizona; another Jurassic formation, the Navajo Sandstone of Utah; parts of the Nubian Sand of North Africa; and Permian sands of northern England have all been described as examples of desert sands.

Overall, there is no single unequivocal indicator of an ancient desert environment. Even aeolian sand deposits may not always be reliable because they frequently are found in nondesert environments, such as along seacoasts. On the other hand, the deposits of the flash floods and **lahars,** so characteristic of modern deserts, do not normally contain any special indicators of arid climate. In practice, it is an association of dune sands with such features as evaporite deposits, caliche crusts, red-colored sediments, and marked lack of fossils that would suggest a desert environment, although each of these features alone would be open to alternative interpretation. So-called **redbeds** were long accepted as an indication of deposition under arid climate conditions. It is now known, however, that a monsoonal climate may be more likely. In some cases, the red color is postdepositional in origin and due to diagenetic changes in the oxidation state of various iron minerals that result in the appearance of hematite within the weathering profile (Walker et al., 1978).

Glacial Sediments

Although glacial ice, both in valley glaciers and major ice caps, is a powerful erosive agent, glaciers are also important transporta-

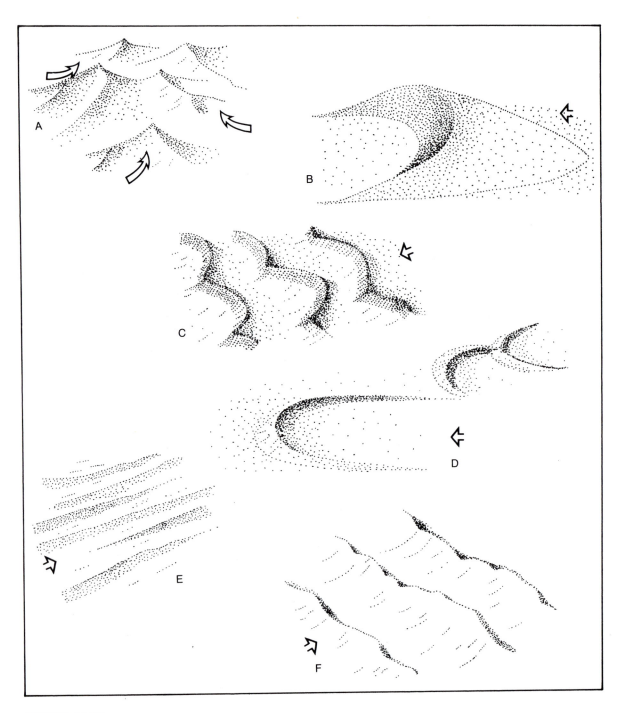

FIGURE 14.30

Types of dunes. A. Star dunes. B. Barchan dune. C. Barchanoid ridges. D. Parabolic (blow-out) dunes. E. Longitudinal (seif) dunes. F. Transverse dunes.

FIGURE 14.31
High star dunes at Sossos Vlei, Great Sand Sea, Namibia. (Photograph courtesy of
I. Watson.)

tional systems and, especially in high latitudes and paleolatitudes, enormous spreads of glacial deposits have been laid down on land surfaces at one time or another. Although much of it is rapidly reworked following the retreat of the ice, enough remains to provide some record of the numerous ice ages from Pleistocene time back to the Precambrian. The most characteristic glacial deposit is **till,** and both the texture and the presence of striated pebbles and boulders are normally the most diagnostic features (Fig. 14.33). In some cases, a preferred orientation in the till fabric gives some sense of ice flow direction but, more usually, there are few internal textural features. Except in the case of Pleistocene drift, the morphology of former

drift sheets is not discernible, although certain ancient fluvioglacial accumulations, such as eskers, have been recognized.

At the present day, the Antarctic and Greenland ice caps cover approximately 10 percent of the earth's surface, but during the Pleistocene glacial maxima, the glaciated area expanded to some 30 percent. It can be assumed that ice caps of earlier times were of similar size, and so the glacial contribution to terrestrial sedimentary veneers was often considerable. In South Africa, for example, glacial deposits comprising the Permo-Carboniferous Dwyka Tillite total nearly 1000 m in thickness. The early Proterozoic Gowganda formation of Ontario, also believed to be a **tillite** (see Fig. 16.14), is even thicker. Because

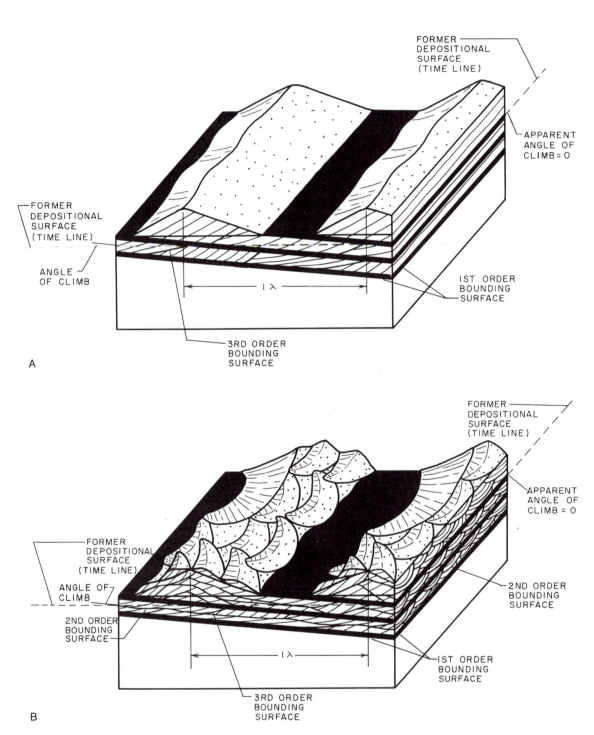

FIGURE 14.32
First-order bounding surfaces represent interdune areas traversed by migrating dunes. In simple dunes (A) internal cross-strata generated by foreset slip faces and represented by third-order bounding surfaces. B. Second-order bounding surfaces separate individual sets of cross-strata. (From Kocurek, G., 1981, Erg reconstruction: The Entrada Sandstone (Jurassic) of northern Utah and Colorado: Palaeogeog., Palaeoclim., Palaeoecol., v. 36, Fig. 6, p. 138. Reprinted by permission of Elsevier Science Publishers B.V.)

A   B

FIGURE 14.33
A. Typical glacial till of Wisconsin age in Bow Valley, Alberta. B. Closeup view of till
overlain by pebbly lag deposit forming a stony soil. (Photographs courtesy of Geological
Survey of Canada, Ottawa.)

pebbly mudstones laid down in a variety of depositional environments may have a very till-like appearance, considerable controversy has surrounded certain of the ancient "tillites," particularly those in the Precambrian. The stratigraphy of glacial deposits is discussed further in Chapter 15.

14.7 CONCLUSION

The discussion of the various tectonic settings has shown them to contain distinctive depositional environments that are, for the most part, unique to their respective settings. In turn, the sedimentary suites within the environments bear individual and recognizable

signatures. What are becoming increasingly well known are the clear linkages from regional tectonic settings, through various field entities of smaller scale, right down to the single rock sample and thin section. As Dott (1978) pointed out, it is the petrologic examination of a sandstone that may provide important information on entire geologic terranes. Numerous workers (e.g., Dickinson and Valloni, 1980; Dickinson, 1982; Potter, 1986) are finding that suites of clastic sediment derived from ancient plate margins contain vital clues in understanding plate tectonic history. In some cases, such clastic sediments may be the only remaining traces of terranes that have long since disappeared, such as is likely with ancient volcanic arcs consumed by subduction.

This chapter has been a summary account of the major tectonic settings of sediments. It is also where stratigraphy, in its deciphering of the history of vanished worlds, comes full circle and parts company with its sister discipline of sedimentology. A study of sediments in any further detail would entail the marshaling of a great body of information on depositional environments and sedimentary associations that is rightly the domain of a quite different kind of book.

REFERENCES

Bagnold, R. A., 1941, The physics of blown sand and desert dunes: London, Methuen, 265 p.

Bates, R. L., and Jackson, J. A., eds., 1987, Glossary of geology, 3rd ed.: Falls Church, VA, Am. Geol. Inst., 788 p.

Berger, W. H., and Winterer, E. L., 1974, Plate stratigraphy and the fluctuating carbonate line, in Hsü, K. J., and Jenkyns, H. C., eds., Pelagic sediments, on land and under the sea: Internat. Assoc. Sedimentologists Spec. Pub., p. 11–48.

Berry, W. B. N., and Boucot, A. J., 1970, Correlation of the North American Silurian rocks: Geol. Soc. America Spec. Paper 102, 289 p.

Bertrand, M., 1897, Structure des Alpes françaises et récurrence de certains facies sédimentaire: Zurich, 6th Internat. Geol. Congr., p. 161–177.

Bradley, W. H., 1964, Geology of the Green River formation and associated Eocene rocks in southwestern Wyoming and adjacent parts of Colorado and Utah: U.S. Geol. Survey Prof. Paper 496A.

Braithwaite, C. J. R., 1973, Reefs: Just a problem of semantics?: Am. Assoc. Petroleum Geologists Bull., v. 57, p. 1100–1116.

Brown, L. F., Jr., and Fisher, W. L., 1977, Seismic-stratigraphic interpretation of depositional systems: Examples from Brazilian rift and pull-apart basins, in Payton, C. E., ed., Seismic stratigraphy—Applications to hydrocarbon ex-

ploration: Am. Assoc. Petroleum Geologists Mem. 26, p. 213–248.

_____ 1979, Interpretation of depositional systems and lithofacies from seismic data: Am. Assoc. Petroleum Geologists, Stratigraphic Interpretation of Seismic Data School, 125 p.

Bubb, J. N., and Hatlelid, W. G., 1977, Seismic stratigraphy and global changes of sea level, Part 10: Seismic recognition of carbonate buildups, in Payton, C. E., ed., Seismic stratigraphy—Applications to hydrocarbon exploration: Am. Assoc. Petroleum Geologists Mem. 26, p. 185–204.

Burke, K., 1976, Development of graben associated with the initial ruptures of the Atlantic Ocean: Tectonophysics, v. 36, p. 93–112.

Chafetz, H. S., 1980, Evidence for an arid to semi-arid climate during deposition of the Cambrian System in central Texas, U.S.A.: Palaeogeog., Paleoclimat., Palaeoecol., v. 30, p. 83–95.

Dapples, E. C., 1955, General lithofacies relationship of St. Peter sandstone and Simpson Group: Am. Assoc. Petroleum Geologists Bull., v. 39, p. 444–467.

Dickinson, W. R., 1982, Composition of sandstones in circum-Pacific subduction complexes and fore-arc basins: Am. Assoc. Petroleum Geologists Bull., v. 66, p. 121–137.

Dickinson, W. R., and Seely, D. R., 1979, Structure and stratigraphy of forearc basins: Am. Assoc. Petroleum Geologists Bull., v. 63, p. 2–31.

Dickinson, W. R., and Valloni, R., 1980, Plate settings and provenance of sands in modern ocean sands: Geology, v. 8, p. 82–86.

Dott, R. H., Jr., 1978, Tectonics and sedimentation a century later: Earth-Science Rev., v. 14, p. 1–34.

Driese, S. A., Byers, C. W., and Dott, R. H., Jr., 1981, Tidal deposition in the basal Upper Cambrian Mt. Simon formation in Wisconsin: Jour. Sed. Petrology, v. 51, p. 367–381.

Dunham, R. J., 1970, Stratigraphic reef versus ecologic reefs: Am. Assoc. Petroleum Geologists Bull., v. 54, p. 1931–1932.

Franks, P. C., 1980, Models of marine transgression—Example from Lower Cretaceous fluvial and paralic deposits, north-central Kansas: Geology, v. 8, p. 56–61.

Frazier, W. J., and Schwimmer, D. R., 1987, Regional stratigraphy of North America: New York, Plenum, 719 p.

Freeman, W. E., and Visher, G. S., 1975, Stratigraphic analysis of the Navajo Sandstone: Jour. Sed. Petrology, v. 45, p. 651–668.

Galloway, W. E., and Brown, L. F., Jr., 1973, Depositional systems and shelf-slope relationships in Upper Pennsylvanian rocks, north-central Texas: Univ. Texas, Bur. Econ. Geology Rept. Inv. 75, 63 p.

Graham, S. A., Dickinson, W. R., and Ingersoll, R. V., 1975, Himalayan-Bengal model for flysch dispersal in the Appalachian-Ouachita system: Geol. Soc. America Bull., v. 86, p. 273–286.

Irwin, M. L., 1965, General theory of epeiric clear water sedimentation: Am. Assoc. Petroleum Geologists Bull., v. 49, p. 445–459.

Kay, M., 1947, Geosynclinal nomenclature and the craton: Am. Assoc. Petroleum Geologists Bull., v. 31, p. 1289–1293.

———— 1951, North American geosynclines: Geol. Soc. America Mem. 48, 143 p.

Klein, G. deV., and Ryer, T. A., 1978, Tidal circulation patterns in Precambrian, Paleozoic and Cretaceous epeiric and mioclinal shelf seas: Geol. Soc. America Bull., v. 89, p. 1050–1058.

Kocurek, G., 1981, Erg reconstruction: The Entrada Sandstone (Jurassic) of northern Utah and Colorado: Palaeogeog., Palaeoclim, Palaeoecol., v. 36, p. 125–153.

Kottlowski, F. E., LeMone, D. V., and Foster, R. W., 1973, Remnant mountains in Early Ordovician seas of the El Paso region, Texas and New Mexico: Geology, v. 1, p. 137–140.

Krumbein, W. C., and Sloss, L. L., 1963, Stratigraphy and sedimentation 2nd. Ed.: San Francisco, Freeman, 660 p.

Lancelot, Y., 1978, Relations entre évolution sédimentaire et tectonique de la plaque Pacifique depuis le Crétacé inférieur: Mem. Soc. Geol. France, No. 1345, 40 p.

Lehner, P., and DeRuiter, P. A. C., 1977, Structural

history of Atlantic margin off Africa: Am. Assoc. Petroleum Geologists Bull., v. 61, p. 961–981.

Miall, A. D., 1977, A review of the braided-river depositional environment: Earth-Sci. Rev., v. 13, p. 1–62.

Moore, T. C., Jr., Van Andel, Tj. H., Sancetta, C., and Pisias, N., 1978, Cenozoic hiatuses in pelagic sediments: Micropaleontology, v. 24, p. 113–138.

Nelson, C. H., 1976, Late Pleistocene and Holocene depositional trends, processes, and history of the Astoria deep sea fan, northeast Pacific: Marine Geology, v. 20, p. 129–173.

Poag, C. W., 1985, Depositional history and stratigraphic reference section for central Baltimore Canyon Trough, in Poag, C. W., ed.: Geologic evolution of the United States Atlantic margin: New York, Van Nostrand, p. 217–264.

Potter, P. E., 1986, South America and a few grains of sand: Part 1—Beach sands: Jour. Geology, v. 94, p. 301–319.

Ryer, T. A., 1977, Patterns of Cretaceous shallow marine sedimentation, Coalville and Rockport areas, Utah: Geol. Soc. America Bull., v. 88, p. 177–188.

Shaw, A. B., 1964, Time in stratigraphy: New York, McGraw-Hill, 365 p.

Sigurdsson, H., Sparks, R. S. J., Carey, S. N., and Huang, T. C., 1980, Volcanogenic sedimentation in the Lesser Antilles arc: Jour. Geology, v. 88, p. 523–540.

Stille, H., 1936, Wege und ergebnisse der geologisch-tektonischen forschung: Wiss. Förh. Gesell. 25 Jahrb. Kaiser Wilhelm, v. 2, p. 77–97.

Swift, D. J. P., 1974, Continental shelf sedimentation, in Burk, C. A., and Drake, C. L., eds., The geology of the continental margins: Berlin, Springer-Verlag, p. 363–371.

Van Houten, F. B., 1973, Meaning of molasse: Geol. Soc. America Bull., v. 84, p. 1973–1976.

Walker, K. R., Ganapathy, S., and Ruppel, S. C., 1983, A model for carbonate to terrigenous clastic sequences: Geol. Soc. America Bull., v. 94, p. 700–712.

Walker, T. R., Waugh, B., and Crowe, A. J., 1978,

Diagenesis in first-cycle desert alluvium of Cenozoic age, southwestern United States and northwestern Mexico: Geol. Soc. America Bull., v. 89, p. 19–32.

Wilson, I. G., 1971, Desert sandflow basins and a model for the development of ergs: Geog. Jour., v. 137, p. 180–199.

_____ 1972, Aeolian bedforms—Their development and origin: Sedimentology, v. 19, p. 173–210.

Wilson, J. T., 1966, Did the Atlantic close and then re-open?: Nature, v. 211, p. 676–681.

PART FOUR
LAST AND FIRST
THINGS

ALTHOUGH SEEMINGLY STRANGE BEDFELLOWS, THE QUATERNARY AND PRECAMBRIAN are brought together, not because they have much in common, but because they each have little in common with any other part of the stratigraphic column. It might be summed up by saying that the Quaternary is too short and the Precambrian too long!

Up to the present, these two parts of the geologic column have been virtually ignored by stratigraphers. However, even a brief perusal of Quaternary and Precambrian literature reveals a wealth of stratigraphic information that rarely is extracted as a separate entity. In this section, the special nature of Quaternary and Precambrian stratigraphic problems will be reviewed. It soon will become obvious that, despite strenuous efforts to use the stratigraphic procedures that work so well elsewhere in the stratigraphic column, a different approach is called for. Far from extending standardized stratigraphic procedure into the Quaternary and Precambrian, it is more likely that the approach used in dealing with the youngest and oldest parts of the geologic record may change the way we look at the 600 million years or so in the middle. There is much to be learned, for example, of ancient glacial episodes from a study of the Quaternary. In the case of the Precambrian, the use of direct dating and the construction of a calendar based on real time will, as dates become more accurate, certainly spread across the entire Phanerozoic.

15
QUATERNARY STRATIGRAPHY

There can be nothing more urgent than finding the way back to a sane interaction between man and his surroundings. And this, I think, is the chief moral to be gained from the study of the Ice Age.

Björn Kurtén

15.1 INTRODUCTION

Much of what has been discussed in earlier chapters cannot always be applied to studies of the Quaternary, so it is the special and modified procedures that have emerged from Quaternary stratigraphy that are the subject of this chapter.

The single most important feature of Quaternary time is that it marked a period when the climatic deterioration that had begun in the Oligocene, or earlier, became more accentuated. It was at this time that the great continental ice sheets that had already existed in Antarctica and Greenland for several million years began to expand to boundaries never reached before. This expansion did not occur suddenly, and its position in time is considered sufficiently ambiguous that some authors (e.g., Berggren and Van Couvering, 1974) consider the distinction between the Tertiary and Quaternary invalid and even artificial. Be that as it may, the chief preoccupation of Quaternary stratigraphers is climatic change. The period is so short that biostratigraphy, based upon zonal indicators as described in Chapter 9, has little contribution to make; instead, fossils are used almost entirely as indicators of past climatic change. It is the more or less regular fluctuation in climate that provides the basis for stratigraphic subdivision and correlation, and this is what makes Quaternary stratigraphy unique. The proper study of the Quaternary requires a multidisciplinary approach, with contributions from many fields, and it is, perhaps, for this reason that it has tended to become a specialist field in its own right.

15.2 THE CLASSICAL MODEL

Glacial gravels, boulder clays, and particularly **erratic boulders** had, in earlier times, excited curiosity, but were usually thought to be evidence of a worldwide flood, the Noachian deluge of biblical scripture. Among those less gullible who sought an alternative explanation was Charles Lyell (1797–1875). He suggested that icebergs, floating in former seas that had once covered the present land surface, could be responsible for transporting the material. This is the reason why glacial deposits came to be known as *drift,* a term still used today. It was the Swiss zoologist Louis Agassiz (1807–1873) who was primarily responsible for showing that erratics and other deposits were, in fact, evidence of a former "ice age" and that valley glaciers and ice sheets, rather than floating icebergs, had transported the drift.

Although, as described in Chapters 3 and 9, the Pleistocene Epoch originally had been defined on the basis of molluscan faunas, it came in general usage to be thought of as virtually synonymous with the glacial period. As the evidence accumulated that the "Great Ice Age" had comprised several glacial episodes, alternating with interglacial episodes, when the climate had been as warm or warmer than it is at present, it was obvious that this provided a ready means of subdividing Pleistocene time. The original, or so-called "classical," division of the Pleistocene was based upon the record of glacial ad-

vances and recessions in the Alps. Albrecht Penck (1858–1945), working in Bavaria on the northern fringe of the Alps, recognized that the **terminal moraines** left behind by former valley glaciers were each associated with outwash gravels deposited by meltwater streams, the ancestors of the Mindel, Würm, Lech, and other rivers flowing north across the German plain (Fig. 15.1). Each glacial episode was marked by a series of terminal moraines (Fig. 15.2) and an associated flood of outwash gravels that choked the valleys downstream. During the ensuing interglacial episode, when the glaciers had retreated many kilometers back into the mountains, the meltwater streams would begin to incise into the outwash gravels, leaving only erosional remnants as terraces perched on the valley sides (Fig. 15.3). This was repeated several times; Penck (1894) eventually concluded that there had been four major glacial advances, which he named Günz, Mindel, Riss, and Würm (in alphabetical order from oldest to youngest) after four of the rivers in his area of study. The interglacial periods he named simply Günz-Mindel, Mindel-Riss, and Riss-Würm.

No basic stratigraphic principles were involved and the evidence was largely geomorphic. The higher terraces were clearly older than lower terraces; fresh, uneroded moraines were younger than those modified by erosion. Although in places well north of the mountains, in areas of alluvial deposition, it was possible to discern some superposition and stratigraphic ordering of the gravels, with the older at the base and the younger overlying them, no "type sections" in the conventional sense were envisaged. The older gravels, those termed the Deckenschotter by Penck, are today found only as scattered erosional remnants capping bluffs and hills, and correlation between them is not easy. Although there was no direct evidence, at least in the form of moraines, of glacial episodes of pre-Günz age, Eberl (1930) described

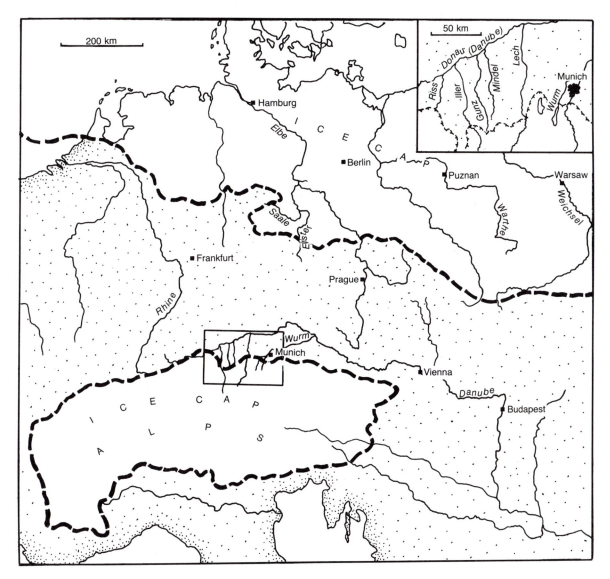

FIGURE 15.1
Location map for the classic Alpine glacial successions described by Albrecht Penck and
Eduard Bruckner for the European lowland successions.

higher and older gravel terrace remnants, the
Otterbeuren. Other gravels marked a so-
called "Donau" glaciation.

Prior to the development of radiometric
and other dating techniques, estimates of
ages for the various glacial and interglacial
episodes were based upon variations in the
degree of weathering of gravels, the thickness of interglacial sediments, geomorphic
changes, and so on. Penck and Bruckner
(1901) decided from such evidence that the
Mindel-Riss interglacial had been four times

FIGURE 15.2
Valley glacier with outwash gravels at terminus, Ball Creek, northern British Columbia.
Note stream downcutting the outwash gravels.

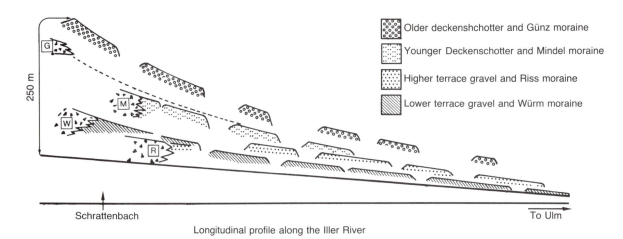

Older deckenshchotter and Günz moraine

Younger Deckenschotter and Mindel moraine

Higher terrace gravel and Riss moraine

Lower terrace gravel and Würm moraine

250 m

Schrattenbach

To Ulm

Longitudinal profile along the Iller River

FIGURE 15.3
Relative positions of the high-level terrace gravels associated with the Günz, Mindel,
Riss, and Würm glaciations of the Alps.

as long as the other two interglacials, and it became known as the "Great Interglacial." Not surprisingly, this conclusion has since been shown to be incorrect. The work of alpine glaciologists was widely recognized and, as a result, the alpine fourfold division became a standard against which the glacial record in many other areas was compared. Thus, the terms Günz, Mindel, Riss, and Würm became, in effect, time terms. Today, they are still widely used in this sense (Bowen, 1978).

In northern Europe, the evidence of former advances and recessions of the Fenno-Scandian Ice Sheet is seen in a series of terminal moraines and associated outwash gravels stretching from the Netherlands to Poland and the USSR. As in the Alps, the glacial-interglacial succession was originally based upon geomorphic evidence (Keihlack, 1926)

along certain rivers, and the standard succession also used the names of rivers, as shown in Table 15.1. In contrast to the Alps, in central and northern Europe some sedimentary record of interglacial times is contained in alluvial and lacustrine deposits and in paleosols. Glacio-isostatic depression of the crust around the ice cap margin resulted in a series of incursions by arms of what is today the Baltic Sea, and so marine deposits also are intercalated in the Pleistocene succession.

In North America, the Laurentide and Greenland ice caps, joined by the Cordilleran ice cap in the west, covered an area nearly three times (18.06×10^6 km^2) that of the Fenno-Scandian ice cap (6.66×10^6 km^2). As in Europe, moraines and outwash deposits are found all across what was formerly the southern margin of the ice cap. The area of the classical studies of the North American

TABLE 15.1

Major divisions of the Pleistocene glacial successions in Europe and North America (interglacials in italics)

Alps	Northern Europe	British Isles	North America
Würm	Weichsel	Weichsel-Devensian	Wisconsin
R/W	*Eem*	*Ipswich*	
Riss	Saale	Wolstonian	
M/R	*Holstein*	*Hoxnian*	*Sangamon*
Mindel	Elster	Anglian	Illinoian
G/M	Cromer	Cromer Beestonian Pastovian	*Yarmouth*
Günz	Menapian		Kansan
D/G	*Waale*		*Aftonian*
Donau	Eburonian		Nebraskan
	Tiglian		
Biber	Praetiglian		

Pleistocene lies in a broad belt to the south and southwest of the Great Lakes, and it was here that Chamberlin (1894) and Leverett (1910) laid the foundations for the nomenclature used to the present day. As in the Alps, four major glaciations were recognized from the evidence of superimposed till sheets (Table 15.1) and loess deposits. Sands, alluvial silts, and soils, in places with fossil invertebrates and vertebrates, plant macrofossils, and also pollen, have provided evidence of environmental conditions during interglacial episodes.

15.3 PLIOCENE-PLEISTOCENE BOUNDARY

Within the stratigraphic divisions that evolved during the nineteenth and twentieth centuries, the Pleistocene Epoch was generally equated with the period of major glaciation, and although, as described earlier, older glaciations such as the Donau were suspected, the beginning of the Pleistocene was usually placed at the time of the first major ice advances. In simple lithostratigraphic terms, therefore, a succession underlying the oldest glacial tills was preglacial, and by definition Pliocene. The difficulties with this approach became increasingly apparent as the evidence for earlier glaciations began to accumulate. It was also clear that any boundary based upon such criteria was bound to be a diachronous one. The need for a properly defined boundary was obvious, but there was considerable controversy as to where it should be. Glacial successions, marine successions, freshwater deposits, invertebrate faunas, microfossils, and mammalian faunas, not to mention paleoclimatic indicators of various kinds, all provided potential markers and had their special advocates.

At the International Geological Congress in 1948, it was agreed that the type area should be in the Calabrian Peninsula in southern Italy (Fig. 15.4). Here was a fossiliferous section ranging in age from Early Pliocene into the Pleistocene and within which three stages had been recognized. At the base, lying unconformably upon Miocene beds, are conglomerates and sands of the Pontian Stage; overlying them are clays and sands of the Piacenzian-Astian Stage, and these are overlain, in turn, by sandstones and conglomerates of the Calabrian stage. Throughout this succession, the marine molluscan faunas have a generally warm-water aspect, but beginning at the base of the Calabrian, a new element begins to make its appearance, and for the first time, cool-water species are found. One of the most characteristic is the bivalve *Arctica (Cyprina) islandica,* which is today found only in northern waters. Other northern forms include *Neptunea despicta, Buccinium undatum, Chlamys islandica,* and *Macoma calcarea.* The foraminiferan *Ilyalina baltica* is also a northern immigrant. Altogether, nearly 10 percent of the total Calabrian foraminiferans are cold-water forms and only 50 percent of the warm-water species from the preceding stage survive into the higher stage.

On the evidence of these faunal changes, it is clear that a marked cooling took place at the beginning of Calabrian time. It was for this reason that the base of the Calabrian stage was selected as the Pliocene-Pleistocene boundary. The boundary stratotype selected by the INQUA (International Association for Quaternary Research) Subcommission on the Pliocene-Pleistocene boundary occurs within a 300-m thick succession of deep-water marine sediments at Vrica (Fig. 15.4), south of the town of Crotone (Calabria). It occurs about 3–6 m above the horizon marking the top of the Olduvai normal polarity subchron. The isotopic age of the Pliocene-Pleistocene boundary in the Vrica section has been a matter of some controversy. According to Colalongo et al. (1980), a volcanic ash just above the first appearance of *Hyalina baltica* has been dated by the K-Ar method as 2.2 ± 0.02 m.y. old, so assuming

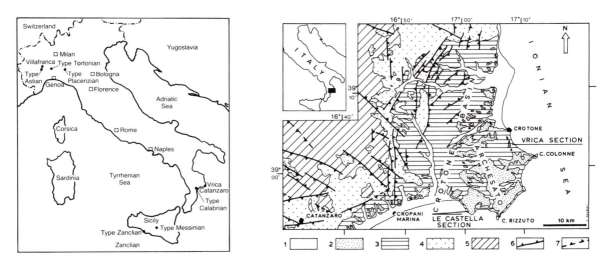

FIGURE 15.4
Locations of type sections of the Pliocene-Pleistocene boundary. Geological map of the Marchesato Peninsula (based on data from Tortorici, L., Analisi delle deformozioni fragili dei sedimenti, postorogeni della Calabria: Bolletino Societa Italia, v. 100, p. 291–308): 1, continental sediments (Pleistocene and Holocene); 2, sediments of upper Pleistocene marine terraces; 3, middle to upper Pliocene and lower Pleistocene sediments; 4, Tortonian to lower Pliocene sediments; 5, basement rocks; 6, main faults (small triangles on downthrown sides); 7, possible faults. (From Aguirre, E., and Pasini, G., 1985, The Pliocene-Pleistocene boundary: Episodes, v. 8, Fig. 1, p. 116.)

average rates of sedimentation, extrapolation down to the *Hyalina baltica* horizon gives about 2.7–2.8 Ma. for the beginning of the Pleistocene. On the other hand, dates from volcanic ashes in the same section, reported by Obradovich et al. (1982), suggest that the older age is incorrect and that the boundary is considerably younger, at about 1.6 m.y. This is a more likely age because, in terms of the magnetostratigraphic time scale, this would place it at the top of the Olduvai Subchron within the Matuyama Chron (Fig. 15.5).

In the deep-sea succession, the Pliocene-Pleistocene boundary has been based upon several biostratigraphic criteria, as shown in Figure 15.6. In general, a boundary set at the top of the Olduvai Subchron is supported by the foraminiferal biostratigraphy, and it also would show agreement with a widely used datum based on the apparent worldwide disappearance of the discoasters (Ericson and Wollin, 1964). Of somewhat doubtful taxonomic affinities, the discoasters have star-shaped calcareous plates, rather similar to those of the Coccolithophoridae; indeed, some taxonomists place them in that family. They are common throughout the Miocene, but show a decline during the Pliocene. According to some workers, the demise of the discoasters at the end of the Pliocene was more apparent than real; they have, for example, been reported in Holocene sediments in New Zealand (Burns, 1972) and also in modern plankton hauls from the Pacific.

In nonmarine successions, the Pliocene-Pleistocene boundary has been established in northern Italy in the succession of fluvial gravels, sands, and peats, ranging in age from late Pliocene to early Pleistocene, that locally interfingers with the marine Calabrian sediments. These are the Villafranchian facies, named after Villa-franca d'Asti in Pied-

FIGURE 15.5
Deep-sea biostratigraphy of the
Pliocene-Pleistocene boundary.

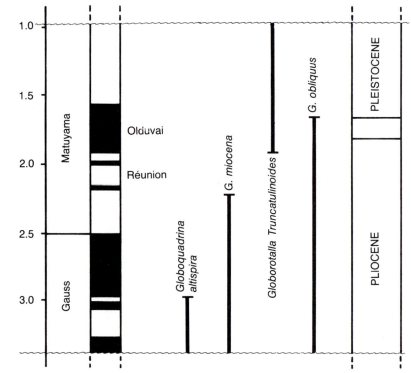

mont (Fig. 15.4), which contain a rich mammalian fauna, including *Elephas, Equus,* and *Leptobos.* The Lower Villafranchian is considered to be Pliocene in age because it contains numerous species that, in general, indicate a relatively warm climate, being essentially holdovers from earlier Tertiary time. Beginning with the Middle Villafranchian, however, new taxa, apparently of Asiatic origin, make their appearance, suggesting a wave of immigration from the east. That a climatic change was responsible for these faunal turnovers is indicated by a concomitant change in the floras. Pollen analyses from the Villafranchian show a marked impoverishment of the warmer-climate trees at about this horizon. Thus, there is little doubt that, just as in the Calabrian marine sequence, the terrestrial fossil evidence points to climatic deterioration and provides an indicator for establishing the Pliocene-Pleistocene boundary in non-

marine successions also, although the precise location of the nonmarine Pliocene-Pleistocene boundary remains a matter for debate. There is no doubt that the deep-ocean succession contains the most complete record, but a boundary stratotype in such a location is not feasible. There are also numerous workers who still emphasize the importance of glacial criteria, and they would prefer an older date—about 2.8 m.y. (Beard et al., 1982).

The work involved in the determination of the Pliocene-Pleistocene boundary described above provides a good example of how practical stratigraphy is conducted. It is not just a question of finding a compromise that keeps everyone happy, but is an exercise in selecting a boundary that, above all, is usable. In other words, this is the "nuts and bolts" of stratigraphy. Thus, in the above discussion, both marine and nonmarine successions are

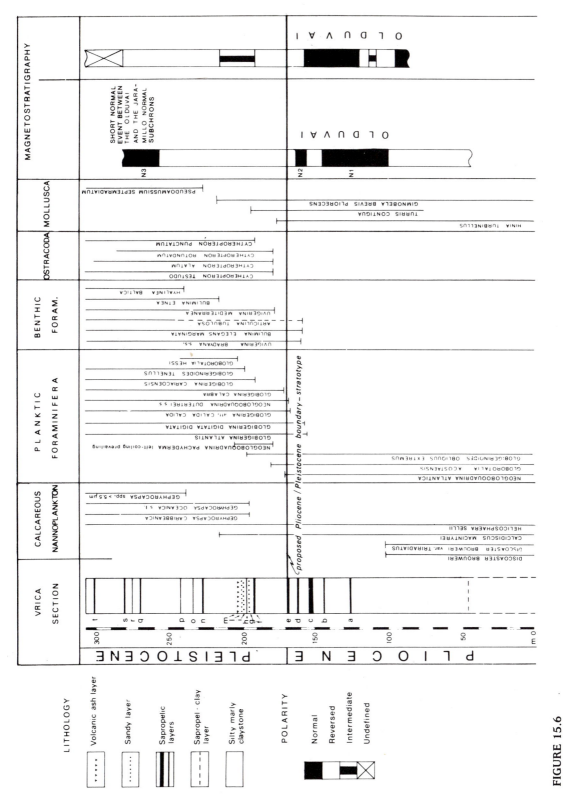

FIGURE 15.6

Biostratigraphy of the Vrica section, Calabria, Italy. (From Aguirre, E., and Pasini, G., 1985, The Pliocene-Pleistocene boundary: Episodes, v. 8, Fig. 3, p. 118.)

involved. This means that the biostratigraphic attributes of selected terrestrial vertebrates, freshwater molluscs, macrofloras and microfloras, marine macrofossils, and marine microfossils are all tested and used to the fullest extent. Essential, also, is the inclusion of a magnetostratigraphic marker, and isotopic dating adds a further dimension. Finally, it should not be forgotten that a marked climatic cooling is involved, so event stratigraphy also makes its contribution.

15.4 QUATERNARY STRATIGRAPHIC TERMINOLOGY

The terminology used in Quaternary stratigraphic studies has, in many ways, developed independently of the stratigraphic terminology in general use for the remainder of the Phanerozoic. Some terms are peculiar to the Quaternary by virtue of their applicability to glacial/interglacial events, and in other instances terms similar to those used in established stratigraphic procedures are used with somewhat different meanings. In defining glacial and interglacial events, the term *stage* has long been used; e.g., Wisconsin Glacial Stage or simply Wisconsin (or Wisconsinan)* Stage. This usage is discouraged by the authors of stratigraphic codes, but the term seems to be firmly established. A subdivision of a glacial stage marking a warmer period, during which there was a secondary recession or stillstand of the ice front, is termed an **interstade** (interstadial). A **stade** (stadial) is a subdivision marking a short glacial readvance.

In an attempt to overcome some of the problems of Quaternary stratigraphy, the

1961 Code of Stratigraphic Nomenclature recommended use of so-called **geologic climate units.** Such a unit was defined as "an inferred widespread climatic episode defined from a subdivision of Quaternary rocks." The boundaries of such units were established by those of the stratigraphic units on which they were based and were, therefore, diachronous. The concept has not been very widely applied and geologic climatic units do not appear in the most recent stratigraphic code. The terms *glaciation* and *interglaciation* are often used informally, and because they describe climatic episodes they are actually geologic climate units. They are usually considered as having stage rank, but clearly not in the sense of conventional usage of that term. Glaciation is often used interchangeably with the term *glacial stage.*

The term *stage* appears again in deep-sea stratigraphy. Here it is used informally by Emiliani (1955) in his oxygen isotope paleotemperature curve, described later in this chapter, in which were recognized 22 oxygen isotope stages, each being either a warm peak or a cool trough on the curve. Boundaries between stages were placed midway between temperature maxima and minima. Smaller peaks on the curve were termed **substages.**

15.5 DEEP-SEA SUCCESSIONS

Oxygen Isotope Stratigraphy

Three isotopes of oxygen occur in nature, O^{16} (ordinary oxygen), O^{17}, and O^{18}; of these, O^{17} occurs in only minute amounts, but O^{18} constitutes about 0.2 percent of natural oxygen. Because the isotopes have different atomic weights, during chemical reactions involving oxygen, a certain fractionation takes place. It was Urey in 1947 who determined that this process was temperature-dependent. Carbonates or carbon dioxide that originate at low temperatures contain a measurably higher

---

*Purists have long sought to standardize the endings of names for the various glacial stages, etc., and this has led to some quite cumbersome words. It now seems generally agreed that inconsistency is preferable; so, for example, the North American glacial stages are now widely referred to as Wisconsin, Illinoian, Kansas (or Kansan), and Nebraskan. In Europe, the same inconsistency seems to have been accepted (see, for example, Table 15.1).

proportion of the heavy isotope O^{18} than do those formed at higher temperatures. It follows that if the oxygen isotope ratio of carbonates forming the shells and tests of fossils could be determined, this value, measured against a standard, could be used to determine the temperature of the water in which the animal lived. The O^{16}/O^{18} ratio in carbonate formed in the oceans is not, however, only temperature-dependent; it is influenced also by the O^{18} content of the water itself, and this, as will be described later, has varied through time.

The oxygen isotope method, as a stratigraphic tool in deep-sea successions, was pioneered by Emiliani (1955). Using the planktonic foraminiferan *Globigerina saculifera*, sampled in cores from the Atlantic and Caribbean, Emiliani constructed a paleotemperature curve that embraced the time back to about 800,000 years B.P. The curve showed that temperature fluctuations had occurred at regular intervals and that there had been at least nine cycles, each comprising a cold interval followed by a warm one (Fig. 15.7). The peaks and lows in the curve, which Emiliani (1955) called stages, were numbered back in time, with number 1 designated the Recent temperature maximum (Table 15.2). It follows, therefore, that all odd-numbered stages represent temperature maxima and the even numbers temperature minima. Later workers (e.g., Shackleton, 1967, 1975) suggested that the isotopic variations, although real enough, could be attributed as much to fluctuations in the isotopic composition of sea water as to the water temperature. It was pointed out that during glacial episodes, when sea levels fell and ocean waters became slightly more saline, they were also relatively enriched in O^{18} (that is, were isotopically positive). A curve showing variations in oxygen isotope abundances was, in fact, a paleoglacial curve as much as a paleotemperature curve. As Figure 15.7 shows, the change from peaks (temperature maxima) to

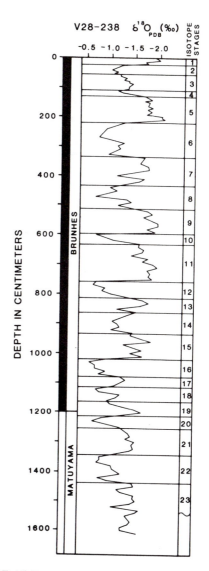

FIGURE 15.7
Succession of oxygen isotope stages as determined in deep-sea core V28-238 from the equatorial Pacific (depth scale in cm), with age estimates for stage boundaries. (From Williams, D. F., 1984, Correlation of Pleistocene marine sediments of the Gulf of Mexico and other basins using oxygen isotope stratigraphy, *in* Healy-Williams, N., ed., Principles of Pleistocene stratigraphy applied to the Gulf of Mexico: Boston, Internat. Human Resources Devel. Corp., Fig. 3–6, p. 80.)

TABLE 15.2

Ages of isotope stage boundaries and terminations

Isotope Stage Boundary	Depth (cm) in Core V28–238	Age (Ka)		Termination
		Constant Sedimentation*	Mean of 3 Spectral Analyses†	
1–2	22	13	11	I
2–3	55	32	27	
3–4	110	64	60	
4–5	128	75	72	
5–6	220	128	127	II
6–7	335	195	188	
7–8	430	251	246	III
8–9	510	303	303	
9–10	595	347	337	IV
10–11	630	367	356	V
11–12	755	440	433	
12–13	810	472	471	VI
13–14	860	502	512	
14–15	930	542	558	VII
15–16	1015	592	620	
16–17	1075	627	654	VIII
17–18	1110	647	676	
18–19	1180	688	719	
B/M		700	729	
19–20	1210	706	736	IX
20/21	1250	729	763	
21/22	1340	782	790	X

*Shackleton, M. J., and Opdyke, N. D. 1973, Oxygen isotope and paleomagnetic stratigraphy of equatorial Pacific core V28–238; Oxygen isotope temperatures and ice volumes on a 10^5 year and 10^6 year scale: Quaternary Research, v. 3, p. 39–55.
†Hays et al., 1976; Kominz et al., 1979; Imbrie et al., 1985.

troughs (temperature minima) on the curve is, in most cases, abrupt, indicating rapid changes from warm- to cool-water conditions. These steep slopes on the curve were labeled *terminations* by Broecker and Van Donk (1970) and given a numbering system of their own, as shown in Table 15.2. As mentioned in Chapter 12, Broecker and Van Donk (1970) described the typical Pleistocene sea-level curve as one showing a relatively protracted fall of sea level and rapid rise. This, in terms of asymmetry, is the opposite to the shape of the surface-water temperature curve seen in

the terminations of the oxygen isotope curve. This apparent contradiction probably is explained by the fact that, as Williams (1984) points out, temperature variations coincident with the glacial/interglacial climatic cycle are only a small part of the Pleistocene isotope signal (typically, less than 25 percent of the total interglacial signal).

Although the oxygen isotope curve (and the timetable of stages derived from it) has come to play a central role in all studies of Late Cenozoic deep-sea successions and to provide a yardstick against which climatic fluctuations can be evaluated, it obviously cannot be used alone. Just as with paleomagnetic reversals, which are also "flip-flop" phenomena, there is nothing unique about one warm/cold cycle to distinguish it from another, and only by counting backward can isotope stages be determined. It is clear that a hiatus in the sedimentary succession that removed one or more complete warm/cold cycles might easily be missed and cause discrepancies in correlation. In practice, various other methods are used in connection with the oxygen isotope tool, including radiometric dating, magnetic stratigraphy, and biostratigraphy. As can be seen in Figure 15.7, the Brunhes/Matuyama paleomagnetic boundary, dated at 700,000 years B.P., occurs at the Stage 19 peak. Stage 2 has been dated at 17,000 years B.P. by radiocarbon, and five further stages by the protactinium-ionium method, as shown. Other stage boundaries have been dated by extrapolation, assuming uniform sedimentation rates. When all the

various cold/warm cycles, as revealed by oxygen isotopes, sea-level changes, ocean temperature, continental glaciations, and so on, are compared, there is not always a very good correlation between them. On the other hand, the fact that they often do show a similar periodicity would indicate that cause-and-effect relationships do exist, although their nature has sometimes remained elusive.

Spectral analysis of the oxygen isotope curve reveals a marked periodicity at 100,000, 41,000, and 23,000 years (Hays et al., 1976); this agrees closely with the insolation cycle of Milankovitch (Fig. 15.8). When the oxygen isotope curve is compared with the other cycles, however, the agreement is sometimes poor. It does not, for example, apparently agree with the continental glacial record or with the eustatic sea-level curve. Similarly, the ocean-temperature curve seems to be out of phase when compared with the continental glacial curve. In

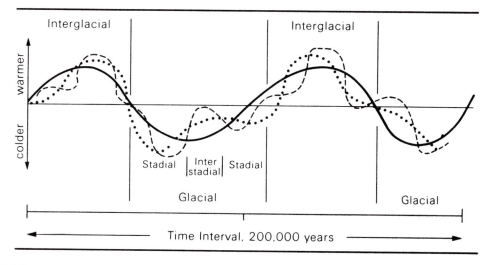

FIGURE 15.8
The 23,000-, 41,000-, and 100,000-year Milankovitch cycles and their relationships to glacial/interglacial stages. (After Lowe, J. J., and Walker, M. J. C., 1984, Reconstructing Quaternary environments: New York, Longmans, Fig. 1.6, p. 13.)

fact, times of continental glaciation seem to correlate with times of relatively warm rather than cold ocean temperatures (Fig. 15.9). One of the problems with trying to match the various cyclical events is that they are often viewed at different resolutions. As Fillon (1984) pointed out, if the resolution of deep-sea and terrestrial records are adjusted appropriately, they often will show quite a good match. For example, if only the most intense interglacial episodes on the paleotemperature curve are highlighted, there is a much closer agreement with the continental glacial timetable. Another factor to be considered is an offset effect when comparing high-latitude and mid-latitude responses to the Milankovitch orbital variations. Mid-latitude ice sheets are controlled by insolation effects that vary primarily in response to 23,000-year and longer cycles, whereas high-latitude ice, including the ice cover of the Arctic Ocean, varies in response to 41,000-year and longer cycles.

Over the late Cenozoic ice age as a whole, ice sheet responses to the Milankovitch mechanism have not been constant. Especially noticeable was the mid-Pleistocene shift from the 41,000-year rhythm of orbital obliquity to the 100,000-year orbital eccentricity rhythm. The cycles of sea-surface temperature, ice sheet fluctuation, and other paleoclimatic indicators during almost all the Matuyama magnetochron (2.5–0.735 Ma) were dominated by the 41,000-year rhythm. After 0.9 Ma, however, the 100,000-year cycle became dominant, although the 41,000- and 23,000-year cycles are still apparent in the record of glacial interstades superimposed on the 100,000-year signal (Ruddiman and Wright, 1987, Chap. 1). The 41,000-year rhythm is controlled by the variation in high-latitude summer insolation effects, whereas the 23,000-year rhythm is a reflection of the expansion of the larger 100,000-year ice sheets into lower latitudes. The reason for the 100,000-year ice sheet rhythm of the later

Pleistocene is still something of a mystery. It has been suggested that an interaction of the 41,000- and 23,000-year rhythms might produce an event analogous to the beat frequency seen in the interaction of ocean swells, although this idea is no more than tentative and probably too simplistic. What, in fact, must also be considered are certain nonlinear effects seen in the marked asymmetry in the ice sheet growth and decay cycle, mentioned in Chapter 12. One reason why ice sheets apparently decay much faster than they accumulate is a feedback mechanism consequent on the lowering of the ice sheet surface. As the surface descends into warmer air, the ablation rate increases and, although this should be partially compensated by crustal rebound under the thinning ice, there is a considerable time lag, variously calculated as 3500 5000 years. Another mechanism, proposed by Denton and Hughes (1983), involves an increase in iceberg calving as ice shelves become buoyant consequent on glacio-eustatic sea-level rise. This, in turn, removed back pressure on inland ice and resulted in an acceleration of the ice streams. The resultant increase in deflation would accelerate sea-level rise, and again there is an obvious feedback effect.

Biostratigraphy

Biostratigraphy, as used in Quaternary deep-sea sediments, differs from its use in older successions. This is because far too few species extinctions occur for any refined division based on zones, as described in Chapter 9. Fossils, however, have been used in other ways, and their main value lies in their role as paleoclimatic indicators. Ericson and Wollin (1956), for example, were able to show that the relative abundance of the foraminiferan *Globorotalia menardii* could be directly correlated with water temperature. The species showed high counts during interglacial episodes and was absent during glacials. A

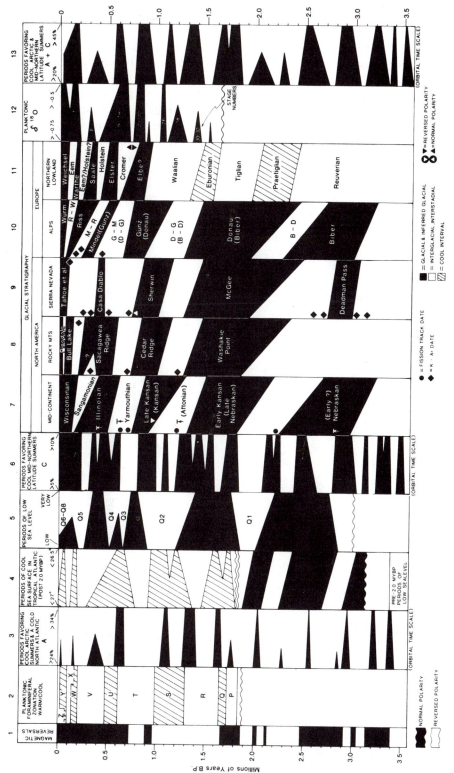

FIGURE 15.9

Suggested chronology for glacial events for the last 3.5 million years. (From Fillon, R. H., 1984, Continental glacial stratigraphy, marine evidence of glaciation, and insights into continental-marine correlations, *in* Healy-Williams, N., ed., Principles of Pleistocene stratigraphy applied to the Gulf of Mexico, Fig. 5.6, p. 164. Reprinted by permission of International Human Resources Development Corp., Boston.)

tenfold division (zones Q–Z) of alternating warm and cold periods originally was equated with the standard North American glacial/interglacial chronology (Fig. 15.10). More recently, the study of the relative abundances of selected species, rather than the simple present/absent method has greatly improved the resolution of the zonation. Even better definition is possible using a whole faunal assemblage rather than single species. With this approach, in cores from the Gulf of Mexico, Kennett and Huddleston (1972) were able to differentiate 18 faunal divisions (Fig. 15.11), with an average duration of less than 15,000 years. The divisions are subzones of the Ericson and Wollin (1956) W–Z zones (Thunell, 1984). Another technique has been based upon the interesting discovery that

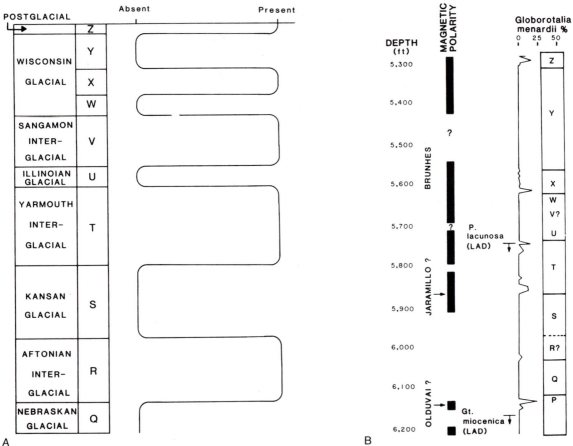

FIGURE 15.10

Interpretation of *Globorotalia menardii* zones by (A) Ericson and Wollin (1968), and (B) Fillon (1984). LAD = last-appearance datum. (From Thunell, R., Pleistocene planktonic Foraminiferal biostratigraphy, *in* Healy-Williams, N., ed., Principles of Pleistocene stratigraphy applied to the Gulf of Mexico: Boston, Internat. Human Resources Devel. Corp., Fig. 2.4, p. 36.)

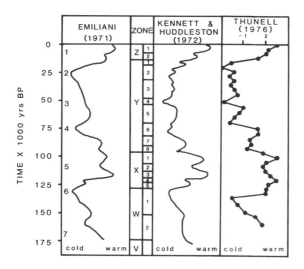

FIGURE 15.11
Subdivision of *Globorotalia menardii zones* by Kennett and Huddleston (From Late Pleistocene paleoclimatology, foraminiferal biostratigraphy and tephrochronology, western Gulf of Mexico: Quarternary Research, v. 2. 1972).

a valuable tool in Quaternary stratigraphy. Provided recognition of individual ash bands is possible, they can be matched from core to core. Because ash bands mark geologically instantaneous events they make excellent time-horizon markers. Various techniques have been developed for the identification of individual ash bands. One approach has been to determine the refractive index of the glass shards as an indicator of the chemical composition of the parent lava. This is not an easy method to use, and it largely has been supplanted by the electron microprobe. This technique has been most successful in fingerprinting different ash layers and even the ash content in sediments beyond the limit of discrete ash falls (Ledbetter, 1984). Dating ash layers usually is done by tracing them into cores with a known stratigraphy based on biostratigraphic or oxygen isotope indicators (Fig. 15.12).

changes in the coiling direction of certain planktonic foraminiferans are apparently temperature-dependent. So, for example, the change from dextral (right) coiling to sinistral (left) coiling in *Globorotalia menardii* indicates the onset of the marked cooling trend at the end of the Pliocene and has been widely used as a Pliocene-Pleistocene boundary indicator. *Neogloboquadrina pachyderma* also is represented by both left- and right-coiled races, again apparently temperature-dependent. In cold water, the sinistral form is characteristic and in warmer water the dextral. Within a sea-bottom core, therefore, the recording of past temperature changes in the water column above is simply a matter of counting variations in the ratio of left- and right-coiled specimens.

Magnetostratigraphy

Within the Pleistocene, four polarity switches occur, as shown in Figure 15.13. Their ages are now fairly well known, so they provide useful real-time dates within oxygen isotope and biostratigraphic successions and also provide the means of global correlation with successions in other depositional environments. The Matuyama Reversed epoch, or chron, contains three short-duration normal polarity events, or subchrons: the Jaramillo, Olduvai, and Réunion. Formerly, another event was recognized and named the Gilsa, but it was found that the dates overlapped with those of the Olduvai, discovered earlier. It now is agreed generally that the Olduvai and Gilsa subchrons are probably one and the same.

Tephrochronology

The recognition and dating of volcanic ash layers in deep-sea sediment has proved to be

15.6 PLEISTOCENE MAMMAL AGES

The remains of fossil mammals are not uncommon in Quaternary terrestrial succes-

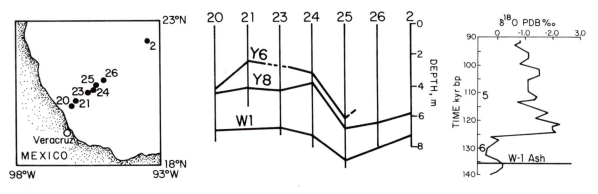

FIGURE 15.12
Three major tephra (Y6, Y8, and W1) horizons identified in the western Gulf of Mexico.
The oxygen isotope stratigraphy of a sediment core was used to date the W1 ash. (After
Ledbetter, M., 1984, Late Pleistocene tephrochronology in the Gulf of Mexico region, *in*
Healy-Williams, N., ed., Principles of Pleistocene stratigraphy applied to the Gulf of Mex-
ico: Boston, Internat. Human Resources Devel. Corp., Fig. 4-3, p. 124 and Fig. 4-13, p.
144.)

sions, and just as with the floras and with the
marine faunas, successive climate changes
are discernible in shifting faunal associations.
In Europe, the terrestrial Villafranchian series
of southern France and northern Italy contains a rich mammalian fauna, and the Pliocene-Pleistocene boundary occurs within
the series. A deterioration in climate is
marked by the appearance of new mammal
groups, and, in general, species from the
Miocene and Pliocene, adapted to life in
warm forest environments, gave way to such
animals as zebrine horses, cattle *(Bos),* and
true elephants, all forms characteristic of
more open, savannah country. This change
was gradual enough that there is no general
agreement as to where precisely the Pliocene-Pleistocene boundary should be
placed. The lower part of the Villafranchian in
Italy grades laterally into marine sediments
of Calabrian and Pliocene age, and this certainly will make it easier to establish the base
of the Pleistocene in terms of terrestrial biostratigraphy.

In North America, Pleistocene mammal
faunas share some features with those of Eu-

rope, but there also are some interesting differences, mainly due to differences of geog-
raphy. Advancing and retreating ice sheets
had less effect on the North American faunas
because, with the main mountain ranges
trending north-south, in contrast with those
in Europe, the response to deteriorating climate was north–south migration. The result
is that there are more Pliocene holdovers
present in North American Pleistocene
faunas. From before the Pleistocene, mammal faunas also were influenced by migration across the Panamanian land bridge, and
during glacial eustatic sea-level lowstands
the emergence of the Bering land bridge resulted in migrations (eventually including
man) from Asia.

Stratigraphically, Pleistocene mammal
faunas can be grouped into three divisions,
or "mammal ages," as they are known. This
is, again, a Quaternary misuse of a stratigraphic term; what is being described are, in
effect, biochrons. From the oldest to the
youngest, they have been named Blancan, after Mount Blanco in northeastern Texas; Irvingtonian, with a type locality at Irvington

FIGURE 15.13
Magnetostratigraphic time scale for the Quaternary. (Black = normal polarity.)

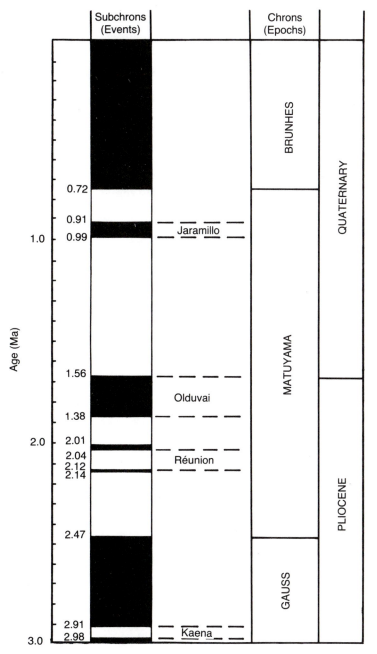

near San Francisco; and Rancholabrean, named for the classic brea, or tarseep, at Rancho La Brea (Hancock Park) in Los Angeles. The term *stage* also is applied to these divisions in the sense that they are "continental stages" as opposed to glacial stages; again, this is at variance with more conventional usage. The boundary between the

Blancan and the underlying Hemphill stage occurs somewhat earlier than the generally accepted Pliocene-Pleistocene boundary. The precise difference in age is unknown, however, because the position of the Pliocene-Pleistocene boundary in North America is itself a matter of some controversy. K-Ar dating of material from numerous Blancan sites give an age range from 1.5 to 3.5 m.y., and these values show that the Blancan is approximately coeval with the Villafranchian of Europe, or at least with its upper two-thirds (Table 15.3). Faunally, the base of the Blancan is marked by numerous extinctions of Pliocene animals and the appearance of new genera such as *Plesippus* and *Nannippus* (horses), *Procastoroides* (beaver), and a species of dog, *Borophagus*. During later Blancan time, the appearance of small boreal mammals suggests a deteriorating climate.

The exchange of mammal species via the Panamanian land bridge in late Blancan time and the incoming of Asian immigrants resulted in the marked change that heralded the Irvingtonian age. One of the most striking features of Irvingtonian faunas was the arrival of the first true elephants *(Mammuthus meridionalis)*. The Rancholabrean fauna, including *Bison latifrons* (giant bison), *Mammuthus columbi* (Columbian mammoth), *Smilodon* (saber tooth), and *Ursus arctos* (brown bear), is a Late Pleistocene fauna of generally cold-climate aspect. The upper boundary of the Rancholabrean coincides roughly with the Pleistocene-Holocene boundary, marked by the disappearance of many of the large mammals. It generally is accepted that the relatively rapid extinction of so many animals, particularly in North America, at the very end of the Pleistocene and set at around 11,000 years B.P. by Martin (1973), was due to the depradations of humans (Fig. 15.14). Sad to relate, our role in what is, in effect, a mass extinction event, continues to the present day.

15.7 PALYNOLOGY

Turning to terrestrial floras, it is again as climatic indicators that fossils are used, in this case **pollen.** Pollen grains, although very small (typically ranging from 10 to 100 micrometers in diameter), have an outer wall, or exine, that is formed of cutin and so is chemically very resistant. Under suitable conditions of burial, therefore, pollen is likely to be readily preserved. Many trees and other plants produce pollen in enormous quantities. There is a close link between the method of pollination and the number of pollen grains produced; trees, such as birch, hazel, and the conifers, and the grasses, which rely on fortuitous dispersal by the wind, are the greatest producers. On the other hand, in plant species with more specialized pollination mechanisms (by insects, nectar-eating birds, or even bats), the need for large numbers of pollen grains is much less.

The outer membrane, or exine, of the pollen grain may be smooth or covered with projections that give the grain a distinctive patterned or sculptured appearance (Fig. 15.15). In some families, the pollen grains are recognizable by species or even subspecies. More often, however, such as in many conifers, classification is possible only by genus. Pollen grains, settling from the pollen "rain" in the air, or carried by water, are a common constituent of many sediments, particularly those in peat bogs or lakes, where an anaerobic environment favors their preservation. Particularly with windblown pollen, there is a direct relationship between the representation of different pollens and the abundance in the immediate area of the plant species that produced them. Because living plants and all habitat variables can be studied at the present time, the bias introduced by overproduction of certain pollen species can be evaluated. Corrections then can be made so as to arrive at a true picture of relative abundances

TABLE 15.3
North American mammal ages and approximate correlation with European mammal faunas

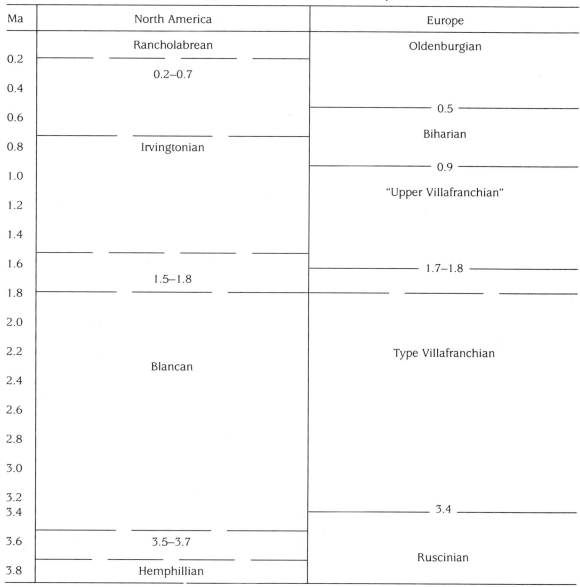

Ma	North America	Europe
	Rancholabrean	Oldenburgian
0.2		
	0.2–0.7	
0.4		
0.6		0.5
0.8	Irvingtonian	Biharian
1.0		0.9
1.2		"Upper Villafranchian"
1.4		
1.6		1.7–1.8
1.8	1.5–1.8	
2.0		
2.2		Type Villafranchian
2.4	Blancan	
2.6		
2.8		
3.0		
3.2		
3.4		3.4
3.6	3.5–3.7	Ruscinian
3.8	Hemphillian	

within the plant associations that existed at the time of pollen deposition.

Pollen analysis is typically carried out with a minimum of 200 grains, although this number is flexible, and for any one stratigraphic horizon what is known as a pollen spectrum is prepared. This shows the distribution of the various taxa in terms of per-

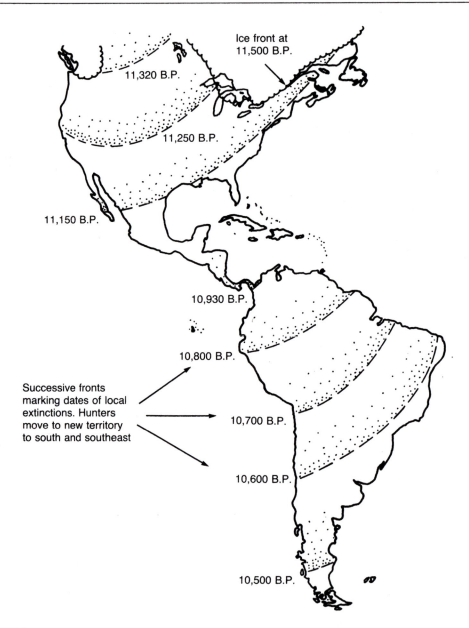

FIGURE 15.14
Map showing the timing of major post-Pleistocene mammal extinctions in North America and South America. (After Martin, P. S., The discovery of America: Science, v. 179, 9 March 1973, Fig. 2 and Fig. 3, p. 972. Copyright 1973 by the AAAS.)

FIGURE 15.15
Pollen from typical Pleistocene-Recent genera: A, *Quercus* (oak); B, *Alnus* (alder); C, *Corylus* (hazel); D, *Acer* (maple); E, *Fagus* (beech); F, *Pinus* (pine). Scale bar is 10μ.

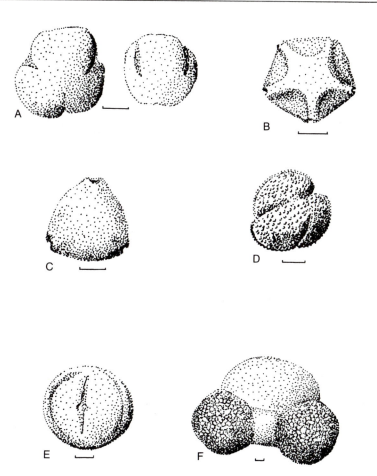

centages of the total pollen count. Normally, three divisions are made, into arboreal (AP), nonarboreal (or herbaceous) flowering plants and grasses (NAP), and spores (Fig. 15.16). As a series of pollen spectra is assembled through a stratigraphic succession, a pollen curve can be constructed to show the variations through time of the proportions of certain plant species. The curves are then combined to produce a pollen diagram plotted along a depth scale, which is also, of course, a relative time scale. Often some calibration in real time is possible using radiocarbon dating. Using suitable graphic symbols, such as bar graphs and pie diagrams, pollen diagrams from different localities can be readily compared and stratigraphic correlations made.

When viewed within a stratigraphic succession, the pollen diagrams from successive horizons provide a record of shifting values in the percentage composition of pollen assemblages. The chief control of these variations is climate, and so the sequential changes through time from cold to warm conditions, wet to dry, and so on, can be determined. Early in a typical interglacial, for example, the forest cover is dominated by *Betula* (birch) and *Pinus,* but as progressive amelioration of the climate occurs, a mixed oak forest with *Quercus, Ulmus* (elm), *Corylus* (hazel), *Fraxinus* (ash), and other trees take its

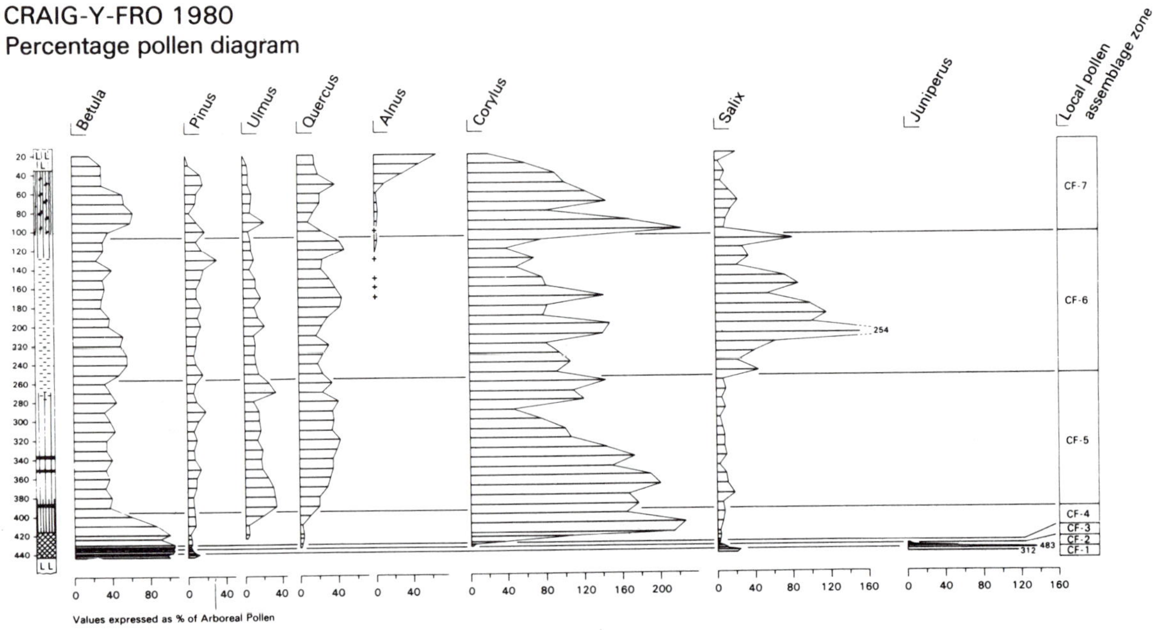

CRAIG-Y-FRO 1980
Percentage pollen diagram

FIGURE 15.16

Typical pollen diagram based on arboreal pollen percentages, from Brecon Beacons, South Wales. (After Walker, M. J. C., 1982b, Early and mid-Flandrian environmental history of the Brecon Beacons, South Wales: New Phytol., v. 91, p. 147–165.)

place. Turner and West (1968) set out a standard zonation table in which four zones are arranged in a cold to warm, and back to cold, climatic cycle. Iverson (1958) and Anderson (1966) used circular diagrams to show essentially the same sequence (Fig. 15.17). Although such standard sequences imply that all interglacials are similar, attention to more subtle changes may reveal features that distinguish one interglacial from another. The order of immigration of species, the times of different species culmination, and the mutual relationships of forest species as the woodlands evolved are all amenable to analysis and apparently significant. Pollen diagrams also provide the means of distinguishing interglacials from interstadials. The climate amelioration during an interstadial falls short of that during a full interglacial

(Table 15.4) and this is normally reflected in the pollen stratigraphy, as it likely would not be if a succession of "interglacial"-type lithologies was the sole source of information. In postglacial successions pollen studies also have proved useful in studies of prehistoric humans in setting the time of the earliest clearing of forests by Neolithic peoples and the introduction of agriculture. Sudden shifts in the proportion of AP and NAP counts and the appearance of pollens from agricultural crops and "weed" plants are easily discernible.

15.8 GLACIAL STRATIGRAPHY

Glacial Tills

The deposits that are most obviously attributable to glaciers are moraines, made up of till

FIGURE 15.17
Glacial/interglacial cycle based on diagrams by Anderson (1966) and Iverson (1958).

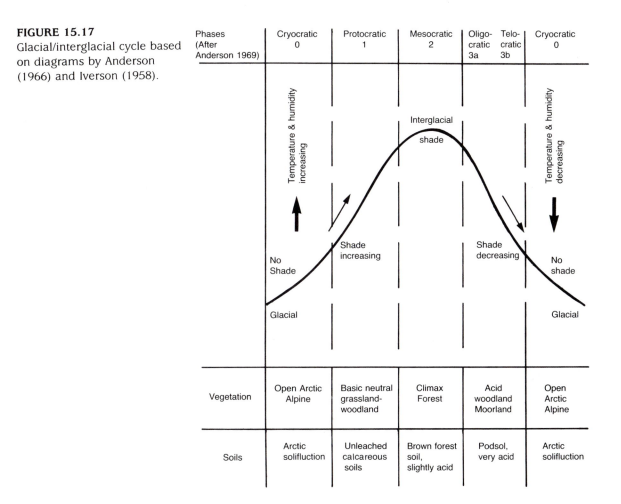

TABLE 15.4
Climatic succession during an interglacial compared with that during an interstadial (After Bowen, 1978, Table 4–5, p. 102.)

Interglacial	Interstadial
Arctic	Arctic
Subarctic	Subarctic
Boreal, with summer temperature at least as high as during the Flandrian climatic optimum of the area in question	Boreal, with summer temperature essentially lower than in the Flandrian climatic optimum of the area in question
Boreal	Subarctic
Subarctic	Arctic
Arctic	

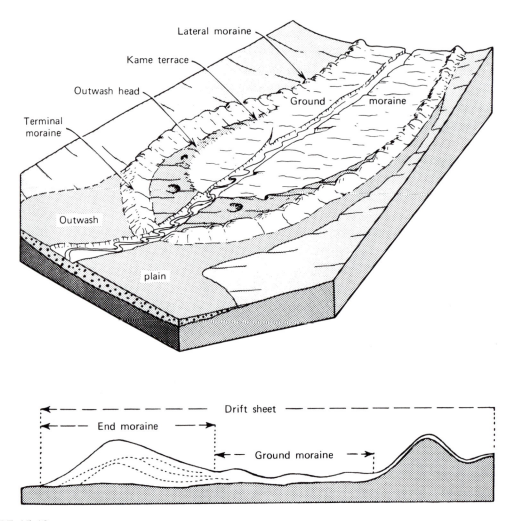

FIGURE 15.18
Diagram showing the relationship between ground moraine and end moraine. (From
Flint, R. F., 1971, Glacial and Quaternary geology, Fig. 9-16, p. 262. Reprinted by per-
mission of John Wiley & Sons.)

transported and deposited directly by the ice.
Ground moraine is defined as being a widespread, often thin and nonpersistent, layer of
glacial drift deposited beneath an ice sheet or
left behind on its melting. End moraines are
masses of till accumulated along the ice margin or at the terminus of a glacier (Fig.
15.18). End moraines may sometimes be
considerable topographic features and, in

areas of former continental glaciation, often
extend for many kilometers as a series of
gently arcuate ridges that mark the former
positions of the lobate front of the ice margin
(Fig. 15.19). Generally speaking, glacial till is
unstratified. Stratified drift is usually indicative of some reworking by meltwater streams.
    In attempting to analyze the stratigraphy of
till sheets and to establish a chronology of

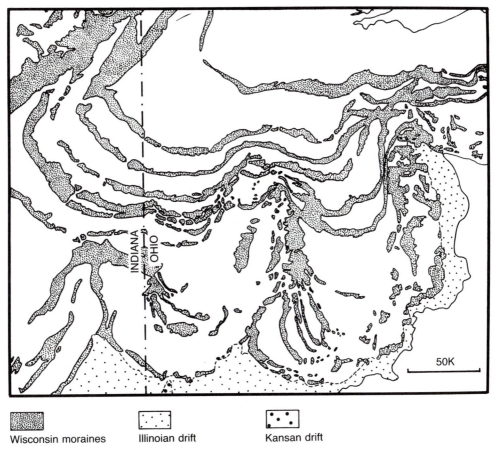

FIGURE 15.19
Major end moraines of the Wisconsin ice sheet in the region south of the Great Lakes.

glacial/interglacial events, it was long as-sumed, in North America at least, that three of the four classical glaciations, the Nebras-kan, Kansan, and Illinoian, were each repre-sented by a single till sheet, and the Wiscon-sin by two, the younger drift features belonging to what used to be called the "clas-sical" Wisconsin. Because of the relatively re-cent age of the Wisconsin deposits, and the fact that the youngest events are accessible to radiocarbon dating, a detailed analysis has been possible, and it has been known for many years that the Wisconsin glaciation was, in fact, multiple in character. Much less

is known about the pre-Wisconsin glaciations because direct evidence is scarce. As can be seen in Figure 15.20, the area covered by the Wisconsin ice sheet extended, except in a few areas, to the limits of and often beyond those areas where the drifts of the earlier glacia-tions were deposited. In the case of the Ne-braskan drift, in particular, it is everywhere masked by younger drifts and is seen only in valley sections or in boreholes. As work con-tinues, it is becoming apparent that, just as with the Wisconsin, the older glaciations were also multiple in character, and so a simple counting of superimposed tills clearly is not

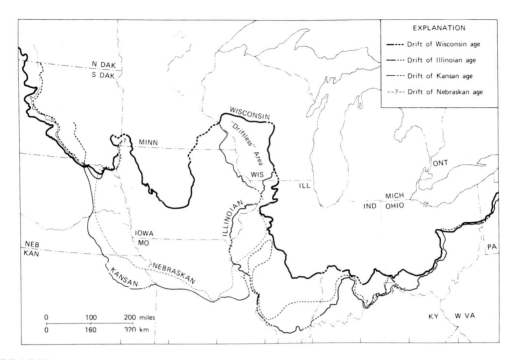

FIGURE 15.20
Limits of drift sheets in the Mississippi basin. The dotted line marks the edge of drift
sheets beneath younger deposits. (From Flint, R. F., 1971, Glacial and Quaternary geology, Fig. 21-1, p. 545. Reprinted by permission of John Wiley & Sons.)

going to provide any meaningful stratigraphic
succession. Work by Boellstorff (1978) and
others is beginning to show the presence of
older tills, including one overlain by a volcanic ash dated at 2.2 Ma.

Although glacial successions are the most
obvious of all indicators of major climatic
swings, taken as a whole they contain only
the most fragmentary record. It seems unlikely, therefore, that all the climatic fluctuations established in deep-sea successions by
oxygen isotope and other methods will ever
be discerned in a complete terrestrial glacial
succession. In attempting to correlate the
North American and European glacial chronologies, it was, until quite recently, generally
accepted that the four classical glacial stages
could be more or less matched up from continent to continent. This now has been shown

as largely incorrect. Improved dating methods and magnetostratigraphic correlations
have revealed a different set of relationships.
The Nebraskan and Kansan glaciations now
are correlated with the Donau and Günz glaciations of the Alps, but there were, apparently, no corresponding lowland glaciations
in central Europe. The oldest major glaciation
there, the Elster, corresponds with the Illinoian and Mindel of the North American and
Alpine successions, respectively (Table 15.5).
As pointed out by Berggren and Van Couvering (1974), earlier correlation models had
failed to take into account the important influences of topography, meteorology, and
other factors. What this means is that although climatic cycles in North America and
Europe generally were synchronous, the effects were not always the same. Although the

TABLE 15.5
Summary of the major glacial stages in North America (interglacials in italics)

Sierra Nevada	Rocky Mountains	Great Lakes			Alps
Tahoe	Pinedale	Wisconsin	Late (main)	Valders Port Huron Tazewell	Würm
			Middle	Farmdale Upper Winnebago	*R/W*
	Bull Lake		Early	Middle Winnebago Lower Winnebago	Riss
		Sangamon			*M/R*
Casa Diablo	Sacajawea Ridge	Illinoian			Mindel
		Yarmouth			*G/M*
Sherwin	Cedar Ridge	Kansan $\begin{cases} \text{II Late} \\ \text{I Early} \end{cases}$			$\begin{cases} \text{Günz II} \\ \text{Günz I} \end{cases}$
		Aftonian			*D/G*
McGee	Washaki Point	Late Nebraskan			Donau
Deadman Pass		Early Nebraskan?			Biber

two earlier cold phases were intense enough in North America to initiate full glacial conditions, in central Europe, apparently, they were not (Fig. 15.21).

Of the North American glaciations, the Wisconsin is naturally the best described, and numerous fluctuations of the ice front have been documented. Three broad divisions into Early, Middle, and Late (Main) Wisconsin are generally recognized, with numerous subdivisions, some of only local significance, particularly in the Late Wisconsin. It was during the Late Wisconsin that the ice sheet apparently reached its maximum extent, dated at approximately 18,000–20,000 years B.P. From that time on there was a net retreat, but interrupted by numerous short readvances, as shown in Figure 15.22.

In Europe, the advances and retreats during the last glaciation (Weichsel) were broadly similar. Particularly well documented by pollen analyses are the various interstadials, and detailed temperature curves have been plotted. For a fuller account of European glacial stratigraphy see Nilsson (1983). In the British Isles, insects, mainly beetles (Coleoptera), have also been used for detailed analyses of climatic change (Coope et al., 1971). The species are all extant and so make excellent paleoecologic indicators, being considered more sensitive to short-term climatic changes than are plants.

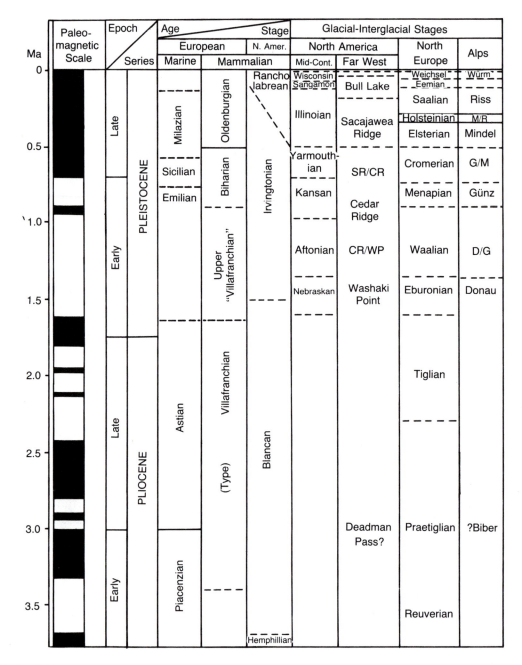

FIGURE 15.21
Summary of the major glacial stages in North America. (Modified from Berggren and Van Couvering, 1974.)

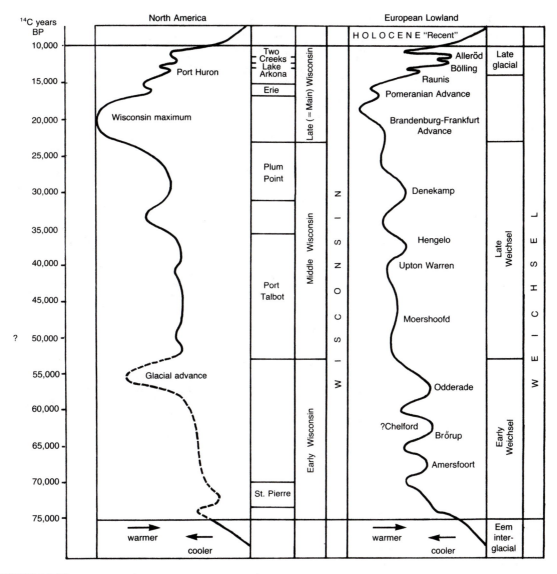

FIGURE 15.22
Climatic fluctuation, ice advances, and recessions during the last glaciation.

Loess Deposits

Loess is a friable, windblown deposit of silt-size grains, dominantly of quartz. It is usually unstratified and varies in color from beige through various shades of yellow and brown, depending upon the presence of iron oxides (Fig. 15.23). Although some loess deposits consist of the finer fractions blown from deserts, the bulk of Quaternary loess sheets was derived from the extensive outwash fans peripheral to the major ice sheets and from areas of drift left bare during glacial retreats. The climate in such areas was naturally cold and dry and only a sparse tundra vegetation

FIGURE 15.23
Loess overlying sands, gravels, and till. Note contrast between the vertical structure within the loess and the horizontal bedding or lack of bedding in the underlying deposits, north end of Trapp Lake, British Columbia. (Photo courtesy of Geological Survey of Canada, Ottawa.)

cover would be present. Such conditions were favorable for the widespread deflation of surface deposits, particularly glacial flour. The areal distribution of loess deposits around the world, and their variation in thickness, suggests a strong climatic control. They are found most commonly in drier areas; the presence of some vegetation, which acts to trap the loess grains, is an added factor. Loess is thin and nonpersistent in moist areas, such as eastern North America and westernmost Europe (Fig. 15.24). Loess deposits are a characteristic of glacial rather than interglacial conditions, and within a stratigraphic sequence they have been widely used for correlation. Commonly, several loess sheets are separated by paleosol horizons.

15.9 SEA-LEVEL FLUCTUATIONS

The general topic of glacio-eustatic sea-level fluctuations has been covered in Chapter 12, and little needs to be added here. As with glacial events, eustatic sea-level changes are particularly well-documented in the period since the Wisconsin maximum. According to Fairbridge (1961), and as described elsewhere, the sea-level rise during the post-Wisconsin marine transgression (also widely known as the **Flandrian transgression**)

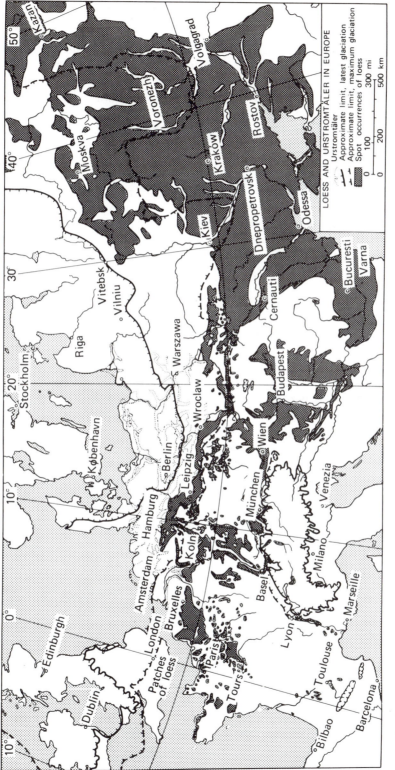

FIGURE 15.24

Map showing the distribution of Pleistocene loess in Europe. (From Flint, R. F., 1971, Glacial and Quaternary geology, Fig. 8-1, p. 201. Reprinted by permission of John Wiley & Sons.)

The map legend reads:

LOESS AND URSTROMTÄLER IN EUROPE

Urstromtäler
Approximate limit, latest glaciation
Approximate limit, maximum glaciation
Spot occurrences of loess

0   100   300 mi
0   200   500 km

shows many fluctuations about a more general background curve (see Fig. 12.26). Inevitably, small-scale sea-level changes sometimes of only a meter or two, are not easy to discern. The erosional or sedimentary imprint from such fleeting stillstands will not only be faint and easily destroyed, but the sea-level "signal" is increasingly difficult to distinguish from the "noise" of other, nonglacial, effects. On some coastlines, for example, wave-cut notches in the splash zone may be mistaken for former sea-level indicators. It is also possible that changes in the shape of the geoid, as suggested by Mörner (1976), may mask small-scale eustatic sea-level changes.

15.10 PLEISTOCENE-HOLOCENE BOUNDARY

The term Holocene, meaning "entirely modern," is a time-stratigraphic term proposed by Gervais (1867–1869). It is synonymous with the term Recent, which is also widely used for the latest division of the Quaternary, in the sense that it denotes postglacial time. Some workers (e.g., Flint, 1971), arguing quite correctly that we are at present in an interglacial episode, see no reason for a "postglacial" division at all and would prefer to use the term Pleistocene for all post-Pliocene time. To define Holocene simply as postglacial time is obviously unwise, because the disappearance of the last major ice sheet occurred at different times in different places. On that basis, Greenland and Antarctica are still in the Pleistocene!

The selection of a Pleistocene-Holocene boundary is not easy because no obvious discontinuities in the geologic or climatic record present themselves. Pollen analyses from European sites, for example, show that since the disappearance of the last ice sheets, there was a progressive warming trend, culminating about 5,000 years B.P. This marked the peak of what has been variously termed the **hypsithermal, altithermal,** or **climatic opti-**

mum (Fig. 15.25). Since that time, there has been a slight cooling trend, with somewhat wetter conditions, down to the present day. Progressive changes in the floras provide the basis for a division of Holocene time into five phases, as shown in Figure 15.25. The placing of the Pleistocene-Holocene boundary at the beginning of the Preboreal, dated at about 10,000 years B.P., is generally favored by European workers. There is markedly less consensus elsewhere. As reported by Hopkins (1975), for example, there is little agreement among North American Quaternary specialists, and boundaries proposed have ranged in age from 18,000 to 4,000 years B.P. The problem remains unresolved, but there is a growing feeling that to arbitrarily select a date of 10,000 years B.P. might be the simplest solution.

15.11 CONCLUSION

It should be clear from what has been said in this chapter that Quaternary stratigraphy is often "different," and it should come as no surprise to learn that there are few workers who have made notable contributions in both Quaternary and pre-Quaternary geology. One of the reasons for this is that, traditionally, a major preoccupation of Quaternary geologists has been, and still is, with glacial phenomena. This, in turn, has meant that much more is known about very late Quaternary history than of events in earlier time, a consequence, naturally, of the fact that glacial activity, by its very nature, is largely destructive.

In making a direct comparison of Quaternary and pre-Quaternary successions, in terms of lithostratigraphic and biostratigraphic documentation, the problem is chiefly one of scale—Quaternary time is just too short for most conventional stratigraphic methodology. Within the larger context of earth history, it could be argued that the biggest contribution of Quaternary studies lies

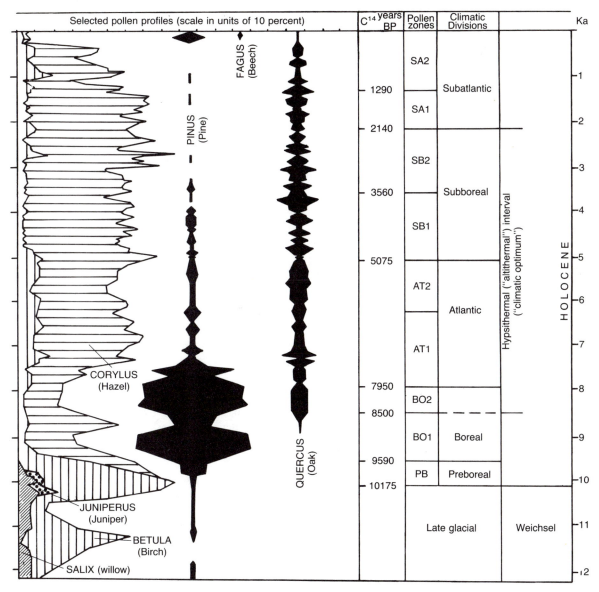

FIGURE 15.25

Pollen zones and climatic subdivisions of the Holocene in Europe, pollen diagram from Littleton Bog, County Tipperary, Ireland. (Redrawn from Mitchell, G. F., 1965, Littleton Bog, Tipperary: An Irish vegetational record: Geol. Soc. America Spec. Paper 84, p. 1– 16.)

in the study of climatic change in the recent geologic past. Some understanding of these changes and the mechanisms involved can be extrapolated into pre-Quaternary successions and, perhaps even more important, may well provide the vital clues in solving the problems of future climatic change induced by human activities on this planet. Continued tinkering with the environment without any in-depth understanding of the consequences may lead to unpleasant, even disastrous changes in the near future. The fact that human activities that already have led to widespread environmental deterioration continue, for the most part, unchecked indicates that there is as yet no real awareness of the changes that may lie ahead. It is in this context that Quaternary studies take on a new meaning. If the study of the past is to have any impact on an understanding of the future, it is uniquely through knowledge of the most recent geologic epoch that this understanding will come.

REFERENCES

Anderson, S. T., 1966, Interglacial vegetational succession and lake development in Denmark: Paleobotanist, v. 15, p. 117–127.

Beard, J. H., Sangree, J. B., and Smith, L. A., 1982, Quaternary, chronology, paleoclimate, depositional sequences, and eustatic cycles: Am. Assoc. Petroleum Geologists Bull., v. 66, p. 158–169.

Berggren, W. A., and Van Couvering, J. A., 1974, The late Neogene, biostratigraphy, geochronology and paleoclimatology of the last 15 million years in marine and continental sequences: Palaeogeog., Palaeoclim., Palaeoecol., v. 16, p. 1–216.

Boellstorff, J., 1978, A need for redefinition of North American Pleistocene stages: Gulf Coast Assoc. Geol. Socs. Trans., v. 28, p. 64–74.

Bowen, D. Q., 1978, Quaternary geology, a stratigraphic framework for multidisciplinary work: New York, Pergamon, 221 p.

Broecker, W. S., and Van Donk, J., 1970, Insolation changes, ice volume and the O^{18} record in deep-sea sediments: Rev. Geophysics and Space Physics, v. 8, p. 169–198.

Burns, D. A., 1972, Discoasters in Holocene sediments, southwest Pacific Ocean: Marine Geology, v. 12, p. 301–306.

Chamberlin, T. C., 1894, Glacial phenomena of North America, in Geikie, J., The Great Ice Age and its relation to the antiquity of man, 3rd ed.: Stanford, Stanford Univ. Press, 850 p.

Colalongo, M. L., Pasini, G., and Sartoni, S., 1980, Remarks on the Neogene/Quaternary boundary and the Vrica section: 26th Internat. Geol. Cong., Preprint Sec. S.08 (Paris), 40 p.

Coope, G. R., Morgan, A., and Osborne, P. J., 1971, Fossil Coleoptera as indicators of climatic fluctuations during the last glaciation in Britain: Palaeogeog., Palaeoclim., Palaeoecol., v. 10, p. 87–101.

Denton, G. H., and Hughes, T., 1983, Milankovitch theory of ice age: Hypothesis of ice-sheet linkage between regional insolation and global climates: Quaternary Research, v. 20, p. 125–144.

Eberl, B., 1930, Die eiszeitenfolge im nordlichen Alpenvorlande: Augsberg, Benno Filzer, 427 p.

Emiliani, C., 1955, Pleistocene temperatures: Jour. Geology, v. 63, p. 538–578.

———— 1978, The cause of the ice ages: Earth and Planetary Sci. Letters, v. 37, p. 349–352.

Ericson, D. B., and Wollin, G., 1956b, Micropaleontological and isotopic determinations of Pleistocene climates: Micropaleontology, v. 2, p. 257–270.

———— 1964, The deep and the past: New York, Knopf, 301 p.

———— 1968, Pleistocene climates and chronology in deep sea sediments: Science, v. 162, p. 1227–1234.

Fairbridge, R. W., 1961, Eustatic changes in sea level, in Ahrens, L. H., Press, F., Rankama, K., and Runcorn, S. K., eds., Physics and chemistry of the earth, v. 4, New York, Pergamon, p. 99–185.

Fillon, R. H., 1984, Continental glacial stratigraphy, marine evidence of glaciation, and insights into continental-marine correlations, in Healy-Williams, N., ed., Principles of Pleistocene stratigraphy applied to the Gulf of Mexico: Boston, Internat. Human Resources Devel. Corp., p. 149–211.

Flint, R. F., 1971, Glacial and Quaternary geology: New York, John Wiley & Sons, 892 p.

Gervais, P., 1867–1869, Zoologie et paléontologie générales: Nouvelle recherches sur les animaux vertébrates vivants et fossiles: Paris, 263 p.

Hays, J. D., Imbrie, J., and Shackleton, N. J., 1976, Variations in the earth's orbit: Pacemaker of the ice ages: Science, v. 194, p. 1121–1132.

Hopkins, D. J., 1975, Time-stratigraphic nomenclature for the Holocene Epoch: Geology, v. 3, p. 10.

Imbrie, J., and 8 others, 1985, The orbital theory of Pleistocene climate: Support from a revised chronology of the marine ^{818}O record, *in* Berger, A. L. et al., eds., Milankovitch and climate: Boston, MA, D. Reidel, p. 269–305.

Iverson, J., 1958, The bearing of glacial and interglacial epochs on the formation and extinction of plant taxa; *in* Hedberg, O., ed., Systematics of today: Acta Univ. Upsaliensis, p. 210–215.

Keihlack, K., 1926, Das Quarter, *in*: Grundzuge der Geologie, v. 2, Ed Salmen.

Kennett, J. P., and Huddleston, P., 1972, Late Pleistocene paleoclimatology, foraminiferal biostratigraphy and tephrochronology, western Gulf of Mexico: Quaternary Research, v. 2, p. 38–69.

Kominz, M. A., Heath, G. R., Ku, T. L., and Pisias, N. G., 1979, Brunhes time scales and the interpretation of climatic change: Earth and Planetary Sci. Letters, v. 45, p. 394–410.

Ledbetter, M. T., 1984, Late Pleistocene tephrochronology in the Gulf of Mexico, *in* Healy-Williams, N., ed., Principles of Pleistocene stratigraphy applied to the Gulf of Mexico: Boston, Internat. Human Resources Devel. Corp., p. 119–148.

Leverett, F., 1910, Comparison of the North American and European glacial deposits: Zeitschr. Gletscherk., v. 4, p. 241–316.

Martin, P. S., 1973, The discovery of America: Science, v. 179, p. 969–974.

Mörner, N. A., 1976, Eustacy and geoid changes: Jour. Geology, v. 84, p. 123–151.

Nilsson, T., 1983, The Pleistocene—Geology and life in the Quaternary ice age: Hingham, MA, Reidel, 651 p.

Obradovich, J. D., Naeser, C. W., Izett, G. A., Pasini, G., and Bigazzi, G., 1982, Age constraints on the proposed Plio-Pleistocene boundary stratotype at Vrica, Italy: Nature, v. 298, p. 55–59.

Penck, A., 1894, Die morphologie de Erdoberfache, 1-2 Stuttgart.

Penck, A., and Bruckner, E., 1901–1909, Die alpen im Eiszeitalter, 1-3 Tauchnitz, p. 83–97.

Ruddiman, W. F., and Wright, H. E., Jr., eds., 1987, North America and adjacent oceans during the last deglaciation, *in* The geology of North America, v. K-3, Geol. Soc. America, Chap. 1, Introduction, p. 1–12.

Shackleton, N. J., 1967, Oxygen isotope analyses and Pleistocene temperatures re-assessed: Nature, v. 215, p. 15–17.

———— 1975, The stratigraphic record of deep-sea cores and its implication for the assessment of glacials, interglacials, stadials and interstadials in the mid-Pleistocene, *in* Butzer, K. W., and Isaacs, G. L., eds., After the australopithecines: The Hague, Mouton, p. 1–24.

Thunell, R. C., 1984, Pleistocene planktonic foraminiferal biostratigraphy and paleoclimatology of the Gulf of Mexico, *in* Healy-Williams, N., ed., Principles of Pleistocene stratigraphy applied to the Gulf of Mexico: Boston, Internat. Human Resources Devel. Corp., p. 25–64.

Turner, C., and West, R. G., 1968, The subdivision and zonation of interglacial periods in northwest Europe: Eiszeitalter und Gegenw., v. 19, p. 93–101.

Urey, H. C., 1947, The thermodynamic properties of isotopic substances: Jour. Chem. Soc., v. 1947, p. 562–581.

Williams, D. F., 1984, Correlation of Pleistocene marine sediments, of the Gulf of Mexico and other basins using oxygen isotope stratigraphy, *in* Principles of Pleistocene stratigraphy applied to the Gulf of Mexico: Boston, Internat. Human Resources Devel. Corp., p. 65–118.

16
PRECAMBRIAN STRATIGRAPHY

In the deep shadow below the Cambrian . . .

John McPhee

16.1 INTRODUCTION

In this chapter, Precambrian events and earth history as such will not be described in detail; rather, the emphasis will be on those aspects of Precambrian research that throw light on age relationships and on what has been the rather neglected field of Precambrian stratigraphy. In viewing the subject of this book, and indeed the literature of stratigraphy as a whole, it is rather discouraging to realize that a great deal of it applies only to the Phanerozoic, less than one-seventh of earth history. Much of our so-called conventional stratigraphy does not work very well in the Precambrian, the first and most obvious reason being that a fossil record that is really usable in stratigraphy does not begin until the start of the Cambrian. Another reason is that because of the extreme age of many of the rocks, especially in the Archean, successive episodes of metamorphism have left little of the original rock textures and minerals intact so that deciphering the rocks in terms of their original mode of formation is often difficult or impossible. A third factor to be considered is that plate tectonic and crustal formational mechanisms were different during the early Precambrian. The composition of the atmosphere was also somewhat different, and there is no doubt that many geologic processes we have come to build into what is still an essentially uniformitarian view of geologic history simply did not operate, or operated in a different way, particularly in the earlier Precambrian. These differences are reflected, for example, in the occurrence of certain rock types that are unique to the Precambrian and in the complete absence of other rocks that are common in later time.

Beyond all these reasons is the possibility that for many stratigraphers, whether they admit it or not, there exists a sort of mystique about the Precambrian. This is partly due to the traditional division that has existed, and still does exist to a considerable extent, between "soft rock" and "hard rock" geologists. It is also due to the fact that a very large number of stratigraphers are engaged directly or indirectly in exploration for hydrocarbons, and **basement** (frequently, although obviously not always, meaning the Precambrian) marks the limit of their interest.

16.2 ARCHEAN AND PROTEROZOIC DIVISIONS

Precambrian rocks can be studied best in those extensive areas where they are exposed at the surface, the shields. The major shield areas of the world are shown in Figure 16.1. It was on the Canadian Shield (sometimes called the Laurentian Shield), particularly in the area adjacent to the Great Lakes and the St. Lawrence River, that some of the very important early work on the Precambrian was carried out by geologists of the Geological Survey of Canada during the latter part of the nineteenth century. In mapping and deciphering the stratigraphy of these rocks, various lithostratigraphic units were recognized on the basis of their relative positions and cross-cutting relationships, and from the very earliest studies a broad twofold division into older, Archean, and younger, Proterozoic, ter-

ranes was recognized. One of the important criteria was considered to be the degree of metamorphism; the higher the metamorphic grade, the older the rock. As it turned out, data from isotopic dating in later years contained many surprises. Some of the highly altered rocks were found to be comparatively young, whereas many unaltered "young-looking" rocks proved to be unexpectedly old.

Within individual structural regions, often bounded by major faults, it was possible to work out the local stratigraphy in some detail on the basis of recognition of metasedimentary sequences, several episodes of granitic intrusion, and the presence of pronounced unconformities overlain by distinctive basal conglomerates. From these local studies of the Canadian Shield and elsewhere, a terminology evolved in which certain names came to be recognized as descriptive of particular lithostratigraphic entities, with characteristic lithologies and having certain field relationships to other named units.

Archean Rocks

Archean rocks have all been metamorphosed to a greater or lesser degree and can be broadly divided into those of high metamorphic rank, the granulites and amphibolite gneisses (Fig. 16.2), and those low-grade rocks collectively known as greenstones. The **greenstones** consist of metamorphosed volcanics and sediments and occur in elongated tracts or belts up to several hundred kilometers long. Typically, they are surrounded, underlain, and intruded by granites and granitic gneisses and have been described as "floating in a sea of granite" (Fig. 16.3). There is some divergence of opinion as to the tectonic setting in which greenstone belts were formed, but one widely held view is that they represent the accumulation of volcanic and sedimentary rocks along the margins of small, relatively thin "protocratons." The marked differentiation between

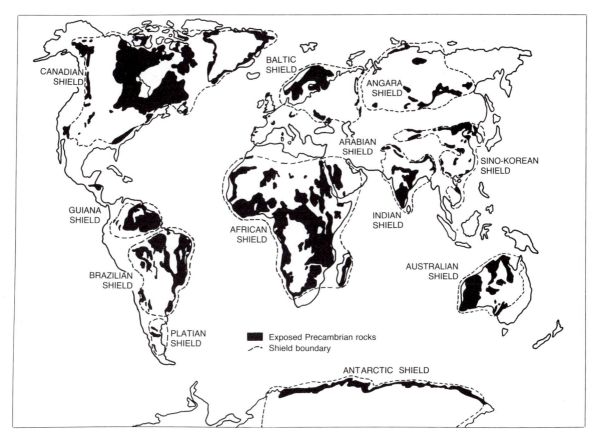

FIGURE 16.1
Precambrian shield areas of the world.

thick continental crust and thin oceanic crust, characteristic of the Phanerozoic, did not exist at this time, so, although apparently there were extensive seas, no ocean basins as such, yet existed. That these seas were, at least in places, quite deep is suggested by the metasediments within greenstone successions; graywackes and mudstones associated with pillow lavas have features interpreted as indicating at least moderately deep water.

Within virtually all greenstone belts, there seems to be a twofold division into a lower volcanic suite and an upper metasedimentary suite (Fig. 16.4), separated by an unconformity. During the early investigations of the

"classic" areas of Ontario, these two divisions were recognized as lithostratigraphic units, with the names Keewatin and Timiskaming, respectively. Although these two names came to be used widely, the implied correlation from one greenstone belt to another could not be further substantiated and it was later found from isotopic dating that there is considerable age variation in the greenstone belts, ranging from 3.4 to 2.3 b.y.

In other greenstone belts of the world, the same twofold division can be recognized, except that in some places on the African shields, a third underlying sequence of ultramafic rocks is present. In South Africa, ultra-

FIGURE 16.2
Archean gneisses, Haviland
Lake, southwestern Colorado.
(Photograph courtesy of D. S.
Robertson.)

mafic and mafic lavas found in the lower part of the Onverwacht Group (the Sebakwian of Zimbabwe) are overlain by the argillaceous Fig Tree Group and in turn by the quartzose Moodies Group (Fig. 16.5). As in the Canadian shield, various names were applied to these rock types, and time correlations from one greenstone belt to another were implied. There is evidence that even in one region there may be two or three sequences re-

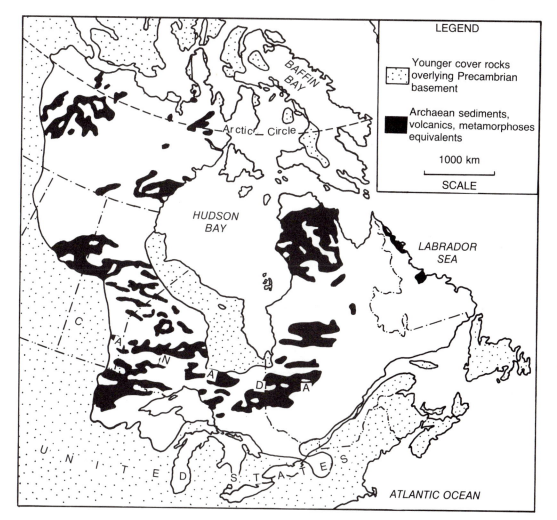

FIGURE 16.3
Greenstone areas of the Canadian Shield.

peated, perhaps reflecting, in Wright's (1985) view, the operation of some kind of "greenstone tectonic cycle." It follows that no direct lithostratigraphic correlation of greenstone belt sequences from one region to another is possible. On the other hand, despite local differences, there is no question, as Anhauesser et al. (1969) pointed out, that there is a worldwide uniformity in the general lithologic, stratigraphic, and structural features of

greenstone belts. They probably formed over an extended time, but they clearly mark an important stage in the evolution of the earth's crust and, as such, have some chronostratigraphic significance.

One of the most striking features of Archean rocks as a whole is that shallow-marine sediments, such as those laid down on modern continental platforms and shelves, are very rare. This strongly suggests that no

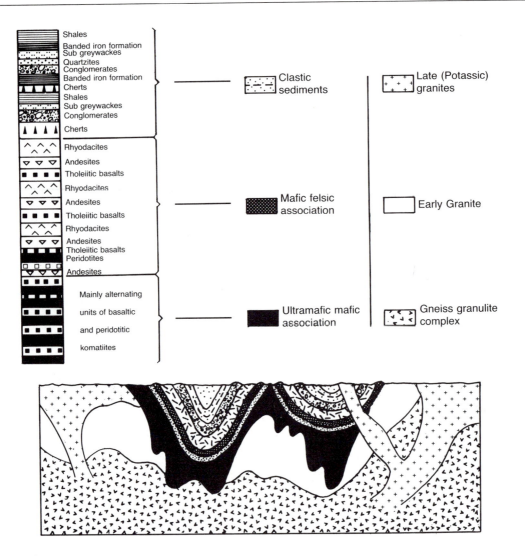

FIGURE 16.4
Typical stratigraphic succession for an Archean greenstone terrane, based on the Barberton model of southern Africa. (After Anhauesser, C. R., et al. 1971, The Barberton Mountain Land: A model of the elements and evolution of an Archaean fold belt: Trans. Geol. Soc. South Africa, Annex. to v. 7, p. 225–254.)

large cratons yet existed. With progressive cooling, convectional recycling in the upper mantle slowed. This caused a thickened and increasingly differentiated crust to gradually form in certain areas. This was an important stage in earth history; it marked the end of early Archean crustal mobility and the beginning of true cratonic successions. It did not, however, occur everywhere at the same time, the earliest occurrence of a true craton being in southeast Africa some 3.0 b.y. ago. The time of appearance of later cratons was

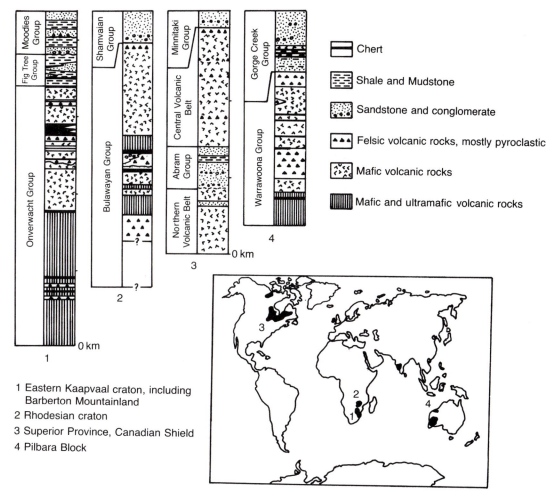

FIGURE 16.5
Stratigraphic successions in typical greenstone belts. (After Lowe, 1980.)

spread over the next 500 m.y. The evidence for this earliest of continental cratons is seen in the great thickness of rocks known as the Witwatersrand Supergroup. The succession contains quartzitic sandstones and siltstones that were clearly laid down in the shallow seas of an extensive continental shelf.

By about 2.5 b.y. ago, it is thought, by some workers at least, that 75 percent or more of the sialic crust was in existence, and this could have resulted only from a fundamental change in the character of crustal processes and a period of relatively rapid continental crustal growth. Opinions vary somewhat as to the age of this very important boundary at the end of the Archean, but the spread of dates is not great and most range from about 2.6 to 2.4 b.y.

Proterozoic Rocks

Proterozoic sequences typically consist of sediments and volcanics laid down on eroded surfaces of Archean gneisses, granites, and greenstones. Compared with Archean sections, Proterozoic rocks consist in many places of compositionally and texturally mature clastic sedimentary suites (Fig. 16.6). Proterozoic sandstones are often almost entirely quartzose, with good sorting and well-rounded grains, all features suggestive of polycyclic reworking under conditions that are reminiscent of the stable shelf environments of Phanerozoic time. These contrasts with Archean environments are further enhanced by the first appearance of limestones and **dolomites.** In the interiors of the Archean blocks the Proterozoic cover rocks are often thin and are usually relatively unaltered, but as they pass laterally into mobile belts, they are seen to increase greatly in thickness. In many places, the thickening and facies changes indicate deposition on relatively shallow platforms marginal to continents. The Huronian and Animikean sediments and volcanics of the Southern Province of the Canadian Shield, for example, thicken markedly to the southeast as a dominantly clastic wedge, similar to the great geosynclinal prisms of Phanerozoic continental margins. Another example of a Proterozoic sequence is seen in the Coronation geosyncline in the northwest part of the Canadian Shield (Hoffman, 1973; Hoffman et al., 1974). Here facies changes and sedimentary associations also are reminiscent of those of Phanerozoic continental plate margins.

Although some Proterozoic mobile belts may have been marginal to continents and thus presumably similar to the orogenic belts of latest Precambrian and Phanerozoic time, others likely had a different setting, and were, in fact, intracratonic. According to Wynne-Edwards (1969), regional stress patterns point to some mechanism of flow folding associated with crustal sagging rather than to a marginal fold-belt setting. A smaller number of marginal fold belts at this time is also supported by evidence that precludes widespread rifting and collision of cratonic

blocks. The close matching of APW curves from several shield areas, and the obviously close match of earlier Precambrian structural traces, dike swarms, and so on, all suggest that the earliest cratons were large and were stable for a long time period. In Piper's (1974, 1976a, 1976b) interpretation, virtually all the continental masses were part of a Proterozoic supercontinent (Fig. 16.7).

16.3 SUBDIVISION OF THE PRECAMBRIAN

One of the clearest indications of the special nature of the Precambrian as compared with the Phanerozoic is seen in the problems associated with stratigraphic classification. Precambrian history covers such immense time spans that in discussing the separate parts of that history, reference is made as much to events as to the rocks themselves. It follows from this that the conventional classifications described earlier in this book have little contribution to make in dealing with Precambrian stratigraphy, and many workers believe that a different approach is called for. Wright (1985), for example, suggested that the orogenic episodes that have proved successful in dividing the Precambrian history of the Canadian Shield and elsewhere be recognized as **chronotectonic cycles,** defined as sequences of sedimentary, metamorphic, intrusive, and tectonic events. As the fundamental unit, a chronotectonic cycle may be subdivided within a hierarchy as listed in Table 16.1. Each of these divisions must be appropriately identified as to its nature; for example, an *episode* may describe a period of sedimentation, metamorphism, folding, and so on. Chronotectonic cycles encompass orogenies that clearly have a beginning and an end so they are, in effect, time units measured in years.

A special classification for dealing with rocks also has been proposed, using so-called **lithotectonic** divisions. Such **divisions** refer

FIGURE 16.6
A. Proterozoic quartzitic sandstones of the Togo Series, Ghana. These were deposited as mature sands in a stable shelf environment. B. Cross-bedded quartzitic sandstones of the Karibib Formation, Damara sequence, Late Proterozoic (Namibian) age, Namibia.

A

B

to lithologic entities that are mappable, but which cannot be defined in terms of the regular lithostratigraphic units such as formations, groups, and so on. As with lithostratigraphic terms, no time connotation is implied, but essential in the definition of such divisions is some degree of tectonic unity. For the largest of such divisions, the term **assemblage** has been proposed and further subdivisions are shown in Table 16.1.

Defined Archean crust

Known or probable
extent of Archean

FIGURE 16.7
Configuration of Proterozoic supercontinents. The shaded areas show the known or
probable extent of Archean terranes. (Redrawn after Piper, 1976b.)

TABLE 16.1

Suggested stratigraphic classification method for the Precambrian (after Wright, 1985)

Chronotectonic Divisions
Cycle
Episode
Phase
Event

Lithotectonic Divisions
Assemblage
Division
Unit

There is, as yet, no internationally accepted scheme for time classification of the Precambrian. The only feasible approach up to the present has been to use chronometric divisions based on isotopically derived dates. Table 16.2 shows one proposed scale; of these divisions only the Archean-Proterozoic boundary, set at 2500 Ma, has been agreed upon. This was the recommendation of the Subcommission on Precambrian Stratigraphy of the International Union of Geological Sciences (IUGS) in 1977. There also is quite wide agreement on a date for the beginning of the Late Proterozoic (Proterozoic III) set at 900 Ma (although Soviet geologists would prefer a date closer to 1000 Ma). These numbers have not been selected arbitrarily and are believed to have some relevance to recognized events in earth history, although there is certainly no consensus on what they are. The Archean-Proterozoic boundary, for example, is supposed to mark an important change in the process of crustal evolution, with the first appearance of major linear orogenic belts and the consolidation of extensive continental cratons. On the other hand, as discussed earlier, there is evidence that the formation of the first true craton actually occurred in southeast Africa some 500 m.y. earlier. It is possible that paleontologic data may eventually give some meaning to these broad divisions, and it is thought that at least three or four major steps in evolutionary progress will eventually be discernible. This topic is discussed later in the chapter.

16.4 TIME DIVISION BY OROGENIES

The granitic terranes of the Archean are much more extensive than the greenstone belts, and from the evidence of cross-cutting field relationships and granite pebbles in basal conglomerates, it was apparent to even the earliest workers that there had been several major episodes of granite formation. In

TABLE 16.2

Chronostratigraphic scale used in the Phanerozoic Eon compared with the chronometric scale used in subdividing the Precambrian

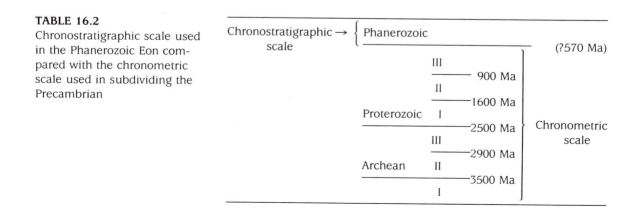

the Lake Superior region of the Canadian Shield, for example, an older granite of post-Keewatin and pre-Timiskaming age was named the Saganagan granite (in other areas the Kenoran or Laurentian granite). A younger post-Timiskaming granite was known as the Algoman granite (Fig. 16.8). In other shield areas also, several periods of granite formation were early noted. All of them, apparently, postdate the greenstones, because nowhere has a pregreenstone basement been determined with any certainty, although this is a matter of some controversy. On the other hand, the presence of granitic pebbles in some conglomerates of greenstone successions indicates that older granites existed somewhere at the time the greenstones were laid down. In the Barberton Mountain region of southeast Africa, three granitic episodes have been described; each one is distinctive in its chemical features and in its structural relationships to the greenstones and other granites. Each of the different episodes of granitic intrusion is considered indicative of a major orogenic phase. An **orogeny** is defined as an episode of mountain building with folding and thrusting of rocks, metamorphism, and usually the intrusion of large masses of granite. Granite batholiths occupy the cores of the mountains and their eroded "roots" are often all that is left of ancient mountain chains. The dating of orogenic events is based upon various criteria. Relative ages can be discerned in angular unconformities and the truncation of regional structures by younger structures. Ages in real time can be determined by isotopic dating methods, although there is typically a considerable spread in the values obtained because the folding and metamorphism that started one set of radiometric clocks may not have been contemporaneous with the granitic intrusions that started others. Because any orogeny obviously had a beginning and an end, it can be said to have some time significance. In the relatively "short" time span of the Phanero-

zoic, orogenies have little direct contribution to make to a time-stratigraphic classification, but within the enormous sweep of Precambrian time, they become significant markers. Each of the several Precambrian orogenies documented on any of the major shields was followed by a long interval of relative quiescence. Such an interval was marked by deposition of sedimentary sequences that lie unconformably on the eroded remnants of the fold mountains formed by the preceding orogeny. The maximum age of the sediments is obviously determined by dating that orogeny. The minimum age is determined by dating the next orogeny that eventually involved the sedimentary sequence, either by folding and metamorphosing it or intruding it with batholiths and dikes, events all of which would reset or start more radiometric clocks. The several Precambrian orogenic episodes provide the means for a natural time-stratigraphic division of the sedimentary and volcanic successions; on the Canadian Shield, for example, four dated orogenies provide the basis for five chronostratigraphic divisions, as shown in Table 16.3. An orogenic cycle not only has a certain range through time, but is seen also as affecting the rocks within a given area, known as a tectonic, or structural, province. Within each province, there is a certain homogeneity in major structural trends that are related to the orogenic cycle, and a significant clustering of isotopic dates derived from the metamorphic and intrusive events associated with the orogeny. The tectonic provinces of North America are shown in Figure 16.9.

Orogenic Cycles

Orogenic cycles, or diastrophic epochs as they are sometimes called, have now been fairly well documented on a worldwide scale; Salop (1983) has recognized no less than 15 Precambrian orogenic cycles, of which five are considered first-order orogenies, as

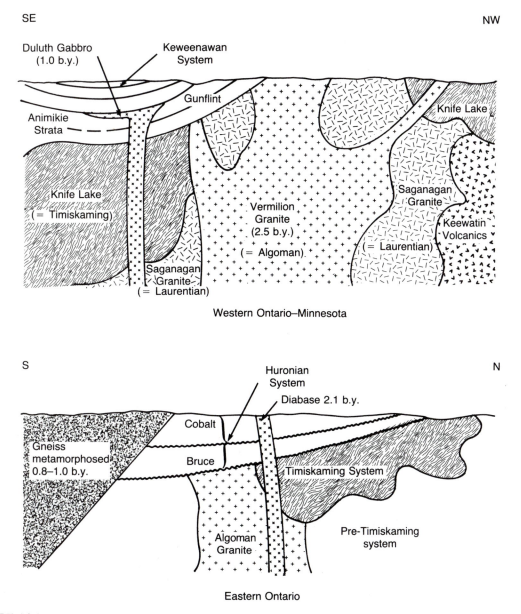

FIGURE 16.8
Field relationships of Archean granites and greenstones, together with Proterozoic successions in the classic areas of the Great Lakes in Ontario and Minnesota. The various names and "systems" were established largely in the 1850s and 1860s by geologists of the Geological Survey of Canada, such as William Logan, Andrew Lawson, Sterry Hunt, and William Dawson.

TABLE 16.3

Time-stratigraphic classification of the Precambrian of the Canadian Shield

Eon	Era	Sub-Era	Orogeny (mean K-Ar mica age, Ma)
Proterozoic	Hadrynian		
Proterozoic	Helikian	Neohelikian	Grenvillian (955)
Proterozoic	Helikian	Paleohelikian	Elsonian (1370)
Proterozoic	Aphebian		Hudsonian (1735)
Archean			Kenoran (2480)

shown in Figure 16.10. As in the Canadian Shield, the orogenies provide the basis for a subdivision into time units of era and subera rank. According to Salop (1983), the orogenic cycles were apparently global in scale, although the mode and intensity must vary from place to place. The time of climax might also vary, and the peak of tectonic activity within a given cycle was not synchronous throughout the regions affected, although it did fall within a limited range of time. As can be seen from Figure 16.10, the orogenies apparently occurred at fairly regular intervals of approximately 200–250 m.y., and this has led to much discussion as to causes. Some workers have suggested, for example, that the duration of the cycles coincided with the galactic year, at least as calculated by some authorities, and that there is some kind of cosmic control at work. The virtual impossibility of ever proving such cause-and-effect relationships inevitably means that topics of this nature are a rich field for speculation, as was described in Chapter 4.

Chelogenic Cycles

As radiometric dates from Precambrian rocks in all parts of the world became increasingly numerous, it was soon apparent that they were not randomly and uniformly distributed through time. Instead, and not surprisingly, they seemed to cluster around certain values related to the timing of the orogenic cycles discussed in the previous section. Dearnley's (1965) study, for example, of approximately 3400 determinations, used graphical plots to demonstrate significant peaks in values at 2,750 m.y., 1,950 m.y. and 1,075 m.y. (Fig. 16.11) and pointed up the particular significance of these times as marking periods of unusual crustal activity. These are the first-order diastrophic cycles of Salop (1983). According to Sutton (1963), these peaks indicate the periodicity of successive long-term cycles that he termed **chelogenic** (literally, "shield-forming") **cycles** and of which he recognized three, as shown in Figure 16.11. According to Sutton (1963), the structural provinces of the world's major shield areas formed during these cycles and each of them experienced a series of four or five smaller orogenic cycles of approximately 200 m.y. duration. These are presumably the second- and third-order cycles recognized by Salop (1983). The most recent of the chelogenic cycles is the Grenville, which, still not com-

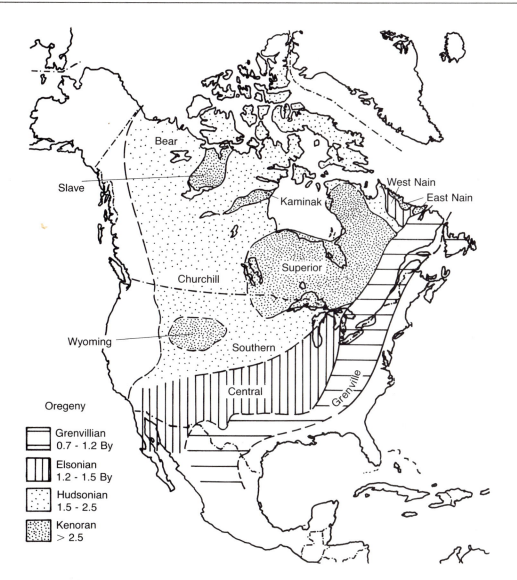

FIGURE 16.9
Tectonic provinces of the Canadian Shield.

pleted, embraces within it the various orogenies of the Phanerozoic.

Superintervals

A division of Precambrian time based on paleomagnetic data was suggested by Irving
and Park (1972), who noted that the path of

the APW curves for a particular continent
during Precambrian time was typically
marked by periodic sharp turns. These bends
they called **hairpins,** and the path of the pole
between the hairpins, they termed tracks.
This topic was mentioned earlier in Chapter
6. Clearly, major changes in the direction of

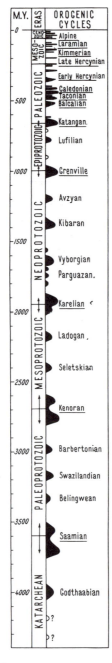

FIGURE 16.10
Diastrophic cycles recognized by Salop (1983, Geological evolution of the earth during the Precambrian, Fig. 77, p. 395, Springer-Verlag, Berlin.)

movement of the lithospheric plates relative to the pole must have occurred at the times marked by hairpins. It was suggested that each sharp bend corresponded to a time of major orogeny, and this is supported by a modest agreement in the timing. The tracks between the hairpins were numbered 1 through 5 consecutively back from the present, and the period of time involved in each case was designated a superinterval, as shown in Figure 16.12.

16.5 BANDED IRON FORMATIONS

Of the two types of **banded iron formation (BIF)** described by Gross (1980), it is the so-called Superior type that is of particular interest in the context of Precambrian time division because of the limited time span during which such deposits occur, ranging from latest Archean to early Proterozoic. Superior-type BIFs have been found on nearly all the Precambrian shields and, according to Ronov (1968), in the early Proterozoic they made up 15 percent of the total sediment thickness. They are not found in any succession older than about 2.5 b.y. or in rocks younger than 1.8 b.y. Their disappearance was apparently quite abrupt, and even, it is claimed by some authorities (Cloud, 1973), instantaneous on a global scale. The internal features of the ironstones have been interpreted as indicating a very uniform depositional environment over extensive areas. The laminations that are such a prominent feature of the formations are often remarkably regular (Fig. 16.13), and distinctive alternations within the sediments can sometimes be correlated over many hundreds of kilometers. Trendall (1968), for example, described microscopic varvelike layers within chert bands which can be traced over a distance of nearly 300 km. Although certain chemical models used to explain the depositional mechanism of BIFs have suggested that they originated in freshwater

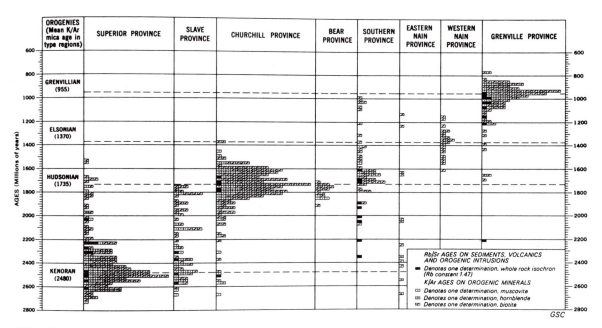

FIGURE 16.11

Histograms based on over 3200 K-Ar age determinations for metamorphic and granitic rocks from six tectonic provinces in the Canadian Shield. (From Stockwell, C. H., 1968, Fourth report on structural provinces, orogenies and time-classification of rocks of the Canadian Precambrian Shield: Geol. Survey Canada Paper 64–17.)

lakes, it is difficult to see anything other than a marine environment as the setting for their wide extent and uniformity of facies. It is believed that the typical Superior-type BIF was deposited in shallow sea water, most likely on continental shelves, or possibly in a lagoonal environment or in intracratonic epeiric seas (Trendall, 1973).

The chemical environment for the deposition of BIFs has long been a topic of lively discussion. Until relatively recently, the generally held view was that BIFs were linked with certain stages reached during the evolution of the atmosphere; more particularly, with the levels of free oxygen present. In the model proposed by Cloud (1972, 1973, 1976), it was assumed that the earth's early atmosphere was essentially devoid of free oxygen, so that iron would presumably remain in a

ferrous state in solution and be widely disseminated in the hydrosphere. Precipitation would have been possible only given a supply of oxygen, and this, it was proposed, was provided by photosynthesizing organisms that floated in shallow-shelf and platform seas. The temporal range of BIFs from late Archean to early Proterozoic was explained in terms of biotic evolution. Their beginning marked the time of first appearance, or at least abundance, of photoautotrophic organisms—cyanobacteria or their progenitors capable of manufacturing their own food by using sunlight in a process of photosynthesis. Because a byproduct of the process was oxygen, and this was toxic to such anaerobic organisms, an oxygen mediator was needed. In the absence of any organic acceptor, the role was apparently filled by ferrous iron, present

FIGURE 16.12

Apparent polar wandering curve for North America (A), showing the tracks and hairpins used by Irving and Park (1972) as the basis for the divisions of Precambrian time shown in the chart (B). (From Irving, E., and Park, J. K., Hairpins and superintervals: Can. Jour. Earth Sci., v. 9, Fig. 3, p. 1320, and Fig. 4, p. 1322.)

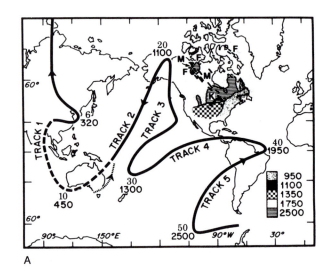

A

Isotopic Ages (Ma)	Orogenies (Canada)	Eon	Era		Hairpins MA	Superinterval
(Precambrian only)	Laramide	Phanerozoic	Cenozoic		⌐ 2	First
	Columbian		Mesozoic		90	
	Appalachian				⌐ 6	
	Acadian		Paleozoic		320	
500	Taconic				⌐ 10 450	
		Proterozoic	Hadrynian			Second
1000	Grenvillian		Neo	Helikean	⌐ 20 1100	Third
	Elsonian				⌐ 30 1300	
1500			Paleo			Fourth
	Hudsonian					
2000			Aphebian		⌐ 40 1900	Fifth
2500	Kenoran				⌐ 50 2500	
		Archean				?

Histogram of isotopic ages (Shaded-Canada. Black-USA)

50 100

B

458

FIGURE 16.13
Folded, interbedded iron formation and metasedimentary rocks of Precambrian age;
Beresford Lake area, Manitoba. (Photo courtesy Geological Survey of Canada, Ottawa.)

in solution in the water, and then precipitated as ferric iron. The final disappearance of BIFs was, according to this model, also of evolutionary significance and was explained as marking the appearance of new, more specialized cyanobacteria with oxygen-mediating enzymes and capable of handling oxygen without the need of an external, nonbiological oxygen sink. It is significant that from about this time on redbeds become a significant feature of nonmarine successions. Their appearance at about the same time as the disappearance of BIFs suggests that oxygen concentrations in the atmosphere had reached a certain threshold and that this was linked with the evolving biota.

It was suggested that the stages in the evolution of algae, as described in the Cloud (1976) model, might be used to subdivide post-Archean time. Acknowledging that only primitive, single-celled life existed for most of this time, three divisions, the Proterophytic, Paleophytic, and Proterozoic, were proposed, the Proterozoic, logically enough, being confined to the latest Precambrian when metazoans had actually appeared (Table 16.4). More recent work has raised doubts as to the validity of the Cloud model, and there is some evidence that the atmosphere of the early earth did, in fact, contain considerable free oxygen. Towe (1983), for example, cited evidence for free atmospheric oxygen in

TABLE 16.4

Subdivision of the Precambrian on the basis of evolving life forms

	Proposed Division	Ga
PHANEROZOIC	Paleozoic (shelly metazoans)	Cambrian
		—0.57
	Proterozoic (soft-bodied metazoans)	Ediacaran
		—0.7
CRYPTOZOIC	Paleophytic (with eucaryotes)	
		—2.0
	Proterophytic (prokaryotes only)	
		—3.4

rocks as old as 3.8 b.y. and suggested that the iron of BIFs may have been derived by upwelling from reservoirs in deep, stagnant ocean basins. For further discussion of various models of BIF origins, see, for example, Kimberly and Dimroth (1976), Holland (1978), Clemmey and Badham (1982), and Towe (1983).

16.6 PRECAMBRIAN GLACIATIONS

In the search for possible marker horizons within Precambrian successions considerable attention has been paid to glacial episodes. The evidence for ice ages in the Precambrian is widespread, and it is obvious that if properly recognized and correlated, they would provide a valuable means of subdividing Precambrian history. The oldest evidence of glacial conditions is seen in the early Proterozoic Gowganda formation, a tillite (Fig. 16.14) in the Huronian of Ontario. Deposits of similar age have been reported in northern Quebec, northern Michigan, and Wyoming. The Gowganda lies unconformably on Archean rocks dated at 2.5 b.y. and is intruded by diabases dated at 2.1 b.y. Paleomagnetic data from the Gowganda (Symons, 1967) suggest a high-latitude location at the time of deposition. This early Proterozoic glacial event has obvious potential as a Precambrian chronostratigraphic indicator, and it is possible that certain tillitelike conglomerates in southwestern Greenland and in Witwatersrand, South Africa, may be correlatable.

In the Late Proterozoic, evidence for an ice age of major magnitude is so widespread as to be literally of global extent. Indeed some authorities claim that the Late Proterozoic glaciation may have been by far the most severe in earth history (in Australia, for example, lasting over 100 m.y.) and apparently extended to the equator (Table 16.5). A

FIGURE 16.14

Gowganda conglomerate, Cobalt, Ontario. The disrupted framework of this polymictic conglomerate generally is interpreted as indicating a glacial origin. (Photo courtesy Geological Survey of Canada, Ottawa.)

TABLE 16.5
Age relationships of the Proterozoic glaciations of Australia

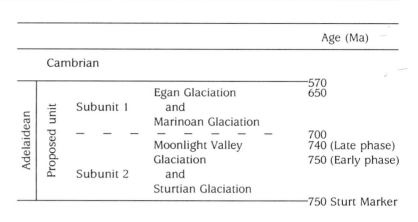

				Age (Ma)
Cambrian				
Adelaidean	Proposed unit	Subunit 1	Egan Glaciation and Marinoan Glaciation	570 650
		Subunit 2	Moonlight Valley Glaciation and Sturtian Glaciation	700 740 (Late phase) 750 (Early phase)
				750 Sturt Marker

puzzling feature of some glacial successions is the association with limestones and dolomites, rocks usually considered indicative of warm-water conditions. An explanation for this was offered by Walter and Bauld (1983), who described carbonate sediments from cold, arid-climate regions, such as certain Antarctic saline lakes. The potential value of Late Proterozoic glaciations as chronostratigraphic markers is a matter of some debate. If, as some authorities maintain, they were spread over a considerable time, the global extent of the tillites might prove to be diachronous and explained in terms of the APW path. This view is supported by McElhinny et al. (1974), whose paleomagnetic data demonstrated large shifts of the apparent pole position during the late Proterozoic. On the other hand, the evidence for low-latitude glacial deposits is also quite convincing. If glacial conditions did, indeed, extend into low latitudes, it must be assumed that Late Proterozoic climatic deterioration was far more extreme than that experienced during the Pleistocene. It would also mean, as Piper (1973) pointed out, that the value of glacial deposits as chronostratigraphic markers would be greatly enhanced.

16.7 PRECAMBRIAN FOSSILS

The earliest definitive evidence of life is seen in simple microscopic forms in rocks of the Onverwacht Group in South Africa and the Warrawoona Group of Western Australia; these rocks range in age from 3.4 to 3.5 b.y. Not surprisingly, the fossil record from these remote times provides only very occasional glimpses of the progress of evolution from the Archean to the Late Proterozoic. It is mainly from microspheres, tiny rods, filaments, and other enigmatic forms present in cherts (Fig. 16.15), such as the 3-b.y.-old Fig Tree Formation of South Africa and the 2-b.y.-old Gunflint Formation of Ontario and Minnesota that some account of the early progress of life has been assembled. Although likely patterns of early evolution have been hypothesized, as described, for example, by Knoll (1985), fossils in the cherts are not easy to interpret as evidence of the timing and development of metabolic processes. That some, at least, of the earliest autotrophs were photosynthesizing, stromatolite-formers is indicated by the earliest fossils of these organo-sedimentary structures, possibly as old as 3.5 b.y., in rocks in the Pilbara region of Western Australia.

All these earliest organisms were probably prokaryotes, belonging to one or the other of the three divisions bacteria, methanogens, or cyanobacteria. The first good evidence of eukaryotic cells comes from rocks dated at about 1.2 to 1.3 b.y., the Beak Springs Dolomite of southeastern California. In such organisms, the cells are larger and with an in-

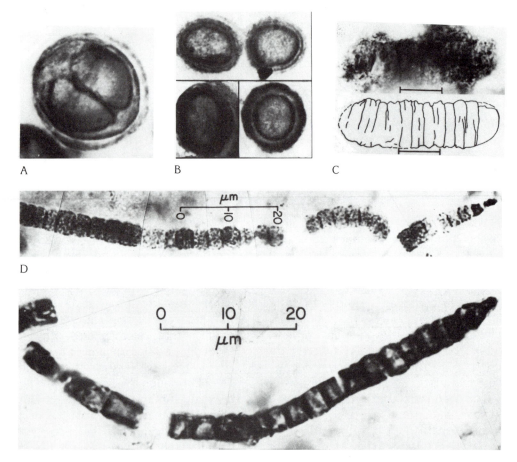

A   B   C

D

E

FIGURE 16.15
Filamentous and spheroidal blue-green algal microfossils from the Bitter Springs Formation, late Precambrian (ca. 850 Ma) of central Australia. (From Schopf, J. W., 1968, Jour. Paleontology, v. 42, p. 651–688, and Schopf, J. W., and Blacic, J. M., 1971, Jour. Paleontology, v. 45, p. 925–960. A *Bigeminococcus lamellosus*, B *Gloediniopsis lamellosa*, C *Palaeolyngbya Barghooniana*, D *Cephalophytarion constrictum*, E *Cephalophytarion grande*. (Photographs by courtesy of J. W. Schopf.)

ternal organization including a cell nucleus. By about 1.0–0.9 b.y. ago, there is evidence that sexual reproduction had evolved: from the Bitter Spring Formation of central Australia come fossils of nucleated cells thought to show stages of meiotic cell division. Table 16.6 summarizes the major biologic and associated events of the Precambrian, but apart from recording a general evolutionary progression, the various microfossils have little contribution to make in a practical division of Precambrian time. It is only in the late Proterozoic that some progress has been made toward establishing biostratigraphic divisions. Stromatolites have been used with some success, as will be described shortly, and some

TABLE 16.6
Development of life forms through the Precambrian

Ga	Major Biogenic Events	More Important Fossiliferous Formations	Lithology and Environment
	First shelly fossils		
	First metazoan fossils	Ediacara	Oxygen content of atmosphere approx. 10% of present day
0.7			
	Origin of sexual reproduction	Bitter Springs (0.9 Ga)	Oxidized cratonic sediments (redbeds)
0.9–1.0			
	First eukaryotes	Beck Springs (1.3 Ga)	
1.3–2.0			
	Advanced oxygen-mediating enzymes	Gunflint Chert (1.9 Ga)	End of Superior-type BIF
1.8			
		Transvaal dolomite (2.25 Ga)	Beginning of free oxygen in atmosphere
2.0			
	Stromatolites becoming abundant	Buluwayan (2.5 Ga)	Earliest Superior-type BIF
2.5			
	Prokaryotes diversify	Fig Tree (3.0 Ga)	Mostly unoxidized cratonic sediments
3.0			
	First stromatolites		K-poor granites, greenstones; no free oxygen; earliest metasediments
3.5			
	First autotrophic prokaryotes		
3.8			
	Chemical evolution leading to biogenesis		Earliest lithosphere and hydrosphere
4.2			
	Origin of earth		Oldest meteorites and terrestrial lead
4.68			

usefulness is claimed also for certain fossil cell structures known as *acritarchs*.

Acritarchs

Unicellular structures of various biological origins and of different ages have collectively come to be termed *acritarchs*. Many are thought to be of algal origin, but their affini-

ties with modern algal species are often in considerable doubt. Some hundreds of "species" have been described, many with supposed biostratigraphic significance in Late Riphean and Vendian rocks, but the reliability of correlation made with such fossils is open to doubt. For example, the use of acritarchs and other microflora by Choubert and Faure-Muret (1980), in support of their attempt at long-distance correlation of lithostratigraphic markers, seems to have been a failure. This is suggested by the fact that the correlations are apparently at variance with isotopically derived age data (Wright, 1985).

Stromatolites

Stromatolites are laminated organosedimentary structures that are formed by so-called **algal mats,** in which cyanobacteria are typically the dominant organisms. The mats form on the surface of sediments in various settings and comprise a loose fabric of intertwined filaments (Fig. 16.16) that secrete a gelatinous material. For this reason, algal mats are extremely effective in trapping and binding loose sedimentary particles, and succeeding generations of algal and bacterial associations build up layer upon layer of sediment. It is these superincumbent layers that are preserved as a stromatolite rather than any fossil remains of the organisms themselves. The characteristic laminated structure of stromatolites is not necessarily regular. The internal fabric of algal mats as they are forming is quite open, and silt and mud are washed in; collapse of the laminae under the superincumbent load results in many irregularities and micro-unconformities in the structure. Where algal mats are exposed to intermittent desiccation, the surface becomes cracked (see Fig. 1.3), and the dried uppermost layers curl up into flakes which, if disturbed, may accumulate in jumbled masses. On submergence, the renewed organic growth over such an irregular surface often leads to dome-, pinnacle-, or biscuitlike forms of the algal mat. Such small-scale topography results in local variations in moisture and salinity and these, in turn, lead to variations in the algal associations and the growth forms they take (Fig. 16.17). One result of these variations is that the growth of the algal and bacterial colonies may become so localized on scattered knoblike protuber-

DAYLIGHT
Upward Growth (S. calcicola)
and Sediment Trapping

DARKNESS
Horizontal Growth (O. submembranacea)
and Sediment Binding

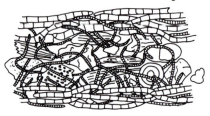

FIGURE 16.16
Day-night cycle of sediment accretion in stromatolite-forming algal mats. (From Gebelein, C. D., 1969, Distribution, morphology, and accretion rate of Recent subtidal stromatolites, Bermuda: Jour. Sed. Petrology, v. 39, Fig. 15, p. 60.)

ances that continued growth results in columnar or pinnaclelike structures.

The numerous variations in growth form, ranging from the simple stratiform or mammillated type to columnar and branching forms, have been described and classified in various ways. Earlier workers applied names based on a Linnaean binomial nomenclature, and many of these, such as the "genera" *Cryptozoan, Collenia,* and *Conophyton,* have become well established in the literature. More recent workers have recognized that the

FIGURE 16.17
Bulbous stromatolites in the intertidal zone at Hamelin pool Shark Bay, Western Australia. (Photo by courtesy of K. J. Mc-Namara, Western Australia Museum, Perth, W.A.)

form of the stromatolite structure is governed by external environmental controls rather than by the particular taxa involved in the algal associations, and opinions vary as to how to handle the classification of stromatolites. Krylov (1976), for example, listed no less than twelve separate classifications for stromatolites. Modern classifications mostly follow the system first introduced by Krylov (1963) based on the gross morphology and fabric, together with laminar shape and microstructure. A binomial nomenclature is used, but it is essentially a morphologic classification only and uses the terms "group" and "form" as roughly analogous to genus and species (Preiss, 1976).

The oldest reported occurrence of stromatolites is from the Pilbara region of Western Australia (Fig. 16.18) in rocks dated at 3.4 to 3.5 b.y. old (Walter, 1972), but it was not until the Proterozoic that they became an important element in sedimentary successions. Studies of stromatolites from Precambrian terranes all over the world, but particularly in the USSR, have demonstrated conclusively that progressive and unidirectional changes in stromatolite assemblages do occur and that they are, therefore, potential biostratigraphic indicators (Fig. 16.19). On the basis of the pioneering work of Soviet geologists, it is clear that the biostratigraphic zonation of the late Proterozoic of the USSR can be ex-

FIGURE 16.18
Section (approx. 30 cm) through a stromatolite from the 3500 m.y. old site at North Pole in the Pilbara region of Western Australia. (Photo by courtesy of K. J. McNamara, Western Australia Museum, Perth, W.A.)

tended to other areas. Up to the present, the most successful correlations have been with Australian successions. Soviet geologists generally use a fourfold subdivision of the Middle and Late Proterozoic, as shown in Table 16.7, although Sokolov (1972), used the term Vendian rather than terminal Riphean for the succession above the highest Precambrian tillites. The applicability of this scheme to intercontinental correlations looks promising, notably in Australia (Preiss, 1976; Walter, 1972; Grey, 1982), North Africa (Bertrand-Sarfati

and Raaben, 1970), India (Raha and Sastry, 1982), and China (Shixing, 1982). Individual "species" are rarely cited and biostratigraphic units are defined mainly on the basis of assemblage zones with various assemblages and subassemblages named by characteristic forms or, alternatively, given a number and letter notation.

Metazoans

The evidence for **metazoan** life in the latest Precambrian has been accumulating rapidly in recent years. Since the initial discovery of the Ediacara fauna (named after the Ediacara Hills in the Flinders Range) in the Pound Sandstone in South Australia (Glaessner and Wade, 1966), similar assemblages have been reported in the USSR, China, Sweden, England, Namibia, Brazil, Argentina, and at a few, but widespread, localities in North America. The assemblage is dominated by coelenterates, or at least coelenterate-like animals, that made up 67 percent of the specimens reported by Glaessner (1971), followed by annelids (25 percent), and arthropods (5 percent) (Fig. 16.20). The age range of these occurrences has been established at 590–700 Ma. It was suggested by Cloud and Glaessner (1982) that an "Ediacaran System" be defined as a chronostratigraphic division on the basis of the first appearance of Ediacara fossils and that it be included in the Phanerozoic, rather than the latest Proterozoic. Although the term Ediacaran fauna has been applied rather loosely to the assemblage as a whole, there are some indications of geographic differences in the distribution of certain elements. McMenamin (1982) reported that benthic forms seem to be restricted to locations on the Proto-Gondwana continent and to have been isolated from the shelly faunas that are apparently confined to the Proto-Laurasian landmass.

Older evidence of metazoan life is seen in the form of tracks, trails, burrows, and other

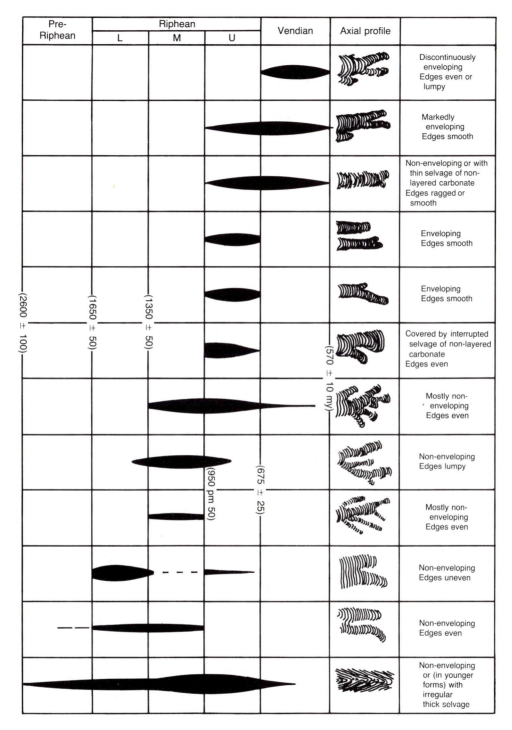

FIGURE 16.19
Variations in the growth forms of stromatolites and their stratigraphic distribution. (Redrawn after Cloud, P. E., Jr., and Semikhatov, M., 1969, Proterozoic stromatolite zonation: Am. Jour. Sci., v. 267, p. 1017–1061.)

TABLE 16.7
Subdivisions of the Late Precambrian in the USSR and characteristic stromatolite assemblages

Subdivision	Characteristic Stromatolites
Cambrian 570 ± 10 Ma	
Vendian or Terminal Riphean 680 ± 20 Ma	*Boxonia grumulosa* and others
Late Riphean 950 ± 50 Ma	*Inzeria, Gymnosolen, Minjaria* etc.
Middle Riphean 1350 ± 50 Ma	*Baicalia baicalica* and others
Early Riphean 1600 ± 50 Ma	*Kussiella kussiensis, Conophyton cylindricus*

trace fossils dated at something less than 1000 Ma; according to Ford (1980), no older than the Vendian, approximately 670 Ma. Prior to this time, there is a significant absence of bioturbation in finely laminated sediments. Although the record is fragmentary, there is a general increase in both variety and complexity of the trace fossils through time.

Dubiofossils

As will be discussed later, the fossils that are present in the latest Precambrian rocks seem to point to a period of rapid evolutionary change that began about 0.7 or 0.8 b.y. ago. The fossil record in older rocks, as described earlier, consists only of stromatolites and microscopic structures in cherts. There are in these older rocks, however, other fossils of a more controversial nature. These are the ichnofossils, or trace fossils, consisting usually of problematical grooves, tubes, sinuous markings, and other bedding-plane features, generally interpreted as tracks, trails, or burrows of metazoan animals of some kind. Many of these fossils are widely known and have been given taxonomic names. Over the years, quite a number of these impressions have failed to stand up to careful scrutiny and have been explained as of inorganic origin, caused by bubbles, escaping gas, sediment desiccation, and so on. Of those traces,

or *dubiofossils* as they are known, that remain unexplained, the oldest so far described are tubelike structures in the Medicine Peak Quartzite of southeastern Wyoming (Kauffman and Steidtmann, 1981), where the quartzite is intruded by a 2.0-b.y.-old granite.

A widely held view is that the late Precambrian fossil record demonstrates very clearly a rapid evolutionary radiation whose causes are explainable in terms of the punctuated equilibrium model of evolutionary change, described in Chapter 8. On the other hand, for those subscribing to the gradualistic model, there is nothing very surprising about indications of metazoans in very old rocks. This, obviously, is an ongoing controversy, yet to be resolved.

16.8 PRECAMBRIAN-CAMBRIAN BOUNDARY

The location of the Precambrian-Cambrian boundary is an example of a problem that has grown rather than decreased with the acquisition of new knowledge. Until comparatively recently, it would have been possible to find fairly general agreement among geologists on the following points:

1. The base of the Cambrian can be placed at the first appearance of shelly faunas, dominated by trilobites.

A

B

C

FIGURE 16.20

Typical forms from the Ediacara fauna of late Proterozoic age. A. Mansonites. B. Spriggina. C. Charniodiscus. (Photos by courtesy of M. Wade, Queensland Museum, Brisbane Queensland.)

2. The base of the Cambrian usually lies above some kind of unconformity.

3. The latest Proterozoic rocks contain fossils, but they are, at best, poorly preserved and problematical.

With these criteria to work with, the placing of the Precambrian-Cambrian boundary should have presented no problem. Unfortunately, none of the above three statements is strictly true.

Perhaps the most remarkable feature of the beginning of the Cambrian, as it appeared to early workers, was the apparent suddenness of appearance and the complexity and diversity of early Cambrian faunas. The absence of any fossil record of the precursors of this fauna was explained by assuming that the ancestral forms did not possess fossilizable hard parts. It was also assumed that a very long and unknown period of evolutionary radiation must have preceded the Cambrian. The lack of Precambrian precursors was also partially explained by the existence of the so-called Lipalian Interval, the name given by Walcott (1910) to the hiatus of the unconformity that was found almost everywhere below the Cambrian, particularly in cratonic successions. The situation has changed considerably in recent years. There is growing evidence that the abrupt appearance of skeletal forms in the early Cambrian was of less significance than had previously been supposed and that the Precambrian-Cambrian boundary simply marks a stage reached during a general but rapid diversification of the metazoans. As Stanley (1976) pointed out, major skeletal taxa made their appearance sequentially, and the apparent richness of faunas of beginning Cambrian age requires no special explanation (Fig. 16.21). Although popularly associated with the beginning Cambrian faunas, trilobites, in fact, did not appear until somewhat later (in the Atdabanian; see Table 16.8). Several other Early Cambrian forms did not appear until post-Tommotian time, and it is clear from Figure 16.21 that numerous taxa, traditionally considered characteristic of the Cambrian, appeared sequentially through the whole period. Certain forms in the Early and Middle Cambrian did not survive into the Late Cambrian. Although it was once supposed that the diversity of forms in the Cambrian necessitated a very long period of Precambrian evolution, it is now believed that the actual time involved was much shorter. As Stanley (1976) suggested, both the traditional view of the sudden appearance of Cambrian faunas and the idea of a long gradualistic climb upward from remote ancestral forms should be abandoned. What, in fact, is surprising is that the appearance of metazoan life forms was so long delayed once life had appeared on earth. Some of the earliest evidence for metazoans is seen in trace fossils and in banded sediments disturbed by bioturbation. Surprisingly, these are considerably younger than 1 b.y.

In approaching the close of the Proterozoic, one of the most important events in terms of stratigraphic markers was the late Precambrian glaciation. This is a logical starting place for a discussion of Late Proterozoic-Paleozoic stratigraphy and particularly of the fossils and their potential in resolving the Cambrian-Precambrian boundary problem. The name Vendian is used by Soviet geologists for the succession in the Podolian area of the Ukraine, beginning below the highest Precambrian tillites up to the base of the Cambrian (i.e., base of the Tommotian; for a fuller account of the Tommotian and the Cambrian-Precambrian boundary, see Raaben, 1981). It is in Vendian rocks and strata of equivalent age from about 670 to 600 Ma (Sepkoski, 1983) that very profound changes in the biotas are recorded. In pre-Vendian rocks, those of the Riphean, the only fossils are those of bacteria, cyanobacteria, and various heterotrophic protoctists, but beginning in the Vendian, trace fossils of trails and

FIGURE 16.21
Stratigraphic ranges of the major invertebrate groups across the Precambrian-Cambrian boundary. (From Stanley, S. M., 1976, Fossil data and the Precambrian-Cambrian evolutionary transition: Am. Jour. Sci., v. 276, Fig. 1, p. 66. Reprinted by permission of American Journal of Science.)

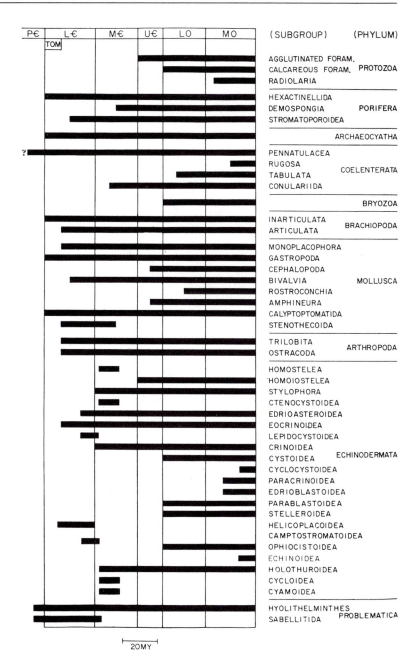

tracks indicate the presence of the first metazoans. No unequivocal burrow structures occur, and this suggests that these first multicelled organisms were relatively simple

forms. The generic term *Planolites* has been given to these simple metozoan trace fossils.

Not far above these trace fossils is found the first direct evidence of metazoans in the

TABLE 16.8
Stratigraphy across the Cambrian-Precambrian boundary based on successions in the USSR (based on data from Trompette, R., 1982, Upper Proterozoic (1800 ± 570 Ma) stratigraphy: A survey of lithostratigraphic, paleontological, radiochronological, and magnetic correlations, Precambrian Research, v. 18, p. 27–52.)

Ma	Stage		
	Botomian		
555			
	Atdabanian		Lower Cambrian (first trilobites)
570			
	Tommotian		(small shelly fossils)
580			
	Upper Vendian		(= Yudomian) (Ediacara fauna)
630			
	Lower Vendian		Late Proterozoic glaciation (Laplandian glaciation, 680–650 Ma)
670–680			
		R4	(Kudashian)
ca. 700			
		R3	(Karatavian)
	Riphean	R2	(Yurmatinian)
		R1	(Burzyanian)
1700			

form of the diverse coelenterate-, annelid-, arthropod-, and echinoderm-like fauna, described earlier. Coincident in time with these Ediacaran fossils were the precursors of the shelly faunas that became increasingly varied in later time. These shelly forms are represented by such fossils as *Cloudina* from Namibia and *Sinotubulites* from southern China, both simple tubelike forms (McMenamin, 1987). Trace fossils are also more complex, consisting of worm tubes made by burrowing organisms, as well as surface tracks and trails of increasing variety. Even higher in the succession, in rocks approximately 570 m.y. old, the Ediacaran forms give way to shelly fossils, the remains of animals capable of secreting skeletal elements and shells of calcium carbonate. Although the disappearance of the soft-bodied Ediacaran forms from the record might seem to reflect evolutionary progression, it is possible that, as Lowenstam (1980) pointed out, the absence of soft parts in post-Ediacaran times may be merely the consequence of the appearance of scavengers. From this time on, there was a progressive increase in the diversity of these animals. None was very big, most are less than one centimeter long, and they have become collectively known as the *small shelly fossils* (Fig. 16.22). In many, the skeletal structures were made of calcium phosphate; this material is particularly characteristic of late Proterozoic and early Cambrian animals. Skeletons of calcium carbonate did not become predominant until considerably later, the chief exception being the Archaeocyatha, a group of primitive spongelike colonial animals that first appeared in the earliest Tommotian and were responsible for the oldest organic reefs.

There has naturally been much speculation as to the cause of what was clearly a marked evolutionary radiation during the period from roughly 700 to 500 m.y. ago, but it is obviously beyond the scope of this work to comment at length. Of particular interest are the reasons that skeletal forms apparently appeared so suddenly. Skeletonization would undoubtedly have provided an adaptive advantage in assisting articulation and leverage for improved musculature and as better circulatory and respiratory systems evolved. It is likely that protection against predators was

FIGURE 16.22

Representative forms from the Tommotian fauna, collectively known as small shelly fossils. A. *Aldanella crassa.* B. *Anabarella plana.* C. *Latouchella* sp. D. *Conotheca mammilata.* E. *Chancelloria* sp. F. *Suchites sacciformis.* G. *Hertzina* sp. H. *Fomitchella infundibuliformis.* I. *Lapworthella dentata.* A, B, and C are probably primitive molluscs, D is a worm tube, E is a sponge spicule, G and H may be conodonts, and F and I are of doubtful affinities.

also a factor because direct evidence in the form of healed injuries comes from early Cambrian fossils. The timing, just after the late Proterozoic ice age, also may be significant, and some workers see widespread marine transgressions as an important factor in extending shallow-marine habitats and providing new niches. Attention has also been focused on possible changes in the chemistry of the oceans.

It has been suggested, for example, that anoxic conditions with increased sulfide accumulations were followed by massive upwelling of ocean water. This so-called Yudonski event must have had profound effects on the biota (Conway Morris, 1987). Cook and Shergold (1984) described a phosphogenic event, perhaps caused by oceanic overturn, that commenced in the Late Proterozoic and reached a climax in the earliest Cambrian. It was suggested by these workers that the appearance of skeletal animals at the beginning of the Cambrian was an opportunistic or crisis reaction to high phosphorus levels in shallow shelf seas. Viewed in its entirety, the evolutionary radiation of late Proterozoic and Cambrian time, described by some workers as an evolutionary "explosion," would seem to provide strong support for the operation of an evolutionary mechanism after the punctuated equilibrium model rather than that of phyletic gradualism.

In terms of the geological time scale, the base of the Cambrian has been variously set between 530 and 590 Ma, but no international agreement on its stratigraphic definition has yet been reached. A pronouncement of the International Geological Congress in 1954 set the boundary at "the base of occurrence of organized fossils." This criterion was reaffirmed in 1983 by the working group on the Precambrian-Cambrian boundary that had been established by the International Commission on Stratigraphy (ICS) in 1972 (International Geological Correlation Project 29). In 1984, and after considering sections in

many parts of the world (Fig. 16.23), it was recommended that the global boundary stratotype section and point for the Precambrian-Cambrian boundary be located at a section in the Kunyang phosphorite mine at Meischucun in Yunnan Province, south China (Cowie, 1985). Two other areas are also under consideration as potential stratotypes; one is on the Burin Peninsula in southeastern Newfoundland, and the other is on the Aldan river in eastern Siberia at a section known as Ula-khan-Sulugur. Which of these locations will be formally adopted as the place for a "golden spike," perhaps the most important one of all, will depend on further studies of the working group. Although the importance of the base of the Cambrian is emphasized by it also being the base of the Paleozoic and of the Phanerozoic, there is, of course, no intrinsic reason why these boundaries should coincide. The fact that they do is simply an artifact of the natural evolution of the geologic time scale. It may be that future studies will result in the moving downward of the Paleozoic or Phanerozoic or both boundaries to embrace the Vendian, widely recognized as the uppermost "system" of the Proterozoic, although its lower boundary has not been properly defined.

16.9 CONCLUSION

Although a large number of classification schemes have been proposed for the Precambrian at one time or another, there are really only two main approaches to the problem. One is to divide Precambrian time on the basis of geochronologic measurements in years, the other is to follow the procedures used in the Phanerozoic and to establish divisions based upon selected type sections or type regions and then attempt to correlate given successions with them. To some extent, opinions are influenced by the availability of good Precambrian exposures. In South Africa, for example, occurs one of the finest Precam-

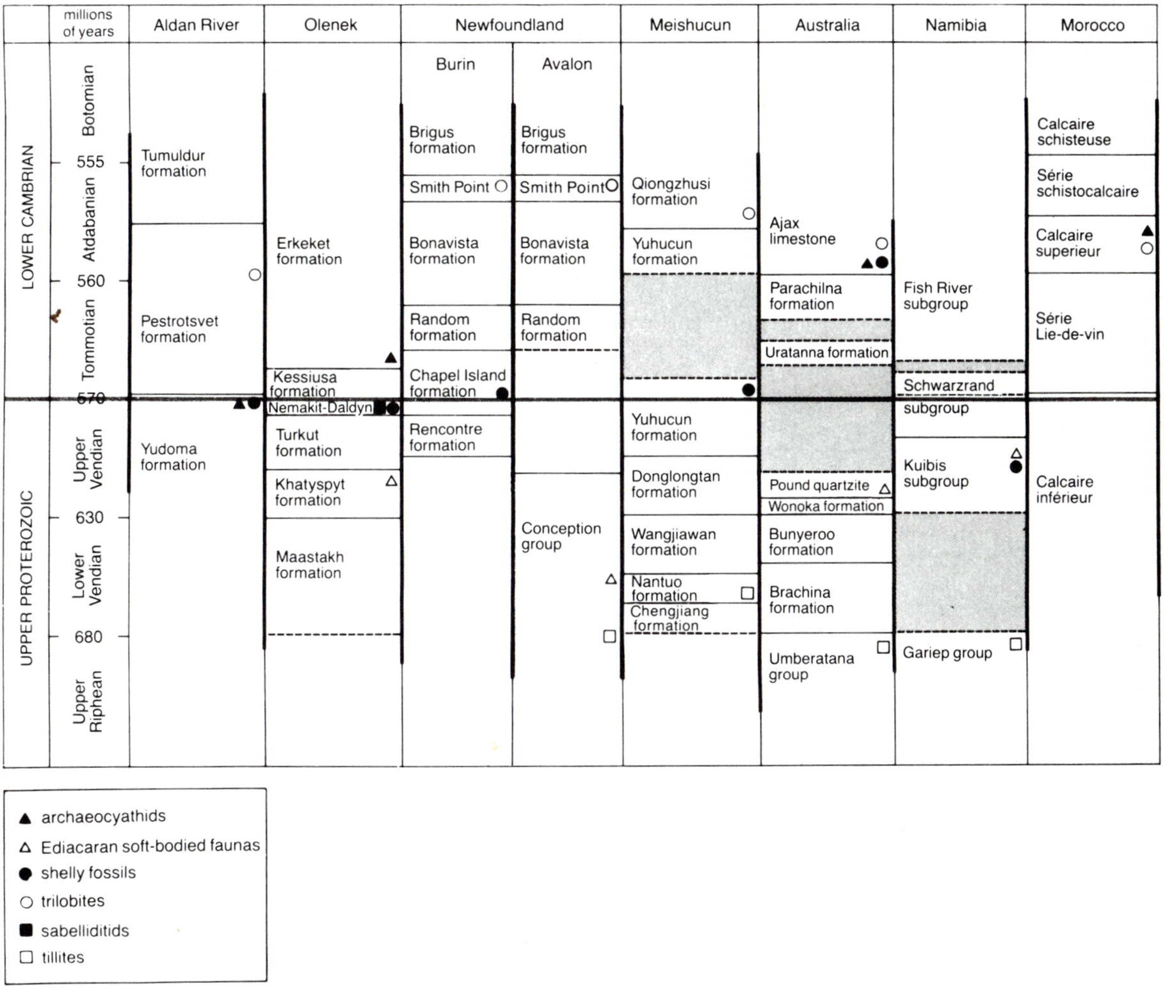

millions of years	Aldan River	Olenek	Newfoundland		Meishucun	Australia	Namibia	Morocco
			Burin	Avalon				
LOWER CAMBRIAN — Botomian / 555	Tumuldur formation	Erkeket formation	Brigus formation	Brigus formation	Qiongzhusi formation	Ajax limestone		Calcaire schisteuse
Atdabanian			Smith Point ○	Smith Point ○				Série schistocalcaire
560			Bonavista formation	Bonavista formation	Yuhucun formation	Parachilna formation	Fish River subgroup	Calcaire superieur ▲ ○
Tommotian	Pestrotsvet formation	Kessiusa formation ▲	Random formation	Random formation		Uratanna formation		Série Lie-de-vin
570	Nemakit-Daldyn ▲○ ■●	Turkut formation	Chapel Island formation ●			Schwarzrand subgroup		
UPPER PROTEROZOIC — Upper Vendian	Yudoma formation	Khatyspyt formation △	Rencontre formation		Yuhucun formation	Pound quartzite △	Kuibis subgroup △ ●	Calcaire inférieur
630 — Lower Vendian		Maastakh formation		Conception group	Donglongtan formation	Wonoka formation		
					Wangjiawan formation	Bunyeroo formation		
680					Nantuo formation □ △	Brachina formation		
					Chengjiang formation □		Gariep group □	
Upper Riphean						Umberatana group □		

Legend:
- ▲ archaeocyathids
- △ Ediacaran soft-bodied faunas
- ● shelly fossils
- ○ trilobites
- ■ sabelliditids
- □ tillites

FIGURE 16.23
Correlation chart of selected stratigraphic successions across the Precambrian-Cambrian boundary. Those at the Aldan River, the Burin Peninsula, Newfoundland, and at Meishucan, China, are the sections presently under consideration by the International Commission on Stratigraphy as possible boundary stratotypes. (From Conway Morris, S., 1987, The search for the Precambrian-Cambrian boundary: Am. Scientist, v. 75, Fig. 4, p. 160.)

brian successions in the world, seen in a succession of largely unmetamorphosed rocks over 65 km thick and ranging in age from 3,750 m.y. to around 570 m.y. The immense thickness and the well-known stratig-raphy of this section persuaded Kent and Hugo (1978) that chronostratigraphic divisions for the whole Precambrian could be recognized, as shown in Table 16.9. Within the succession, however, are several major

TABLE 16.9

Stratigraphic subdivision of South African Precambrian (From Kent, L. E., and Hugo, P. J., 1978, Aspects of the revised South African stratigraphic classification and a proposal for the chronostratigraphic subdivision of the Precambrian, *in* Cohee, G. V., Glaessner, M. F., and Hedberg, H. D., eds., Contributions to the geological time scale: AAPG Studies in Geology, no. 6, reprinted by permission of American Assocation of Petroleum Geologists.)

Approx. age (Ma)	Chronostratigraphic Unit		Lithostratigraphic Unit		Thickness (from Anhaeusser, 1973	
		Era	Group	Supergroup/ Sequence		
	Paleozoic		Nama/Malmesbury Gariep. Nosib			
1080	Namibian			Damara		
2070	Mogolian		Koras Waterberg/Soutpansberg		6½ km	5 km sed 1½ km volc
	Vaalian		Rooiberg Oilfantshoek Pretoria/Postmasburg Chuniespoort/Campbell/ Griquatown Woikberg	Transvaal/ Griqualand West	11 km	9 km sed 2 km volc
2630	Randian		Pniel Platberg Klipriviersberg	Ventersdorp	5 km	1 km sed 4 km volc
			Central Rand West Rand	Witwatersrand	11 km	9 km sed 2 km volc
2800			Dominion Limpopo (Beit Bridge)			
2900	Swazian		Pongola		10 km	5 km sed 5 km volc
3750			Moodies Fig Tree Onverwacht	Swaziland	21 km	5 km sed 16 km volc

TABLE 16.10

Selected Precambrian time and time-rock classifications

Time Scale (Ma)	CANADA Stockwell (1964)	USA James (1972)	USSR Semikhatov (1974)	AUSTRALIA Dunn, Plumb and Roberts (1966)	Cloud (1976)	SOUTH AFRICA Geological Survey (1970)	SOUTH AFRICA Kent and Hugo (1978)
	PHANEROZOIC	PHANEROZOIC			V PHANEROZOIC	PHANEROZOIC	PHANEROZOIC
500	PROTEROZOIC EON — Hadrynian Era	PRECAMBRIAN Z	PROTEROZOIC — Riphean	PROTEROZOIC — Adelaidean System	IV PALEOPHYTIC	PRECAMBRIAN	NAMIBIAN
1000	Helikian Era	PRECAMBRIAN Y		Carpentarian System			MOGOLIAN
1500					III PROTEROPHYTIC		
2000	Aphebian Era	PRECAMBRIAN X	Aphebian	Lower Proterozoic "Nullaginian System"			VAALIAN
2500	ARCHEAN EON	PRECAMBRIAN W	ARCHEAN	ARCHEAN	II ARCHEAN	"ARCHEAN COMPLEX"	RANDIAN
3000							SWAZIAN
3500							
4000					I HADEAN		

unconformities, so how much of the 3200-m.y. time span is not represented is unknown.

In view of the complexity of Precambrian geology and the immense time spans involved, it seems likely that the most straightforward approach lies in the use of numbers. As James (1972) put it, there is little "that is inherently more stable than numbers," and there is no doubt that the only proper understanding of Precambrian events will come from the growing number of age determinations. This is the view of Snelling (1985), who pointed out that whereas intracratonic correlations of stratigraphic successions and/or major plutonic/metamorphic and tectonic events may be feasible, even over considerable distances, global correlation is a different matter. In other words, to attempt worldwide correlations extending from a single stratotype somewhere, as is done in the Phanerozoic, is not feasible. Such a suggestion overlooks one very important point, and that is the existence in the Phanerozoic of a highly sophisticated biostratigraphic data base that is unavailable in the Precambrian. In Snelling's (1985) view, Precambrian age determinations are now commonly available, and it is they that can provide the basis for correlation, as do fossils in the Phanerozoic. It follows from this that a time scale for the Precambrian could be based on geochronometric units with arbitrarily chosen boundaries or ages in years. There is, quite naturally, some disagreement with this view, and many workers are convinced that the only correct procedure is to define all Precambrian time divisions in terms of boundary stratotypes, as advocated by Hedberg (1974). Others, while accepting the need for geochronometric measurement, feel that the times must be event-related (e.g., Douglas, 1980) and that a proper assessment of the synchroneity or otherwise of geologic events and an understanding of the sequential aspects of geological data can come only from

such a relationship. It is true that in any one region it may be possible to establish time divisions that are event-related as, for example, in the divisions of the Canadian shield, based upon orogenic cycles (Stockwell, 1968). In terms of worldwide subdivisions, however, this approach can be used only if such cycles can be recognized elsewhere. Salop (1983), as mentioned earlier, has claimed that his diastrophic cycles are global in their effect, but this has not found universal acceptance. Examples of several Precambrian classifications are given in Table 16.10. Which of these, if any, will provide the basis for an internationally acceptable global classification, only time will tell.

REFERENCES

Anhauesser, C. R., Roering, C., Viljoen, M. J., and Viljoen, R. P., 1969, The Barberton Mountain Land: A model of the elements and evolution of an Archaean fold belt: Trans. Geol. Soc. South Africa, Annex. to v. 71, p. 225–254.

Bertrand-Sarfati, J., and Raaben, M. E., 1970, Comparison de ensembles stromatolitiques du Precambrien supérieur du Sabana occidental de l'Oural: Bull. Soc. geol. de France, v. XII, p. 563–577.

Choubert, G., and Faure-Muret, A., 1980, The Precambrian in north peri-Atlantic and South Mediterranean mobile zones: General results: Earth Sci. Rev., v. 16, p. 85–219.

Clemmey, H., and Badham, N., 1982, Oxygen in the Precambrian atmosphere: An evaluation of the geological evidence: Geology, v. 10, p. 141–146.

Cloud, P., 1972, A working model of the primitive earth: Am. Jour. Sci., v. 272, p. 537–548.

———— 1973, Paleoecological significance of the banded iron formation: Econ. Geology, v. 68, p. 1135–1143.

———— 1976, Major features of crustal evolution, Alex L. DuToit Memorial Lecture No. 14: Trans. Geol. Soc. South Africa, Annex. to v. 79, 33 p.

Cloud, P., and Glaessner, M. F., 1982, The Ediacaran Period and System: Metazoa inherit the earth: Science, v. 217, p. 783–792.

Conway Morris, S., 1987, The search for the Precambrian-Cambrian boundary: Am. Scientist, v. 75, p. 157–167.

Cook, P. J., and Shergold, J. H., 1984, Phosphorus, phosphorites and skeletal evolution at the Precambrian-Cambrian boundary: Nature, v. 308, p. 231–236.

Cowie, J. W., 1985, Continuing work on the Precambrian/Cambrian boundary: Episodes, v. 8, p. 93–97.

Dearnley, R., 1965, Orogenic fold belts and continental drift: Nature, v. 206, p. 1083–1087, 1284–1290.

Douglas, R. J. W., 1980, On the age of rocks and Precambrian time scales: Geology, v. 8, p. 167–171.

Dunn, P. R., Plumb, K. A., and Roberts, H. G., 1966, A proposal for time-stratigraphic subdivision of the Australian Precambrian: Jour. Geol. Soc. Australia, v. 13, p. 593–608.

Ford, T. D., 1980, Life in the Precambrian: Nature, v. 285, p. 193–194.

Glaessner, M. F., 1971, Geographic distribution and time range of the Ediacara Precambrian faunas: Geol. Soc. America Bull., v. 82, p. 509–514.

Glaessner, M. F., and Wade, M. 1966, The late Precambrian fossils from Ediacara, South Australia: Paleontology, v. 9, p. 599–628.

Grey, K., 1982, Aspects of Proterozoic stromatolite biostratigraphy in Western Australia: Precambrian Research, v. 18, p. 347–365.

Gross, G. A., 1980, A classification of iron-formations based on depositional environment: Canadian Mineralogist, v. 18, p. 215–222.

Hedberg, H. D., 1974, Basis for chronostratigraphic classification of the Precambrian: Precambrian Research, v. 1, p. 165–177.

Hoffman, P., 1973, The Coronation geosyncline; Lower Proterozoic analogue of the Cordilleran geosyncline in the north-western Canadian shield: Philos. Trans. Royal Soc. [London], Ser. A, v. 273, p. 547.

Hoffman, P., Dewey, J. F., and Burke, K., 1974, Aulacogens and their genetic relation to geosynclines, with a Proterozoic example from Great Slave Lake, Canada, in Dott, R. H., Jr., and Sharer, H., eds., Modern and ancient geosynclinal sediments: Soc. Econ. Paleontologists and Mineralogists Spec. Pub. 19, p. 38–55.

Holland, H. D., 1978, The chemistry of the atmosphere and oceans: New York, John Wiley & Sons, 351 p.

Irving, E., and Park, J. X., 1972, Hairpins and superintervals: Can. Jour. Earth Sci., v. 9, p. 1318–1324.

James, H. L., 1972, Subdivision of Precambrian: Reply: Am. Assoc. Petroleum Geologists Bull., v. 56, p. 2084–2086.

Kauffman, E. G., and Steidtmann, J. R., 1981, Are these the oldest metazoan trace fossils?: Jour. Paleontology, v. 55, p. 923–947.

Kent, L. E., and Hugo, P. J., 1978, Aspects of the revised South African stratigraphic classification and a proposal for the chronostratigraphic subdivision of the Precambrian; in Cohee, G. V., Glaessner, M. F., and Hedberg, H. D., eds., Contributions to the geological time scale: Am. Assoc. Petroleum Geologists Studies in Geology, No. 6, p. 367–379.

Kimberly, M. M., and Dimroth, E., 1976, Basic similarity of Archaean to subsequent atmospheric and hydrospheric compositions as evidence in the distribution of sedimentary carbon, sulphur, uranium and iron, in Windley, B. F., ed., The early history of the earth: New York, John Wiley & Sons, p. 579–585.

Knoll, A. H., 1985, Patterns of evolution in the Archaean and Proterozoic eons: Paleobiology, v. 11, p. 53–64.

Krylov, I. N., 1963, Columnar branching stromatolites of Riphean beds of the southern Urals and their significance for the stratigraphy of the Upper Precambrian: Akad. Nauk. U.S.S.R., Trudy Inst. Geol., v. 69, 133 p.

———— 1976, Approaches to the classification of stromatolites; in Walter, M. R., ed., Stromatolites: New York, Elsevier, p. 31–43.

Lowenstam, H. A., 1980, What, if anything, happened at the transition from the Precambrian to the Phanerozoic?: Precambrian Research, v. 11, p. 89–91.

McElhinny, M. W., Gidding, J. W., and Embleton, B. J. J., 1974, Paleomagnetic results and late Precambrian glaciations: Nature, v. 248, p. 557–561.

McMenamin, M. A. S., 1982, A case for two late Proterozoic–earliest Cambrian faunal province loci: Geology, v. 10, p. 290–292.

—————— 1987, The emergence of animals: Sci. American, v. 256, p. 94–102.

Piper, J. D. A., 1973, Latitudinal extent of late Precambrian glaciations: Nature, v. 244, p. 342–344.

—————— 1974, Proterozoic crustal distribution, mobile belts and apparent polar movement: Nature, v. 251, p. 381–384.

—————— 1976a, Definition of pre-2000 m.y. apparent polar movement: Earth and Planetary Sci. Letters, v. 28, p. 470–478.

—————— 1976b, Paleomagnetic evidence for a Proterozoic supercontinent: Philos. Trans. Royal Soc. [London], Ser. A, v. 280, p. 469–490.

Preiss, W. V., 1976, Intercontinental correlations, in Walter, N. R., ed., Stromatolites: New York, Elsevier, p. 359–370.

Raaben, M. E., ed., 1981, The Tommotian stage and the Cambrian lower boundary problem (trans. from Russian, 1969): Amerind Pub., 359 p.

Raha, P. K., and Sastry, M. V. A., 1982, Stromatolites and Precambrian stratigraphy in India: Precambrian Research, v. 18, p. 293–318.

Ronov, A. B., 1968, Probable changes in the composition of sea water during the course of geological time: Sedimentology, v. 10, p. 25–43.

Salop, L. J., 1983, Geological evolution of the earth during the Precambrian (trans. V. P. Grudina): Berlin, Springer-Verlag, 459 p.

Semikhatov, M. A., 1974, Stratigraphy and geochronology of the Proterozoic: Nauka, Moscow.

Sepkoski, J. J., Jr., 1983, Precambrian-Cambrian boundary: The spike is driven and the monolith crumbles: Paleobiology, v. 9, p. 199–206.

Shixing, Z., 1982, An outline of studies on the Precambrian stromatolites of China: Precambrian Research, v. 18, p. 367–396.

Snelling, N. J., 1985, Geochronology and the geological record, in Snelling, N. J., ed., The chronology of the geological record: Geol. Soc. London Mem. 10, p. 3–9.

Sokolov, B. S., 1972, The Vendian stage in earth history: 24th Internat. Geol. Congr. (Montreal), sec. 1, Precambrian geology, p. 78–84.

Stanley, S. M., 1976, Fossil data and the Precambrian-Cambrian evolutionary transition: Am. Jour. Sci., v. 276, p. 56–76.

Stockwell, C. H., 1964, Fourth report on structural provinces, orogenies and time-classification of rocks of the Canadian Precambrian Shield: Geol. Surv. Canada Paper 64–17.

—————— 1968, Geochronology of stratified rocks, Canadian Shield: Can. Jour. Earth Sci., v. 5, p. 693–698.

Sutton, J., 1963, Long-term cycles in the evolution of the continents: Nature, v. 198, p. 731–735.

Symons, D. T. A., 1967, Paleomagnetism of Precambrian rocks near Cobalt, Ont.: Can. Jour. Earth Sci., v. 4, p. 1161–1170.

Towe, K. M., 1983, Precambrian atmospheric oxygen and banded iron formations: A delayed ocean model: Precambrian Research, v. 20, p. 161–170.

Trendall, A. F., 1968, Three great basins of Precambrian banded iron formations: A systematic comparison: Geol. Soc. America Bull., v. 79, p. 1527–1544.

—————— 1973, Varve cycles in the Weeli Wolli formation of the Precambrian Hammersley Group: Western Australia: Econ. Geology, v. 68, p. 1089–1097.

Walcott, C. D., 1910, Abrupt appearance of the Cambrian fauna on the North American continent: Smithsonian Misc. Coll., v. 57, No. 1.

Walter, M. R., 1972, Stromatolites and biostratigraphy of the Australian Precambrian and Cambrian: Palaeont. Assoc. London Spec. Pub. 11, 190 p.

Walter, M. R., and Bauld, J., 1983, The association of sulphates, evaporites, stromatolites, carbonates and glacial sediments: Examples from the Proterozoic of Australia and the Cainozoic of Antarctica: Precambrian Research, v. 21, p. 129–148.

Wright, A. E., 1985, Subdivision of the Precambrian, *in* Snelling, N. J., ed., The chronology of the geological record: Geol. Soc. London Mem. 10, p. 29–40.

Wynne-Edwards, H. R., 1969, Tectonic overprinting in the Grenville Province, southwest Quebec: Geol. Assoc. Canada Spec. Paper 5, p. 163–182.

APPENDIXES

APPENDIX 1
TIME SCALES

APPENDIX 2
LIST OF COMMON ABBREVIATIONS

APPENDIX 3
NORTH AMERICAN STRATIGRAPHIC CODE

APPENDIX 1
TIME SCALES

GLOBAL CHRONOSTRATIGRAPHIC UNITS						NORTH AMERICAN CHRONOSTRATIGRAPHIC UNITS	NUMERICAL TIME SCALE (Ma)
ERATHEM	SYSTEMS		SERIES / STAGES			SERIES / STAGES	
CENOZOIC	QUATERNARY		HOLOCENE			NORTH AMERICAN PLEISTOCENE GLACIAL STAGES ONLY WHEN APPLICABLE AND NECESSARY	0.01
			PLEISTOCENE				1.7 to 2.8
	TERTIARY	NEOGENE	PLIOCENE		PIACENZIAN	PACIFIC AREA STAGES OR MAMMALIAN STAGES ONLY WHEN APPLICABLE AND NECESSARY	4.6
					ZANCLEAN		5.3
			MIOCENE	UPPER	MESSINIAN		6.7
					TORTONIAN		10.8
				MIDDLE	SERRAVALLIAN		15.4
					LANGHIAN		17
				LOWER	BURDIGALIAN		23
					AQUITANIAN		25
		PALEOGENE	OLIGOCENE	UPPER	CHATTIAN		33
				LOWER	RUPELIAN		38
			EOCENE	UPPER	PRIABONIAN		41
				MIDDLE	BARTONIAN		45
					LUTETIAN		50
				LOWER	YPRESIAN		55
			PALEOCENE	UPPER	THANETIAN		62
				LOWER	DANIAN		67

FIGURE A.1
Chronostratigraphic and geochronometric scales used in the COSUNA charts, 1985.

ERATHEM	SYSTEMS	SERIES / STAGES		SERIES / STAGES	NUMERICAL TIME SCALE (Ma)	
		GLOBAL CHRONOSTRATIGRAPHIC UNITS		**NORTH AMERICAN CHRONOSTRATIGRAPHIC UNITS**		
M E S O Z O I C	CRETACEOUS	UPPER	MAASTRICHTIAN	SAME AS GLOBAL	67 / 72 / 80 / 85 / 90 / 92	65 / 70 / 80 / 90
			CAMPANIAN			
			SANTONIAN			
			CONIACIAN			
			TURONIAN			
			CENOMANIAN		100	100
		LOWER	ALBIAN		108	110
			APTIAN		115	
			BARREMIAN			120
			HAUTERIVIAN		125	
			VALANGINIAN		130	130
			BERRIASIAN		135	
	JURASSIC	UPPER	TITHONIAN	SAME AS GLOBAL	140 / 145	140
			KIMMERIDGIAN			150
			OXFORDIAN		155	
		MIDDLE	CALLOVIAN		160	160
			BATHONIAN		165	
			BAJOCIAN		170	170
			AALENIAN		175	
		LOWER	TOARCIAN		180	180
			PLIENSBACHIAN		185	
			SINEMURIAN		190	190
			HETTANGIAN		195	
	TRIASSIC	UPPER	RHAETIAN	SAME AS GLOBAL	200	200 / 210
			NORIAN		215	
			CARNIAN		220	220
		MIDDLE	LADINIAN		230	230
			ANISIAN		240	240
		LOWER	SCYTHIAN		245 / 250	250

FIGURE A.1
(continued)

GLOBAL CHRONOSTRATIGRAPHIC UNITS				NORTH AMERICAN CHRONOSTRATIGRAPHIC UNITS			NUMERICAL TIME SCALE (Ma)	
ERATHEM	SYSTEMS	SERIES / STAGES		SERIES / STAGES				
P A L E O Z O I C	PERMIAN	UPPER	TATARIAN	OCHOAN			250 — 255	250 — 260
			KAZANIAN	GUADALUPIAN				
			KUNGURIAN				270 — 275	270
		LOWER	ARTINSKIAN	LEONARDIAN				280
			SAKMARIAN	WOLFCAMPIAN			285	
			ASSELIAN				290	290
	CARBONIFEROUS	UPPER	STEPHANIAN	GZHELIAN	PENNSYLVANIAN SUB-SYSTEM	VIRGILIAN		300
				KASIMOVIAN		MISSOURIAN		
			WESTPHALIAN	MOSCOVIAN		DESMOINESIAN	310	310
		MIDDLE				ATOKAN	315	315
				BASHKIRIAN		MORROWAN		320
			"NAMURIAN"				330	330
		LOWER		SERPUKHOVIAN	MISSISSIPPIAN SUB-SYSTEM	CHESTERIAN		
			VISEAN			MERAMECIAN	340	340
						OSAGEAN		350
			TOURNAISIAN			KINDERHOOKIAN	355	355
								360
	DEVONIAN	UPPER	FAMENNIAN	CHAUTAUQUAN	CONEWANGOAN		365	370
					CASSADAGAN			
			FRASNIAN	SENECAN	CHEMUNGIAN		380	380
		MIDDLE	GIVETIAN	ERIAN	FINGERLAKESIAN		385	385
			EIFELIAN				390	390
		LOWER	EMSIAN	ULSTERIAN	ESOPUSIAN		395	
			SIEGENIAN		DEERPARKIAN		400	400
			GEDINNIAN		HELDERBERGIAN		405	
	SILURIAN	UPPER	PRIDOLIAN	CAYUGAN				410
			LUDLOVIAN	NIAGARAN	LOCKPORTIAN		415	415
			WENLOCKIAN		CLIFTONIAN			
		LOWER	LLANDOVERIAN		CLINTONIAN		420	420
				ALEXANDRIAN			425	
	ORDOVICIAN	UPPER	ASHGILLIAN	CINCINNATIAN	RICHMONDIAN			430
					MAYSVILLIAN			440
			CARADOCIAN		EDENIAN		455	450
				SHERMANIAN KIRKFIELDIAN ROCKLANDIAN	BLACKRIVERIAN		460	455
		MIDDLE	LLANDEILIAN		CHAZYAN			460
				CHAMPLAINIAN			475	470
			LLANVIRNIAN		WHITEROCKIAN			480
		LOWER	ARENIGIAN	CANADIAN			485	485
			TREMADOCIAN				490	490
	CAMBRIAN	UPPER		TREMPEALEAUAN			500	500
				FRANCONIAN				510
				DRESBACHIAN			515	515
		MIDDLE						520
								530
								540
		LOWER					540	550
								560
							570	570

FIGURE A.1

(continued)

Left column table

Era	Sub-era / Period / Sub-period	Epoch	Age	Ma age	Age abbrev.	Ma intervals
Cenozoic (Cz) / Tertiary (TT) / Neogene (Ng)	Quaternary of Pleistogene	Holocene		0-01	Hol	.01
		Pleistocene		2-0	Ple	1.99 → 2
		Pliocene 2	Piacenzian		Pia	3.1
		Pliocene 1	Zanclian	5-1	Zan	
		Miocene 3	Messinian		Mes	6.2
			Tortonian		Tor	
		Miocene 2	Serravallian		Srv	3.1
			Langhian-Late	14-4	Lan2	3.1
			Langhian-Early		Lan1	→ 22-6
		Miocene 1	Burdigalian		Bur	10.2
			Aquitanian	24-6	Aqt	
Cenozoic (Cz) / Tertiary (TT) / Paleogene (Pg)		Oligocene 2	Chattian	32-8	Cht	8.2
		Oligocene 1	Rupelian	38-0	Rup	5.2
		Eocene 3	Priabonian	42-0	Prb	4
		Eocene 2	Bartonian		Brt	8.5
			Lutetian	50-5	Lut	→ 40-4
		Eocene 1	Ypresian	54-9	Ypr	4.4
		Paleocene 2	Thanetian	60-2	Tha	5.3
		Paleocene 1	Danian	65	Dan	4.8
Mesozoic (Mz)	Cretaceous (K) / K2 Senonian		Maastrichtian	73	Maa	8
			Campanian	83	Cmp	10
			Santonian	87-5	San	4.5
			Coniacian	88-5	Con	1
			Turonian	91	Tur	2.5
			Cenomanian	97-5	Cen	6.5 → 79
	K1		Albian	113	Alb	15.5
			Aptian	119	Apt	6
			Barremian	125	Brm	6
	Neocomian		Hauterivian	131	Hau	6
			Valanginian	138	Vlg	7
			Berriasian	144	Ber	6
	Jurassic (J) / Malm (J3)		Tithonian	150	Tth	6
			Kimmeridgian	156	Kim	6
			Oxfordian	163	Oxf	7
	Dogger (J2)		Callovian	169	Clv	6
			Bathonian	175	Bth	6 → 69
			Bajocian	181	Baj	6
			Aalenian	188	Aal	7
	Lias (J1)		Toarcian	194	Toa	6
			Pliensbachian	200	Plb	6
			Sinemurian	206	Sin	6
			Hettangian	213	Het	7
	Triassic (Tr) / Tr3		Rhaetian	219	Rht	6
			Norian	225	Nor	6
			Carnian	231	Crn	6
	Tr2		Ladinian	238	Lad	7 → 35
			Anisian	243	Ans	5
	Scythian (Tr1)		Spathian		Spa	1¼
			Smithian		Smi	1¼
			Dienerian		Die	1¼
			Griesbachian	248	Gri	1¼
Paleozoic	Permian (P) / P2		Tatarian	253	Tat	5
			Kazanian		Kaz	2-5
			Ufimian	258	Ufi	2-5 → 38
	P1		Kungurian	263	Kun	5
			Artinskian	268	Art	5
			Sakmarian		Sak	9
			Asselian	286	Ass	9
	Carboniferous / Pennsylvanian / Gzelian		Noginskian		Nog	
			Klazminskian		Kla	
	Kasimovian		Dorogomilovsk.		Dor	
			Chamovnichesk.		Chv	
			Krevyakinskian		Kre	
	Moscovian		Myachkovskian	296	Mya	
			Podolskian		Pod	→ 34
			Kashirskian		Ksk	
			Vereiskian		Vrk	
	Bashkirian		Melekesskian		Mel	
			Cheremshansk.	315	Che	
			Yeadonian		Yea	
			Marsdenian	320	Mrd	
			Kinderscoutian		Kin	
	C1 Serpukhovian		Alportian		Alp	
			Chokierian		Cho	40

Right column table

Era	Sub-era / Period / Sub-period	Epoch	Age	Ma age	Age abbrev.	Ma intervals
Paleozoic (Pz)	Carboniferous / Mississippian / C2 Bashkirian		Marsdenian		Mrd	
			Kinderscoutian	320	Kin	34
	Serpukhovian		Alportian		Alp	
			Chokerian		Cho	13
			Arnsbergian		Arn	
			Pendleian	333	Pnd	
	Visean		Brigantian		Bri	
			Asbian		Asb	40
			Holkerian		Hlk	19
			Arundian		Aru	
			Chadian	352	Cha	
	C1 Tournaisian		Ivorian		Ivo	8
			Hastarian	360	Has	
	Devonian (D) / D3		Famennian	367	Fam	7
			Frasnian	374	Frs	7
	D2		Givetian	380	Giv	6 → 48
			Eifelian	387	Eif	7
	D1		Emsian	394	Ems	7
			Siegenian	401	Sig	7
			Gedinnian	408	Ged	7
	Silurian (S)	Pridoli		414	Prd	6
		Ludlow	Ludfordian		Ldf	7
			Gorstian	421	Gor	
		Wenlock	Gleedon		Gle	30
			Whitwell		Whi	7
			Sheinwoodian	428	She	
		Llandovery	Telychian		Tel	
			Fronian		Fro	10
			Idwian		Idw	
			Rhuddanian	438	Rhu	
	Ordovician / O3	Ashgill	Hirnantian		Hir	
			Rawtheyan		Raw	10
			Cautleyan		Cau	
			Pusgillian	448	Pus	
		Caradoc	Onnian		Onn	
			Actonian		Act	
			Marshbrookian		Mrb	
			Longvillian		Lon	10
			Soudleyan		Sou	→ 67
			Harnagian		Har	
			Costonian	458	Cos	
	O2	Llandeilo	Late		Llo3	
			Middle		Llo2	10
			Early	468	Llo1	
		Llanvirn	Late		Lln2	10
			Early	478	Lln1	
	O1	Arenig		488	Arg	10
		Tremadoc		505	Tre	17
	Cambrian	Merioneth	Dolgellian		Dol	9
			Maentwrogian	523	Mnt	9
		St David's	Menevian		Men	9 → 85
			Solvan	540	Sol	8
		Caerfai	Lenian		Len	15
			Atdabanian		Atb	15
			Tommotian	590	Tom	20
Sinian	Vendian / Ediacaran		Poundian		Pou	
			Wonokian	630	Won	40
			Mortensnes	650	Mor	20 → 80
	Varangian		Smalfjord	670	Sma	20

			Ma
Z	Sturtian	U	800
	Riphean	R	1050
		Yurmatin Y	1350
		Burzyan B	1650
	Huronian	H	2100 / 2400
	Randian	Ran	2630 / 2800
	Swazian	Sw	3750
	Isuan	I	
	Hadean	Hde	3900

Harland, Cox, Llewellyn, Pickton, Smith, Walters 1982. A geologic time scale. Cambridge University Press

FIGURE A.2 Geologic time scale after Harland et al., 1982.

APPENDIX 2
LIST OF COMMON
ABBREVIATIONS

AAPG	American Association of Petroleum Geologists	DNAG	Decade of North American Geology
ACSN	American Commission on Stratigraphic Nomenclature	DRM	Detrital remanent magnetism
		DSDP	Deep Sea Drilling Project
AGI	American Geological Institute	FAD	First-appearance datum
API	American Petroleum Institute	Ga	Billions of years (giga-ans)
AP	Arboreal pollen	GBSSP	Global boundary stratotype section and point
APW	Apparent polar wandering		
BHT	Bottom-hole temperature	GR	Gamma ray (log)
BIF	Banded iron formation	GSA	Geological Society of America
B.P.	Before present	GSC	Geological Survey of Canada
b.y.	Billion years (10^9 yrs.)	GY	Billion years (giga years)
CA	Cluster analysis	HO	Highest occurrence
CCD	Calcium (calcite) compensation depth	ICS	International Commission on Stratigraphy
COST	Continental offshore stratigraphic test	IGC	International Geological Congress
COSUNA	Correlation of Stratigraphic Units of North America	IGCP	International Geological Correlation Program
CRM	Chemical remanent magnetism	INQUA	International Association for Quaternary Research
CSRS	Composite standard reference section	ISSC	International Subcommission on Stratigraphic Classification
DF	Derrick floor		

IUGS	International Union of Geological Sciences	NASC	North American Stratigraphic Code
JOIDES	Joint Oceanographic Institutes Deep Earth Sampling	PTS	Phanerozoic time scale
		RBV	Relative biostratigraphic value
ka	Thousands of years (kilo-ans)	SEG	Society of Exploration Geophysicists
KB	Kelly bushing		
LAD	Last-appearance datum	SEPM	Society of Economic Paleontologists and Mineralogists
LO	Lowest occurrence		
LOC	Line of correlation	SN	Short normal
Ma	Millions of years (mega ans)	SP	Self (spontaneous) potential (log)
m.y.	Million years	SRS	Standard reference section
NACSN	North American Commission on Stratigraphic Nomenclature	TD	Total depth
		TRM	Thermoremanent magnetism
NAP	Nonarboreal pollen	VRM	Viscous remanent magnetism

APPENDIX 3
NORTH AMERICAN STRATIGRAPHIC CODE[1]

NORTH AMERICAN COMMISSION ON STRATIGRAPHIC NOMENCLATURE

FOREWORD

This code of recommended procedures for classifying and naming stratigraphic and related units has been prepared during a four-year period, by and for North American earth scientists, under the auspices of the North American Commission on Stratigraphic Nomenclature. It represents the thought and work of scores of persons, and thousands of hours of writing and editing. Opportunities to participate in and review the work have been provided throughout its development, as cited in the Preamble, to a degree unprecedented during preparation of earlier codes.

Publication of the International Stratigraphic Guide in 1976 made evident some insufficiencies of the American Stratigraphic Codes of 1961 and 1970. The Commission considered whether to discard our codes, patch them over, or rewrite them fully, and chose the last. We believe it desirable to sponsor a code of stratigraphic practice for use in North America, for we can adapt to new methods and points of view more rapidly than a worldwide body. A timely example was the recognized need to develop modes of establishing formal nonstratiform (igneous and high-grade metamorphic) rock units, an objective which is met in this Code, but not yet in the Guide.

The ways in which this Code differs from earlier American codes are evident from the Contents. Some categories have disappeared and others are new, but this Code has evolved from earlier codes and from the International Stratigraphic Guide. Some new units have not yet stood the test of long practice, and conceivably may not, but they are introduced toward meeting recognized and defined needs of the profession. Take this Code, use it, but do not condemn it because it contains something new or not of direct interest to you. Innovations that prove unacceptable to the profession will expire without damage to other concepts and procedures, just as did the geologic-climate units of the 1961 Code.

This Code is necessarily somewhat innovative because of: (1) the decision to write a new code, rather than to revise the old; (2) the open invitation to members of the geologic profession to

offer suggestions and ideas, both in writing and orally; and (3) the progress in the earth sciences since completion of previous codes. This report strives to incorporate the strength and acceptance of established practice, with suggestions for meeting future needs perceived by our colleagues; its authors have attempted to bring together the good from the past, the lessons of the Guide, and carefully reasoned provisions for the immediate future.

Participants in preparation of this Code are listed in Appendix I, but many others helped with their suggestions and comments. Major contributions were made by the members, and especially the chairmen, of the named subcommittees and advisory groups under the guidance of the Code Committee, chaired by Steven S. Oriel, who also served as principal, but not sole, editor. Amidst the noteworthy contributions by many, those of James D. Aitken have been outstanding. The work was performed for and supported by the Commission, chaired by Malcolm P. Weiss from 1978 to 1982.

This Code is the product of a truly North American effort. Many former and current commissioners representing not only the ten organizational members of the North American Commission on Stratigraphic Nomenclature (Appendix II), but other institutions as well, generated the product. Endorsement by constituent organizations is anticipated, and scientific communication will be fostered if Canadian, United States, and Mexican scientists, editors, and administrators consult Code recommendations for guidance in scientific reports. The Commission will appreciate reports of formal adoption or endorsement of the Code, and asks that they be transmitted to the Chairman of the Commission (c/o American Association of Petroleum Geologists, Box 979, Tulsa, Oklahoma 74101, U.S.A.).

Any code necessarily represents but a stage in the evolution of scientific communication. Suggestions for future changes of, or additions to, the North American Stratigraphic Code are welcome. Suggested and adopted modifications will be announced to the profession, as in the past, by serial Notes and Reports published in the *Bulletin* of the American Association of Petroleum Geologists. Suggestions may be made to representatives of your association or agency who are current commissioners, or directly to the Commission itself. The Commission meets annually, during the national meetings of the Geological Society of America.

1982 NORTH AMERICAN COMMISSION
ON STRATIGRAPHIC NOMENCLATURE

[1]Reprinted by permission from American Association of Petroleum Geologists Bulletin, v. 67, no. 5 (May 1983), p 841–875.

Copies are available at $1.00 per copy postpaid. Order from American Association of Petroleum Geologists, Box 979, Tulsa, Oklahoma 74101.

CONTENTS

PART I. PREAMBLE

BACKGROUND

PERSPECTIVE

Codes of Stratigraphic Nomenclature prepared by the American Commission on Stratigraphic Nomenclature (ACSN, 1961) and its predecessor (Committee on Stratigraphic Nomenclature, 1933) have been used widely as a basis for stratigraphic terminology. Their formulation was a response to needs recognized during the past century by government surveys (both national and local) and by editors of scientific journals for uniform standards and common procedures in defining and classifying formal rock bodies, their fossils, and the time spans represented by them. The most recent Code (ACSN, 1970) is a slightly revised version of that published in 1961, incorporating some minor amendments adopted by the Commission between 1962 and 1969. The Codes have served the profession admirably and have been drawn upon heavily for codes and guides prepared in other parts of the world (ISSC, 1976, p. 104-106). The principles embodied by any code, however, reflect the state of knowledge at the time of its preparation, and even the most recent code is now in need of revision.

New concepts and techniques developed during the past two decades have revolutionized the earth sciences. Moreover, increasingly evident have been the limitations of previous codes in meeting some needs of Precambrian and Quaternary geology and in classification of plutonic, high-grade metamorphic, volcanic, and intensely deformed rock assemblages. In addition, the important contributions of numerous international stratigraphic organizations associated with both the International Union of Geological Sciences (IUGS) and UNESCO, including working groups of the International Geological Correlation Program (IGCP), merit recognition and incorporation into a North American code.

For these and other reasons, revision of the American Code has been undertaken by committees appointed by the North American Commission on Stratigraphic Nomenclature (NACSN). The Commission, founded as the American Commission on Stratigraphic Nomenclature in 1946 (ACSN, 1947), was renamed the NACSN in 1978 (Weiss, 1979b) to emphasize that delegates from ten organizations in Canada, the United States, and Mexico represent the geological profession throughout North America (Appendix II).

Although many past and current members of the Commission helped prepare this revision of the Code, the participation of all interested geologists has been sought (for example, Weiss, 1979a). Open forums were held at the national meetings of both the Geological Society of America at San Diego in November, 1979, and the American Association of Petroleum Geologists at Denver in June, 1980, at which comments and suggestions were offered by more than 150 geologists. The resulting draft of this report was printed, through the courtesy of the Canadian Society of Petroleum Geologists, on October 1, 1981, and additional comments were invited from the profession for a period of one year before submittal of this report to the Commission for adoption. More than 50 responses were received with sufficient suggestions for improvement to prompt moderate revision of the printed draft (NACSN, 1981). We are particularly indebted to Hollis D. Hedberg and Amos Salvador for their exhaustive and perceptive reviews of early drafts of this Code, as well as to those who responded to the request for comments. Participants in the preparation and revisions of this report, and conferees, are listed in Appendix I.

Some of the expenses incurred in the course of this work were defrayed by National Science Foundation Grant EAR 7919845,
for which we express appreciation. Institutions represented by the participants have been especially generous in their support.

SCOPE

The North American Stratigraphic Code seeks to describe explicit practices for classifying and naming all formally defined geologic units. *Stratigraphic procedures* and principles, although developed initially to bring order to strata and the events recorded therein, are applicable to all earth materials, not solely to strata. They promote systematic and rigorous study of the composition, geometry, sequence, history, and genesis of rocks and unconsolidated materials. They provide the framework within which time and space relations among rock bodies that constitute the Earth are ordered systematically. Stratigraphic procedures are used not only to reconstruct the history of the Earth and of extra-terrestrial bodies, but also to define the distribution and geometry of some commodities needed by society. *Stratigraphic classification* systematically arranges and partitions bodies of rock or unconsolidated materials of the Earth's crust into units based on their inherent properties or attributes.

A *stratigraphic code* or guide is a formulation of current views on stratigraphic principles and procedures designed to promote standardized classification and formal nomenclature of rock materials. It provides the basis for formalization of the language used to denote rock units and their spatial and temporal relations. To be effective, a code must be widely accepted and used; geologic organizations and journals may adopt its recommendations for nomenclatural procedure. Because any code embodies only current concepts and principles, it should have the flexibility to provide for both changes and additions to improve its relevance to new scientific problems.

Any system of nomenclature must be sufficiently explicit to enable users to distinguish objects that are embraced in a class from those that are not. This stratigraphic code makes no attempt to systematize structural, petrographic, paleontologic, or physiographic terms. Terms from these other fields that are used as part of formal stratigraphic names should be sufficiently general as to be unaffected by revisions of precise petrographic or other classifications.

The objective of a system of classification is to promote unambiguous communication in a manner not so restrictive as to inhibit scientific progress. To minimize ambiguity, a code must promote recognition of the distinction between observable features (reproducible data) and inferences or interpretations. Moreover, it should be sufficiently adaptable and flexible to promote the further development of science.

Stratigraphic classification promotes understanding of the *geometry* and *sequence* of rock bodies. The development of stratigraphy as a science required formulation of the Law of Superposition to explain sequential stratal relations. Although superposition is not applicable to many igneous, metamorphic, and tectonic rock assemblages, other criteria (such as crosscutting relations and isotopic dating) can be used to determine sequential arrangements among rock bodies.

The term *stratigraphic unit* may be defined in several ways. Etymological emphasis requires that it be a stratum or assemblage of adjacent strata distinguished by any or several of the many properties that rocks may possess (ISSC, 1976, p. 13). The scope of stratigraphic classification and procedures, however, suggests a broader definition: a naturally occurring body of rock or rock material distinguished from adjoining rock on the basis of some stated property or properties. Commonly used properties include composition, texture, included fossils, magnetic signature, radioactivity, seismic velocity, and age. Sufficient care is required in defining the boundaries of a unit to enable others to distinguish the material body from those adjoining it. Units based on one property commonly do not coincide with those based on another and, therefore, distinctive terms are needed to identify the property used in defining each unit.

The adjective *stratigraphic* is used in two ways in the remainder of this report. In discussions of lithic (used here as synonymous with "lithologic") units, a conscious attempt is made to restrict the term to lithostratigraphic or layered rocks and sequences that obey the Law of Superposition. For nonstratiform rocks (of plutonic or tectonic origin, for example), the term *lithodemic* (see Article 27) is used. The adjective *stratigraphic* is also used in a broader sense to refer to those procedures derived from stratigraphy which are now applied to all classes of earth materials.

An assumption made in the material that follows is that the reader has some degree of familiarity with basic principles of stratigraphy as outlined, for example, by Dunbar and Rodgers (1957), Weller (1960), Shaw (1964), Matthews (1974), or the International Stratigraphic Guide (ISSC, 1976).

RELATION OF CODES TO INTERNATIONAL GUIDE

Publication of the International Stratigraphic Guide by the International Subcommission on Stratigraphic Classification (ISSC, 1976), which is being endorsed and adopted throughout the world, played a part in prompting examination of the American Stratigraphic Code and the decision to revise it.

The International Guide embodies principles and procedures that had been adopted by several national and regional stratigraphic committees and commissions. More than two decades of effort by H. D. Hedberg and other members of the Subcommission (ISSC, 1976, p. VI, 1, 3) developed the consensus required for preparation of the Guide. Although the Guide attempts to cover all kinds of rocks and the diverse ways of investigating them, it is necessarily incomplete. Mechanisms are needed to stimulate individual innovations toward promulgating new concepts, principles, and practices which subsequently may be found worthy of inclusion in later editions of the Guide. The flexibility of national and regional committees or commissions enables them to perform this function more readily than an international subcommission, even while they adopt the Guide as the international standard of stratigraphic classification.

A guiding principle in preparing this Code has been to make it as consistent as possible with the International Guide, which was endorsed by the ACSN in 1976, and at the same time to foster further innovations to meet the expanding and changing needs of earth scientists on the North American continent.

OVERVIEW

CATEGORIES RECOGNIZED

An attempt is made in this Code to strike a balance between serving the needs of those in evolving specialties and resisting the proliferation of categories of units. Consequently, more formal categories are recognized here than in previous codes or in the International Guide (ISSC, 1976). On the other hand, no special provision is made for formalizing certain kinds of units (deep oceanic, for example) which may be accommodated by available categories.

Four principal categories of units have previously been used widely in traditional stratigraphic work; these have been termed lithostratigraphic, biostratigraphic, chronostratigraphic, and geochronologic and are distinguished as follows:

1. A *lithostratigraphic unit* is a stratum or body of strata, generally but not invariably layered, generally but not invariably tabular, which conforms to the Law of Superposition and is distinguished and delimited on the basis of lithic characteristics and stratigraphic position. Example: Navajo Sandstone.

2. A *biostratigraphic unit* is a body of rock defined and characterized by its fossil content. Example: *Discoaster multiradiatus* Interval Zone.

3. A *chronostratigraphic unit* is a body of rock established to serve as the material reference for all rocks formed during the same span of time. Example: Devonian System. Each boundary of a chronostratigraphic unit is synchronous. Chronostratigraphy provides a means of organizing strata into units based on their age relations. A chronostratigraphic body also serves as the basis for defining the specific interval of geologic time, or geochronologic unit, represented by the referent.

4. A *geochronologic unit* is a division of time distinguished on the basis of the rock record preserved in a chronostratigraphic unit. Example: Devonian Period.

The first two categories are comparable in that they consist of material units defined on the basis of content. The third category differs from the first two in that it serves primarily as the standard for recognizing and isolating materials of a specific age. The fourth, in contrast, is not a material, but rather a conceptual, unit; it is a division of time. Although a geochronologic unit is not a stratigraphic body, it is so intimately tied to chronostratigraphy that the two are discussed properly together.

Properties and procedures that may be used in distinguishing geologic units are both diverse and numerous (ISSC, 1976, p. 1, 96; Harland, 1977, p. 230), but all may be assigned to the following principal classes of categories used in stratigraphic classification (Table 1), which are discussed below:

I. Material categories based on content, inherent attributes, or physical limits,

II. Categories distinguished by geologic age:

    A. Material categories used to define temporal spans, and

    B. Temporal categories.

Table 1. Categories of Units Defined*

MATERIAL CATEGORIES BASED ON CONTENT OR PHYSICAL LIMITS

    Lithostratigraphic (22)
    Lithodemic (31)**
    Magnetopolarity (44)
    Biostratigraphic (48)
    Pedostratigraphic (55)
    Allostratigraphic (58)

CATEGORIES EXPRESSING OR RELATED TO GEOLOGIC AGE

    Material Categories Used to Define Temporal Spans
        Chronostratigraphic (66)
        Polarity-Chronostratigraphic (83)
    Temporal (Non-Material) Categories
        Geochronologic (80)
        Polarity-Chronologic (88)
        Diachronic (91)
        Geochronometric (96)

*Numbers in parentheses are the numbers of the Articles where units are defined.
**Italicized categories are those introduced or developed since publication of the previous code (ACSN, 1970).

Material Categories Based on Content or Physical Limits

The basic building blocks for most geologic work are rock bodies defined on the basis of composition and related lithic characteristics, or on their physical, chemical, or biologic content or properties. Emphasis is placed on the relative objectivity and reproducibility of data used in defining units within each category.

Foremost properties of rocks are composition, texture, fabric, structure, and color, which together are designated *lithic characteristics*. These serve as the basis for distinguishing and defining the most fundamental of all formal units. Such units based primarily on composition are divided into two categories (Henderson and others, 1980): lithostratigraphic (Article 22) and lithodemic (defined here in Article 31). A lithostratigraphic unit obeys the Law of Superposition, whereas a lithodemic unit does not. A *lithodemic unit* is a defined body of predominantly intrusive, highly metamorphosed, or intensely deformed rock that, because it is intrusive or has lost primary structure through metamorphism or tectonism, generally does not conform to the Law of Superposition.

Recognition during the past several decades that remanent magnetism in rocks records the Earth's past magnetic characteristics (Cox, Doell, and Dalrymple, 1963) provides a powerful new tool encompassed by magnetostratigraphy (McDougall, 1977; McElhinny, 1978). *Magnetostratigraphy* (Article 43) is the study of remanent magnetism in rocks; it is the record of the Earth's magnetic polarity (or field reversals), dipole-field-pole position (including apparent polar wander), the non-dipole component (secular variation), and field intensity. Polarity is of particular utility and is used to define a *magnetopolarity unit* (Article 44) as a body of rock identified by its remanent magnetic polarity (ACSN, 1976; ISSC, 1979). Empirical demonstration of uniform polarity does not necessarily have direct temporal connotations because the remanent magnetism need not be related to rock deposition or crystallization. Nevertheless, polarity is a physical attribute that may characterize a body of rock.

Biologic remains contained in, or forming, strata are uniquely important in stratigraphic practice. First, they provide the means of defining and recognizing material units based on fossil content (biostratigraphic units). Second, the irreversibility of organic evolution makes it possible to partition enclosing strata temporally. Third, biologic remains provide important data for the reconstruction of ancient environments of deposition.

Composition also is important in distinguishing pedostratigraphic units. A *pedostratigraphic unit* is a body of rock that consists of one or more pedologic horizons developed in one or more lithic units now buried by a formally defined lithostratigraphic or allostratigraphic unit or units. A pedostratigraphic unit is the part of a buried soil characterized by one or more clearly defined soil horizons containing pedogenically formed minerals and organic compounds. Pedostratigraphic terminology is discussed below and in Article 55.

Many upper Cenozoic, especially Quaternary, deposits are distinguished and delineated on the basis of content, for which lithostratigraphic classification is appropriate. However, others are delineated on the basis of criteria other than content. To facilitate the reconstruction of geologic history, some compositionally similar deposits in vertical sequence merit distinction as separate stratigraphic units because they are the products of different processes; others merit distinction because they are of demonstrably different ages. Lithostratigraphic classification of these units is impractical and a new approach, allostratigraphic classification, is introduced here and may prove applicable to older deposits as well. An *allostratigraphic unit* is a mappable stratiform body of sedimentary rock defined and identified on the basis of bounding discontinuities (Article 58 and related Remarks).

Geologic-Climate units, defined in the previous Code (ACSN, 1970, p. 31), are abandoned here because they proved to be of dubious utility. Inferences regarding climate are subjective and too tenuous a basis for the definition of formal geologic units. Such inferences commonly are based on deposits assigned more appropriately to lithostratigraphic or allostratigraphic units and may be expressed in terms of diachronic units (defined below).

Categories Expressing or Related to Geologic Age

Time is a single, irreversible continuum. Nevertheless, various categories of units are used to define intervals of geologic time, just as terms having different bases, such as Paleolithic, Renaissance, and Elizabethan, are used to designate specific periods of human history. Different temporal categories are established to express intervals of time distinguished in different ways.

Major objectives of stratigraphic classification are to provide a basis for systematic ordering of the time and space relations of rock bodies and to establish a time framework for the discussion of geologic history. For such purposes, units of geologic time traditionally have been named to represent the span of time during which a well-described sequence of rock, or a chronostratigraphic unit, was deposited ("time units based on material referents," Fig. 1). This procedure continues, to the exclusion of other possible approaches, to be standard practice in studies of Phanerozoic rocks. Despite admonitions in previous American codes and the International Stratigraphic Guide (ISSC, 1976, p. 81) that similar procedures should be applied to the Precambrian, no comparable chronostratigraphic units, or geochronologic units derived therefrom, proposed for the Precambrian have yet been accepted worldwide. Instead, the IUGS Subcommission on Precambrian Stratigraphy (Sims, 1979) and its Working Groups (Harrison and Peterman, 1980) recommend division of Precambrian time into *geochronometric units* having no material referents.

A distinction is made throughout this report between *isochronous* and *synchronous*, as urged by Cumming, Fuller, and Porter (1959, p. 730), although the terms have been used synonymously by many. *Isochronous* means of equal duration; *synchronous* means simultaneous, or occurring at the same time. Although two rock bodies of very different ages may be formed during equal durations of time, the term *isochronous* is not applied to them in the earth sciences. Rather, isochronous bodies are those bounded by synchronous surfaces and formed during the same span of time. *Isochron*, in contrast, is used for a line connecting points of equal age on a graph representing physical or chemical phenomena; the line represents the same or equal time. The adjective *diachronous* is applied either to a rock unit with one or two bounding surfaces which are not synchronous, or to a boundary which is not synchronous (which "transgresses time").

Two classes of time units based on material referents, or stratotypes, are recognized (Fig. 1). The first is that of the traditional and conceptually isochronous units, and includes *geochronologic units*, which are based on *chronostratigraphic units*, and *polarity-geochronologic units*. These isochronous units have worldwide applicability and may be used even in areas lacking a material record of the named span of time. The second class of time units, newly defined in this Code, consists of

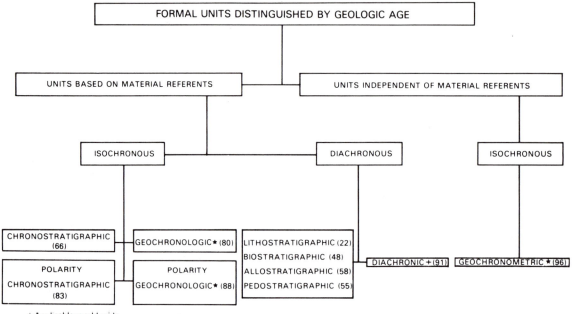

FIG. 1.—Relation of geologic time units to the kinds of rock-unit referents on which most are based.

diachronic units (Article 91), which are based on rock bodies known to be diachronous. In contrast to isochronous units, a diachronic term is used only where a material referent is present; a diachronic unit is coextensive with the material body or bodies on which it is based.

A *chronostratigraphic unit*, as defined above and in Article 66, is a body of rock established to serve as the material reference for all rocks formed during the same span of time; its boundaries are synchronous. It is the referent for a *geochronologic unit*, as defined above and in Article 80. Internationally accepted and traditional chronostratigraphic units were based initially on the time spans of lithostratigraphic units, biostratigraphic units, or other features of the rock record that have specific durations. In sum, they form the Standard Global Chronostratigraphic Scale (ISSC, 1976, p. 76-81; Harland, 1978), consisting of established systems and series.

A *polarity-chronostratigraphic unit* is a body of rock that contains a primary magnetopolarity record imposed when the rock was deposited or crystallized (Article 83). It serves as a material standard or referent for a part of geologic time during which the Earth's magnetic field had a characteristic polarity or sequence of polarities; that is, for a *polarity-chronologic unit* (Article 88).

A *diachronic unit* comprises the unequal spans of time represented by one or more specific diachronous rock bodies (Article 91). Such bodies may be lithostratigraphic, biostratigraphic, pedostratigraphic, allostratigraphic, or an assemblage of such units. A diachronic unit is applicable only where its material referent is present.

A *geochronometric* (or chronometric) *unit* is an isochronous direct division of geologic time expressed in years (Article 96). It has no material referent.

Pedostratigraphic Terms

The definition and nomenclature for pedostratigraphic[2] units in this Code differ from those for soil-stratigraphic units in the previous Code (ACSN, 1970, Article 18), by being more specific with regard to content, boundaries, and the basis for determining stratigraphic position.

The term "soil" has different meanings to the geologist, the soil scientist, the engineer, and the layman, and commonly has no stratigraphic significance. The term *paleosol* is currently used in North America for any soil that formed on a landscape of the past; it may be a buried soil, a relict soil, or an exhumed soil (Ruhe, 1965; Valentine and Dalrymple, 1976).

A *pedologic soil* is composed of one or more soil horizons.[3] A *soil horizon* is a layer within a pedologic soil that (1) is approximately parallel to the soil surface, (2) has distinctive physical, chemical, biological, and morphological properties that differ from those of adjacent, genetically related, soil horizons, and (3) is distinguished from other soil horizons by objective compositional properties that can be observed or measured in the field. The physical boundaries of buried pedologic horizons are objective traceable boundaries with stratigraphic significance. A buried pedologic soil provides the material basis for definition of a stratigraphic unit in pedostratigraphic classification (Article 55), but a buried pedologic soil may be somewhat more inclusive than a pedostratigraphic unit. A pedologic soil may contain both an

---

[2]From Greek, *pedon*, ground or soil.
[3]As used in a geological sense, a *horizon* is a surface or line. In pedology, however, it is a body of material, and such usage is continued here.

0-horizon and the entire C-horizon (Fig. 6), whereas the former is excluded and the latter need not be included in a pedostratigraphic unit.

The definition and nomenclature for pedostratigraphic units in this Code differ from those of soil stratigraphic units proposed by the International Union for Quaternary Research and International Society of Soil Science (Parsons, 1981). The pedostratigraphic unit, geosol, also differs from the proposed INQUA-ISSS soil-stratigraphic unit, pedoderm, in several ways, the most important of which are: (1) a geosol may be in any part of the geologic column, whereas a pedoderm is a surficial soil; (2) a geosol is a buried soil, whereas a pedoderm may be a buried, relict, or exhumed soil; (3) the boundaries and stratigraphic position of a geosol are defined and delineated by criteria that differ from those for a pedoderm; and (4) a geosol may be either all or only a part of a buried soil, whereas a pedoderm is the entire soil.

The term *geosol*, as defined by Morrison (1967, p. 3), is a laterally traceable, mappable, geologic weathering profile that has a consistent stratigraphic position. The term is adopted and redefined here as the fundamental and only unit in formal pedostratigraphic classification (Article 56).

FORMAL AND INFORMAL UNITS

Although the emphasis in this Code is necessarily on formal categories of geologic units, informal nomenclature is highly useful in stratigraphic work.

Formally named units are those that are named in accordance with an established scheme of classification; the fact of formality is conveyed by capitalization of the initial letter of the *rank* or *unit* term (for example, Morrison Formation). Informal units, whose unit terms are ordinary nouns, are not protected by the stability provided by proper formalization and recommended classification procedures. Informal terms are devised for both economic and scientific reasons. Formalization is appropriate for those units requiring stability of nomenclature, particularly those likely to be extended far beyond the locality in which they were first recognized. Informal terms are appropriate for casually mentioned, innovative, and most economic units, those defined by unconventional criteria, and those that may be too thin to map at usual scales.

Casually mentioned geologic units not defined in accordance with this Code are informal. For many of these, there may be insufficient need or information, or perhaps an inappropriate basis, for formal designations. Informal designations as beds or lithozones (the pebbly beds, the shaly zone, third coal) are appropriate for many such units.

Most economic units, such as aquifers, oil sands, coal beds, quarry layers, and ore-bearing "reefs," are informal, even though they may be named. Some such units, however, are so significant scientifically and economically that they merit formal recognition as beds, members, or formations.

Innovative approaches in regional stratigraphic studies have resulted in the recognition and definition of units best left as informal, at least for the time being. Units bounded by major regional unconformities on the North American craton were designated "sequences" (example: Sauk sequence) by Sloss (1963). Major unconformity-bounded units also were designated "synthems" by Chang (1975), who recommended that they be treated formally. Marker-defined units that are continuous from one lithofacies to another were designated "formats" by Forgotson (1957). The term "chronosome" was proposed by Schultz (1982) for rocks of diverse facies corresponding to geographic variations in sedimentation during an interval of deposition identified on the basis of bounding stratigraphic markers. Successions of faunal zones containing evolutionarily

related forms, but bounded by non-evolutionary biotic discontinuities, were termed "biomeres" (Palmer, 1965). The foregoing are only a few selected examples to demonstrate how informality provides a continuing avenue for innovation.

The terms *magnafacies* and *parvafacies*, coined by Caster (1934) to emphasize the distinction between lithostratigraphic and chronostratigraphic units in sequences displaying marked facies variation, have remained informal despite their impact on clarifying the concepts involved.

Tephrochronologic studies provide examples of informal units too thin to map at conventional scales but yet invaluable for dating important geologic events. Although some such units are named for physiographic features and places where first recognized (e.g., Guaje pumice bed, where it is not mapped as the Guaje Member of the Bandelier Tuff), others bear the same name as the volcanic vent (e.g., Huckleberry Ridge ash bed of Izett and Wilcox, 1981).

Informal geologic units are designated by ordinary nouns, adjectives or geographic terms and lithic or unit-terms that are not capitalized (chalky formation or beds, St. Francis coal).

No geologic unit should be established and defined, whether formally or informally, unless its recognition serves a clear purpose.

CORRELATION

Correlation is a procedure for demonstrating correspondence between geographically separated parts of a geologic unit. The term is a general one having diverse meanings in different disciplines. Demonstration of temporal correspondence is one of the most important objectives of stratigraphy. The term "correlation" frequently is misused to express the idea that a unit has been identified or recognized.

Correlation is used in this Code as the demonstration of correspondence between two geologic units in both some defined property and relative stratigraphic position. Because correspondence may be based on various properties, three kinds of correlation are best distinguished by more specific terms. *Lithocorrelation* links units of similar lithology and stratigraphic position (or sequential or geometric relation, for lithodemic units). *Biocorrelation* expresses similarity of fossil content and biostratigraphic position. *Chronocorrelation* expresses correspondence in age and in chronostratigraphic position.

Other terms that have been used for the similarity of content and stratal succession are homotaxy and chronotaxy. *Homotaxy* is the similarity in separate regions of the serial arrangement or succession of strata of comparable compositions or of included fossils. The term is derived from *homotaxis*, proposed by Huxley (1862, p. xlvi) to emphasize that similarity in succession does not prove age equivalence of comparable units. The term *chronotaxy* has been applied to similar stratigraphic sequences composed of units which are of equivalent age (Henbest, 1952, p. 310).

Criteria used for ascertaining temporal and other types of correspondence are diverse (ISSC, 1976, p. 86-93) and new criteria will emerge in the future. Evolving statistical tests, as well as isotopic and paleomagnetic techniques, complement the traditional paleontologic and lithologic procedures. Boundaries defined by one set of criteria need not correspond to those defined by others.

PART II. ARTICLES

INTRODUCTION

Article 1.—**Purpose.** This Code describes explicit stratigraphic procedures for classifying and naming geologic units accorded formal status. Such procedures, if widely adopted, assure consistent and uniform usage in classification and terminology and therefore promote unambiguous communication.

Article 2.—**Categories.** Categories of formal stratigraphic units, though diverse, are of three classes (Table 1). The first class is of rock-material categories based on inherent attributes or content and stratigraphic position, and includes lithostratigraphic, lithodemic, magnetopolarity, biostratigraphic, pedostratigraphic, and allostratigraphic units. The second class is of material categories used as standards for defining spans of geologic time, and includes chronostratigraphic and polarity-chronostratigraphic units. The third class is of non-material temporal categories, and includes geochronologic, polarity-chronologic, geochronometric, and diachronic units.

GENERAL PROCEDURES

DEFINITION OF FORMAL UNITS

Article 3.—**Requirements for Formally Named Geologic Units.** Naming, establishing, revising, redefining, and abandoning formal geologic units require publication in a recognized scientific medium of a comprehensive statement which includes: (i) intent to designate or modify a formal unit; (ii) designation of category and rank of unit; (iii) selection and derivation of name; (iv) specification of stratotype (where applicable); (v) description of unit; (vi) definition of boundaries; (vii) historical background; (viii) dimensions, shape, and other regional aspects; (ix) geologic age; (x) correlations; and possibly (xi) genesis (where applicable). These requirements apply to subsurface and offshore, as well as exposed, units.

Article 4.—**Publication.**[4] "Publication in a recognized scientific medium" in conformance with this Code means that a work, when first issued, must (1) be reproduced in ink on paper or by some method that assures numerous identical copies and wide distribution; (2) be issued for the purpose of scientific, public, permanent record; and (3) be readily obtainable by purchase or free distribution.

Remarks. (a) **Inadequate publication.**—The following do not constitute publication within the meaning of the Code: (1) distribution of microfilms, microcards, or matter reproduced by similar methods; (2) distribution to colleagues or students of a note, even if printed, in explanation of an accompanying illustration; (3) distribution of proof sheets; (4) open-file release; (5) theses, dissertations, and dissertation abstracts; (6) mention at a scientific or other meeting; (7) mention in an abstract, map explanation, or figure caption; (8) labeling of a rock specimen in a collection; (9) mere deposit of a document in a library; (10) anonymous publication; or (11) mention in the popular press or in a legal document.

(b). **Guidebooks.**—A guidebook with distribution limited to participants of a field excursion does not meet the test of availability. Some organizations publish and distribute widely large editions of serial guidebooks that include refereed regional papers; although these do meet the tests of scientific purpose and availability, and therefore constitute valid publication, other media are preferable.

Article 5.—**Intent and Utility.** To be valid, a new unit must serve a clear purpose and be duly proposed and duly described, and the intent to establish it must be specified. Casual mention of a unit, such as "the granite exposed near the Middleville

schoolhouse," does not establish a new formal unit, nor does mere use in a table, columnar section, or map.

Remark. (a) **Demonstration of purpose served.**—The initial definition or revision of a named geologic unit constitutes, in essence, a proposal. As such, it lacks status until use by others demonstrates that a clear purpose has been served. A unit becomes established through repeated demonstration of its utility. The decision not to use a newly proposed or a newly revised term requires a full discussion of its unsuitability.

Article 6.—**Category and Rank.** The category and rank of a new or revised unit must be specified.

Remark. (a) **Need for specification.**—Many stratigraphic controversies have arisen from confusion or misinterpretation of the category of a unit (for example, lithostratigraphic vs. chronostratigraphic). Specification and unambiguous description of the category is of paramount importance. Selection and designation of an appropriate rank from the distinctive terminology developed for each category help serve this function (Table 2).

Article 7.—**Name.** The name of a formal geologic unit is compound. For most categories, the name of a unit should consist of a geographic name combined with an appropriate rank (Wasatch Formation) or descriptive term (Viola Limestone). Biostratigraphic units are designated by appropriate biologic forms (*Exus albus* Assemblage Biozone). Worldwide chronostratigraphic units bear long established and generally accepted names of diverse origins (Triassic System). The first letters of all words used in the names of formal geologic units are capitalized (except for the trivial species and subspecies terms in the name of a biostratigraphic unit).

Remarks. (a) **Appropriate geographic terms.**—Geographic names derived from permanent natural or artificial features at or near which the unit is present are preferable to those derived from impermanent features such as farms, schools, stores, churches, crossroads, and small communities. Appropriate names may be selected from those shown on topographic, state, provincial, county, forest service, hydrographic, or comparable maps, particularly those showing names approved by a national board for geographic names. The generic part of a geographic name, e.g., river, lake, village, should be omitted from new terms, unless required to distinguish between two otherwise identical names (e.g., Redstone Formation and Redstone River Formation). Two names should not be derived from the same geographic feature. A unit should not be named for the source of its components; for example, a deposit inferred to have been derived from the Keewatin glaciation center should not be designated the "Keewatin Till."

(b) **Duplication of names.**—Responsibility for avoiding duplication, either in use of the same name for different units (homonymy) or in use of different names for the same unit (synonymy), rests with the proposer. Although the same geographic term has been applied to different categories of units (example: the lithostratigraphic Word Formation and the chronostratigraphic Wordian Stage) now entrenched in the literature, the practice is undesirable. The extensive geologic nomenclature of North America, including not only names but also nomenclatural history of formal units, is recorded in compendia maintained by the Committee on Stratigraphic Nomenclature of the Geological Survey of Canada, Ottawa, Ontario; by the Geologic Names Committee of the United States Geological Survey, Reston, Virginia; by the Instituto de Geología, Ciudad Universitaria, México, D.F.; and by many state and provincial geological surveys. These organizations respond to inquiries regarding the availability of names, and some are prepared to reserve names for units likely to be defined in the next year or two.

(c) **Priority and preservation of established names.**—Stability of nomenclature is maintained by use of the rule of priority and by preservation of well-established names. Names should not be modified without explaining the need. Priority in publication is to be respected, but priority alone does not justify displacing a well-established name by one

---

[4]This article is modified slightly from a statement by the International Commission of Zoological Nomenclature (1964, p. 7-9).

Table 2. Categories and Ranks of Units Defined in This Code*

A. Material Units

LITHOSTRATIGRAPHIC	LITHODEMIC	MAGNETOPOLARITY	BIOSTRATIGRAPHIC	PEDOSTRATIGRAPHIC	ALLOSTRATIGRAPHIC
Supergroup	Supersuite				
Group	Suite	Polarity Superzone			Allogroup
Formation	Lithodeme	Polarity zone	Biozone (Interval, Assemblage or Abundance)	Geosol	Alloformation
Member (or Lens, or Tongue)		Polarity Subzone	Subbiozone		Allomember
Bed(s) or Flow(s)					

(Complex — spanning Supersuite/Suite in the LITHODEMIC column)

B. Temporal and Related Chronostratigraphic Units

CHRONO-STRATIGRAPHIC	GEOCHRONOLOGIC GEOCHRONOMETRIC	POLARITY CHRONO-STRATIGRAPHIC	POLARITY CHRONOLOGIC	DIACHRONIC
Eonothem	Eon	Polarity Superchronozone	Polarity Superchron	
Erathem (Supersystem)	Era (Superperiod)			
System (Subsystem)	Period (Subperiod)	Polarity Chronozone	Polarity Chron	Episode
Series	Epoch			Phase
Stage (Substage)	Age (Subage)	Polarity Subchronozone	Polarity Subchron	Span
Chronozone	Chron			Cline

(Diachron — spanning Episode/Phase/Span/Cline in the DIACHRONIC column)

*Fundamental units are italicized.

neither well-known nor commonly used; nor should an inadequately established name be preserved merely on the basis of priority. Redefinitions in precise terms are preferable to abandonment of the names of well-established units which may have been defined imprecisely but nonetheless in conformance with older and less stringent standards.

(d) **Differences of spelling and changes in name.**—The geographic component of a well-established stratigraphic name is not changed due to differences in spelling or changes in the name of a geographic feature. The name Bennett Shale, for example, used for more than half a century, need not be altered because the town is named Bennet. Nor should the Mauch Chunk Formation be changed because the town has been renamed Jim Thorpe. Disappearance of an impermanent geographic feature, such as a town, does not affect the name of an established geologic unit.

(e) **Names in different countries and different languages.**—For geologic units that cross local and international boundaries, a single name for each is preferable to several. Spelling of a geographic name commonly conforms to the usage of the country and linguistic group involved. Although geographic names are not translated (Cuchillo is not translated to Knife), lithologic or rank terms are (Edwards Limestone, Caliza Edwards; Formación La Casita, La Casita Formation).

Article 8.—**Stratotypes.** The designation of a unit or boundary stratotype (type section or type locality) is essential in the definition of most formal geologic units. Many kinds of units are best defined by reference to an accessible and specific sequence of rock that may be examined and studied by others. A stratotype is the standard (original or subsequently designated) for a named geologic unit or boundary and constitutes the basis for definition or recognition of that unit or boundary; therefore, it must be

illustrative and representative of the concept of the unit or boundary being defined.

Remarks. (a) **Unit stratotypes.**—A unit stratotype is the type section for a stratiform deposit or the type area for a nonstratiform body that serves as the standard for definition and recognition of a geologic unit. The upper and lower limits of a unit stratotype are designated points in a specific sequence or locality and serve as the standards for definition and recognition of a stratigraphic unit's boundaries.

(b) **Boundary stratotype.**—A boundary stratotype is the type locality for the boundary reference point for a stratigraphic unit. Both boundary stratotypes for any unit need not be in the same section or region. Each boundary stratotype serves as the standard for definition and recognition of the base of a stratigraphic unit. The top of a unit may be defined by the boundary stratotype of the next higher stratigraphic unit.

(c) **Type locality.**—A type locality is the specified geographic locality where the stratotype of a formal unit or unit boundary was originally defined and named. A type area is the geographic territory encompassing the type locality. Before the concept of a stratotype was developed, only type localities and areas were designated for many geologic units which are now long- and well-established. Stratotypes, though now mandatory in defining most stratiform units, are impractical in definitions of many large nonstratiform rock bodies whose diverse major components may be best displayed at several reference localities.

(d) **Composite-stratotype.**—A composite-stratotype consists of several reference sections (which may include a type section) required to demonstrate the range or totality of a stratigraphic unit.

(e) **Reference sections.**—Reference sections may serve as invaluable standards in definitions or revisions of formal geologic units. For those well-established stratigraphic units for which a type section never was specified, a principal reference section (lectostratotype of ISSC, 1976, p. 26) may be designated. A principal reference section (neostratotype of ISSC, 1976, p. 26) also may be designated for those units or boundaries whose stratotypes have been destroyed, covered, or otherwise made inaccessible. Supplementary reference sections often are designated to illustrate the diversity or heterogeneity of a defined unit or some critical feature not evident or exposed in the stratotype. Once a unit or boundary stratotype section is designated, it is never abandoned or changed; however, if a stratotype proves inadequate, it may be supplemented by a principal reference section or by several reference sections that may constitute a composite-stratotype.

(f) **Stratotype descriptions.**—Stratotypes should be described both geographically and geologically. Sufficient geographic detail must be included to enable others to find the stratotype in the field, and may consist of maps and/or aerial photographs showing location and access, as well as appropriate coordinates or bearings. Geologic information should include thickness, descriptive criteria appropriate to the recognition of the unit and its boundaries, and discussion of the relation of the unit to other geologic units of the area. A carefully measured and described section provides the best foundation for definition of stratiform units. Graphic profiles, columnar sections, structure-sections, and photographs are useful supplements to a description; a geologic map of the area including the type locality is essential.

Article 9.—**Unit Description.** A unit proposed for formal status should be described and defined so clearly that any subsequent investigator can recognize that unit unequivocally. Distinguishing features that characterize a unit may include any or several of the following: composition, texture, primary structures, structural attitudes, biologic remains, readily apparent mineral composition (e.g., calcite vs. dolomite), geochemistry, geophysical properties (including magnetic signatures), geomorphic expression, unconformable or cross-cutting relations, and age. Although all distinguishing features pertinent to the unit category should be described sufficiently to characterize the unit, those not pertinent to the category (such as age and inferred genesis for lithostratigraphic units, or lithology for biostratigraphic units) should not be made part of the definition.

Article 10.—**Boundaries.** The criteria specified for the recognition of boundaries between adjoining geologic units are of paramount importance because they provide the basis for scientific reproducibility of results. Care is required in describing the criteria, which must be appropriate to the category of unit involved.

Remarks. (a) **Boundaries between intergradational units.**—Contacts between rocks of markedly contrasting composition are appropriate boundaries of lithic units, but some rocks grade into, or intertongue with, others of different lithology. Consequently, some boundaries are necessarily arbitrary as, for example, the top of the uppermost limestone in a sequence of interbedded limestone and shale. Such arbitrary boundaries commonly are diachronous.

(b) **Overlaps and gaps.**—The problem of overlaps and gaps between long-established adjacent chronostratigraphic units is being addressed by international IUGS and IGCP working groups appointed to deal with various parts of the geologic column. The procedure recommended by the Geological Society of London (George and others, 1969; Holland and others, 1978), of defining only the basal boundaries of chronostratigraphic units, has been widely adopted (e.g., McLaren, 1977) to resolve the problem. Such boundaries are defined by a carefully selected and agreed-upon boundary-stratotype (marker-point type section or "golden spike") which becomes the standard for the base of a chronostratigraphic unit. The concept of the mutual-boundary stratotype (ISSC, 1976, p. 84-86), based on the assumption of continuous deposition in selected sequences, also has been used to define chronostratigraphic units.

Although international chronostratigraphic units of series and higher rank are being redefined by IUGS and IGCP working groups, there may be a continuing need for some provincial series. Adoption of the basal boundary-stratotype concept is urged.

Article 11.—**Historical Background.** A proposal for a new name must include a nomenclatorial history of rocks assigned to the proposed unit, describing how they were treated previously and by whom (references), as well as such matters as priorities, possible synonymy, and other pertinent considerations. Consideration of the historical background of an older unit commonly provides the basis for justifying definition of a new unit.

Article 12.—**Dimensions and Regional Relations.** A perspective on the magnitude of a unit should be provided by such information as may be available on the geographic extent of a unit; observed ranges in thickness, composition, and geomorphic expression; relations to other kinds and ranks of stratigraphic units; correlations with other nearby sequences; and the bases for recognizing and extending the unit beyond the type locality. If the unit is not known anywhere but in an area of limited extent, informal designation is recommended.

Article 13.—**Age.** For most formal material geologic units, other than chronostratigraphic and polarity-chronostratigraphic, inferences regarding geologic age play no proper role in their definition. Nevertheless, the age, as well as the basis for its assignment, are important features of the unit and should be stated. For many lithodemic units, the age of the protolith should be distinguished from that of the metamorphism or deformation. If the basis for assigning an age is tenuous, a doubt should be expressed.

Remarks. (a) **Dating.**—The geochronologic ordering of the rock record, whether in terms of radioactive-decay rates or other processes, is generally called "dating." However, the use of the noun "date" to mean "isotopic age" is not recommended. Similarly, the term "absolute age" should be suppressed in favor of "isotopic age" for an age determined on the basis of isotopic ratios. The more inclusive term "numerical age" is recommended for all ages determined from isotopic ratios, fission tracks, and other quantifiable age-related phenomena.

(b) **Calibration**—The dating of chronostratigraphic boundaries in terms of numerical ages is a special form of dating for which the word "calibration" should be used. The geochronologic time-scale now in use has been developed mainly through such calibration of chronostratigraphic sequences.

(c) **Convention and abbreviations.**—The age of a stratigraphic unit or the time of a geologic event, as commonly determined by numerical dating or by reference to a calibrated time-scale, may be expressed in years before the present. The unit of time is the modern year as presently recognized worldwide. Recommended (but not mandatory) abbreviations for such ages are SI (International System of Units) multipliers coupled with "a" for annum: ka, Ma, and Ga[5] for kilo-annum (10^3 years), Mega-annum (10^6 years), and Giga-annum (10^9 years), respectively. Use of these terms after the age value follows the convention established in the field of C-14 dating. The "present" refers to 1950 AD, and such qualifiers as "ago" or "before the present" are omitted after the value because measurement of the duration from the present to the past is implicit in the designation. In contrast, the duration of a remote interval of geologic time, as a number of years, should not be expressed by the same symbols. Abbreviations for numbers of years, without reference to the present, are informal (e.g., y or yr for years; my, m.y., or m.yr. for

---

[5]Note that the initial letters of Mega- and Giga- are capitalized, but that of kilo- is not, by SI convention.

millions of years; and so forth, as preference dictates). For example, boundaries of the Late Cretaceous Epoch currently are calibrated at 63 Ma and 96 Ma, but the interval of time represented by this epoch is 33 m.y.

(d) **Expression of "age" of lithodemic units.**—The adjectives "early," "middle," and "late" should be used with the appropriate geochronologic term to designate the age of lithodemic units. For example, a granite dated isotopically at 510 Ma should be referred to using the geochronologic term "Late Cambrian granite" rather than either the chronostratigraphic term "Upper Cambrian granite" or the more cumbersome designation "granite of Late Cambrian age."

Article 14.—Correlation. Information regarding spatial and temporal counterparts of a newly defined unit beyond the type area provides readers with an enlarged perspective. Discussions of criteria used in correlating a unit with those in other areas should make clear the distinction between data and inferences.

Article 15.—Genesis. Objective data are used to define and classify geologic units and to express their spatial and temporal relations. Although many of the categories defined in this Code (e.g., lithostratigraphic group, plutonic suite) have genetic connotations, inferences regarding geologic history or specific environments of formation may play no proper role in the definition of a unit. However, observations, as well as inferences, that bear on genesis are of great interest to readers and should be discussed.

Article 16.—Subsurface and Subsea Units. The foregoing procedures for establishing formal geologic units apply also to subsurface and offshore or subsea units. Complete lithologic and paleontologic descriptions or logs of the samples or cores are required in written or graphic form, or both. Boundaries and divisions, if any, of the unit should be indicated clearly with their depths from an established datum.

Remarks. (a) **Naming subsurface units.**—A subsurface unit may be named for the borehole (Eagle Mills Formation), oil field (Smackover Limestone), or mine which is intended to serve as the stratotype, or for a nearby geographic feature. The hole or mine should be located precisely, both with map and exact geographic coordinates, and identified fully (operator or company, farm or lease block, dates drilled or mined, surface elevation and total depth, etc).

(b) **Additional recommendations.**—Inclusion of appropriate borehole geophysical logs is urged. Moreover, rock and fossil samples and cores and all pertinent accompanying materials should be stored, and available for examination, at appropriate federal, state, provincial, university, or museum depositories. For offshore or subsea units (Clipperton Formation of Tracey and others, 1971, p. 22; Argo Salt of McIver, 1972, p. 57), the names of the project and vessel, depth of sea floor, and pertinent regional sampling and geophysical data should be added.

(c) **Seismostratigraphic units.**—High-resolution seismic methods now can delineate stratal geometry and continuity at a level of confidence not previously attainable. Accordingly, seismic surveys have come to be the principal adjunct of the drill in subsurface exploration. On the other hand, the method identifies rock types only broadly and by inference. Thus, formalization of units known only from seismic profiles is inappropriate. Once the stratigraphy is calibrated by drilling, the seismic method may provide objective well-to-well correlations.

REVISION AND ABANDONMENT OF FORMAL UNITS

Article 17.—Requirements for Major Changes. Formally defined and named geologic units may be redefined, revised, or abandoned, but revision and abandonment require as much justification as establishment of a new unit.

Remark. (a) **Distinction between redefinition and revision.**—Redefinition of a unit involves changing the view or emphasis on the content of the unit without changing the boundaries or rank, and differs only slightly from redescription. Neither redefinition nor redescription is considered revision. A redescription corrects an inadequate or inaccurate description, whereas a redefinition may change a descriptive (for example, lithologic) designation. Revision involves either minor changes in the definition of one or both boundaries or in the rank of a unit (normally, elevation to a higher rank). Correction of a misidentification of a unit outside its type area is neither redefinition nor revision.

Article 18.—Redefinition. A correction or change in the descriptive term applied to a stratigraphic or lithodemic unit is a redefinition which does not require a new geographic term.

Remarks. (a) **Change in lithic designation.**—Priority should not prevent more exact lithic designation if the original designation is not everywhere applicable; for example, the Niobrara Chalk changes gradually westward to a unit in which shale is prominent, for which the designation "Niobrara Shale" or "Formation" is more appropriate. Many carbonate formations originally designated "limestone" or "dolomite" are found to be geographically inconsistent as to prevailing rock type. The appropriate lithic term or "formation" is again preferable for such units.

(b) **Original lithic designation inappropriate.**—Restudy of some long-established lithostratigraphic units has shown that the original lithic designation was incorrect according to modern criteria; for example, some "shales" have the chemical and mineralogical composition of limestone, and some rocks described as felsic lavas now are understood to be welded tuffs. Such new knowledge is recognized by changing the lithic designation of the unit, while retaining the original geographic term. Similarly, changes in the classification of igneous rocks have resulted in recognition that rocks originally described as quartz monzonite now are more appropriately termed granite. Such lithic designations may be modernized when the new classification is widely adopted. If heterogeneous bodies of plutonic rock have been misleadingly identified with a single compositional term, such as "gabbro," the adoption of a neutral term, such as "intrusion" or "pluton," may be advisable.

Article 19.—Revision. Revision involves either minor changes in the definition of one or both boundaries of a unit, or in the unit's rank.

Remarks. (a) **Boundary change.**—Revision is justifiable if a minor change in boundary or content will make a unit more natural and useful. If revision modifies only a minor part of the content of a previously established unit, the original name may be retained.

(b) **Change in rank.**—Change in rank of a stratigraphic or temporal unit requires neither redefinition of its boundaries nor alteration of the geographic part of its name. A member may become a formation or vice versa, a formation may become a group or vice versa, and a lithodeme may become a suite or vice versa.

(c) **Examples of changes from area to area.**—The Conasauga Shale is recognized as a formation in Georgia and as a group in eastern Tennessee; the Osgood Formation, Laurel Limestone, and Waldron Shale in Indiana are classed as members of the Wayne Formation in a part of Tennessee; the Virgelle Sandstone is a formation in western Montana and a member of the Eagle Sandstone in central Montana; the Skull Creek Shale and the Newcastle Sandstone in North Dakota are members of the Ashville Formation in Manitoba.

(d) **Example of change in single area.**—The rank of a unit may be changed without changing its content. For example, the Madison Limestone of early work in Montana later became the Madison Group, containing several formations.

(e) **Retention of type section.**—When the rank of a geologic unit is changed, the original type section or type locality is retained for the newly ranked unit (see Article 22c).

(f) **Different geographic name for a unit and its parts.**—In changing the rank of a unit, the same name may not be applied both to the unit as a whole and to a part of it. For example, the Astoria Group should not contain an Astoria Sandstone, nor the Washington Formation, a Washington Sandstone Member.

(g) **Undesirable restriction.**—When a unit is divided into two or more of the same rank as the original, the original name should not be used for any of the divisions. Retention of the old name for one of the units precludes use of the name in a term of higher rank. Furthermore, in order to understand an author's meaning, a later reader would have to know about the modification and its date, and whether the author is following the original or the modified usage. For these reasons, the normal practice is to raise the rank of an established unit when units of the same rank are recognized and mapped within it.

Article 20.—Abandonment. An improperly defined or obsolete stratigraphic, lithodemic, or temporal unit may be formally abandoned, provided that (a) sufficient justification is presented to demonstrate a concern for nomenclatural stability, and (b) recommendations are made for the classification and nomenclature to be used in its place.

Remarks. (a) **Reasons for abandonment.**—A formally defined unit may be abandoned by the demonstration of synonymy or homonymy, of assignment to an improper category (for example, definition of a lithostratigraphic unit in a chronostratigraphic sense), or of other direct violations of a stratigraphic code or procedures prevailing at the time of the original definition. Disuse, or the lack of need or useful purpose for a unit, may be a basis for abandonment; so, too, may widespread misuse in diverse ways which compound confusion. A unit also may be abandoned if it proves impracticable, neither recognizable nor mappable elsewhere.

(b) **Abandoned names.**—A name for a lithostratigraphic or lithodemic unit, once applied and then abandoned, is available for some other unit only if the name was introduced casually, or if it has been published only once in the last several decades and is not in current usage, and if its reintroduction will cause no confusion. An explanation of the history of the name and of the new usage should be a part of the designation.

(c) **Obsolete names.**—Authors may refer to national and provincial records of stratigraphic names to determine whether a name is obsolete (see Article 7b).

(d) **Reference to abandoned names.**—When it is useful to refer to an obsolete or abandoned formal name, its status is made clear by some such term as "abandoned" or "obsolete," and by using a phrase such as "La Plata Sandstone of Cross (1898)". (The same phrase also is used to convey that a named unit has not yet been adopted for usage by the organization involved.)

(e) **Reinstatement.**—A name abandoned for reasons that seem valid at the time, but which subsequently are found to be erroneous, may be reinstated. Example: the Washakie Formation, defined in 1869, was abandoned in 1918 and reinstated in 1973.

CODE AMENDMENT

Article 21.—Procedure for Amendment. Additions to, or changes of, this Code may be proposed in writing to the Commission by any geoscientist at any time. If accepted for consideration by a majority vote of the Commission, they may be adopted by a two-thirds vote of the Commission at an annual meeting not less than a year after publication of the proposal.

FORMAL UNITS DISTINGUISHED BY CONTENT, PROPERTIES, OR PHYSICAL LIMITS

LITHOSTRATIGRAPHIC UNITS

Nature and Boundaries

Article 22.—Nature of Lithostratigraphic Units. A lithostratigraphic unit is a defined body of sedimentary, extrusive igneous, metasedimentary, or metavolcanic strata which is distinguished and delimited on the basis of lithic characteristics and stratigraphic position. A lithostratigraphic unit generally conforms to the Law of Superposition and commonly is stratified and tabular in form.

Remarks. (a) **Basic units.**—Lithostratigraphic units are the basic units of general geologic work and serve as the foundation for delineating strata, local and regional structure, economic resources, and geologic history in regions of stratified rocks. They are recognized and defined by observable rock characteristics; boundaries may be placed at clearly distinguished contacts or drawn arbitrarily within a zone of gradation. Lithification or cementation is not a necessary property; clay, gravel, till, and other unconsolidated deposits may constitute valid lithostratigraphic units.

(b) **Type section and locality.**—The definition of a lithostratigraphic unit should be based, if possible, on a stratotype consisting of readily accessible rocks in place, e.g., in outcrops, excavations, and mines, or of rocks accessible only to remote sampling devices, such as those in drill holes and underwater. Even where remote methods are used, definitions must be based on lithic criteria and not on the geophysical characteristics of the rocks, nor the implied age of their contained fossils. Definitions must be based on descriptions of actual rock material. Regional validity must be demonstrated for all such units. In regions where the stratigraphy has been established through studies of surface exposures, the naming of new units in the subsurface is justified only where the subsurface section differs materially from the surface section, or where there is doubt as to the equivalence of a subsurface and a surface unit. The establishment of subsurface reference sections for units originally defined in outcrop is encouraged.

(c) **Type section never changed.**—The definition and name of a lithostratigraphic unit are established at a type section (or locality) that, once specified, must not be changed. If the type section is poorly designated or delimited, it may be redefined subsequently. If the originally specified stratotype is incomplete, poorly exposed, structurally complicated, or unrepresentative of the unit, a principal reference section or several reference sections may be designated to supplement, but not to supplant, the type section (Article 8e).

(d) **Independence from inferred geologic history.**—Inferred geologic history, depositional environment, and biological sequence have no place in the definition of a lithostratigraphic unit, which must be based on composition and other lithic characteristics; nevertheless, considerations of well-documented geologic history properly may influence the choice of vertical and lateral boundaries of a new unit. Fossils may be valuable during mapping in distinguishing between two lithologically similar, noncontiguous lithostratigraphic units. The fossil content of a lithostratigraphic unit is a legitimate lithic characteristic; for example, oyster-rich sandstone, coquina, coral reef, or graptolitic shale. Moreover, otherwise similar units, such as the Formación Mendez and Formación Velasco mudstones, may be distinguished on the basis of coarseness of contained fossils (foraminifera).

(e) **Independence from time concepts.**—The boundaries of most lithostratigraphic units may transgress time horizons, but some may be approximately synchronous. Inferred time-spans, however measured, play no part in differentiating or determining the boundaries of any lithostratigraphic unit. Either relatively short or relatively long intervals of time may be represented by a single unit. The accumulation of material assigned to a particular unit may have begun or ended earlier in some localities than in others; also, removal of rock by erosion, either within the time-span of deposition of the unit or later, may reduce the time-span represented by the unit locally. The body in some places may be entirely younger than in other places. On the other hand, the establishment of formal units that straddle known, identifiable, regional disconformities is to be avoided, if at all possible. Although concepts of time or age play no part in defining lithostratigraphic units nor in determining their boundaries, evidence of age may aid recognition of similar lithostratigraphic units at localities far removed from the type sections or areas.

(f) **Surface form.**—Erosional morphology or secondary surface form may be a factor in the recognition of a lithostratigraphic unit, but properly should play a minor part at most in the definition of such units.

Because the surface expression of lithostratigraphic units is an important aid in mapping, it is commonly advisable, where other factors do not countervail, to define lithostratigraphic boundaries so as to coincide with lithic changes that are expressed in topography.

(g) **Economically exploited units.**—Aquifers, oil sands, coal beds, and quarry layers are, in general, informal units even though named. Some such units, however, may be recognized formally as beds, members, or formations because they are important in the elucidation of regional stratigraphy.

(h) **Instrumentally defined units.**—In subsurface investigations, certain bodies of rock and their boundaries are widely recognized on borehole geophysical logs showing their electrical resistivity, radioactivity, density, or other physical properties. Such bodies and their boundaries may or may not correspond to formal lithostratigraphic units and their boundaries. Where other considerations do not countervail, the boundaries of subsurface units should be defined so as to correspond to useful geophysical markers; nevertheless, units defined exclusively on the basis of remotely sensed physical properties, although commonly useful in stratigraphic analysis, stand completely apart from the hierarchy of formal lithostratigraphic units and are considered informal.

(i) **Zone.**—As applied to the designation of lithostratigraphic units, the term "zone" is informal. Examples are "producing zone," "mineralized zone," "metamorphic zone," and "heavy-mineral zone." A zone may include all or parts of a bed, a member, a formation, or even a group.

(j) **Cyclothems.**—Cyclic or rhythmic sequences of sedimentary rocks, whose repetitive divisions have been named cyclothems, have been recognized in sedimentary basins around the world. Some cyclothems have been identified by geographic names, but such names are considered informal. A clear distinction must be maintained between the division of a stratigraphic column into cyclothems and its division into groups, formations, and members. Where a cyclothem is identified by a geographic name, the word *cyclothem* should be part of the name, and the geographic term should not be the same as that of any formal unit embraced by the cyclothem.

(k) **Soils and paleosols.**—Soils and paleosols are layers composed of the in-situ products of weathering of older rocks which may be of diverse composition and age. Soils and paleosols differ in several respects from lithostratigraphic units, and should not be treated as such (see "Pedostratigraphic Units," Articles 55 et seq).

(l) **Depositional facies.**—Depositional facies are informal units, whether objective (conglomeratic, black shale, graptolitic) or genetic and environmental (platform, turbiditic, fluvial), even when a geographic term has been applied, e.g., Lantz Mills facies. Descriptive designations convey more information than geographic terms and are preferable.

Article 23.—Boundaries.

Boundaries of lithostratigraphic units are placed at positions of lithic change. Boundaries are placed at distinct contacts or may be fixed arbitrarily within zones of gradation (Fig. 2a). Both vertical and lateral boundaries are based on the lithic criteria that provide the greatest unity and utility.

Remarks. (a) **Boundary in a vertically gradational sequence.**—A named lithostratigraphic unit is preferably bounded by a single lower and a single upper surface so that the name does not recur in a normal stratigraphic succession (see Remark b). Where a rock unit passes vertically into another by intergrading or interfingering of two or more kinds of rock, unless the gradational strata are sufficiently thick to warrant designation of a third, independent unit, the boundary is necessarily arbitrary and should be selected on the basis of practicality (Fig. 2b). For example, where a shale unit overlies a unit of interbedded limestone and shale, the boundary commonly is placed at the top of the highest readily traceable limestone bed. Where a sandstone unit grades upward into shale, the boundary may be so gradational as to be difficult to place even arbitrarily; ideally it should be drawn at the level where the rock is composed of one-half of each component. Because of creep in outcrops and caving in boreholes, it is generally best to define such arbitrary boundaries by the highest occurrence of a particular rock type, rather than the lowest.

(b) **Boundaries in lateral lithologic change.**—Where a unit changes laterally through abrupt gradation into, or intertongues with, a markedly different kind of rock, a new unit should be proposed for the different rock type. An arbitrary lateral boundary may be placed between the two equivalent units. Where the area of lateral intergradation or intertonguing is sufficiently extensive, a transitional interval of interbedded rocks may constitute a third independent unit (Fig. 2c). Where tongues (Article 25b) of formations are mapped separately or otherwise set apart without being formally named, the unmodified formation name should not be repeated in a normal stratigraphic sequence, although the modified name may be repeated in such phrases as "lower tongue of Mancos Shale" and "upper tongue of Mancos Shale." To show the order of superposition on maps and cross sections, the unnamed tongues may be distinguished informally (Fig. 2d) by number, letter, or other means. Such relationships may also be dealt with informally through the recognition of depositional facies (Article 22-1).

(c) **Key beds used for boundaries.**—Key beds (Article 26b) may be used as boundaries for a formal lithostratigraphic unit where the internal lithic characteristics of the unit remain relatively constant. Even though bounding key beds may be traceable beyond the area of the diagnostic overall rock type, geographic extension of the lithostratigraphic unit bounded thereby is not necessarily justified. Where the rock between key beds becomes drastically different from that of the type locality, a new name should be applied (Fig. 2e), even though the key beds are continuous (Article 26b). Stratigraphic and sedimentologic studies of stratigraphic units (usually informal) bounded by key beds may be very informative and useful, especially in subsurface work where the key beds may be recognized by their geophysical signatures. Such units, however, may be a kind of chronostratigraphic, rather than lithostratigraphic, unit (Article 75, 75c), although others are diachronous because one, or both, of the key beds are also diachronous.

(d) **Unconformities as boundaries.**—Unconformities, where recognizable objectively on lithic criteria, are ideal boundaries for lithostratigraphic units. However, a sequence of similar rocks may include an obscure unconformity so that separation into two units may be desirable but impractical. If no lithic distinction adequate to define a widely recognizable boundary can be made, only one unit should be recognized, even though it may include rock that accumulated in different epochs, periods, or eras.

(e) **Correspondence with genetic units.**—The boundaries of lithostratigraphic units should be chosen on the basis of lithic changes and, where feasible, to correspond with the boundaries of genetic units, so that subsequent studies of genesis will not have to deal with units that straddle formal boundaries.

Ranks of Lithostratigraphic Units

Article 24.—Formation.

The formation is the fundamental unit in lithostratigraphic classification. A formation is a body of rock identified by lithic characteristics and stratigraphic position; it is prevailingly but not necessarily tabular and is mappable at the Earth's surface or traceable in the subsurface.

Remarks. (a) **Fundamental unit.**—Formations are the basic lithostratigraphic units used in describing and interpreting the geology of a region. The limits of a formation normally are those surfaces of lithic change that give it the greatest practicable unity of constitution. A formation may represent a long or short time interval, may be composed of materials from one or several sources, and may include breaks in deposition (see Article 23d).

(b) **Content.**—A formation should possess some degree of internal lithic homogeneity or distinctive lithic features. It may contain between its upper and lower limits (i) rock of one lithic type, (ii) repetitions of two or more lithic types, or (iii) extreme lithic heterogeneity which in itself may constitute a form of unity when compared to the adjacent rock units.

(c) **Lithic characteristics.**—Distinctive lithic characteristics include chemical and mineralogical composition, texture, and such supplementary features as color, primary sedimentary or volcanic structures, fossils (viewed as rock-forming particles), or other organic content (coal, oil-shale). A unit distinguishable only by the taxonomy of its fossils is not a lithostratigraphic but a biostratigraphic unit (Article 48). Rock type may be distinctively represented by electrical, radioactive, seismic, or other

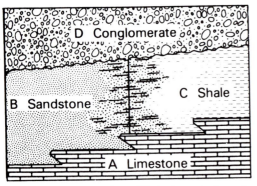

A.--Boundaries at sharp lithologic contacts and in laterally gradational sequence.

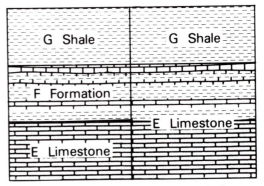

B.--Alternative boundaries in a vertically gradational or interlayered sequence.

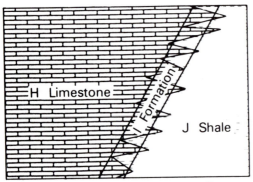

C.--Possible boundaries for a laterally intertonguing sequence.

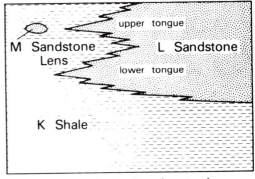

D.--Possible classification of parts of an intertonguing sequence.

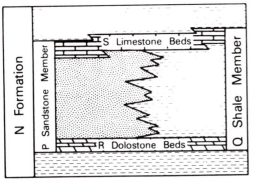

E.--Key beds, here designated the R Dolostone Beds and the S Limestone Beds, are used as boundaries to distinguish the Q Shale Member from the other parts of the N Formation. A lateral change in composition between the key beds requires that another name, P Sandstone Member, be applied. The key beds are part of each member.

EXPLANATION

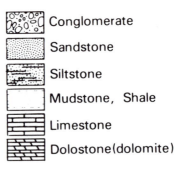

Conglomerate

Sandstone

Siltstone

Mudstone, Shale

Limestone

Dolostone(dolomite)

FIG. 2.—Diagrammatic examples of lithostratigraphic boundaries and classification.

properties (Article 22h), but these properties by themselves do not describe adequately the lithic character of the unit.

(d) **Mappability and thickness.**—The proposal of a new formation must be based on tested mappability. Well-established formations commonly are divisible into several widely recognizable lithostratigraphic units; where formal recognition of these smaller units serves a useful purpose, they may be established as members and beds, for which the requirement of mappability is not mandatory. A unit formally recognized as a formation in one area may be treated elsewhere as a group, or as a member of another formation, without change of name. Example: the Niobrara is mapped at different places as a member of the Mancos Shale, of the Cody Shale, or of the Colorado Shale, and also as the Niobrara Formation, as the Niobrara Limestone, and as the Niobrara Shale.

Thickness is not a determining parameter in dividing a rock succession into formations; the thickness of a formation may range from a feather edge at its depositional or erosional limit to thousands of meters elsewhere. No formation is considered valid that cannot be delineated at the scale of geologic mapping practiced in the region when the formation is proposed. Although representation of a formation on maps and cross sections by a labeled line may be justified, proliferation of such exceptionally thin units is undesirable. The methods of subsurface mapping permit delineation of units much thinner than those usually practicable for surface studies; before such thin units are formalized, consideration should be given to the effect on subsequent surface and subsurface studies.

(e) **Organic reefs and carbonate mounds.**—Organic reefs and carbonate mounds ("buildups") may be distinguished formally, if desirable, as formations distinct from their surrounding, thinner, temporal equivalents. For the requirements of formalization, see Article 30f.

(f) **Interbedded volcanic and sedimentary rock.**—Sedimentary rock and volcanic rock that are interbedded may be assembled into a formation under one name which should indicate the predominant or distinguishing lithology, such as Mindego Basalt.

(g) **Volcanic rock.**—Mappable distinguishable sequences of stratified volcanic rock should be treated as formations or lithostratigraphic units of higher or lower rank. A small intrusive component of a dominantly stratiform volcanic assemblage may be treated informally.

(h) **Metamorphic rock.**—Formations composed of low-grade metamorphic rock (defined for this purpose as rock in which primary structures are clearly recognizable) are, like sedimentary formations, distinguished mainly by lithic characteristics. The mineral facies may differ from place to place, but these variations do not require definition of a new formation. High-grade metamorphic rocks whose relation to established formations is uncertain are treated as lithodemic units (see Articles 31 et seq).

Article 25.—Member. A member is the formal lithostratigraphic unit next in rank below a formation and is always a part of some formation. It is recognized as a named entity within a formation because it possesses characteristics distinguishing it from adjacent parts of the formation. A formation need not be divided into members unless a useful purpose is served by doing so. Some formations may be divided completely into members; others may have only certain parts designated as members; still others may have no members. A member may extend laterally from one formation to another.

Remarks. (a) **Mapping of members.**—A member is established when it is advantageous to recognize a particular part of a heterogeneous formation. A member, whether formally or informally designated, need not be mappable at the scale required for formations. Even if all members of a formation are locally mappable, it does not follow that they should be raised to formational rank, because proliferation of formation names may obscure rather than clarify relations with other areas.

(b) **Lens and tongue.**—A geographically restricted member that terminates on all sides within a formation may be called a lens (lentil). A wedging member that extends outward beyond a formation or wedges ("pinches") out within another formation may be called a tongue.

(c) **Organic reefs and carbonate mounds.**—Organic reefs and carbonate mounds may be distinguished formally, if desirable, as members

within a formation. For the requirements of formalization, see Article 30f.

(d) **Division of members.**—A formally or informally recognized division of a member is called a bed or beds, except for volcanic flow-rocks, for which the smallest formal unit is a flow. Members may contain beds or flows, but may never contain other members.

(e) **Laterally equivalent members.**—Although members normally are in vertical sequence, laterally equivalent parts of a formation that differ recognizably may also be considered members.

Article 26.—Bed(s). A bed, or beds, is the smallest formal lithostratigraphic unit of sedimentary rocks.

Remarks. (a) **Limitations.**—The designation of a bed or a unit of beds as a formally named lithostratigraphic unit generally should be limited to certain distinctive beds whose recognition is particularly useful. Coal beds, oil sands, and other beds of economic importance commonly are named, but such units and their names usually are not a part of formal stratigraphic nomenclature (Articles 22g and 30g).

(b) **Key or marker beds.**—A key or marker bed is a thin bed of distinctive rock that is widely distributed. Such beds may be named, but usually are considered informal units. Individual key beds may be traced beyond the lateral limits of a particular formal unit (Article 23c).

Article 27.—Flow. A flow is the smallest formal lithostratigraphic unit of volcanic flow rocks. A flow is a discrete, extrusive, volcanic body distinguishable by texture, composition, order of superposition, paleomagnetism, or other objective criteria. It is part of a member and thus is equivalent in rank to a bed or beds of sedimentary-rock classification. Many flows are informal units. The designation and naming of flows as formal rock-stratigraphic units should be limited to those that are distinctive and widespread.

Article 28.—Group. A group is the lithostratigraphic unit next higher in rank to formation; a group may consist entirely of named formations, or alternatively, need not be composed entirely of named formations.

Remarks. (a) **Use and content.**—Groups are defined to express the natural relationships of associated formations. They are useful in small-scale mapping and regional stratigraphic analysis. In some reconnaissance work, the term "group" has been applied to lithostratigraphic units that appear to be divisible into formations, but have not yet been so divided. In such cases, formations may be erected subsequently for one or all of the practical divisions of the group.

(b) **Change in component formations.**—The formations making up a group need not necessarily be everywhere the same. The Rundle Group, for example, is widespread in western Canada and undergoes several changes in formational content. In southwestern Alberta, it comprises the Livingstone, Mount Head, and Etherington Formations in the Front Ranges, whereas in the foothills and subsurface of the adjacent plains, it comprises the Pekisko, Shunda, Turner Valley, and Mount Head Formations. However, a formation or its parts may not be assigned to two vertically adjacent groups.

(c) **Change in rank.**—The wedge-out of a component formation or formations may justify the reduction of a group to formation rank, retaining the same name. When a group is extended laterally beyond where it is divided into formations, it becomes in effect a formation, even if it is still called a group. When a previously established formation is divided into two or more component units that are given formal formation rank, the old formation, with its old geographic name, should be raised to group status. Raising the rank of the unit is preferable to restricting the old name to a part of its former content, because a change in rank leaves the sense of a well-established unit unchanged (Articles 19b, 19g).

Article 29.—Supergroup. A supergroup is a formal assemblage of related or superposed groups, or of groups and formations. Such units have proved useful in regional and provincial syntheses. Supergroups should be named only where their recognition serves a clear purpose.

Remark. (a) **Misuse of "series" for group or supergroup.**—Although "series" is a useful general term, it is applied formally only to a chronostratigraphic unit and should not be used for a lithostratigraphic unit. The term "series" should no longer be employed for an assemblage of formations or an assemblage of formations and groups, as it has been, especially in studies of the Precambrian. These assemblages are groups or supergroups.

Lithostratigraphic Nomenclature

Article 30.—**Compound Character.** The formal name of a lithostratigraphic unit is compound. It consists of a geographic name combined with a descriptive lithic term or with the appropriate rank term, or both. Initial letters of all words used in forming the names of formal rock-stratigraphic units are capitalized.

Remarks. (a) **Omission of part of a name.**—Where frequent repetition would be cumbersome, the geographic name, the lithic term, or the rank term may be used alone, once the full name has been introduced; as "the Burlington," "the limestone," or "the formation," for the Burlington Limestone.

(b) **Use of simple lithic terms.**—The lithic part of the name should indicate the predominant or diagnostic lithology, even if subordinate lithologies are included. Where a lithic term is used in the name of a lithostratigraphic unit, the simplest generally acceptable term is recommended (for example, limestone, sandstone, shale, tuff, quartzite). Compound terms (for example, clay shale) and terms that are not in common usage (for example, calcirudite, orthoquartzite) should be avoided. Combined terms, such as "sand and clay," should not be used for the lithic part of the names of lithostratigraphic units, nor should an adjective be used between the geographic and the lithic terms, as "Chattanooga Black Shale" and "Biwabik Iron-Bearing Formation."

(c) **Group names.**—A group name combines a geographic name with the term "group," and no lithic designation is included; for example, San Rafael Group.

(d) **Formation names.**—A formation name consists of a geographic name followed by a lithic designation or by the word "formation." Examples: Dakota Sandstone, Mitchell Mesa Rhyolite, Monmouth Formation, Halton Till.

(e) **Member names.**—All member names include a geographic term and the word "member;" some have an intervening lithic designation, if useful; for example, Wedington Sandstone Member of the Fayetteville Shale. Members designated solely by lithic character (for example, siliceous shale member), by position (upper, lower), or by letter or number, are informal.

(f) **Names of reefs.**—Organic reefs identified as formations or members are formal units only where the name combines a geographic name with the appropriate rank term, e.g., Leduc Formation (a name applied to the several reefs enveloped by the Ireton Formation), Rainbow Reef Member.

(g) **Bed and flow names.**—The names of beds or flows combine a geographic term, a lithic term, and the term "bed" or "flow;" for example, Knee Hills Tuff Bed, Ardmore Bentonite Beds, Negus Variolitic Flows.

(h) **Informal units.**—When geographic names are applied to such informal units as oil sands, coal beds, mineralized zones, and informal members (see Articles 22g and 26a), the unit term should not be capitalized. A name is not necessarily formal because it is capitalized, nor does failure to capitalize a name render it informal. Geographic names should be combined with the terms "formation" or "group" only in formal nomenclature.

(i) **Informal usage of identical geographic names.**—The application of identical geographic names to several minor units in one vertical sequence is considered informal nomenclature (lower Mount Savage coal, Mount Savage fireclay, upper Mount Savage coal, Mount Savage rider coal, and Mount Savage sandstone). The application of identical geographic names to the several lithologic units constituting a cyclothem likewise is considered informal.

(j) **Metamorphic rock.**—Metamorphic rock recognized as a normal stratified sequence, commonly low-grade metavolcanic or metasedimentary rocks, should be assigned to named groups, formations, and members, such as the Deception Rhyolite, a formation of the Ash Creek Group, or the Bonner Quartzite, a formation of the Missoula Group. High-grade metamorphic and metasomatic rocks are treated as lithodemes and suites (see Articles 31, 33, 35).

(k) **Misuse of well-known name.**—A name that suggests some well-known locality, region, or political division should not be applied to a unit typically developed in another less well-known locality of the same name. For example, it would be inadvisable to use the name "Chicago Formation" for a unit in California.

LITHODEMIC UNITS

Nature and Boundaries

Article 31.—**Nature of Lithodemic Units.** A lithodemic[6] unit is a defined body of predominantly intrusive, highly deformed, and/or highly metamorphosed rock, distinguished and delimited on the basis of rock characteristics. In contrast to lithostratigraphic units, a lithodemic unit generally does not conform to the Law of Superposition. Its contacts with other rock units may be sedimentary, extrusive, intrusive, tectonic, or metamorphic (Fig. 3).

Remarks. (a) **Recognition and definition.**—Lithodemic units are defined and recognized by observable rock characteristics. They are the practical units of general geological work in terranes in which rocks generally lack primary stratification; in such terranes they serve as the foundation for studying, describing, and delineating lithology, local and regional structure, economic resources, and geologic history.

(b) **Type and reference localities.**—The definition of a lithodemic unit should be based on as full a knowledge as possible of its lateral and vertical variations and its contact relationships. For purposes of nomenclatural stability, a type locality and, wherever appropriate, reference localities should be designated.

(c) **Independence from inferred geologic history.**—Concepts based on inferred geologic history properly play no part in the definition of a lithodemic unit. Nevertheless, where two rock masses are lithically similar but display objective structural relations that preclude the possibility of their being even broadly of the same age, they should be assigned to different lithodemic units.

(d) **Use of "zone."**—As applied to the designation of lithodemic units, the term "zone" is informal. Examples are: "mineralized zone," "contact zone," and "pegmatitic zone."

Article 32.—**Boundaries.** Boundaries of lithodemic units are placed at positions of lithic change. They may be placed at clearly distinguished contacts or within zones of gradation. Boundaries, both vertical and lateral, are based on the lithic criteria that provide the greatest unity and practical utility. Contacts with other lithodemic and lithostratigraphic units may be depositional, intrusive, metamorphic, or tectonic.

Remark. (a) **Boundaries within gradational zones.**—Where a lithodemic unit changes through gradation into, or intertongues with, a rock-mass with markedly different characteristics, it is usually desirable to propose a new unit. It may be necessary to draw an arbitrary boundary within the zone of gradation. Where the area of intergradation or intertonguing is sufficiently extensive, the rocks of mixed character may constitute a third unit.

Ranks of Lithodemic Units

Article 33.—**Lithodeme.** The lithodeme is the fundamental unit in lithodemic classification. A lithodeme is a body of intru-

---

[6]From the Greek *demas*, *-os:* "living body, frame".

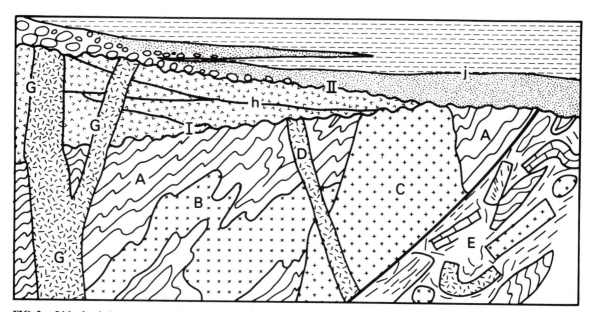

FIG. 3.—Lithodemic (upper case) and lithostratigraphic (lower case) units. A *lithodeme* of *gneiss* (A) contains an *intrusion* of diorite (B) that was deformed with the gneiss. A and B may be treated jointly as a *complex*. A younger *granite* (C) is cut by a dike of *syenite* (D), that is cut in turn by unconformity I. All the foregoing are in fault contact with a *structural complex* (E). A *volcanic complex* (G) is built upon unconformity I, and its feeder dikes cut the unconformity. Laterally equivalent volcanic strata in orderly, mappable succession (h) are treated as lithostratigraphic units. A *gabbro* feeder (G′), to the volcanic complex, where surrounded by gneiss is readily distinguished as a separate lithodeme and named as a *gabbro* or an *intrusion*. All the foregoing are overlain, at unconformity II, by sedimentary rocks (j) divided into formations and members.

sive, pervasively deformed, or highly metamorphosed rock, generally non-tabular and lacking primary depositional structures, and characterized by lithic homogeneity. It is mappable at the Earth's surface and traceable in the subsurface. For cartographic and hierarchical purposes, it is comparable to a formation (see Table 2).

Remarks. (a) **Content.**—A lithodeme should possess distinctive lithic features and some degree of internal lithic homogeneity. It may consist of (i) rock of one type, (ii) a mixture of rocks of two or more types, or (iii) extreme heterogeneity of composition, which may constitute in itself a form of unity when compared to adjoining rock-masses (see also "complex," Article 37).

(b) **Lithic characteristics.**—Distinctive lithic characteristics may include mineralogy, textural features such as grain size, and structural features such as schistose or gneissic structure. A unit distinguishable from its neighbors only by means of chemical analysis is informal.

(c) **Mappability.**—Practicability of surface or subsurface mapping is an essential characteristic of a lithodeme (see Article 24d).

Article 34.—**Division of Lithodemes.** Units below the rank of lithodeme are informal.

Article 35.—**Suite.** A *suite* (metamorphic suite, intrusive suite, plutonic suite) is the lithodemic unit next higher in rank to lithodeme. It comprises two or more associated lithodemes of the same class (e.g., plutonic, metamorphic). For cartographic and hierarchical purposes, suite is comparable to group (see Table 2).

Remarks. (a) **Purpose.**—Suites are recognized for the purpose of expressing the natural relations of associated lithodemes having signifi-

cant lithic features in common, and of depicting geology at compilation scales too small to allow delineation of individual lithodemes. Ideally, a suite consists entirely of named lithodemes, but may contain both named and unnamed units.

(b) **Change in component units.**—The named and unnamed units constituting a suite may change from place to place, so long as the original sense of natural relations and of common lithic features is not violated.

(c) **Change in rank.**—Traced laterally, a suite may lose all of its formally named divisions but remain a recognizable, mappable entity. Under such circumstances, it may be treated as a lithodeme but retain the same name. Conversely, when a previously established lithodeme is divided into two or more mappable divisions, it may be desirable to raise its rank to suite, retaining the original geographic component of the name. To avoid confusion, the original name should not be retained for one of the divisions of the original unit (see Article 19g).

Article 36.—**Supersuite.** A supersuite is the unit next higher in rank to a suite. It comprises two or more suites or complexes having a degree of natural relationship to one another, either in the vertical or the lateral sense. For cartographic and hierarchical purposes, supersuite is similar in rank to supergroup.

Article 37.—**Complex.** An assemblage or mixture of rocks of *two or more genetic classes*, i.e., igneous, sedimentary, or metamorphic, with or without highly complicated structure, may be named a *complex*. The term "complex" takes the place of the lithic or rank term (for example, Boil Mountain Complex, Franciscan Complex) and, although unranked, commonly is comparable to suite or supersuite and is named in the same manner (Articles 41, 42).

Remarks (a) **Use of "complex."**—Identification of an assemblage of diverse rocks as a complex is useful where the mapping of each separate lithic component is impractical at ordinary mapping scales. "Complex" is unranked but commonly comparable to suite or supersuite; therefore, the term may be retained if subsequent, detailed mapping distinguishes some or all of the component lithodemes or lithostratigraphic units.

(b) **Volcanic complex.**—Sites of persistent volcanic activity commonly are characterized by a diverse assemblage of extrusive volcanic rocks, related intrusions, and their weathering products. Such an assemblage may be designated a *volcanic complex*.

(c) **Structural complex.**—In some terranes, tectonic processes (e.g., shearing, faulting) have produced heterogeneous mixtures or disrupted bodies of rock in which some individual components are too small to be mapped. *Where there is no doubt that the mixing or disruption is due to tectonic processes,* such a mixture may be designated as a structural complex, whether it consists of two or more classes of rock, or a single class only. A simpler solution for some mapping purposes is to indicate intense deformation by an overprinted pattern.

(d) **Misuse of "complex".**—Where the rock assemblage to be united under a single, formal name consists of diverse types of a *single class* of rock, as in many terranes that expose a variety of either intrusive igneous or high-grade metamorphic rocks, the term "intrusive suite," "plutonic suite," or "metamorphic suite" should be used, rather than the unmodified term "complex." Exceptions to this rule are the terms *structural complex* and *volcanic complex* (see Remarks c and b, above).

Article 38.—Misuse of "Series" for Suite, Complex, or Supersuite. The term "series" has been employed for an assemblage of lithodemes or an assemblage of lithodemes and suites, especially in studies of the Precambrian. This practice now is regarded as improper; these assemblages are suites, complexes, or supersuites. The term "series" also has been applied to a sequence of rocks resulting from a succession of eruptions or intrusions. In these cases a different term should be used; "group" should replace "series" for volcanic and low-grade metamorphic rocks, and "intrusive suite" or "plutonic suite" should replace "series" for intrusive rocks of group rank.

Lithodemic Nomenclature

Article 39.—General Provisions. The formal name of a lithodemic unit is compound. It consists of a geographic name combined with a descriptive or appropriate rank term. The principles for the selection of the geographic term, concerning suitability, availability, priority, etc., follow those established in Article 7, where the rules for capitalization are also specified.

Article 40.—Lithodeme Names. The name of a lithodeme combines a geographic term with a lithic or descriptive term, e.g., Killarney Granite, Adamant Pluton, Manhattan Schist, Skaergaard Intrusion, Duluth Gabbro. The term *formation* should not be used.

Remarks. (a) **Lithic term.**—The lithic term should be a common and familiar term, such as schist, gneiss, gabbro. Specialized terms and terms not widely used, such as websterite and jacupirangite, and compound terms, such as graphitic schist and augen gneiss, should be avoided.

(b) **Intrusive and plutonic rocks.**—Because many bodies of intrusive rock range in composition from place to place and are difficult to characterize with a single lithic term, and because many bodies of plutonic rock are considered not to be intrusions, latitude is allowed in the choice of a lithic or descriptive term. Thus, the descriptive term should preferably be compositional (e.g., gabbro, granodiorite), but may, if necessary, denote form (e.g., dike, sill), or be neutral (e.g., intrusion, pluton[7]). In any event, specialized compositional terms not widely used are to be avoided, as are form terms that are not widely used, such as bysmalith and chonolith. Terms implying genesis should be avoided as much as possible, because interpretations of genesis may change.

Article 41.—Suite Names. The name of a suite combines a geographic term, the term "suite," and an adjective denoting the fundamental character of the suite; for example, Idaho Springs Metamorphic Suite, Tuolumne Intrusive Suite, Cassiar Plutonic Suite. The geographic name of a suite may not be the same as that of a component lithodeme (see Article 19f). Intrusive assemblages, however, may share the same geographic name if an intrusive lithodeme is representative of the suite.

Article 42.—Supersuite Names. The name of a supersuite combines a geographic term with the term "supersuite."

MAGNETOSTRATIGRAPHIC UNITS

Nature and Boundaries

Article 43.—Nature of Magnetostratigraphic Units. A magnetostratigraphic unit is a body of rock unified by specified remanent-magnetic properties and is distinct from underlying and overlying magnetostratigraphic units having different magnetic properties.

Remarks. (a) **Definition.**—Magnetostratigraphy is defined here as all aspects of stratigraphy based on remanent magnetism (paleomagnetic signatures). Four basic paleomagnetic phenomena can be determined or inferred from remanent magnetism: polarity, dipole-field-pole position (including apparent polar wander), the non-dipole component (secular variation), and field intensity.

(b) **Contemporaneity of rock and remanent magnetism.**—Many paleomagnetic signatures reflect earth magnetism at the time the rock formed. Nevertheless, some rocks have been subjected subsequently to physical and/or chemical processes which altered the magnetic properties. For example, a body of rock may be heated above the blocking temperature or Curie point for one or more minerals, or a ferromagnetic mineral may be produced by low-temperature alteration long after the enclosing rock formed, thus acquiring a component of remanent magnetism reflecting the field at the time of alteration, rather than the time of original rock deposition or crystallization.

(c) **Designations and scope.**—The prefix *magneto* is used with an appropriate term to designate the aspect of remanent magnetism used to define a unit. The terms "magnetointensity" or "magnetosecular-variation" are possible examples. This Code considers only polarity reversals, which now are recognized widely as a stratigraphic tool. However, apparent-polar-wander paths offer increasing promise for correlations within Precambrian rocks.

Article 44.—Definition of Magnetopolarity Unit. A magnetopolarity unit is a body of rock unified by its remanent magnetic polarity and distinguished from adjacent rock that has different polarity.

Remarks. (a) **Nature.**—Magnetopolarity is the record in rocks of the polarity history of the Earth's magnetic-dipole field. Frequent past reversals of the polarity of the Earth's magnetic field provide a basis for magnetopolarity stratigraphy.

(b) **Stratotype.**—A stratotype for a magnetopolarity unit should be designated and the boundaries defined in terms of recognized lithostratigraphic and/or biostratigraphic units in the stratotype. The formal definition of a magnetopolarity unit should meet the applicable specific requirements of Articles 3 to 16.

(c) **Independence from inferred history.**—Definition of a magnetopolarity unit does not require knowledge of the time at which the unit acquired its remanent magnetism; its magnetism may be primary or sec-

---

[7]Pluton—a mappable body of plutonic rock.

ondary. Nevertheless, the unit's present polarity is a property that may be ascertained and confirmed by others.

(d) **Relation to lithostratigraphic and biostratigraphic units.**— Magnetopolarity units resemble lithostratigraphic and biostratigraphic units in that they are defined on the basis of an objective recognizable property, but differ fundamentally in that most magnetopolarity unit boundaries are thought not to be time transgressive. Their boundaries may coincide with those of lithostratigraphic or biostratigraphic units, or be parallel to but displaced from those of such units, or be crossed by them.

(e) **Relation of magnetopolarity units to chronostratigraphic units.**— Although transitions between polarity reversals are of global extent, a magnetopolarity unit does not contain within itself evidence that the polarity is primary, or criteria that permit its unequivocal recognition in chronocorrelative strata of other areas. Other criteria, such as paleontologic or numerical age, are required for both correlation and dating. Although polarity reversals are useful in recognizing chronostratigraphic units, magnetopolarity alone is insufficient for their definition.

Article 45.—Boundaries. The upper and lower limits of a magnetopolarity unit are defined by boundaries marking a change of polarity. Such boundaries may represent either a depositional discontinuity or a magnetic-field transition. The boundaries are either polarity-reversal horizons or polarity transition-zones, respectively.

Remark. (a) **Polarity-reversal horizons and transition-zones.**—A polarity-reversal horizon is either a single, clearly definable surface or a thin body of strata constituting a transitional interval across which a change in magnetic polarity is recorded. Polarity-reversal horizons describe transitional intervals of 1 m or less; where the change in polarity takes place over a stratigraphic interval greater than 1 m, the term "polarity transition-zone" should be used. Polarity-reversal horizons and polarity transition-zones provide the boundaries for polarity zones, although they may also be contained within a polarity zone where they mark an internal change subsidiary in rank to those at its boundaries.

Ranks of Magnetopolarity Units

Article 46.—Fundamental Unit. A polarity zone is the fundamental unit of magnetopolarity classification. A polarity zone is a unit of rock characterized by the polarity of its magnetic signature. Magnetopolarity zone, rather than polarity zone, should be used where there is risk of confusion with other kinds of polarity.

Remarks. (a) **Content.**—A polarity zone should possess some degree of internal homogeneity. It may contain rocks of (1) entirely or predominantly one polarity, or (2) mixed polarity.

(b) **Thickness and duration.**—The thickness of rock of a polarity zone or the amount of time represented should play no part in the definition of the zone. The polarity signature is the essential property for definition.

(c) **Ranks.**—When continued work at the stratotype for a polarity zone, or new work in correlative rocks elsewhere, reveals smaller polarity units, these may be recognized formally as polarity subzones. If it should prove necessary or desirable to group polarity zones, these should be termed polarity superzones. The rank of a polarity unit may be changed when deemed appropriate.

Magnetopolarity Nomenclature

Article 47.—Compound Name. The formal name of a magnetopolarity zone should consist of a geographic name and the term *Polarity Zone.* The term may be modified by *Normal, Reversed,* or *Mixed* (example: Deer Park Reversed Polarity Zone). In naming or revising magnetopolarity units, appropriate parts of Articles 7 and 19 apply. The use of informal designations, e.g., numbers or letters, is not precluded.

BIOSTRATIGRAPHIC UNITS

Nature and Boundaries

Article 48.—Nature of Biostratigraphic Units. A biostratigraphic unit is a body of rock defined or characterized by its fossil content. The basic unit in biostratigraphic classification is the biozone, of which there are several kinds.

Remarks. (a) **Enclosing strata.**—Fossils that define or characterize a biostratigraphic unit commonly are contemporaneous with the body of rock that contains them. Some biostratigraphic units, however, may be represented only by their fossils, preserved in normal stratigraphic succession (e.g., on hardgrounds, in lag deposits, in certain types of remanié accumulations), which alone represent the rock of the biostratigraphic unit. In addition, some strata contain fossils derived from older or younger rocks or from essentially coeval materials of different facies; such fossils should not be used to define a biostratigraphic unit.

(b) **Independence from lithostratigraphic units.**—Biostratigraphic units are based on criteria which differ fundamentally from those for lithostratigraphic units. Their boundaries may or may not coincide with the boundaries of lithostratigraphic units, but they bear no inherent relation to them.

(c) **Independence from chronostratigraphic units.**—The boundaries of most biostratigraphic units, unlike the boundaries of chronostratigraphic units, are both characteristically and conceptually diachronous. An exception is an abundance biozone boundary that reflects a mass-mortality event. The vertical and lateral limits of the rock body that constitutes the biostratigraphic unit represent the limits in distribution of the defining biotic elements. The lateral limits never represent, and the vertical limits rarely represent, regionally synchronous events. Nevertheless, biostratigraphic units are effective for interpreting chronostratigraphic relations.

Article 49.—Kinds of Biostratigraphic Units. Three principal kinds of biostratigraphic units are recognized: *interval, assemblage,* and *abundance* biozones.

Remark: (a) **Boundary definitions.**—Boundaries of interval zones are defined by lowest and/or highest occurrences of single taxa; boundaries of some kinds of assemblage zones (Oppel or concurrent range zones) are defined by lowest and/or highest occurrences of more than one taxon; and boundaries of abundance zones are defined by marked changes in relative abundances of preserved taxa.

Article 50.—Definition of Interval Zone. An interval zone (or subzone) is the body of strata between two specified, documented lowest and/or highest occurrences of single taxa.

Remarks. (a) **Interval zone types.**—Three basic types of interval zones are recognized (Fig. 4). These include the range zones and interval zones of the International Stratigraphic Guide (ISSC, 1976, p. 53, 60) and are:

1. The interval between the documented lowest and highest occurrences of a single taxon (Fig. 4A). This is the *taxon range zone* of ISSC (1976, p. 53).

2. The interval included between the documented lowest occurrence of one taxon and the documented highest occurrence of another taxon (Fig. 4B). When such occurrences result in stratigraphic overlap of the taxa (Fig. 4B-1), the interval zone is the *concurrent range zone* of ISSC (1976, p. 55), that involves only two taxa. When such occurrences do not result in stratigraphic overlap (Fig. 4B-2), but are used to partition the range of a third taxon, the interval is the *partial range zone* of George and others (1969).

3. The interval between documented successive lowest occurrences or successive highest occurrences of two taxa (Fig. 4C). When the interval is between successive documented lowest occurrences within an evolutionary lineage (Fig. 4C-1), it is the *lineage zone* of ISSC (1976, p. 58). When

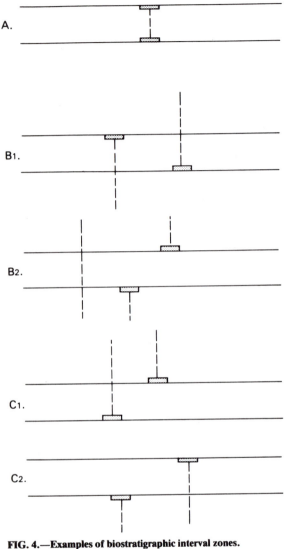

FIG. 4.—Examples of biostratigraphic interval zones.
Vertical broken lines indicate ranges of taxa; bars indicate lowest or highest documented occurrences.

the interval is between successive lowest occurrences of unrelated taxa or between successive highest occurrences of either related or unrelated taxa (Fig. 4C-2), it is a kind of *interval zone* of ISSC (1976, p. 60).

(b) **Unfossiliferous intervals.**—Unfossiliferous intervals between or within biozones are the *barren interzones* and *intrazones* of ISSC (1976, p. 49).

Article 51.—**Definition of Assemblage Zone.** An assemblage zone is a biozone characterized by the association of three or more taxa. It may be based on all kinds of fossils present, or restricted to only certain kinds of fossils.

Remarks. (a) **Assemblage zone contents.**—An assemblage zone may consist of a geographically or stratigraphically restricted assemblage, or may incorporate two or more contemporaneous assemblages with shared characterizing taxa (*composite assemblage zones* of Kauffman, 1969) (Fig. 5c).

(b) **Assemblage zone types.**—In practice, two assemblage zone concepts are used:

1. The *assemblage zone* (or cenozone) of ISSC (1976, p. 50), which is characterized by taxa without regard to their range limits (Fig. 5a). Recognition of this type of assemblage zone can be aided by using techniques of multivariate analysis. Careful designation of the characterizing taxa is especially important.

2. The *Oppel zone,* or the *concurrent range zone* of ISSC (1976, p. 55, 57), a type of zone characterized by more than two taxa and having boundaries based on two or more documented first and/or last occurrences of the included characterizing taxa (Fig. 5b).

Article 52.—**Definition of Abundance Zone.** An abundance zone is a biozone characterized by quantitatively distinctive maxima of relative abundance of one or more taxa. This is the *acme zone* of ISSC (1976, p. 59).

Remark. (a) **Ecologic controls.**—The distribution of biotic assemblages used to characterize some assemblage and abundance biozones may reflect strong local ecological control. Biozones based on such assemblages are included within the concept of ecozones (Vella, 1964), and are informal.

Ranks of Biostratigraphic Units

Article 53.—**Fundamental Unit.** The fundamental unit of biostratigraphic classification is a biozone.

Remarks. (a) **Scope.**—A single body of rock may be divided into various kinds and scales of biozones or subzones, as discussed in the International Stratigraphic Guide (ISSC, 1976, p. 62). Such usage is recommended if it will promote clarity, but only the unmodified term *biozone* is accorded formal status.

(b) **Divisions.**—A biozone may be completely or partly divided into formally designated sub-biozones (subzones), if such divisions serve a useful purpose.

Biostratigraphic Nomenclature

Article 54.—**Establishing Formal Units.** Formal establishment of a biozone or subzone must meet the requirements of Article 3 and requires a unique name, a description of its content and its boundaries, reference to a stratigraphic sequence in which the zone is characteristically developed, and a discussion of its spatial extent.

Remarks. (a) **Name.**—The name, which is compound and designates the kind of biozone, may be based on:

1. One or two characteristic and common taxa that are restricted to the biozone, reach peak relative abundance within the biozone, or have their total stratigraphic overlap within the biozone. These names most commonly are those of genera or subgenera, binomial designations of species, or trinomial designations of subspecies. If names of the nominate taxa change, names of the zones should be changed accordingly. Generic or subgeneric names may be abbreviated. Trivial species or subspecies names should not be used alone because they may not be unique.

2. Combinations of letters derived from taxa which characterize the biozone. However, alpha-numeric code designations (e.g., N1, N2, N3...) are informal and not recommended because they do not lend themselves readily to subsequent insertions, combinations, or eliminations. Biozonal

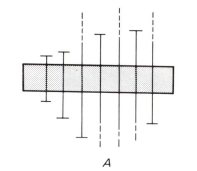

A

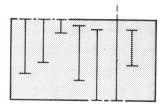

B

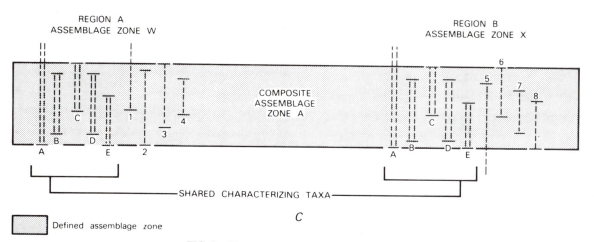

C

FIG. 5.—Examples of assemblage zone concepts.

systems based *only* on simple progressions of letters or numbers (e.g., A, B, C, or 1, 2, 3) are also not recommended.

(b) **Revision.**—Biozones and subzones are established empirically and may be modified on the basis of new evidence. Positions of established biozone or subzone boundaries may be stratigraphically refined, new characterizing taxa may be recognized, or original characterizing taxa may be superseded. If the concept of a particular biozone or subzone is substantially modified, a new unique designation is required to avoid ambiguity in subsequent citations.

(c) **Specifying kind of zone.**—Initial designation of a formally proposed biozone or subzone as an abundance zone, or as one of the types of interval zones, or assemblage zones (Articles 49-52), is strongly recommended. Once the type of biozone is clearly identified, the designation may be dropped in the remainder of a text (e.g., *Exus albus* taxon range zone to *Exus albus* biozone).

(d) **Defining taxa.**—Initial description or subsequent emendation of a biozone or subzone requires designation of the defining and characteristic taxa, and/or the documented first and last occurrences which mark the biozone or subzone boundaries.

(e) **Stratotypes.**—The geographic and stratigraphic position and boundaries of a formally proposed biozone or subzone should be defined precisely or characterized in one or more designated reference sections. Designation of a stratotype for each new biostratigraphic unit and of reference sections for emended biostratigraphic units is required.

PEDOSTRATIGRAPHIC UNITS

Nature and Boundaries

Article 55.—**Nature of Pedostratigraphic Units.** A pedostratigraphic unit is a body of rock that consists of one or more pedologic horizons developed in one or more lithostratigraphic, allostratigraphic, or lithodemic units (Fig. 6) and is overlain by one or more formally defined lithostratigraphic or allostratigraphic units.

Remarks. (a) **Definition.**—A pedostratigraphic[8] unit is a buried, traceable, three-dimensional body of rock that consists of one or more differentiated pedologic horizons.

(b) **Recognition.**—The distinguishing property of a pedostratigraphic unit is the presence of one or more distinct, differentiated, pedologic horizons. Pedologic horizons are products of soil development (pedogenesis) which occurred subsequent to formation of the lithostrati-

---

[8]Terminology related to pedostratigraphic classification is summarized on page 850.

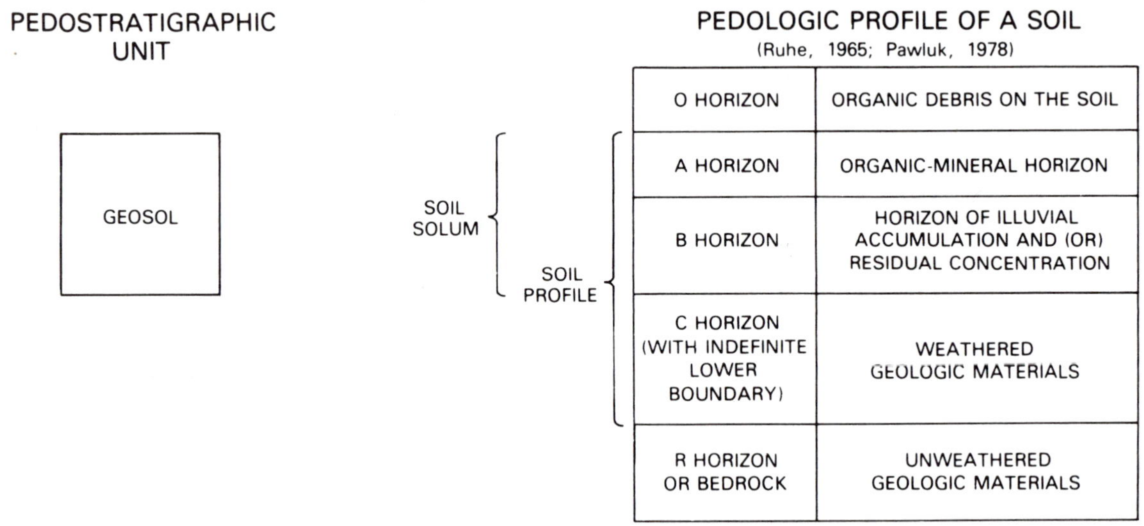

PEDOSTRATIGRAPHIC
UNIT

PEDOLOGIC PROFILE OF A SOIL

(Ruhe, 1965; Pawluk, 1978)

GEOSOL		O HORIZON	ORGANIC DEBRIS ON THE SOIL

SOIL
SOLUM

SOIL
PROFILE

O HORIZON	ORGANIC DEBRIS ON THE SOIL
A HORIZON	ORGANIC-MINERAL HORIZON
B HORIZON	HORIZON OF ILLUVIAL ACCUMULATION AND (OR) RESIDUAL CONCENTRATION
C HORIZON (WITH INDEFINITE LOWER BOUNDARY)	WEATHERED GEOLOGIC MATERIALS
R HORIZON OR BEDROCK	UNWEATHERED GEOLOGIC MATERIALS

FIG. 6.—Relationship between pedostratigraphic units and pedologic profiles.
The base of a geosol is the lowest clearly defined physical boundary of a pedologic horizon in a buried soil profile. In this example it is the lower boundary of the B horizon because the base of the C horizon is not a clearly defined physical boundary. In other profiles the base may be the lower boundary of a C horizon.

graphic, allostratigraphic, or lithodemic unit or units on which the buried soil was formed; these units are the parent materials in which pedogenesis occurred. Pedologic horizons are recognized in the field by diagnostic features such as color, soil structure, organic-matter accumulation, texture, clay coatings, stains, or concretions. Micromorphology, particle size, clay mineralogy, and other properties determined in the laboratory also may be used to identify and distinguish pedostratigraphic units.

(c) **Boundaries and stratigraphic position.**—The upper boundary of a pedostratigraphic unit is the top of the uppermost pedologic horizon formed by pedogenesis in a buried soil profile. The lower boundary of a pedostratigraphic unit is the lowest *definite* physical boundary of a pedologic horizon within a buried soil profile. The stratigraphic position of a pedostratigraphic unit is determined by its relation to overlying and underlying stratigraphic units (see Remark d).

(d) **Traceability.**—Practicability of subsurface tracing of the upper boundary of a buried soil is essential in establishing a pedostratigraphic unit because (1) few buried soils are exposed continuously for great distances, (2) the physical and chemical properties of a specific pedostratigraphic unit may vary greatly, both vertically and laterally, from place to place, and (3) pedostratigraphic units of different stratigraphic significance in the same region generally do not have unique identifying physical and chemical characteristics. Consequently, extension of a pedostratigraphic unit is accomplished by lateral tracing of the contact between a buried soil and an overlying, formally defined lithostratigraphic or allostratigraphic unit, or between a soil and two or more demonstrably correlative stratigraphic units.

(e) **Distinction from pedologic soils.**—Pedologic soils may include organic deposits (e.g., litter zones, peat deposits, or swamp deposits) that overlie or grade laterally into differentiated buried soils. The organic deposits are not products of pedogenesis, and O horizons are not included in a pedostratigraphic unit (Fig. 6); they may be classified as biostratigraphic or lithostratigraphic units. Pedologic soils also include the entire C horizon of a soil. The C horizon in pedology is not rigidly defined; it is merely the part of a soil profile that underlies the B horizon. The base of the C horizon in many soil profiles is gradational or unidentifiable; commonly it is placed arbitrarily. The need for clearly defined and easily recognized physical boundaries for a stratigraphic unit requires that the

lower boundary of a pedostratigraphic unit be defined as the lowest *definite* physical boundary of a pedologic horizon in a buried soil profile, and part or all of the C horizon may be excluded from a pedostratigraphic unit.

(f) **Relation to saprolite and other weathered materials.**—A material derived by in situ weathering of lithostratigraphic, allostratigraphic, and(or) lithodemic units (e.g., saprolite, bauxite, residuum) may be the parent material in which pedologic horizons form, but is not a pedologic soil. A pedostratigraphic unit may be based on the pedologic horizons of a buried soil developed in the product of in-situ weathering, such as saprolite. The parents of such a pedostratigraphic unit are both the saprolite and, indirectly, the rock from which it formed.

(g) **Distinction from other stratigraphic units.**—A pedostratigraphic unit differs from other stratigraphic units in that (1) it is a product of surface alteration of one or more older material units by specific processes (pedogenesis), (2) its lithology and other properties differ markedly from those of the parent material(s), and (3) a single pedostratigraphic unit may be formed in situ in parent material units of diverse compositions and ages.

(h) **Independence from time concepts.**—The boundaries of a pedostratigraphic unit are time-transgressive. Concepts of time spans, however measured, play no part in defining the boundaries of a pedostratigraphic unit. Nonetheless, evidence of age, whether based on fossils, numerical ages, or geometrical or other relationships, may play an important role in distinguishing and identifying non-contiguous pedostratigraphic units at localities away from the type areas. The name of a pedostratigraphic unit should be chosen from a geographic feature in the type area, and not from a time span.

Pedostratigraphic Nomenclature and Unit

Article 56.—**Fundamental Unit.** The fundamental and only unit in pedostratigraphic classification is a geosol.

Article 57.—**Nomenclature.**—The formal name of a pedostratigraphic unit consists of a geographic name combined with the term "geosol." Capitalization of the initial letter in each word

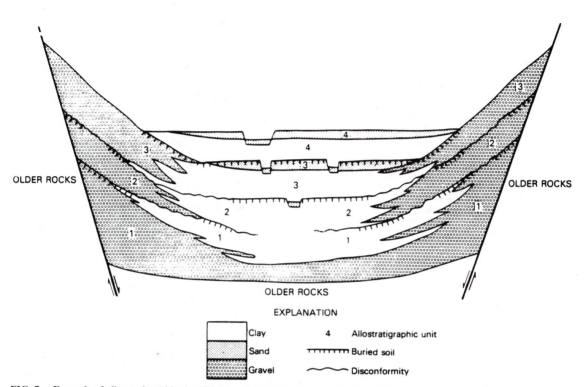

FIG. 7.—**Example of allostratigraphic classification of alluvial and lacustrine deposits in a graben.**

The alluvial and lacustrine deposits may be included in a single formation, or may be separated laterally into formations distinguished on the basis of contrasting texture (gravel, clay). Textural changes are abrupt and sharp, both vertically and laterally. The gravel deposits and clay deposits, respectively, are lithologically similar and thus cannot be distinguished as members of a formation. Four allostratigraphic units, each including two or three textural facies, may be defined on the basis of laterally traceable discontinuities (buried soils and disconformities).

serves to identify formal usage. The geographic name should be selected in accordance with recommendations in Article 7 and should not duplicate the name of another formal geologic unit. Names based on subjacent and superjacent rock units, for example the super-Wilcox–sub-Claiborne soil, are informal, as are those with time connotations (post-Wilcox–pre-Claiborne soil).

Remarks. (a) **Composite geosols.**—Where the horizons of two or more merged or "welded" buried soils can be distinguished, formal names of pedostratigraphic units based on the horizon boundaries can be retained. Where the horizon boundaries of the respective merged or "welded" soils cannot be distinguished, formal pedostratigraphic classification is abandoned and a combined name such as Hallettville-Jamesville geosol may be used informally.

(b) **Characterization.**—The physical and chemical properties of a pedostratigraphic unit commonly vary vertically and laterally throughout the geographic extent of the unit. A pedostratigraphic unit is characterized by the *range* of physical and chemical properties of the unit in the type area, rather than by "typical" properties exhibited in a type section. Consequently, a pedostratigraphic unit is characterized on the basis of a composite stratotype (Article 8d).

(c) **Procedures for establishing formal pedostratigraphic units.**—A formal pedostratigraphic unit may be established in accordance with the

applicable requirements of Article 3, and additionally by describing major soil horizons in each soil facies.

ALLOSTRATIGRAPHIC UNITS

Nature and Boundaries

Article 58.—**Nature of Allostratigraphic Units.** An allostratigraphic[9] unit is a mappable stratiform body of sedimentary rock that is defined and identified on the basis of its bounding discontinuities.

Remarks. (a) **Purpose.**—Formal allostratigraphic units may be defined to distinguish between different (1) superposed discontinuity-bounded deposits of similar lithology (Figs. 7, 9), (2) contiguous discontinuity-bounded deposits of similar lithology (Fig. 8), or (3) geographically separated discontinuity-bounded units of similar lithology (Fig. 9), or to distinguish as single units discontinuity-bounded deposits characterized by lithic heterogeneity (Fig. 8).

(b) **Internal characteristics.**—Internal characteristics (physical, chemical, and paleontological) may vary laterally and vertically throughout the unit.

(c) **Boundaries.**—Boundaries of allostratigraphic units are laterally traceable discontinuities (Figs. 7, 8, and 9).

(d) **Mappability.**—A formal allostratigraphic unit must be mappable at the scale practiced in the region where the unit is defined.

[9]From the Greek *allo*: "other, different."

(e) **Type locality and extent.**—A type locality and type area must be designated; a composite stratotype or a type section and several reference sections are desirable. An allostratigraphic unit may be laterally contiguous with a formally defined lithostratigraphic unit; a vertical cut-off between such units is placed where the units meet.

(f) **Relation to genesis.**—Genetic interpretation is an inappropriate basis for defining an allostratigraphic unit. However, genetic interpretation may influence the choice of its boundaries.

(g) **Relation to geomorphic surfaces.**—A geomorphic surface may be used as a boundary of an allostratigraphic unit, but the unit should not be given the geographic name of the surface.

(h) **Relation to soils and paleosols.**—Soils and paleosols are composed of products of weathering and pedogenesis and differ in many respects from allostratigraphic units, which are depositional units (see "Pedostratigraphic Units," Article 55). The upper boundary of a surface or buried soil may be used as a boundary of an allostratigraphic unit.

(i) **Relation to inferred geologic history.**—Inferred geologic history is not used to define an allostratigraphic unit. However, well-documented geologic history may influence the choice of the unit's boundaries.

(j) **Relation to time concepts.**—Inferred time spans, however measured, are not used to define an allostratigraphic unit. However, age relationships may influence the choice of the unit's boundaries.

(k) **Extension of allostratigraphic units.**—An allostratigraphic unit is extended from its type area by tracing the boundary discontinuities or by tracing or matching the deposits between the discontinuities.

Ranks of Allostratigraphic Units

Article 59.—**Hierarchy.** The hierarchy of allostratigraphic units, in order of decreasing rank, is allogroup, alloformation, and allomember.

Remarks. (a) **Alloformation.**—The alloformation is the fundamental unit in allostratigraphic classification. An alloformation may be completely or only partly divided into allomembers, if some useful purpose is served, or it may have no allomembers.

(b) **Allomember.**—An allomember is the formal allostratigraphic unit next in rank below an alloformation.

(c) **Allogroup.**—An allogroup is the allostratigraphic unit next in rank above an alloformation. An allogroup is established only if a unit of that rank is essential to elucidation of geologic history. An allogroup may consist entirely of named alloformations or, alternatively, may contain one or more named alloformations which jointly do not comprise the entire allogroup.

(d) **Changes in rank.**—The principles and procedures for elevation and reduction in rank of formal allostratigraphic units are the same as those in Articles 19b, 19g, and 28.

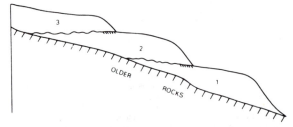

FIG. 8.—**Example of allostratigraphic classification of contiguous deposits of similar lithology.**
Allostratigraphic units 1, 2, and 3 are physical records of three glaciations. They are lithologically similar, reflecting derivation from the same bedrock, and constitute a single lithostratigraphic unit.

Allostratigraphic Nomenclature

Article 60.—**Nomenclature.** The principles and procedures for naming allostratigraphic units are the same as those for naming of lithostratigraphic units (see Articles 7, 30).

Remark. (a) **Revision.**—Allostratigraphic units may be revised or otherwise modified in accordance with the recommendations in Articles 17 to 20.

FORMAL UNITS DISTINGUISHED BY AGE

GEOLOGIC-TIME UNITS

Nature and Types

Article 61.—**Types.** Geologic-time units are conceptual, rather than material, in nature. Two types are recognized: those based on material standards or referents (specific rock sequences or bodies), and those independent of material referents (Fig. 1).

Units Based on Material Referents

Article 62.—**Types Based on Referents.** Two types of formal geologic-time units based on material referents are recognized: they are isochronous and diachronous units.

Article 63.—**Isochronous Categories.** Isochronous time units and the material bodies from which they are derived are twofold: geochronologic units (Article 80), which are based on corresponding material chronostratigraphic units (Article 66), and polarity-geochronologic units (Article 88), based on corresponding material polarity-chronostratigraphic units (Article 83).

Remark. (a) **Extent.**—Isochronous units are applicable worldwide; they may be referred to even in areas lacking a material record of the named span of time. The duration of the time may be represented by a unit-stratotype referent. The beginning and end of the time are represented by point-boundary-stratotypes either in a single stratigraphic sequence or in separate stratotype sections (Articles 8b, 10b).

Article 64.—**Diachronous Categories.** Diachronic units (Article 91) are time units corresponding to diachronous material allostratigraphic units (Article 58), pedostratigraphic units (Article 55), and most lithostratigraphic (Article 22) and biostratigraphic (Article 48) units.

Remarks. (a) **Diachroneity.**—Some lithostratigraphic and biostratigraphic units are clearly diachronous, whereas others have boundaries which are not demonstrably diachronous within the resolving power of available dating methods. The latter commonly are treated as isochronous and are used for purposes of chronocorrelation (see biochronozone, Article 75). However, the assumption of isochroneity must be tested continually.

(b) **Extent.**—Diachronic units are coextensive with the diachronous material stratigraphic units on which they are based and are not used beyond the extent of their material referents.

Units Independent of Material Referents

Article 65.—**Numerical Divisions of Time.** Isochronous geologic-time units based on numerical divisions of time in years are geochronometric units (Article 96) and have no material referents.

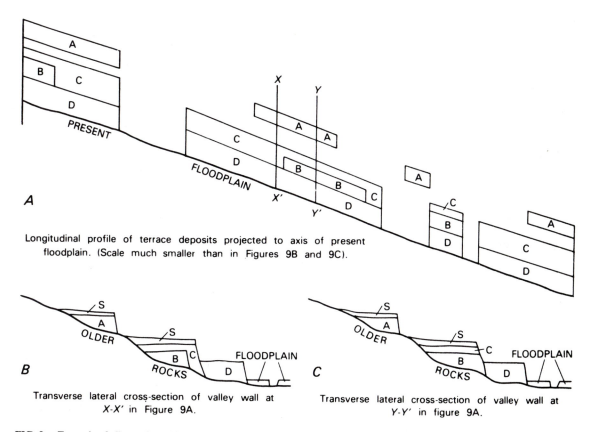

Longitudinal profile of terrace deposits projected to axis of present floodplain. (Scale much smaller than in Figures 9B and 9C).

Transverse lateral cross-section of valley wall at X-X′ in Figure 9A.

Transverse lateral cross-section of valley wall at Y-Y′ in figure 9A.

FIG. 9.—Example of allostratigraphic classification of lithologically similar, discontinuous terrace deposits.
A, B, C, and D are terrace gravel units of similar lithology at different topographic positions on a valley wall. The deposits may be defined as separate formal allostratigraphic units if such units are useful and if bounding discontinuities can be traced laterally. Terrace gravels of the same age commonly are separated geographically by exposures of older rocks. Where the bounding discontinuities cannot be traced continuously, they may be extended geographically on the basis of objective correlation of internal properites of the deposits other than lithology (e.g., fossil content, included tephras), topographic position, numerical ages, or relative-age criteria (e.g., soils or other weathering phenomena). The criteria for such extension should be documented. Slope deposits and eolian deposits (S) that mantle terrace surfaces may be of diverse ages and are not included in a terrace-gravel allostratigraphic unit. A single terrace surface may be underlain by more than one allostratigraphic unit (units B and C in sections b and c).

CHRONOSTRATIGRAPHIC UNITS

Nature and Boundaries

Article 66.—**Definition.** A chronostratigraphic unit is a body of rock established to serve as the material reference for all rocks formed during the same span of time. Each of its boundaries is synchronous. The body also serves as the basis for defining the specific interval of time, or geochronologic unit (Article 80), represented by the referent.

Remarks. (a) **Purposes.**—Chronostratigraphic classification provides a means of establishing the temporally sequential order of rock bodies. Principal purposes are to provide a framework for (1) temporal correlation of the rocks in one area with those in another, (2) placing the rocks of the Earth's crust in a systematic sequence and indicating their relative position and age with respect to earth history as a whole, and (3) constructing an internationally recognized Standard Global Chronostratigraphic Scale.

(b) **Nature.**—A chronostratigraphic unit is a material unit and consists of a body of strata formed during a specific time span. Such a unit represents all rocks, and only those rocks, formed during that time span.

(c) **Content.**—A chronostratigraphic unit may be based upon the time span of a biostratigraphic unit, a lithic unit, a magnetopolarity unit, or any other feature of the rock record that has a time range. Or it may be any arbitrary but specified sequence of rocks, provided it has properties allowing chronocorrelation with rock sequences elsewhere.

Article 67.—**Boundaries.** Boundaries of chronostratigraphic units should be defined in a designated stratotype on the basis of observable paleontological or physical features of the rocks.

Remark. (a) **Emphasis on lower boundaries of chronostratigraphic units.**—Designation of point boundaries for both base and top of chronostratigraphic units is not recommended, because subsequent information on relations between successive units may identify overlaps or gaps. One means of minimizing or eliminating problems of duplication or gaps in chronostratigraphic successions is to define formally as a point-boundary stratotype only the base of the unit. Thus, a chronostratigraphic unit with its base defined at one locality, will have its top defined by the base of an overlying unit at the same, but more commonly another, locality (Article 8b).

Article 68.—Correlation. Demonstration of time equivalence is required for geographic extension of a chronostratigraphic unit from its type section or area. Boundaries of chronostratigraphic units can be extended only within the limits of resolution of available means of chronocorrelation, which currently include paleontology, numerical dating, remanent magnetism, thermoluminescence, relative-age criteria (examples are superposition and cross-cutting relations), and such indirect and inferential physical criteria as climatic changes, degree of weathering, and relations to unconformities. Ideally, the boundaries of chronostratigraphic units are independent of lithology, fossil content, or other material bases of stratigraphic division, but, in practice, the correlation or geographic extension of these boundaries relies at least in part on such features. Boundaries of chronostratigraphic units commonly are intersected by boundaries of most other kinds of material units.

Ranks of Chronostratigraphic Units

Article 69.—Hierarchy. The hierarchy of chronostratigraphic units, in order of decreasing rank, is eonothem, erathem, system, series, and stage. Of these, system is the primary unit of worldwide major rank; its primacy derives from the history of development of stratigraphic classification. All systems and units of higher rank are divided completely into units of the next lower rank. Chronozones are non-hierarchical and commonly lower-rank chronostratigraphic units. Stages and chronozones in sum do not necessarily equal the units of next higher rank and need not be contiguous. The rank and magnitude of chronostratigraphic units are related to the time interval represented by the units, rather than to the thickness or areal extent of the rocks on which the units are based.

Article 70.—Eonothem. The unit highest in rank is eonothem. The Phanerozoic Eonothem encompasses the Paleozoic, Mesozoic, and Cenozoic Erathems. Although older rocks have been assigned heretofore to the Precambrian Eonothem, they also have been assigned recently to other (Archean and Proterozoic) eonothems by the IUGS Precambrian Subcommission. The span of time corresponding to an eonothem is an *eon*.

Article 71.—Erathem. An erathem is the formal chronostratigraphic unit of rank next lower to eonothem and consists of several adjacent systems. The span of time corresponding to an erathem is an *era*.

Remark. (a) **Names.**—Names given to traditional Phanerozoic erathems were based upon major stages in the development of life on Earth: Paleozoic (old), Mesozoic (intermediate), and Cenozoic (recent) life. Although somewhat comparable terms have been applied to Precambrian units, the names and ranks of Precambrian divisions are not yet universally agreed upon and are under consideration by the IUGS Subcommission on Precambrian Stratigraphy.

Article 72.—System. The unit of rank next lower to erathem is the system. Rocks encompassed by a system represent a time-span and an episode of Earth history sufficiently great to serve as a worldwide chronostratigraphic reference unit. The temporal equivalent of a system is a *period*.

Remark. (a) **Subsystem and supersystem.**—Some systems initially established in Europe later were divided or grouped elsewhere into units ranked as systems. *Subsystems* (Mississippian Subsystem of the Carboniferous System) and *supersystems* (Karoo Supersystem) are more appropriate.

Article 73.—Series. Series is a conventional chronostratigraphic unit that ranks below a system and always is a division of a system. A series commonly constitutes a major unit of chronostratigraphic correlation within a province, between provinces, or between continents. Although many European series are being adopted increasingly for dividing systems on other continents, provincial series of regional scope continue to be useful. The temporal equivalent of a series is an *epoch*.

Article 74.—Stage. A stage is a chronostratigraphic unit of smaller scope and rank than a series. It is most commonly of greatest use in intra-continental classification and correlation, although it has the potential for worldwide recognition. The geochronologic equivalent of stage is *age*.

Remark. (a) **Substage.**—Stages may be, but need not be, divided completely into substages.

Article 75.—Chronozone. A chronozone is a non-hierarchical, but commonly small, formal chronostratigraphic unit, and its boundaries may be independent of those of ranked units. Although a chronozone is an isochronous unit, it may be based on a biostratigraphic unit (example: *Cardioceras cordatum* Biochronozone), a lithostratigraphic unit (Woodbend Lithochronozone), or a magnetopolarity unit (Gilbert Reversed-Polarity Chronozone). Modifiers (litho-, bio-, polarity) used in formal names of the units need not be repeated in general discussions where the meaning is evident from the context, e.g., *Exus albus* Chronozone.

Remarks. (a) **Boundaries of chronozones.**—The base and top of a *chronozone* correspond in the unit's stratotype to the observed, defining, physical and paleontological features, but they are extended to other areas by any means available for recognition of synchroneity. The temporal equivalent of a chronozone is a chron.

(b) **Scope.**—The scope of the non-hierarchical chronozone may range markedly, depending upon the purpose for which it is defined either formally or informally. The informal "biochronozone of the ammonites," for example, represents a duration of time which is enormous and exceeds that of a system. In contrast, a biochronozone defined by a species of limited range, such as the *Exus albus* Chronozone, may represent a duration equal to or briefer than that of a stage.

(c) **Practical utility.**—Chronozones, especially thin and informal biochronozones and lithochronozones bounded by key beds or other "markers," are the units used most commonly in industry investigations of selected parts of the stratigraphy of economically favorable basins. Such units are useful to define geographic distributions of lithofacies or biofacies, which provide a basis for genetic interpretations and the selection of targets to drill.

Chronostratigraphic Nomenclature

Article 76.—Requirements. Requirements for establishing a formal chronostratigraphic unit include: (i) statement of intention to designate such a unit; (ii) selection of name; (iii) statement of kind and rank of unit; (iv) statement of general concept of unit including historical background, synonymy, previous treatment, and reasons for proposed establishment; (v) description of characterizing physical and/or biological features; (vi) designation and description of boundary type sections, stratotypes, or other kinds of units on which it is based; (vii) correlation and age relations; and (viii) publication in a recognized scientific medium as specified in Article 4.

Article 77.—Nomenclature. A formal chronostratigraphic unit is given a compound name, and the initial letter of all words, except for trivial taxonomic terms, is capitalized. Except for chronozones (Article 75), names proposed for new chronostratigraphic units should not duplicate those for other stratigraphic units. For example, naming a new chronostratigraphic unit simply by adding "-an" or "-ian" to the name of a lithostratigraphic unit is improper.

Remarks. (a) **Systems and units of higher rank.**—Names that are generally accepted for systems and units of higher rank have diverse origins, and they also have different kinds of endings (Paleozoic, Cambrian, Cretaceous, Jurassic, Quaternary).

(b) **Series and units of lower rank.**—Series and units of lower rank are commonly known either by geographic names (Virgilian Series, Ochoan Series) or by names of their encompassing units modified by the capitalized adjectives Upper, Middle, and Lower (Lower Ordovician). Names of chronozones are derived from the unit on which they are based (Article 75). For series and stage, a geographic name is preferable because it may be related to a type area. For geographic names, the adjectival endings -an or -ian are recommended (Cincinnatian Series), but it is permissible to use the geographic name without any special ending, if more euphonious. Many series and stage names already in use have been based on lithic units (groups, formations, and members) and bear the names of these units (Wolfcampian Series, Claibornian Stage). Nevertheless, a stage preferably should have a geographic name not previously used in stratigraphic nomenclature. Use of internationally accepted (mainly European) stage names is preferable to the proliferation of others.

Article 78.—Stratotypes. An ideal stratotype for a chronostratigraphic unit is a completely exposed unbroken and continuous sequence of fossiliferous stratified rocks extending from a well-defined lower boundary to the base of the next higher unit. Unfortunately, few available sequences are sufficiently complete to define stages and units of higher rank, which therefore are best defined by boundary-stratotypes (Article 8b).

Boundary-stratotypes for major chronostratigraphic units ideally should be based on complete sequences of either fossiliferous monofacial marine strata or rocks with other criteria for chronocorrelation to permit widespread tracing of synchronous horizons. Extension of synchronous surfaces should be based on as many indicators of age as possible.

Article 79.—Revision of units. Revision of a chronostratigraphic unit without changing its name is allowable but requires as much justification as the establishment of a new unit (Articles 17, 19, and 76). Revision or redefinition of a unit of system or higher rank requires international agreement. If the definition of a chronostratigraphic unit is inadequate, it may be clarified by establishment of boundary stratotypes in a principal reference section.

GEOCHRONOLOGIC UNITS

Nature and Boundaries

Article 80.—Definition and Basis. Geochronologic units are divisions of time traditionally distinguished on the basis of the rock record as expressed by chronostratigraphic units. A geochronologic unit is not a stratigraphic unit (i.e., it is not a material unit), but it corresponds to the time span of an established chronostratigraphic unit (Articles 65 and 66), and its beginning and ending corresponds to the base and top of the referent.

Ranks and Nomenclature of Geochronologic Units

Article 81.—Hierarchy. The hierarchy of geochronologic units in order of decreasing rank is *eon, era, period, epoch,* and *age.* Chron is a non-hierarchical, but commonly brief, geochronologic unit. Ages in sum do not necessarily equal epochs and need not form a continuum. An eon is the time represented by the rocks constituting an eonothem; era by an erathem; period by a system; epoch by a series; age by a stage; and chron by a chronozone.

Article 82.—Nomenclature. Names for periods and units of lower rank are identical with those of the corresponding chronostratigraphic units; the names of some eras and eons are independently formed. Rules of capitalization for chronostratigraphic units (Article 77) apply to geochronologic units. The adjectives Early, Middle, and Late are used for the geochronologic epochs equivalent to the corresponding chronostratigraphic Lower, Middle, and Upper series, where these are formally established.

POLARITY-CHRONOSTRATIGRAPHIC UNITS

Nature and Boundaries

Article 83.—Definition. A polarity-chronostratigraphic unit is a body of rock that contains the primary magnetic-polarity record imposed when the rock was deposited, or crystallized, during a specific interval of geologic time.

Remarks. (a) **Nature.**—Polarity-chronostratigraphic units depend fundamentally for definition on actual sections or sequences, or measurements on individual rock units, and without these standards they are meaningless. They are based on material units, the polarity zones of magnetopolarity classification. Each polarity-chronostratigraphic unit is the record of the time during which the rock formed and the Earth's magnetic field had a designated polarity. Care should be taken to define polarity-chronologic units in terms of polarity-chronostratigraphic units, and not vice versa.

(b) **Principal purposes.**—Two principal purposes are served by polarity-chronostratigraphic classification: (1) correlation of rocks at one place with those of the same age and polarity at other places; and (2) delineation of the polarity history of the Earth's magnetic field.

(c) **Recognition.**—A polarity-chronostratigraphic unit may be extended geographically from its type locality only with the support of physical and/or paleontologic criteria used to confirm its age.

Article 84.—Boundaries. The boundaries of a polarity chronozone are placed at polarity-reversal horizons or polarity transition-zones (see Article 45).

Ranks and Nomenclature of Polarity-Chronostratigraphic Units

Article 85.—Fundamental Unit. The polarity chronozone consists of rocks of a specified primary polarity and is the fundamental unit of worldwide polarity-chronostratigraphic classification.

Remarks. (a) **Meaning of term.**—A polarity chronozone is the worldwide body of rock strata that is collectively defined as a polarity-chronostratigraphic unit.

(b) **Scope.**—Individual polarity zones are the basic building blocks of polarity chronozones. Recognition and definition of polarity chronozones may thus involve step-by-step assembly of carefully dated or correlated individual polarity zones, especially in work with rocks older than the oldest ocean-floor magnetic anomalies. This procedure is the method by which the Brunhes, Matuyama, Gauss, and Gilbert Chronozones were recognized (Cox, Doell, and Dalrymple, 1963) and defined originally (Cox, Doell, and Dalrymple, 1964).

(c) **Ranks.**—Divisions of polarity chronozones are designated polarity subchronozones. Assemblages of polarity chronozones may be termed polarity superchronozones.

Article 86.—Establishing Formal Units. Requirements for establishing a polarity-chronostratigraphic unit include those specified in Articles 3 and 4, and also (1) definition of boundaries of the unit, with specific references to designated sections and data; (2) distinguishing polarity characteristics, lithologic descriptions, and included fossils; and (3) correlation and age relations.

Article 87.—Name. A formal polarity-chronostratigraphic unit is given a compound name beginning with that for a named geographic feature; the second component indicates the normal, reversed, or mixed polarity of the unit, and the third component is *chronozone*. The initial letter of each term is capitalized. If the same geographic name is used for both a magnetopolarity zone and a polarity-chronostratigraphic unit, the latter should be distinguished by an -an or -ian ending. Example: Tetonian Reversed-Polarity Chronozone.

Remarks: (a) **Preservation of established name.**—A particularly well-established name should not be displaced, either on the basis of priority, as described in Article 7c, or because it was not taken from a geographic feature. Continued use of Brunhes, Matuyama, Gauss, and Gilbert, for example, is endorsed so long as they remain valid units.

(b) **Expression of doubt.**—Doubt in the assignment of polarity zones to polarity-chronostratigraphic units should be made explicit if criteria of time equivalence are inconclusive.

POLARITY-CHRONOLOGIC UNITS

Nature and Boundaries

Article 88.—Definition. Polarity-chronologic units are divisions of geologic time distinguished on the basis of the record of magnetopolarity as embodied in polarity-chronostratigraphic units. No special kind of magnetic time is implied; the designations used are meant to convey the parts of geologic time during which the Earth's magnetic field had a characteristic polarity or sequence of polarities. These units correspond to the time spans represented by polarity chronozones, e.g., Gauss Normal Polarity Chronozone. They are not material units.

Ranks and Nomenclature of Polarity-Chronologic Units

Article 89.—Fundamental Unit. The polarity chron is the fundamental unit of geologic time designating the time span of a polarity chronozone.

Remark. (a) **Hierarchy.**—Polarity-chronologic units of decreasing hierarchical ranks are polarity superchron, polarity chron, and polarity subchron.

Article 90.—Nomenclature. Names for polarity chronologic units are identical with those of corresponding polarity-chronostratigraphic units, except that the term chron (or superchron, etc) is substituted for chronozone (or superchronozone, etc).

DIACHRONIC UNITS

Nature and Boundaries

Article 91.—Definition. A diachronic unit comprises the unequal spans of time represented either by a specific lithostratigraphic, allostratigraphic, biostratigraphic, or pedostratigraphic unit, or by an assemblage of such units.

Remarks. (a) **Purposes.**—Diachronic classification provides (1) a means of comparing the spans of time represented by stratigraphic units with diachronous boundaries at different localities, (2) a basis for broadly establishing in time the beginning and ending of deposition of diachronous stratigraphic units at different sites, (3) a basis for inferring the rate of change in areal extent of depositional processes, (4) a means of determining and comparing rates and durations of deposition at different localities, and (5) a means of comparing temporal and spatial relations of diachronous stratigraphic units (Watson and Wright, 1980).

(b) **Scope.**—The scope of a diachronic unit is related to (1) the relative magnitude of the transgressive division of time represented by the stratigraphic unit or units on which it is based and (2) the areal extent of those units. A diachronic unit is not extended beyond the geographic limits of the stratigraphic unit or units on which it is based.

(c) **Basis.**—The basis for a diachronic unit is the diachronous referent.

(d) **Duration.**—A diachronic unit may be of equal duration at different places despite differences in the times at which it began and ended at those places.

Article 92.—Boundaries. The boundaries of a diachronic unit are the times recorded by the beginning and end of deposition of the material referent at the point under consideration (Figs. 10, 11).

Remark. (a) **Temporal relations.**—One or both of the boundaries of a diachronic unit are demonstrably time-transgressive. The varying time significance of the boundaries is defined by a series of boundary reference sections (Article 8b, 8e). The duration and age of a diachronic unit differ from place to place (Figs. 10, 11).

Ranks and Nomenclature of Diachronic Units

Article 93.—Ranks. A diachron is the fundamental and non-hierarchical diachronic unit. If a hierarchy of diachronic units is needed, the terms episode, phase, span, and cline, in order of decreasing rank, are recommended. The rank of a hierarchical unit is determined by the scope of the unit (Article 91 b), and not by the time span represented by the unit at a particular place.

Remarks. (a) **Diachron.**—Diachrons may differ greatly in magnitude because they are the spans of time represented by individual or grouped lithostratigraphic, allostratigraphic, biostratigraphic, and(or) pedostratigraphic units.

(b) **Hierarchical ordering permissible.**—A hierarchy of diachronic units may be defined if the resolution of spatial and temporal relations of diachronous stratigraphic units is sufficiently precise to make the hierarchy useful (Watson and Wright, 1980). Although all hierarchical units of rank lower than episode are part of a unit next higher in rank, not all parts of an episode, phase, or span need be represented by a unit of lower rank.

(c) **Episode.**—An episode is the unit of highest rank and greatest scope in hierarchical classification. If the "Wisconsinan Age" were to be redefined as a diachronic unit, it would have the rank of episode.

Article 94.—Name. The name for a diachronic unit should be compound, consisting of a geographic name followed by the term diachron or a hierarchical rank term. Both parts of the com-

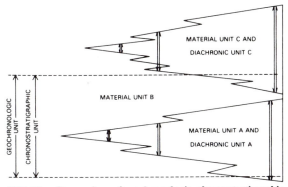

FIG. 10.—Comparison of geochronologic, chronostratigraphic, and diachronic units.

pound name are capitalized to indicate formal status. If the diachronic unit is defined by a single stratigraphic unit, the geographic name of the unit may be applied to the diachronic unit. Otherwise, the geographic name of the diachronic unit should not duplicate that of another formal stratigraphic unit. Genetic terms (e.g., alluvial, marine) or climatic terms (e.g., glacial, interglacial) are not included in the names of diachronic units.

Remarks. (a) **Formal designation of units.**—Diachronic units should be formally defined and named only if such definition is useful.

(b) **Inter-regional extension of geographic names.**—The geographic name of a diachronic unit may be extended from one region to another if the stratigraphic units on which the diachronic unit is based extend across the regions. If different diachronic units in contiguous regions eventually

prove to be based on laterally continuous stratigraphic units, one name should be applied to the unit in both regions. If two names have been applied, one name should be abandoned and the other formally extended. Rules of priority (Article 7d) apply. Priority in publication is to be respected, but priority alone does not justify displacing a well-established name by one not well-known or commonly used.

(c) **Change from geochronologic to diachronic classification.**—Lithostratigraphic units have served as the material basis for widely accepted chronostratigraphic and geochronologic classifications of Quaternary nonmarine deposits, such as the classifications of Frye et al (1968), Willman and Frye (1970), and Dreimanis and Karrow (1972). In practice, time-parallel horizons have been extended from the stratotypes on the basis of markedly time-transgressive lithostratigraphic and pedostratigraphic unit boundaries. The time ("geochronologic") units, defined on the basis of the stratotype sections but extended on the basis of diachronous stratigraphic boundaries, are diachronic units. Geographic names established for such "geochronologic" units may be used in diachronic classification if (1) the chronostratigraphic and geochronologic classifications are formally abandoned and diachronic classifications are proposed to replace the former "geochronologic" classifications, and (2) the units are redefined as formal diachronic units. Preservation of well-established names in these specific circumstances retains the intent and purpose of the names and the units, retains the practical significance of the units, enhances communication, and avoids proliferation of nomenclature.

Article 95.—Establishing Formal Units. Requirements for establishing a formal diachronic unit, in addition to those in Article 3, include (1) specification of the nature, stratigraphic relations, and geographic or areal relations of the stratigraphic unit or units that serve as a basis for definition of the unit, and (2) specific designation and description of multiple reference sections that illustrate the temporal and spatial relations of the defining stratigraphic unit or units and the boundaries of the unit or units.

Remark. (a) **Revision or abandonment.**—Revision or abandonment of the stratigraphic unit or units that serve as the material basis for defini-

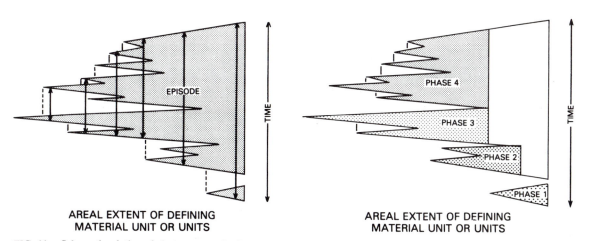

AREAL EXTENT OF DEFINING
MATERIAL UNIT OR UNITS

AREAL EXTENT OF DEFINING
MATERIAL UNIT OR UNITS

FIG. 11.—Schematic relation of phases to an episode.
    Parts of a phase similarly may be divided into spans, and spans into clines. Formal definition of spans and clines is unnecessary in most diachronic unit hierarchies.

tion of a diachronic unit may require revision or abandonment of the diachronic unit. Procedure for revision must follow the requirements for establishing a new diachronic unit.

GEOCHRONOMETRIC UNITS

Nature and Boundaries

Article 96.—**Definition.** Geochronometric units are units established through the direct division of geologic time, expressed in years. Like geochronologic units (Article 80), geochronometric units are abstractions, i.e., they are not material units. Unlike geochronologic units, geochronometric units are not based on the time span of designated chronostratigraphic units (stratotypes), but are simply time divisions of convenient magnitude for the purpose for which they are established, such as the development of a time scale for the Precambrian. Their boundaries are arbitrarily chosen or agreed-upon ages in years.

Ranks and Nomenclature of Geochronometric Units

Article 97.—**Nomenclature.** Geochronologic rank terms (eon, era, period, epoch, age, and chron) may be used for geochronometric units when such terms are formalized. For example, Archean Eon and Proterozoic Eon, as recognized by the IUGS Subcommission on Precambrian Stratigraphy, are formal geochronometric units in the sense of Article 96, distinguished on the basis of an arbitrarily chosen boundary at 2.5 Ga. Geochronometric units are not defined by, but may have, corresponding chronostratigraphic units (eonothem, erathem, system, series, stage, and chronozone).

PART III: REFERENCES[10]

American Commission on Stratigraphic Nomenclature, 1947, Note 1—Organization and objectives of the Stratigraphic Commission: American Association of Petroleum Geologists Bulletin, v. 31, no. 3, p. 513-518.

——— ,1961, Code of Stratigraphic Nomenclature: American Association of Petroleum Geologists Bulletin, v. 45, no. 5, p. 645-665.

——— ,1970, Code of Stratigraphic Nomenclature (2d ed.): American Association of Petroleum Geologists, Tulsa, Okla., 45 p.

——— ,1976, Note 44—Application for addition to code concerning magnetostratigraphic units: American Association of Petroleum Geologists Bulletin, v. 60, no. 2, p. 273-277.

Caster, K. E., 1934, The stratigraphy and paleontology of northwestern Pennsylvania, Part 1, Stratigraphy: Bulletins of American Paleontology, v. 21, 185 p.

Chang, K. H., 1975, Unconformity-bounded stratigraphic units: Geological Society of America Bulletin, v. 86, no. 11, p. 1544-1552.

Committee on Stratigraphic Nomenclature, 1933, Classification and nomenclature of rock units: Geological Society of America Bulletin, v. 44, no. 2, p. 423-459, and American Association of Petroleum Geologists Bulletin, v. 17, no. 7, p. 843-868.

Cox, A. V., R. R. Doell, and G. B. Dalrymple, 1963, Geomagnetic polarity epochs and Pleistocene geochronometry: Nature, v. 198, p. 1049-1051.

——— ,1964, Reversals of the Earth's magnetic field: Science, v. 144, no. 3626, p. 1537-1543.

Cross, C. W., 1898, Geology of the Telluride area: U.S. Geological Survey 18th Annual Report, pt. 3, p. 759.

Cumming, A. D., J. G. C. M. Fuller, and J. W. Porter, 1959, Separation of strata: Paleozoic limestones of the Williston basin: American Journal of Science, v. 257, no. 10, p. 722-733.

Dreimanis, Aleksis, and P. F. Karrow, 1972, Glacial history of the Great Lakes–St. Lawrence region, the classification of the Wisconsin(an) Stage, and its correlatives: International Geologic Congress, 24th Session, Montreal, 1972, Section 12, Quaternary Geology, p. 5-15.

Dunbar, C. O., and John Rodgers, 1957, Principles of stratigraphy: Wiley, New York, 356 p.

Forgotson, J. M., Jr., 1957, Nature, usage and definition of marker-defined vertically segregated rock units: American Association of Petroleum Geologists Bulletin, v. 41, no. 9, p. 2108-2113.

Frye, J. C., H. B. Willman, Meyer Rubin, and R. F. Black, 1968, Definition of Wisconsinan Stage: U.S. Geological Survey Bulletin 1274-E, 22 p.

George, T. N., and others, 1969, Recommendations on stratigraphical usage: Geological Society of London, Proceedings no. 1656, p. 139-166.

Harland, W. B., 1977, Essay review [of] International Stratigraphic Guide, 1976: Geology Magazine, v. 114, no. 3, p. 229-235.

——— ,1978, Geochronologic scales, in G. V. Cohee et al, eds., Contributions to the Geologic Time Scale: American Association of Petroleum Geologists, Studies in Geology, no. 6, p. 9-32.

Harrison, J. E., and Z. E. Peterman, 1980, North American Commission on Stratigraphic Nomenclature Note 52—A preliminary proposal for a chronometric time scale for the Precambrian of the United States and Mexico: Geological Society of America Bulletin, v. 91, no. 6, p. 377-380.

Henbest, L. G., 1952, Significance of evolutionary explosions for diastrophic division of Earth history: Journal of Paleontology, v. 26, p. 299-318.

Henderson, J. B., W. G. E. Caldwell, and J. E. Harrison, 1980, North American Commission on Stratigraphic Nomenclature, Report 8—Amendment of code concerning terminology for igneous and high-grade metamorphic rocks: Geological Society of America Bulletin, v. 91, no. 6, p. 374-376.

Holland, C. H., and others, 1978, A guide to stratigraphical procedure: Geological Society of London, Special Report 10, p. 1-18.

Huxley, T. H., 1862, The anniversary address: Geological Society of London, Quarterly Journal, v. 18, p. xl-liv.

International Commission on Zoological Nomenclature, 1964: International Code of Zoological Nomenclature adopted by the XV International Congress of Zoology: International Trust for Zoological Nomenclature, London, 176 p.

International Subcommission on Stratigraphic Classification (ISSC), 1976, International Stratigraphic Guide (H. D. Hedberg, ed.): John Wiley and Sons, New York, 200 p.

International Subcommission on Stratigraphic Classification, 1979, Magnetostratigraphy polarity units—a supplementary chapter of the ISSC International Stratigraphic Guide: Geology, v. 7, p. 578-583.

Izett, G. A., and R. E. Wilcox, 1981, Map showing the distribution of the Huckleberry Ridge, Mesa Falls, and Lava Creek volcanic ash beds (Pearlette family ash beds) of Pliocene and Pleistocene age in the western United States and southern Canada: U. S. Geological Survey Miscellaneous Geological Investigations Map I-1325.

Kauffman, E. G., 1969, Cretaceous marine cycles of the Western Interior: Mountain Geologist: Rocky Mountain Association of Geologists, v. 6, no. 4, p. 227-245.

Matthews, R. K., 1974, Dynamic stratigraphy—an introduction to sedimentation and stratigraphy: Prentice-Hall, New Jersey, 370 p.

McDougall, Ian, 1977, The present status of the geomagnetic polarity time scale: Research School of Earth Sciences, Australian National University, Publication no. 1288, 34 p.

McElhinny, M. W., 1978, The magnetic polarity time scale; prospects and possibilities in magnetostratigraphy, in G. V. Cohee et al, eds., Contributions to the Geologic Time Scale, American Association of Petroleum Geologists, Studies in Geology, no. 6, p. 57-65.

---

[10]Readers are reminded of the extensive and noteworthy bibliography of contributions to stratigraphic principles, classification, and terminology cited by the International Stratigraphic Guide (ISSC, 1976, p. 111-187).

McIver, N. L., 1972, Cenozoic and Mesozoic stratigraphy of the Nova Scotia shelf: Canadian Journal of Earth Science, v. 9, p. 54-70.

McLaren, D. J., 1977, The Silurian-Devonian Boundary Committee. A final report, *in* A. Martinsson, ed., The Silurian-Devonian boundary: IUGS Series A, no. 5, p. 1-34.

Morrison, R. B., 1967, Principles of Quaternary soil stratigraphy, *in* R. B. Morrison and H. E. Wright, Jr., eds., Quaternary soils: Reno, Nevada, Center for Water Resources Research, Desert Research Institute, Univ. Nevada, p. 1-69.

North American Commission on Stratigraphic Nomenclature, 1981, Draft North American Stratigraphic Code: Canadian Society of Petroleum Geologists, Calgary, 63 p.

Palmer, A. R., 1965, Biomere-a new kind of biostratigraphic unit: Journal of Paleontology, v. 39, no. 1, p. 149-153.

Parsons, R. B., 1981, Proposed soil-stratigraphic guide, *in* International Union for Quaternary Research and International Society of Soil Science: INQUA Commission 6 and ISSS Commission 5 Working Group, Pedology, Report, p. 6-12.

Pawluk, S., 1978, The pedogenic profile in the stratigraphic section, *in* W. C. Mahaney, ed., Quaternary soils: Norwich, England, GeoAbstracts, Ltd., p. 61-75.

Ruhe, R. V., 1965, Quaternary paleopedology, *in* H. E. Wright, Jr., and D. G. Frey, eds., The Quaternary of the United States: Princeton, N.J., Princeton University Press, p. 755-764.

Schultz, E. H., 1982, The chronosome and supersome--terms proposed for low-rank chronostratigraphic units: Canadian Petroleum Geology, v. 30, no. 1, p. 29-33.

Shaw, A. B., 1964, Time in stratigraphy: McGraw-Hill, New York, 365 p.

Sims, P. K., 1979, Precambrian subdivided: Geotimes, v. 24, no. 12, p. 15.

Sloss, L. L., 1963, Sequences in the cratonic interior of North America: Geological Society of America Bulletin, v. 74, no. 2, p. 94-114.

Tracey, J. I., Jr., and others, 1971, Initial reports of the Deep Sea Drilling Project, v. 8: U.S. Government Printing Office, Washington, 1037 p.

Valentine, K. W. G., and J. B. Dalrymple, 1976, Quaternary buried paleosols: A critical review: Quaternary Research, v. 6, p. 209-222.

Vella, P., 1964, Biostratigraphic units: New Zealand Journal of Geology and Geophysics, v. 7, no. 3, p. 615-625.

Watson, R. A., and H. E. Wright, Jr., 1980, The end of the Pleistocene: A general critique of chronostratigraphic classification: Boreas, v. 9, p. 153-163.

Weiss, M. P., 1979a, Comments and suggestions invited for revision of American Stratigraphic Code: Geological Society of America, News and Information, v. 1, no. 7, p. 97-99.

———— ,1979b, Stratigraphic Commission Note 50--Proposal to change name of Commission: American Association of Petroleum Geologists Bulletin, v. 63, no. 10, p. 1986.

Weller, J. M., 1960, Stratigraphic principles and practice: Harper and Brothers, New York, 725 p.

Willman, H. B., and J. C. Frye, 1970, Pleistocene stratigraphy of Illinois: Illinois State Geological Survey Bulletin 94, 204 p.

GLOSSARY

As far as possible, the definitions have been kept brief and simple. Where there are alternative uses in other fields of geology, only that used in stratigraphy or related areas is cited. Most of the terms are listed in the *Glossary of Geology* (3rd ed.) by R. L. Bates and J. A. Jackson (1987), published by the American Geological Institute, and the reader is referred to that work for alternative or more detailed definitions.

Words italicized are themselves defined elsewhere in the glossary.

Absolute time scale Geologic time scale measured in real time; i.e., in years before the present (B.P.; i.e., before A.D. 1950).

Abundance zone *Biozone* characterized by a relative abundance or quantitative maximum of a species or other taxon or taxa.

Abyssal plain Flattest region of the deep-ocean floor, formed by the blanketing effect of *turbidity current* and *pelagic* deposits over older topographic features.

Acme zone Syn. with *abundance zone*.

Acrozone Syn. with *range zone*.

Active continental margin Collision plate margin adjacent or close to a subduction zone. Characterized by compressional and thrust tectonics and vulcanism.

Aeolian (eolian) Pertaining to the wind; e.g., aeolian dunes of windblown sand.

Age Geochronologic (time) division, smaller than an *epoch*.

Algal mat Interwoven mass of various species of algae, mostly *cyanobacteria* (blue-green algae) that form a coating on sediment surfaces in various environments (shallow-marine, intertidal, lacustrine, etc.). Layers of sediment trapped by algal mats form *stromatolites*.

Allochthonous Originated elsewhere, typically in reference to clastic sediments; cf. *autochthonous*.

Allopatric species Species geographically isolated from other species.

Allostratigraphic unit Mappable body of stratified rock defined by its bounding unconformities.

Alluvial Pertaining to the deposits of streams.

Alpha particle Helium nucleus (of two protons and two neutrons) emitted during radioactive decay.

Altithermal Period from about 7500 to

4000 B.P. when the climate was warmer than at present.

Anagenesis   Linear and progressive evolution.

Angiosperm   Flowering plant, with seed protected within fruit.

Angular unconformity   Unconformity between younger strata deposited across older strata that were folded and eroded previously.

Anoxic   Oxygen-deficient.

Apparent polar wandering curve (APW curve)   Sinuous line joining up the apparent positions of the poles in the geologic past as they can be plotted from various continents. The movement was on the part of the continental blocks rather than the poles.

Arboreal pollen   Pollen from trees.

Archeomagnetism   *Thermoremanent* magnetism exhibited in magnetic minerals contained in archeological artifacts of baked clay.

Arkose   Sandstone in which feldspar is the dominant mineral. Typically derived from weathering of a nearby granite.

Assemblage   Highest ranking unit in the hierarchy of *lithotectonic* divisions, suggested by Wright (1985) in classifying the Precambrian.

Assemblage zone   Biozone characterized by the association of three or more species or other taxa.

Astrobleme   Meteorite crater.

Atomic number   Number of protons in the nucleus of an atom.

Autochthonous   Originated or formed in the place where now located. Refers usually to sediments such as biogenic limestones, coals and evaporites; cf. *Allochthonous*.

Backarc   In an arc-trench system, refers to structural features on the side of the volcanoes away from the trench.

Back reef   Sheltered, shallow lagoon on the shoreward side of a barrier reef.

Banded iron formation (BIF)   Interlaminations of ferric iron oxide (dominantly hematite) and chert.

Barchan dune   Migrating crescentic dune, formed of sand transported by unidirectional wind. The concave (downwind) side of the dune has a slip face.

Barren interzone   Unfossiliferous interval between two *biozones*.

Barrier beach   Sandy beach ridge separated from the mainland by a narrow and shallow lagoon or by tidal marshes.

Barrier reef   Elongate organic reef that lies roughly parallel to the shore and separated from it by a quiet lagoon.

Baselap   General term used in seismic stratigraphy to describe the relationship between the lower boundary of strata where they terminate against other strata in either an *onlap* or *downlap* relationship.

Basement   Oldest rocks in an area, often with complex structures, intruded and metamorphosed and genetically unrelated to overlying rocks; in many cases, of Precambrian age.

Basin   Site of deposition of great thicknesses of sediment, often with indications of structural downsagging.

Beach   Sloping zone between low-water mark and the highest point affected by wave action. Usually covered by sand, although beach material may consist of pebbles or cobbles (shingle).

Beach face   Foreshore; the zone of beach exposed to wave action.

Bed   Layer or stratum of sedimentary rock. The smallest formal lithostratigraphic unit.

Bedding plane   Surface of a bed, the plane of stratification.

Benthonic   (benthic) Adjective referring to *benthos*.

Benthos   Bottom-dwelling organisms that live on a seafloor or lake bed *(epifauna)* or that burrow below the sediment–water interface *(infauna)*. Sessile benthos are attached in their adult life (e.g., corals); vagrant benthos move about.

Bentonite Clay derived from alteration of volcanic ash. Commonly a montmorillonite clay.

Beta particle Electron emitted during radioactive decay.

Biochron Time represented by a *biozone*.

Bioclast Fragment of fossil skeletal material.

Biocoenosis Life assemblage; a group of animals that lived together.

Biofacies That aspect of a rock concerned only with its fossil content rather than its lithologic features.

Bioherm Reefoid rock mass built from the skeletal remains of organisms and having a moundlike form.

Biostratigraphic unit Body of rock defined on the basis of its fossil content; the basic unit is a *biozone*.

Biotope *Habitat* of an animal or plant community to which that community is adapted.

Bioturbation Mixing and disturbance of sediment by organisms, typically burrowing *infaunal* species.

Biozone Succession of sediments characterized by the presence of one or more species or other taxa. The fundamental biostratigraphic unit.

Bolide Exploding meteor.

Bottomset beds Beds deposited beyond the slope of a delta front and buried by the encroaching *foreset beds* as the delta front advances.

Bouma cycle Vertical succession of sediment types laid down by a turbidity current. At the base is a massive and coarse-grained graded bed, deposited as the turbidity current passes. Overlying this is a laminated sand, a rippled sand and silt, and finally, a layer of mud, all reflecting the waning carrying capacity of the current.

Boundary stratotype Specific horizon that serves as the standard in defining the boundary (usually the base) of a stratigraphic unit.

Carbonate compensation depth (CCD) In the ocean, the depth at which the calcium carbonate content of deep-sea sediments decreases rapidly. If the rate of solution of calcium carbonate equals the rate of deposition of calcareous sediment, this depth will correspond with the *lysocline*. In areas of high organic production at the surface, the CCD will occur below the lysocline.

Carbonate rock General term for limestone, consisting of calcite and/or aragonite, $CaCO_3$, and *dolomite* (dolostone), consisting of dolomite, $CaMg(CO_3)_2$.

Cenozone Syn. with *assemblage zone*

Chelogenic cycle Cycle of *orogenies* that resulted in the accretion of a craton. Literally, a *shield*-forming cycle.

Chert Amorphous or cryptocrystalline silica; most typically occurs as nodules in limestones or as beds in shale successions.

Chron Smallest *geochronologic* (geochronometric) unit. The time span of a *chronozone*.

Chronostratigraphic unit Time-rock unit. A body of rock formed during a specific time span and bounded by *isochronous* surfaces. Used as the basis for defining a specific time interval.

Chronotectonic Division Division suggested by Wright (1985) to be used for the subdivision of Precambrian time; based on sedimentary, metamorphic, intrusive, or tectonic events. A chronotectonic cycle is the fundamental unit.

Chronozone Defined as the smallest division in the hierarchy of chronostratigraphic terms. Used more generally in a nonhierarchial sense to embrace all rocks formed anywhere during the time span of a designated geologic unit.

Clade Group of species or other taxa with ancestor-descendant (phylogenetic) relationships, as demonstrated by shared derived characters.

Cladogenesis Phyletic speciation.

Clast Piece or fragment of an older rock, derived by weathering and breakdown.

Clastic rock Sedimentary rock formed by the *lithification* of fragments or mineral grains derived from the weathering and breakdown of an earlier rock.

Clastic wedge Mass of clastic sediments derived from a tectonic land mass or orogenic belt and filling an adjacent basin, such as a foreland basin.

Clay General term for hydrous aluminum silicate minerals related to the micas. Typically the crystals are small (av. 2–4 μm).

Climatic optimum Period in the Holocene from about 9000 to 2500 B.P. when mean annual temperatures were higher than at present day.

Cline 1. Continuous and gradational change in morphological character in a species population across its geographic range.
2. Smallest division in a hierarchy of *diachronic* units.

Clinoform Sloping subaqueous surface on the margin of a sedimentary basin, separating the shallow *(undaform)* surface of the basin rim from the deep *(fondoform)* surface of the basin floor.

Clinothem Sedimentary rocks formed by deposition on the *clinoform* surface; i.e., in the clino environment.

Coastal onlap See *onlap.*

Commensalism Relationship between two organisms when the first benefits from the second, but the benefit is not reciprocated, the second organism being unaffected and uninvolved.

Community Total aggregation of living things within an ecosystem; a composite group of organisms with some degree of homogeneity. Grouped into larger entities known as *provinces,* which, in turn, form parts of *regions* and *realms,* within a loosely defined hierarchy.

Complex Assemblage or association of rocks of different types and/or ages that are involved together in a structural relationship that renders difficult their differentiation as separate mappable units.

Composite section *Stratigraphic section* compiled from adding together two or more sequential (and preferably overlapping) local sections.

Concordant Describing strata that are parallel and structurally *conformable* but within which there may possibly be a *hiatus.*

Concordia Time curve joining graphed points plotted for ages derived from the ratios of Pb^{208}/U^{238} and Pb^{207}/U^{235} values.

Concurrent range zone Stratigraphic interval between the first appearance of one taxon and the last appearance of a second or additional taxa; i.e., the interval of overlap of the two or more ranges, hence also called an *overlap zone.* This concept appears to match that of Alfred Oppel so it is sometimes referred to as an *Oppel zone.*

Conformable Said of strata when they lie one above the other in parallel and apparently continuous succession. There may be evidence of some interruption in sedimentation, provided no erosion or disturbance of the earlier strata occurred.

Connate water Water trapped in the interstices of sedimentary grains at the time they were first deposited in a subaqueous environment.

Conodont Small, toothlike fossil derived from an animal believed to have been a pelagic hemichordate. Range in age from possibly late Precambrian to latest Triassic.

Continental rise Gentle sloping surface that extends up from the *abyssal plain* to the foot of the *continental slope.*

Continental shelf Gentle inclined submerged margin of the continent; extends from the shoreline to where there is a sudden increase in slope (the shelf break), typically at about 200-m depth.

Continental slope Sloping surface lying between the *continental shelf* and the *continental rise*.

Correlation coefficient Measure of similarity between two samples. Syn. with *similarity coefficient*.

Craton Central stable core of the ancient basement rocks of a continental block, consisting of exposed *shield* areas and adjacent *platforms* that are covered by younger sediments and/or modern shallow-marine waters.

Cross-bedding Inclined beds of sediment laid down by water currents or the wind, each bed lying downstream (or downwind) from the previously laid bed.

Curie point Temperature at which a magnetic mineral acquires *thermoremanent* magnetism on cooling.

Cyanobacteria Unicellular or multicellular photosynthesizing organisms containing chlorophyll. Formerly termed blue-green algae, they are the group mainly responsible for the formation of *stromatolites*.

Cyclothem Sedimentary unit deposited during a single cycle of deposition. Each cycle is one of a series of rhythmically repetitive changes in the environment, such as water depth, alternation of brackish and marine conditions, sediment supply, etc. The presence of coal was implicit in earlier definitions based on Pennsylvanian examples.

Daughter element (or daughter nuclide) Element that is the end product of radioactive decay.

Decay constant Constant value for each radioactive isotope that is a measure of the rate of spontaneous radioactive decay.

Declination Angle between magnetic and true north.

Dendrochronology Dating of the recent past by means of the study of annual growth rings in trees.

Dendrogram Graphic presentation of numerical data on the degrees of similarity between samples and groups of samples, set out in a treelike pattern.

Density log Well log of formation density as determined by the backscatter of gamma rays from a source such as cobalt-60 or cesium-137 contained in a *sonde* passed up the borehole.

Derived fossil Fossil originating in an older formation and which has been weathered out and redeposited in a younger formation.

Deviation Informal term used for a minor or short-lived paleomagnetic reversal, perhaps lasting less than 10,000 years.

Diachron Fundamental *diachronic unit*.

Diachronic unit Geologic time unit whose boundaries vary in age from place to place. It is defined by a particular stratigraphic unit whose boundaries are *diachronous*.

Diachronous Said of a rock unit or geologic boundary that is of different ages in different places; i.e., is time-transgressive.

Diagenesis Encompasses all changes, physical, chemical, and biological, that occur in a sediment after deposition but prior to metamorphism or weathering.

Diapir Anticline or dome formed in rocks overlying less dense, plastic material, such as salt or clay, which has intruded upward. Similar structures may be formed by igneous intrusions.

Diastem Relatively short break, or interruption, in sedimentation, typically with no erosion of previously deposited material before deposition is resumed. Physically, a diastem might appear to be little more than a bedding surface, but it may be indicated by missing paleontologic indicators.

Diatom Unicellular plant of the class Bacillariophyceae. Of microscopic size, diatoms secrete supporting structures (frustules) of silica. They grow in both marine and fresh water.

Diffusion threshold Temperature above which radiogenic argon is likely to leak from the mineral during the potassium-argon decay process.

Disconformity *Unconformity* that demonstrates a significant hiatus within a sedimentary succession but in which the beds below and above the break are parallel

Distal downlap In seismic stratigraphy, the relationship where downlap is in a direction basinward from the source of the sediments.

Distal onlap Onlap in a direction away from the source of sediment; i.e., across the basin.

Ditch sample Sample of drill cuttings taken from the drilling mud as it circulates to the surface.

Divergent plate boundary Plate boundary at an ocean-floor spreading center (mid-ocean ridge) or at a zone of continental rifting, at which the surfaces of lithospheric plates are moving apart.

Division Unit within the hierarchy of *lithotectonic units* proposed by Wright (1985) for stratigraphic classification in the Precambrian.

Dolomite Rhombohedral mineral of composition $CaMg(CO_3)_2$. The term also refers to the rock (syn. dolostone) composed dominantly of this mineral.

Downlap Relationship between inclined strata that terminate downdip against an initially horizontal surface.

Draa Dune complex with dune bodies having smaller dunes superimposed on them.

Drape fold Most commonly used to describe compaction folding in which more or less compressible strata, such as shales, are "draped" and differentially compacted over a relatively incompressible mass, such as a bioherm or basement feature.

Drilling mud Mud circulated in the borehole during rotary drilling to bring drill cuttings to the surface.

Drill string (syn. drill stem) In rotary drilling, the assemblage of drill bit and drill pipe rotated in the borehole by the drilling rig at the surface.

Electric log Generic term for logs that measure the electrical properties of rocks; e.g., spontaneous potential, resistivity, etc.

Endemic Said of an organism that is restricted in its geographic range or confined to a specific environment or region.

Eon Largest geochronologic (time) unit.

Eonothem Sequence of rocks formed during a specific *eon* of geologic time. The largest chronostratigraphic unit.

Epeiric sea Shallow sea that covered the continental interior.

Epeirogeny Large-scale and relatively slow uplift or downwarping of the crust causing little folding or faulting.

Epibole Sediments accumulated during a *hemera.*

Epicontinental sea Syn. with *epeiric sea.*

Epifauna *Benthonic* organisms that live on the surface of the seafloor or lake bed.

Episode 1. Highest ranking *diachronic unit.*
2. One of the hierarchy of terms used in the *chronotectonic* classification of the Precambrian by Wright (1985).

Epoch 1. Geologic time (geochronologic) unit of lesser rank than a *period.*
2. Time interval of about 1 m.y. characterized by normal or reversed magnetic polarity.

Era Geologic time unit of lesser rank than an *eon* and including several *periods.*

Erathem Sequence of rocks formed during a specific *era.*

Erg Extensive area of sand dunes; a sand sea.

Erratic boulder Boulder transported from a distant source by glacial ice.

Eugeosyncline Outer (oceanward) part of a geosynclinal belt marginal to a craton, typified by deep-water clastics *(turbidites,* etc.) and associated volcanics.

Eurytopic species Species having a tolerance for a wide range of environmental conditions.

Eustatic sea-level change Worldwide change in the level of the ocean.

Evaporite Sediment or mineral formed by the precipitation of salt crystals.

Event 1. Polarity switch of short duration (perhaps 10,000 to 100,000 years) during a *polarity epoch*.
2. Smallest unit in the hierarchy of *chronotectonic* terms of Wright (1985).

Excursion Short polarity event of 10,000 years or less.

Exotic terrane Syn. with *suspect terrane.*

Facies Overall aspect or set of characteristics of a rock that reflect its particular depositional environment. The features that set it off from adjacent facies within the same rock unit.

Fan delta Broadly arcuate and gently sloping accumulation of alluvial sands and gravels formed where a mountain stream emerges onto the lowland.

Faunal province Geographic region containing a specific assemblage of animals.

Faunizone Syn. with *assemblage zone.*

Flandrian transgression Postglacial rise of sea level that has occurred since the end of the Pleistocene.

Floating chronology Chronology of geologic events based upon a counting of *rhythmites,* annual varves, etc. that may cover a considerable time span (up to tens or hundreds of thousands of years) but which cannot be tied in with the geologic time scale in years B.P.

Flood plain Flat area of a river valley that is covered by water when the river floods.

Flushed zone That part of the rock formation immediately adjacent to the borehole in which the original pore-space fluid is replaced by the liquid filtrate of the drilling mud.

Fluvial Pertaining to a river or stream.

Flysch Marine shales and turbidites characteristic of deposition in a *foredeep* adjacent to a rising mountain belt during the early stages of an *orogeny.*

Fondoform Floor of a sedimentary basin.

Fondothem Deposits laid down on the floor of a basin, i.e., in the fondo environment.

Food chain Transfer of energy within an ecosystem from producers to consumers.

Foraminifera Single-celled marine protozoans that secrete calcareous tests, mostly of microscopic size. Many are *planktonic.*

Forearc basin Basin formed by crustal sagging above a descending lithospheric plate between a trench and volcanic belt.

Foredeep Downwarped trench adjacent to an *island arc* or orogenic belt.

Foreland Stable portion of a craton adjacent to a collision margin and toward which the rising orogenic belt is being thrust.

Fore reef Seaward side of a reef.

Foreset beds Inclined beds laid down on the advancing slope at the front of a delta or a migrating sand dune. In a delta successive foreset beds come to cover the *bottomset beds* and are themselves covered (or truncated) by *topset* beds of the delta surface. In a migrating sand dune, the foreset beds are laid down on the slip face.

Formation Fundamental *lithostratigraphic* unit. It is distinguished from adjacent formations by specific physical characters and is large enough to be mappable.

Fringing reef Elongate organic reef that lies along a coast and normally has no lagoon behind it.

Gamma ray log Log of natural gamma radiation from rocks. Sources are dominantly minerals containing thorium, potassium, and uranium.

Gene pool Sum total of genetic material of a species population.

Genotype Genetic constitution (the particular set of genes in each and all cells) of an individual organism.

Geochronology Science of geologic time, especially its measurement in terms of real time in years B. P.

Geochronometry Branch of *geochronology* concerned with the quantitative measurement of geologic time in years B.P.

Geoid Ideal figure of the earth considered as the sea-level surface and ignoring the continents. The surface of reference of geodetic surveys.

Geologic climate unit Widespread climatic episode as interpreted from Quaternary successions; e.g., *glaciations, interglacials, stades,* etc.

Geophone Electrical detecting device laid on the ground and which transmits an electrical signal to record ground vibrations.

Geosyncline Large elongated trough marginal to a continent; subsides over long periods and has accumulated great thicknesses of sediment (see also *eugeosyncline* and *miogeosyncline*).

Glacial refugium Typically an upland area to which cold-climate species of animals and plants retreated as continental ice sheets withdrew and the postglacial climate ameliorated.

Glaciation Informal geologic climatic unit, usually accorded the status of a *stage,* as that term is used in Quaternary stratigraphy.

Gondwana Ancient supercontinent that comprised the modern southern continents of Africa, South America, Australia, and Antarctica, together with peninsula India. Named after the Gondwana System of India, the name Gondwana meaning "land of the Gonds."

Graded bed Individual bed of sediment having an upward gradual change from relatively coarse to fine-grained texture; typical of *turbidites.*

Granulite High-grade metamorphic rock with coarse granular texture.

Graptolite Colonial marine organism of the class Graptolithina of the phylum Hemichordata. Many species were planktonic or pseudoplanktonic and became widely disseminated. Found typically in black shales from Middle Cambrian to Carboniferous.

Graywacke Poorly sorted sandstone with abundant feldspar, lithic fragments, and clay; typically dark in color.

Greenstone belt Belt of associated metavolcanics and metasediments; found in all Archean terranes.

Group Lithostratigraphic unit next in rank above a *formation.*

Growth fault Fault cutting sedimentary rocks and that continues to move while sediments are being deposited. The result is that the throw increases with depth and sediments are thicker on the downthrow side.

Gymnosperm Naked seed plant; i.e., seeds commonly in a cone and not enclosed in an ovary.

Habitat Environment to which an organism is best adapted.

Hairpin On an *apparent polar wandering curve,* a sudden change in direction.

Half-life Time required for half of a given amount of radioactive isotope to decay. Half-life is inversely related to the *decay constant.*

Hemera Time span of an *acme zone.*

Hemicyclothem Half of a cyclothem; typically used in Pennsylvanian cyclothems associated with coals, applied either to the lower (nonmarine) part or the upper (marine) part.

Heritable trait Physical trait in an organism that has a genetic cause and that is potentially inheritable by succeeding generations.

Hiatus Within a stratigraphic succession the interval of time not represented by sediments; caused by erosion, nondeposition, or both.

Highstand Refers to sea level when it is above the edge of the continental shelf.

Holostratotype Original *stratotype* (type section) selected by the author of a given stratigraphic unit or boundary as the standard.

Homeomorph Species that closely resembles another species but which is not related below the family level.

Horizon   1. Interface, parting, or possibly a very thin lamina within a bedded succession.

2. In British usage, a *zonule,* a subdivision of a *subzone.*

Hydrocarbon   Organic compound, gaseous, liquid, or solid, made up of carbon and hydrogen atoms arranged in rings or chains.

Hypercyclothem   Grouping of *megacyclothems.*

Hypostratotype   Additional stratotype intended to supplement or augment the *holostratotype* and to extend its recognition into other areas or facies.

Hypsithermal   Syn. with *climatic optimum.*

Hypsometric curve   Graph showing relative areas of the earth's surface at different elevations above and below sea level.

Iapetus   Proto-Atlantic ocean that existed roughly from the late Precambrian until its closure in the Early Paleozoic.

Inclination (magnetic)   Angle at which the lines of force of the magnetic field dip; the magnetic dip.

Index fossil   Fossil used as a zonal indicator to correlate or date the strata in which it occurs (syn. guide fossil).

Induction Log   Log that measures the secondary magnetic field induced in the rock formation by alternating currents transmitted from electrical coils in the logging tool.

Infauna   *Benthic* organisms that live within the seafloor or lake-bottom sediments.

Infiltrated fossil   In borehole logging, a fossil that originated higher in the succession than the horizon from which samples are currently being logged. This is usually a result of caving from the walls of the hole.

Interglacial   Time between two glacial advances.

Interstade (interstadial)   Short-lived recession of the ice during a period of glacial advance, a warm substage.

Interval zone   Comprises the strata between two specified lowest and/or highest occurrences of a species or other taxon.

Interzone   Syn. with *barren interzone.*

Intracratonic basin   Basin on a craton surface.

Intrazone   Syn. with *barren interzone.*

Invaded zone   That part of the rock formation adjacent to a borehole that is penetrated by the mud filtrate.

Island arc   Curving belt of islands near a continental margin and typically volcanic in origin. The curve is convex toward the ocean basin and beyond it lies a trench; e.g., the Aleutian and Kurile Islands.

Isochron   Line on a map connecting points of equal age.

Isochronous   Being everywhere the same age (syn. with synchronous).

Isostasy   Mechanism whereby areas of the crust rise or sink until they reach a buoyant equilibrium in "floating" on the denser mantle.

Joint   Crack or fracture in a rock across which no relative displacement has occurred.

Karst topography   Topography typical of areas underlain by limestones. Characterized by caverns, dry valleys, sink holes, and subsurface drainage.

Kelly   Steel pipe, square or hexagonal in cross section and approximately 12 m long. It is attached to the *drill string* and fits into the rotary table, transmitting the rotary motion of the table to the drill string.

Kelly bushing   Rotating journal box (shaft and bearing) in the center of the rotary table through which the kelly passes.

Kerogen   Hydrocarbon substance (bitumen) found in oil shale and other fine-grained sediments. A source of petroleum on heating.

Lacuna   Gap in the stratigraphic succession represented by an unconformity. The chronostratigraphic equivalent of a time of no deposition (hiatus) and/or erosion (vacuity).

Lacustrine   Pertaining to lakes.

Lag deposit   More-resistant and/or coarser rock and mineral detritus left behind after finer and/or less-resistant material has been removed by wind or water currents; e.g., lag gravels on a stream bed.

Lagoon   Shallow body of water along a coast behind a barrier island or organic reef.

Lahar   Mudflow of clastic or pyroclastic material.

Lectostratotype   Stratotype selected as a standard section for a previously described stratigraphic unit and for which no specific type section had ever been selected.

Levee   Natural low embankment along a stream formed by deposition of *alluvial* sediment during periods of flooding of the river.

Lineage   Chronologic succession of evolutionary descent from ancestral to descendant forms.

Lineage zone   Range zone consisting of the stratigraphic interval between successively documented first appearances of forms within an evolutionary *lineage*. Syn. with *phylozone*.

Lithification   Process whereby an unconsolidated sediment is converted into a solid rock.

Lithodemic unit   Body of rock, typically igneous or metamorphic, that has irregular contacts with adjacent rock units and that does not normally conform to the law of superposition.

Lithofacies   Characteristics or aspect of a rock that are defined in terms of its composition, mineralogy, texture, color, appearance, etc.

Lithosome   Body of sediment deposited under essentially uniform conditions and, therefore, having uniform lithologic characters. The contacts with adjacent lithosomes of different lithology are typically irregular.

Lithostratigraphic unit   Rock unit. A body of rock defined on the basis of its lithic characteristics (lithology). It conforms to the law of superposition and is usually stratified; the fundamental unit is the *formation*.

Lithotectonic unit   Mappable lithologic unit that cannot be defined in terms of regular lithostratigraphic divisions such as formations. Primarily applicable in Precambrian terranes (Wright, 1985).

Lithotope   Sedimentary rock body containing within it a *biotope* and, therefore, reflecting the uniformity of sedimentary environment implied by the presence of the biotope.

Loess   Windblown silt-size material derived from the finer fractions from deserts or glacial *outwash deposits*.

Lowstand   Refers to sea level when it is below the shelf edge.

Lysocline   Ocean depth at which there is a marked increase in the dissolution rate of calcite. In the central Pacific, for example, this is about 3800 m.

Magnafacies   Sedimentary rock body of uniform and homogeneous lithologic and paleontologic characteristics that over its geographic extent is *diachronous*. Comprised of two or more *parvafacies*.

Magnetic anomaly   Measurement of the polarity or intensity of the earth's magnetic field that is a departure from the normal, average, or anticipated direction or value at that time and place.

Magnetic reversal   A 180° change of polarity of the earth's magnetic field: from normal to reverse or reverse to normal.

Magnetostratigraphic unit   Body of rock characterized by normal, reversed, or alternating polarities that set it apart from magnetostratigraphic units above and below.

Marl Calcareous clay or clayey limestone, typically soft and unconsolidated. Most often deposited in freshwater lakes.

Megacyclothem Grouping of cyclothems (Moore, 1936).

Member Lithostratigraphic unit of lower rank than a *formation*. Division of a formation.

Mesothem Grouping of cyclothems or a division of a *synthem* (Ramsbottom, 1977).

Metasediment Metamorphosed sedimentary rock.

Metazoa Multicellular animal life forms.

Mid-Plate margin Passive, aseismic, or Atlantic-type continental margin that has moved away from a spreading center.

Milankovitch cycle Cycle of world climatic changes caused by fluctuation in the amount of the sun's energy received by the earth. These fluctuations are due to variations in the tilt of the rotational axis, the eccentricities of the earth's orbit, and the time of the year when the earth is at perihelion.

Miogeosyncline Geosynclinal trough adjacent to a craton. It is filled with a wedge of sediments, typically including well-sorted clastics derived from the continent, together with limestones and evaporites. Volcanic rocks are absent.

Miosynthem Lowest ranked unit in a hierarchy of unconformity-bounded units suggested by Salvador (1987).

Molasse Sediments, typically nonmarine or sometimes shallow-marine, that accumulate in front of a mountain chain and fill a subsiding *foredeep*.

Moment 1. Smallest detectable geologic time interval.
2. Time span of a biozone.

Moraine Accumulation of unsorted, unstratified till deposited by a glacier.

Morphogenetic zone Syn. with *lineage zone*.

Morphospecies Supposed species defined on the basis of its external features, assuming that they reflect the genetic features that would normally differentiate one species from another.

Morphostratigraphic unit Informal unit defined as a body of rock described on the basis of its surface morphology; typically applies to surficial deposits such as beach ridges, alluvial fans, and glacial moraines.

Mud cake Layer of drilling mud that becomes plastered to the wall of the borehole as the liquid fraction passes into the formation; i.e., into the *invaded zone*.

Mud Log Log of drilling mud and drill cuttings from rotary drilling.

Natural selection Evolutionary mechanism whereby the environmental effect on a population is seen in the higher survival rates of certain varieties within a species so that they contribute more offspring to the succeeding generation. If the characteristics of the variant are heritable, there is a gradual change in the composition of the population.

Nekton Free-swimming organisms.

Neostratotype New stratotype established after the original *holostratotype* has been destroyed or become inaccessible.

Neutron Elementary particle in the atomic nucleus having the mass of one proton and no electrical charge.

Neutron log Log of the changing intensity of radiation produced by rock formations as they are bombarded by neutrons from a source contained in a *sonde* drawn up the borehole.

Niche Ecologic position of an organism within its environment and to which it is adapted.

Nonangular unconformity Syn. with *disconformity*.

Nonarboreal pollen Pollen from herbaceous plants.

Nonconformity That type of unconformity

where sedimentary rocks overlie an eroded surface of igneous or metamorphic rock.

Nuclide   Atom defined by the number of protons and neutrons in its nucleus.

Offlap   Relationship between strata as they are deposited during a marine regression, as successively younger beds are deposited basinward of older beds and leave the landward extent of older beds progressively emergent.

Onlap   Basal relationship of beds laid down during a marine transgression as they terminate progressively farther inland, each younger bed encroaching successively landward over previously deposited beds.

Oolith   Spherical grain of calcium carbonate, approximately 0.5 to 1.0 mm in diameter and formed initially by the precipitation or accretion of aragonite crystals around a nucleus, typically a small bioclastic or quartz grain. Ooliths form in warm, shallow-marine, moderately agitated water.

Oort cloud   Supposed ringlike swarm of comets that lies around the sun in an orbit beyond that of the outer planets.

Ooze   Fine-grained deep-sea sediment, derived from the calcareous and siliceous skeletons of microscopic planktonic foraminiferans, radiolarians, diatoms, etc.

Oppel zone   Syn. with *concurrent range zone.*

Orogeny   Mountain-building episode.

Ostracode (ostracod)   Aquatic crustacean of the subclass Ostracoda. The majority are of microscopic size (0.5–1.5 mm) and secrete a bivalved, generally calcified carapace. Occur in both marine and freshwater environments. Cambrian to Recent.

Outer arc ridge   Shallow-water or emergent ridge that forms as oceanic sediments are scraped off the subducting plate along a subduction zone.

Outwash deposits   Fluvioglacial deposits laid down by meltwater streams issuing from glaciers.

Overbank deposits   Fine-grained alluvial silt and mud deposited over a flood plain by a river in flood.

Overlap zone   Syn. with *concurrent range zone.*

Pandemic   Said of an organism that has an extensive geographic range.

Pangaea   Supercontinent that supposedly existed through approximately the Upper Paleozoic and which split into a northern continent (Laurasia) and a southern continent *(Gondwana).*

Paraconformity   Obscure unconformity with the discontinuity apparently of no greater magnitude than a bedding plane, but across which biostratigraphic evidence points to a time gap of some magnitude.

Paracontinuity   Small-scale hiatus that is both a lithostratigraphic and biostratigraphic break or discontinuity. A geographically widespread diastem.

Paracycle   Term coined by Mitchum (1977) for the time interval during which sea level rises to a stillstand and then rises still further with no intervening relative fall.

Parastratotype   Stratotype used by the original author to supplement the *holostratotype.*

Partial range zone   Biozone in which a larger biozone is partitioned by reference to two other taxa where ranges do not overlap. The lower boundary of the partition is the highest occurrence of the second taxon, and the upper boundary is the lowest occurrence of the third taxon.

Parvafacies   Body of rock defined as that part of a *magnafacies* that lies between designated time horizons.

Passage beds   Beds that are transitional between a formation of one lithology below to one with a contrasting lithology above.

Passive margin   Continental margin located

within a plate. Tectonic movements are dominantly of vertical type; some initial uplift occurs at rifting, followed mainly by a history of subsidence.

Patch reef Small, isolated reef mound or pinnacle growing in a lagoon or on a shallow platform.

Peak Zone Syn. with *abundance zone.*

Peat Unconsolidated swamp deposit of dead and partially decayed vegetation; represents the first stage in the formation of coal.

Pedostratigraphic unit Body of rock consisting of one or more pedalogic (paleosol, or fossil soil) horizons.

Pelagic Referring to open-ocean environments.

Peneplain Extensive low, almost featureless land surface, supposedly the result of long-continued subaerial erosion.

Period Fundamental geochronologic (time) unit, a subdivision of an *era.*

Peripheral isolate Member of a population near the geographic limits of the species range and which is likely to become isolated from the main gene pool.

Permeability Ability of a rock or sediment to transport fluids through pores and cracks.

Phase 1. Minor variant of facies of local extent and/or brief time.
2. Second unit in a descending series of *diachronic* divisions.
3. Third unit in a hierarchy of *chronotectonic* units suggested for use in Precambrian stratigraphy by Wright (1985).

Phosphorite Sedimentary rock with a high concentration of phosphate minerals.

Phyletic gradualism Evolutionary model in which a new species gradually evolves from an ancestral form over a long period of time.

Phylogenesis Evolution within a lineage.

Phylogenetic zone Syn. with *lineage zone.*

Phylozone Syn. with *phylogenetic zone.*

Pinnacle reef Isolated, stromatoporoid-algal reef in the Middle Silurian of the Michigan Basin.

Plankton Organisms that drift with currents, either passively or with only very limited swimming ability.

Platform That part of a craton where the basement rocks are covered by mainly flat-lying sedimentary rocks.

Playa lake Temporary lake in an arid or semiarid region with internal drainage. When such lakes dry up, evaporitic salts are left.

Point bar Accumulation of sand or gravel on the inside of a bend in a meandering stream.

Polarity chron Fundamental time unit designating the time span of a polarity chronozone.

Polarity zone Fundamental unit of magnetopolarity classification. A rock unit characterized by its magnetic signature.

Pollen Microgametophytes (microspores) of seed plants *(angiosperms* and *gymnosperms);* each contains a microscopic male gamete.

Porosity Percentage of the total rock volume that constitutes pore spaces.

Prodelta Gentle slope beyond the front of the delta. The surface is below wave base.

Progradation Outward building of the coastline by sediment accumulation, such as in deltas, mangrove swamps, or beaches.

Proton Elementary particle in the atomic nucleus; has a positive charge.

Province (Faunal or floral) An area defined by climatic and topographic parameters and containing certain spatially and temporally associated animal or plant communities.

Proximal onlap Onlap in the direction of the source of clastic sediment, as with marine advance onto the land from which the sediment is derived.

Pyroclastic sediment Accumulation of volcanic ejecta (blocks, cinders, ash, etc.).

Quiet zone Zone on the seafloor where the magnetic anomaly record shows a relatively long period with no reversals.

Race Population of animals that exhibits certain differences from other portions of the *gene pool* but which are not sufficiently distinctive to be considered a separate subspecies or species.

Radiolaria Planktonic marine protozoans of microscopic size that secrete skeletons of silica.

Range zone Biozone based on the total stratigraphic range of a species, or other taxon, or of an assemblage of taxa.

Realm Large area of the earth's surface of continental (or oceanic) or larger size delimited by major geographic barriers of land or water. Within each realm may be distinguished numerous *regions.*

Redbeds Sediments of any type but typically sandstones, siltstones, and shales, in a nonmarine environment, that are reddish in color. The color is due to the presence of ferric iron. Typically, such sediments are deposited in a continental environment.

Reef Rocklike mass made up of skeletons of sedentary calcareous marine organisms and built up as a wave-resistant structure from the seafloor. Internally, the reef consists of the remains of frame-building organisms such as corals, together with calcareous algae and many other reef dwellers.

Region Major area of the earth's surface distinguished by topographic and climatic characteristics. Subdivisions of regions, often climatically delimited, are known as *provinces.*

Regression Retreat of the sea from the land. This may be due to relative fall in sea level or to depositional *progradation.*

Relict sediments Sediments characteristic of or originally deposited in a different environment from that in which they presently occur.

Resistivity log Well log that records the changing values of the resistance in the rock to an electric current sent into the ground by two electrodes.

Retroarc basin Basin on the *foreland* caused by underthrusting of continental lithosphere under the flank of the island arc.

Retrogradation Landward retreat of a shoreline as a consequence of marine erosion.

Rhythmite Single unit of a succession of rhythmically repeated sediments.

Rotary drilling Drilling method employing a drill bit at the bottom of a rotating drill pipe. The drill bit is lubricated and cooled, and the drill cuttings are brought to the surface by circulating *drilling mud.*

Sabkha Extensive supratidal flat in an arid region and on which accumulates a variety of evaporite minerals.

Sand line Line drawn on an SP log through the maximum deflections for thick, clean sands, provided the salinity of the pore space water does not vary.

Sechron Time interval involved in the deposition of a sequence.

Secule Time span of an *assemblage zone.*

Sequence Major unconformity-bounded unit of interregional extent.

Series Chronostratigraphic unit of lesser rank than a *system.* The time unit equivalent is an *epoch.*

Shale line Line drawn through the minimum deflections marking impermeable shales on an SP log.

Shelf margin reef Organic barrier-reeflike structure formed at the outer margin of the shelf.

Shield Extensive area of exposed basement rocks, typically of Precambrian age, in the interior of a craton.

Shoestring sand In the subsurface, a long and relatively narrow mass of sand, frequently a buried river-channel sand or beach ridge.

Shothole In seismic surveying, a shallow borehole in which explosive charges are placed so as to generate seismic waves.

Similarity coefficient Syn. with *Correlation coefficient.*

Sonde Cylindrical downhole logging tool that houses electrical and radioactivity measuring instruments.

Span Unit within the hierarchy of *diachronic* units.

Speleothem Cave deposit, typically of calcite; e.g., stalagmites and stalactites.

Spontaneous-potential log Type of electric log that records the changes in the natural potential between different parts of a stratigraphic succession and the drilling mud.

Stade (stadial) Substage of a glacial stage when a climatic cooling caused a brief readvance of the ice.

Stage 1. Chronostratigraphic unit of lesser rank than a *series* and, theoretically, the smallest unit recognizable on a world scale.

2. In Quaternary stratigraphy, a glacial or interglacial episode.

3. Unit in deep-sea oxygen isotope stratigraphy.

Stenotopic Refers to organisms that are relatively intolerant and can survive only in a narrow range of ecologic conditions.

Stratigraphic section Informal term applied to a documented, more or less local succession of sequential strata. It could be compiled from surface exposures and/or well logs.

Stratotype Type section for a specific stratigraphic unit. It constitutes the standard in defining the unit and/or its boundaries.

Stromatolite Organosedimentary structure consisting of fine sedimentary layers trapped by successive generations of algal mats; produced by *cyanobacteria;* may be more or less horizontal, mammilated, columnar, or variable in gross form.

Substage Subdivision of a *stage.*

Subsynthem Subdivision of a *synthem,* according to Salvador (1987).

Subzone Subdivision of a *biozone;* may itself be divided into *zonules.*

Supergroup Assemblage of contiguous groups that have features in common.

Supersequence According to Mitchum (1977), a group of contiguous sequences, roughly similar in magnitude to the *sequence* of Sloss (1963).

Supersynthem Assemblage of *synthems* in Salvador's (1987) classification of unconformity-bounded units.

Suspect terrane (allochthonous terrane) Region consisting of rocks of markedly different type and/or age from adjacent terranes. Believed to have been transported as a segment of ocean floor, island arc, or so-called microplate fragment by lithospheric plate movements and accreted to a stable craton. Syn. with *Exotic terrane.*

Synthem Unconformity-bounded stratigraphic unit. A grouping of *mesothems* in Ramsbottom's (1977) hierarchy.

System Fundamental and original chronostratigraphic unit. Equivalent to the geochronologic (time) division of *period;* i.e., it was formed during a period.

Taphonomy Study of fossils, particularly as concerned with postmortem events.

Taxon range zone *Biozone* between the lowest and highest occurrence of a given taxon.

Teilchron Locally recognized time span of a species or other taxon. The time span of a teilzone.

Teilzone Local range zone of a species or other taxon.

Tempestite Sediment laid down under storm conditions, showing evidence of stirring and redeposition.

Tephra Generic term for all pyroclastics.

Tephrochronology Study of stratigraphic correlation and dating of tephra; invariably this means ash falls.

Terminal moraine Moraine that forms at the margin of the glacier or ice sheet marking its farthest advance.

Terrestrial sediment Sediment that is deposited on the continental surface, as opposed to marine sediments. Includes freshwater deposits in lakes, swamps and streams, and glacial and desert deposits.

Terrigenous sediment Sediment derived from the land and deposited in shallow-marine waters.

Thanatocoenosis Death assemblage. A group of fossils brought together after death simply as clasts by sedimentary processes.

Thermohaline circulation Large-scale vertical movements in the oceans due to convective mechanisms consequent on variations in temperature and salinity.

Thermoremanent magnetisn (TRM) Magnetism acquired by certain minerals in igneous rocks as they cool through the *Curie point* in the presence of the earth's magnetic field.

Till Unconsolidated, unstratified, and poorly sorted accumulation of boulders, pebbles, and other clasts, together with claylike rock flour, transported and deposited by glacial ice.

Tillite Lithified till.

Topset beds Horizontal beds that form on the top surface of a delta and include the subaerial and subaqueous portions of the delta.

Transgression Rise in relative sea level causing marine sediment to encroach on the land.

Trench Elongated and deep trough (typically deeper than 5000 m) in the ocean floor, the topographic manifestation of a subduction zone.

Turbidite Sedimentary deposit of a *Turbidity current,* typically with *graded bedding.*

Turbidity current Type of density current that flows down subaqueous slopes carrying a mixed mass of sediment.

Type section Syn. with *stratotype.*

Unconformity Surface between a lower succession of rocks and an upper succession and which represents a period of erosion.

Undaform Shallow subaqueous surface affected by wave action at the margin of a sedimentary basin.

Undathem Sedimentary rocks formed by deposition on the *undaform* surface; i.e., in the unda environment.

Unit Smallest of the *lithotectonic* divisions proposed by Wright (1985) for use in Precambrian stratigraphy.

Uplap In seismic stratigraphy, the term used to describe the relationship of strata to a relatively small basin that is becoming filled with sediment derived from both sides of the basin. A particular example might be sediment accumulating between shore and offshore barrier.

Varve Annual (or possibly monthly or daily) sedimentary increment, typically a thin lamina deposited on a lake bed. A glacial varve consists of a couplet containing a relatively coarse-grained summer layer overlain by a fine-grained winter layer.

Walther's law Principle that when depth-related depositional environments migrate laterally, as, for example, during a marine transgression or regression the sediments of a particular environment come to lie above the sediments of an adjacent environment.

Wave base Maximum depth at which bottom sediment is stirred by normal wave action, typically about 10 m.

Wilson cycle Cycle of plate tectonic events from continental rifting to produce an ocean basin, followed by closing of the basin and continental collision.

Wireline log Well log derived from instruments in a sonde lowered down the borehole on flexible steel cable; e.g., resistivity, sonic gamma ray logs, etc.

Zone Body of rock characterized by the presence of distinctive and particular unifying features; e.g., fossils, heavy minerals, etc. Should be used only with a modifying term to indicate the kind of zone referred to.

Zonule In biostratigraphy, a subdivision of a *subzone*.

AUTHOR INDEX

SUBJECT INDEX